Handbook of Energy-Aware and Green Computing
Volume 2

CHAPMAN & HALL/CRC
COMPUTER and INFORMATION SCIENCE SERIES

Series Editor: Sartaj Sahni

PUBLISHED TITLES

ADVERSARIAL REASONING: COMPUTATIONAL
APPROACHES TO READING THE OPPONENT'S MIND
Alexander Kott and William M. McEneaney

DISTRIBUTED SENSOR NETWORKS
S. Sitharama Iyengar and Richard R. Brooks

DISTRIBUTED SYSTEMS: AN ALGORITHMIC APPROACH
Sukumar Ghosh

ENERGY EFFICIENT HARDWARE-SOFTWARE
CO-SYNTHESIS USING RECONFIGURABLE HARDWARE
Jingzhao Ou and Viktor K. Prasanna

FUNDAMENTALS OF NATURAL COMPUTING: BASIC
CONCEPTS, ALGORITHMS, AND APPLICATIONS
Leandro Nunes de Castro

HANDBOOK OF ALGORITHMS FOR WIRELESS
NETWORKING AND MOBILE COMPUTING
Azzedine Boukerche

HANDBOOK OF APPROXIMATION ALGORITHMS
AND METAHEURISTICS
Teofilo F. Gonzalez

HANDBOOK OF BIOINSPIRED ALGORITHMS
AND APPLICATIONS
Stephan Olariu and Albert Y. Zomaya

HANDBOOK OF COMPUTATIONAL MOLECULAR BIOLOGY
Srinivas Aluru

HANDBOOK OF DATA STRUCTURES AND APPLICATIONS
Dinesh P. Mehta and Sartaj Sahni

HANDBOOK OF DYNAMIC SYSTEM MODELING
Paul A. Fishwick

HANDBOOK OF ENERGY-AWARE AND GREEN
COMPUTING
Ishfaq Ahmad and Sanjay Ranka

HANDBOOK OF PARALLEL COMPUTING: MODELS,
ALGORITHMS AND APPLICATIONS
Sanguthevar Rajasekaran and John Reif

HANDBOOK OF REAL-TIME AND EMBEDDED SYSTEMS
Insup Lee, Joseph Y-T. Leung, and Sang H. Son

HANDBOOK OF SCHEDULING: ALGORITHMS, MODELS,
AND PERFORMANCE ANALYSIS
Joseph Y.-T. Leung

HIGH PERFORMANCE COMPUTING IN REMOTE SENSING
Antonio J. Plaza and Chein-I Chang

INTRODUCTION TO NETWORK SECURITY
Douglas Jacobson

LOCATION-BASED INFORMATION SYSTEMS:
DEVELOPING REAL-TIME TRACKING APPLICATIONS
Miguel A. Labrador, Alfredo J. Pérez, and
Pedro M. Wightman

METHODS IN ALGORITHMIC ANALYSIS
Vladimir A. Dobrushkin

PERFORMANCE ANALYSIS OF QUEUING AND COMPUTER
NETWORKS
G. R. Dattatreya

THE PRACTICAL HANDBOOK OF INTERNET COMPUTING
Munindar P. Singh

SCALABLE AND SECURE INTERNET SERVICES AND
ARCHITECTURE
Cheng-Zhong Xu

SPECULATIVE EXECUTION IN HIGH PERFORMANCE
COMPUTER ARCHITECTURES
David Kaeli and Pen-Chung Yew

VEHICULAR NETWORKS: FROM THEORY TO PRACTICE
Stephan Olariu and Michele C. Weigle

Handbook of Energy-Aware and Green Computing

Volume 2

Edited by

Ishfaq Ahmad
Sanjay Ranka

CRC Press
Taylor & Francis Group
Boca Raton London New York

CRC Press is an imprint of the
Taylor & Francis Group, an **informa** business

A CHAPMAN & HALL BOOK

CRC Press
Taylor & Francis Group
6000 Broken Sound Parkway NW, Suite 300
Boca Raton, FL 33487-2742

First issued in hardback 2019

© 2012 by Taylor & Francis Group, LLC
CRC Press is an imprint of Taylor & Francis Group, an Informa business

No claim to original U.S. Government works

ISBN 13: 978-1-4665-0112-6 (hbk)

Library of Congress Cataloging-in-Publication Data

Handbook of energy-aware and green computing / edited by Ishfaq Ahmad and Sanjay Ranka.
    p. cm. -- (Chapman & Hall/CRC computer and information science series)
  Includes bibliographical references and index.
  ISBN 978-1-4398-5040-4
  1. Electronic digital computers--Power supply. 2. Computer systems--Energy conservation. 3. Green technology. 4. Low voltage systems. I. Ahmad, Ishfaq. II. Ranka, Sanjay.

TK7895.P68H36 2012
621.39'5--dc23   2011029259

Visit the Taylor & Francis Web site at
http://www.taylorandfrancis.com

and the CRC Press Web site at
http://www.crcpress.com

Contents

PART VI Monitoring, Modeling, and Evaluation

PART VII Software Systems

PART VIII   Data Centers and Large-Scale Systems

PART IX   Green Applications

PART X Social and Environmental Issues

Preface

Green computing is an emerging interdisciplinary research area spanning the fields of computer science and engineering, electrical engineering, and other engineering disciplines. Green computing or sustainable computing is the study and practice of using computing resources efficiently, which in turn can impact a spectrum of economic, ecological, and social objectives. Such practices include the implementation of energy-efficient central processing unit processors and peripherals as well as reduced resource consumption. During the last several decades and in particular in the last five years, the area has produced a prodigious amount of knowledge that needs to be consolidated in the form of a comprehensive book.

Researchers and engineers are now considering energy as a first-class resource and are inventing means to manage it along with performance, reliability, and security. Thus, a considerable amount of knowledge has emerged, as is evident by numerous tracks in leading conferences in a wide variety of areas such as mobile and pervasive computing, circuit design, architecture, real-time systems, and software. Active research is going on in power and thermal management at the component, software, and system level, as well as on defining power management standards for servers and devices and operating systems. Heat dissipation control is equally important, forcing circuit designers and processor architects to consider not only performance issues but also factors such as packaging, reliability, dynamic power consumption, and the distribution of heat. Thus, research growth in this area has been explosive.

The aim of this edited handbook is to provide basic and fundamental knowledge in all related areas, including, but not limited to, circuit and component design, software, operating systems, networking and mobile computing, and data centers. As a comprehensive reference, the book provides the readers with the state of the art of various aspects of energy-aware computing at the component, software, and system level. It also provides a broad range of topics dealing with power-, energy-, and temperature-related research areas of current importance.

The *Handbook of Energy-Aware and Green Computing* is divided into two volumes, and its 52 chapters are categorized into 10 parts:

Volume 1

> Part I: Components, Platforms, and Architectures
> Part II: Energy-Efficient Storage
> Part III: Green Networking
> Part IV: Algorithms
> Part V: Real-Time Systems

Volume 2

> Part VI: Monitoring, Modeling, and Evaluation
> Part VII: Software Systems

Part VIII: Data Centers and Large-Scale Systems
Part IX: Green Applications
Part X: Social and Environmental Issues

Readers interested in architecture, networks, circuit design, software, and applications can find useful information on the topics of their choice. For some of the topics, there are multiple chapters that essentially address the same issue but provide a different and unique problem solving and presentation by different research groups. Rather than trying to have a single unified view, we have tried to encourage this diversity so that the reader has the benefit of a variety of perspectives.

We hope the book provides a good compendium of ideas and information to a wide range of individuals from industry, academia, national labs and industry. It should be of interest not just to computer science researchers but to several other areas as well, due to the inherent interdisciplinary nature of green computing.

Ishfaq Ahmad
University of Texas at Arlington
Arlington, Texas

Sanjay Ranka
University of Florida
Gainesville, Florida

For MATLAB® and Simulink® product information, please contact:

The MathWorks, Inc.
3 Apple Hill Drive
Natick, MA, 01760-2098 USA
Tel: 508-647-7000
Fax: 508-647-7001
E-mail: info@mathworks.com
Web: www.mathworks.com

Editors

Ishfaq Ahmad received his BSc in electrical engineering from the University of Engineering and Technology, Pakistan, in 1985, and his MS in computer engineering and PhD in computer science from Syracuse University, New York in 1987 and 1992, respectively. Since 2002, he has been a professor of computer science and engineering at the University of Texas at Arlington. Prior to this, he was an associate professor of computer science at the Hong Kong University of Science and Technology. His research focus is on the broader areas of parallel and distributed computing systems and their applications, optimization algorithms, multimedia systems, video compression, and energy-aware green computing.

Dr. Ahmad has received numerous research awards, including three best paper awards at leading conferences and the best paper award for *IEEE Transactions on Circuits and Systems for Video Technology*, the IEEE Service Appreciation Award, and the Outstanding Area Editor Award from *IEEE Transactions on Circuits and Systems for Video Technology*. He was elevated to the IEEE Fellow grade in 2008.

Dr. Ahmad's current research is funded by the U.S. Department of Justice, the National Science Foundation, SRC, the Department of Education, and several companies. He is the founding editor in chief of a new journal, *Sustainable Computing: Informatics and Systems*, and a cofounder of the *International Green Computing Conference*. He is an editor of the *Journal of Parallel and Distributed Computing, IEEE Transactions on Circuits and Systems for Video Technology, IEEE Transactions on Parallel and Distributed Systems*, and *Hindawi Journal of Electrical and Computer Engineering*. He has guest edited several special issues and has been a member of the editorial boards of the *IEEE Transactions on Multimedia* and *IEEE Concurrency*.

Sanjay Ranka is a professor in the Department of Computer Information Science and Engineering at the University of Florida, Gainesville, Florida. His current research interests are energy-efficient computing, high-performance computing, data mining, and informatics. Most recently, he was the chief technology officer at Paramark, where he developed real-time optimization software for optimizing marketing campaigns. He has also held a tenured faculty position at Syracuse University and has been a researcher/visitor at IBM T.J. Watson Research Labs and Hitachi America Limited.

Dr. Ranka received his PhD and BTech in computer science from the University of Minnesota and from IIT, Kanpur, India, respectively. He has coauthored 2 books, *Elements of Neural Networks*

(MIT Press) and *Hypercube Algorithms* (Springer-Verlag), 70 journal articles, and 110 refereed conference articles. His recent work has received a student best paper award at ACM-BCB 2010, best paper runner-up award at KDD-2009, a nomination for the Robbins Prize for the best paper in the journal *Physics in Medicine and Biology* for 2008, and a best paper award at ICN 2007.

Dr. Ranka is a fellow of the IEEE and AAAS, and a member of the IFIP Committee on System Modeling and Optimization. He is also the associate editor in chief of the *Journal of Parallel and Distributed Computing* and an associate editor for *IEEE Transactions on Parallel and Distributed Computing*; *Sustainable Computing: Systems and Informatics*; *Knowledge and Information Systems*; and the *International Journal of Computing*.

Dr. Ranka was a past member of the Parallel Compiler Runtime Consortium, the Message Passing Initiative Standards Committee, and the Technical Committee on Parallel Processing. He is the program chair for the 2010 International Conference on Contemporary Computing and co-general chair for the *2009 International Conference on Data Mining* and the *2010 International Conference on Green Computing*.

Dr. Ranka has had consulting assignments with a number of companies (e.g., AT&T Wireless, IBM, Hitachi) and has served as an expert witness in patent disputes.

Contributors

Sherif Abdelwahed
Department of Electrical and Computer
  Engineering
Mississippi State University
Mississippi State, Mississippi

Tarek F. Abdelzaher
Department of Computer Science
University of Illinois at Urbana–Champaign
Champaign, Illinois

Sheikh Iqbal Ahamed
Department of Mathematics, Statistics
  and Computer Science
Marquette University
Milwaukee, Wisconsin

Swarup Bhunia
Electrical Engineering and Computer Science
  Department
Case Western Reserve University
Cleveland, Ohio

Mingsong Bi
The University of Arizona
Tucson, Arizona

Paul Brenner
University of Notre Dame
Notre Dame, Indiana

Laurent Broto
University of Toulouse
Toulouse, France

Aimee Buccellato
University of Notre Dame
Notre Dame, Indiana

Yu Cai
Michigan Technological University
Houghton, Michigan

Hui Chen
Department of Computer Science
Wayne State University
Detroit, Michigan

Wei Chen
School of Information Technologies
University of Sydney
Sydney, New South Wales, Australia

Song Ci
Department of Computer and Electronics
  Engineering
University of Nebraska–Lincoln
Lincoln, Nebraska

Sylvain Contassot-Vivier
AlGorille Project Team
National Institute for Research in Computer
  Science and Control (INRIA)
Henri Poincaré University (Nancy 1)
Nancy, France

Julita Corbalan
Barcelona Supercomputing Center
Barcelona, Spain

Georges Da Costa
University of Toulouse
Toulouse, France

Igor Crk
IBM Systems and Technology Group
Tucson, Arizona

Noel Depalma
Joseph Fourier University
Grenoble, France

Abhishek Dubey
Institute for Software Integrated Systems
Vanderbilt University
Nashville, Tennessee

Maja Etinski
Barcelona Supercomputing Center
Barcelona, Spain

Jeffrey J. Evans
Purdue University
West Lafayette, Indiana

Geoffrey C. Fox
Indiana University
Bloomington, Indiana

Aeiman Gadafi
University of Toulouse
Toulouse, France

Chris Gniady
The University of Arizona
Tucson, Arizona

David B. Go
University of Notre Dame
Notre Dame, Indiana

Francesc Guim
Intel Barcelona
Barcelona, Spain

Sandeep Gupta
School of Computing, Informatics and Decision
    Systems Engineering
Arizona State University
Tempe, Arizona

Daniel Hagimont
University of Toulouse
Toulouse, France

Anna Haywood
School of Engineering for Matter, Transport, and
    Energy
Arizona State University
Tempe, Arizona

Helmut Hlavacs
University of Vienna
Vienna, Austria

Chung-Hsing Hsu
Oak Ridge National Laboratory
Oak Ridge, Tennessee

Karin Anna Hummel
University of Vienna
Vienna, Austria

Khaled Z. Ibrahim
Lawrence Berkeley National Laboratory
Berkeley, California

Lennart Johnsson
Department of Computer Science
University of Houston
Houston, Texas

and

School of Computer Science and Communications
Royal Institute of Technology
Stockholm, Sweden

Siny Joseph
School of Business
Newman University
Wichita, Kansas

Thomas Jost
ALICE and AlGorille Project Teams
National Institute for Research in Computer
    Science and Control (INRIA)
Henri Poincaré University (Nancy 1)
Nancy, France

Krishna Kant
Center for Secure Information Systems
George Mason University
Fairfax, Virginia

and

Intel Corporation
Hillsboro, Oregon

Karen L. Karavanic
Portland State University
Portland, Oregon

Michael Knobloch
Jülich Supercomputing Centre
Forschungszentrum Jülich
Jülich, Germany

Michael Kuhn
Department of Informatics
University of Hamburg
Hamburg, Germany

Julian Kunkel
Department of Informatics
University of Hamburg
Hamburg, Germany

Jesus Labarta
Barcelona Supercomputing Center
Barcelona, Spain

Gregor von Laszewski
Indiana University
Bloomington, Indiana

Young Choon Lee
School of Information Technologies
University of Sydney
Sydney, New South Wales, Australia

Laurent Lefèvre
National Institute for Research in Computer
    Science and Control (INRIA)
University of Lyon
Lyon, France

Lin Liu
School of Software
Tsinghua University
Beijing, China

Thomas Ludwig
Department of Informatics
University of Hamburg
Hamburg, Germany

Andres Marquez
Pacific Northwest National Laboratory
Richland, Washington

Rajat Mehrotra
Department of Electrical and Computer
    Engineering
Mississippi State University
Mississippi State, Mississippi

Timo Minartz
Department of Informatics
University of Hamburg
Hamburg, Germany

Daniel Molka
Center for Information Services and High
    Performance Computing (ZIH)
Technical University of Dresden
Dresden, Germany

Vasily G. Moshnyaga
Department of Electronics Engineering
    and Computer Science
Fukuoka University
Fukuoka, Japan

Tridib Mukherjee
Xerox Corporation
Rochester, New York

Vinod Namboodiri
Department of Electrical Engineering
    and Computer Science
Wichita State University
Wichita, Kansas

Seetharam Narasimhan
Electrical Engineering and Computer Science
    Department
Case Western Reserve University
Cleveland, Ohio

Casey O'Brien
Department of Mathematics, Statistics
    and Computer Science
Marquette University
Milwaukee, Wisconsin

Anne-Cécile Orgerie
Ecole Normale Supérieure de Lyon
University of Lyon
Lyon, France

Manish Parashar
Center for Autonomic Computing
Rutgers University
Piscataway, New Jersey

Patrick Phelan
School of Engineering for Matter, Transport and
    Energy
Arizona State University
Tempe, Arizona

Jean-Marc Pierson
University of Toulouse
Toulouse, France

Stephen W. Poole
Oak Ridge National Laboratory
Oak Ridge, Tennessee

Andres Quiroz
Xerox Research
Xerox Corporation
Webster, New York

Farzana Rahman
Department of Mathematics, Statistics
    and Computer Science
Marquette University
Milwaukee, Wisconsin

Ivan Rodero
Center for Autonomic Computing
Rutgers University
Piscataway, New Jersey

Weisong Shi
Department of Computer Science
Wayne State University
Detroit, Michigan

Jianxin Sun
Department of Computer and Electronics
    Engineering
University of Nebraska–Lincoln
Lincoln, Nebraska

Asser N. Tantawi
IBM Thomas J. Watson Research Center
Yorktown, New York

Douglas Thain
University of Notre Dame
Notre Dame, Indiana

Georgios Varsamopoulos
School of Computing, Informatics and Decision
    Systems Engineering
Arizona State University
Tempe, Arizona

Stephane Vialle
GT-CNRS UMI 2958 and AlGorille INRIA
    Project Team
L'École Supérieure d'Électricité (Supélec)
Metz, France

Lizhe Wang
Indiana University
Bloomington, Indiana

Xueyi Wang
Department of Computer and Electronics
    Engineering
University of Nebraska–Lincoln
Lincoln, Nebraska

Dalei Wu
Department of Computer and Electronics
    Engineering
University of Nebraska–Lincoln
Lincoln, Nebraska

Andrew J. Younge
Indiana University
Bloomington, Indiana

He Zhang
School of Software
Tsinghua University
Beijing, China

Jiucai Zhang
Department of Computer and Electronics
    Engineering
University of Nebraska–Lincoln
Lincoln, Nebraska

Xinying Zheng
Michigan Technological University
Houghton, Michigan

Albert Y. Zomaya
School of Information Technologies
University of Sydney
Sydney, New South Wales, Australia

VI

Monitoring, Modeling, and Evaluation

27

Power-Aware Modeling and Autonomic Management Framework for Distributed Computing Systems

Rajat Mehrotra
Mississippi State University

Abhishek Dubey
Vanderbilt University

Sherif Abdelwahed
Mississippi State University

Asser N. Tantawi
IBM Thomas J. Watson Research Center

27.1 Introduction

According to two related recent studies [1,2], energy consumption cost contributes to more than 12% of overall data center cost and is the fastest growing component of the operating cost. The same studies also point out that data centers consume only 15% energy in actual processing, while the rest of the energy is used by the cooling equipment. Cooling infrastructure adds a significant overhead to the operating cost in terms of power consumption as well as system management. In addition, according to [2], the data center industry accounts for the 2% of global CO_2 emission, which is at the same level as the emission introduced by the aviation industry. Therefore, it is clear that improving power consumption will have a significant influence on the cost effectiveness, reliability, and environmental impact of current and future distributed systems. Consequently, extensive research effort has been recently directed toward developing power-efficient computing systems, also referred to as "power-aware" systems.

Power awareness requires identifying the main factors contributing to power consumption as well as the mechanisms that can be used to control (reduce) this consumption, as well as the effect of using these control mechanisms on other quality of service (QoS) aspects of the system [10]. Adjustments in power consumption need to be made within the tolerance limit of the system QoS. Typically, power management requires expert administrator knowledge to identify workload pattern, system behavior, capacity planning, and resource allocation. However, with increasing size and complexity of computing systems, effective administration is not only tedious but also error-prone and in many cases infeasible. Autonomic computing [31] is a new strategy aiming at replacing manual management with a more systematic approach based on well-founded approaches in AI and systems theory. Such approaches rely on a model that defines the relationship among the system performance, various measurements, and operating parameters of the system. An effective system model is therefore essential to achieve the power awareness in computing systems infrastructure.

In this chapter, a model-based power management framework is presented for distributed multitier systems. This framework implements a predictive control approach to minimize power consumption in a representative system while keeping the QoS parameters at the desired level. The proposed approach starts by experimentally identifying the various system parameters that impact system performance. The dependency relationships among identified parameters are determined and then used to develop a mathematical model structure of the system. Off-line regression techniques are used to estimate the parameters of the power consumption model while an online Bayesian method (exponential extended Kalman filter) is used to estimate the state of the system modeled as an equivalent limited processor sharing queue system. Experiments show that the developed model captures the system behavior accurately in varying environmental conditions with small error variance. Finally, we apply a predictive control approach to optimize power consumption while maintaining a desired level of response time at a negligible overhead for a representative multitier system. Some preliminary research and results have been previously published in [19,42].

This chapter is organized as follows. Preliminary concepts related to the proposed approach and related research work is presented in Section 27.2. System setup is discussed in Section 27.4 and identified system parameters are presented in Section 27.5. The system modeling approach is outlined in Section 27.6. The predictive controller is presented in Section 27.7 and a case study related to managing power consumption and response time using the developed system model and controller is described in Section 27.7.1.

27.2 Background and Related Research

27.2.1 Power Consumption in Computing Systems

Most modern electronic components are built using CMOS (complementary metal-oxide semiconductor) technology. Advances made in the last decade have led to increased clock rates and narrower feature length of the CMOS transistor. This in turn has allowed chip developers to stack more transistors on the die, increasing the available computational power. However, these advancements have come at the cost of increased power consumption.

At the level of a transistor, power consumption can be attributed to three factors. These factors are applicable to all electronic systems of the computer, including CPU, memory, and even the hard drive. In the hard drive, there are some other mechanical factors that lead to increased power consumption. We will discuss them later in this section.

1. Switching (dynamic) power consumption: The working principle of a CMOS FET (field effect transistor) is based upon modulation of the electric charge stored by the capacitance between the gate and the body of the transistor. This capacitor actually charges and discharges during one cycle, i.e., turning the switch on and then off. Effectively, this causes a drain on power, which goes toward charging the capacitor. This power loss is also called the switching or the dynamic loss.
2. Leakage (static) current power consumption: This is due to leakage current flowing through the transistor while being in OFF state. Previously static power consumption was negligible due to low number of transistors per inch and high resistance of wires used on chip. Currently, power loss due to leakage current is 40% of the total power budget. Lowering the voltage across the chip increases the leakage current by making transistors too leaky, which in turn increases the power consumption of the microprocessor [47]. Additionally, high operating temperature of microprocessor increases the leakage current power consumption significantly.
3. Short circuit power consumption: A small amount of power consumption is present in CMOS due to short circuit current on short circuit path between supply rails and ground.

Dynamic power loss has been the main component of total power loss in the past. However, lately the percentage of static power loss is increasing as feature sizes have been decreasing.

27.2.2 Power Consumption Modeling

A nonintrusive but accurate real-time power consumption model is presented in [20], which generates a power model with the help of AC power measurements and user-level utilization metrics. It provides a power consumption model with high accuracy for both temporal and overall power consumption in the servers. A microprocessor-level power consumption estimation technique is described in [25] that first examines the hardware performance counters and then uses relevant counters to estimate the power consumption through sampling-based approaches. Another approach to modeling hard disk power consumption is shown in [53] that extracts performance information from the hard disk itself and predicts the power model. Additionally, it shows that the modeling of idle periods is an important step in predicting the power consumption model of a hard disk. Another approach for power consumption modeling in embedded multimedia applications is presented in [23] that takes various image, speech, and video coding algorithms into account with supplied frequency and voltage to predict the power consumption behavior. A highly scalable power modeling approach is described in [28] for high-performance computing systems by linearly extrapolating the power consumed by a single node to complete large-scale system using various electrical equipment. A micro-architecture-level temperature and voltage-aware performance and leakage power model is introduced in [35] that shows variation of leakage current and energy consumption with varying temperature. An approach for accurately estimating power consumption in embedded systems is presented in [44] while running a software application and considering pipeline stall, inter-instructions effect, and cache misses. A power modeling approach for smart phones is described in [33] that models the power consumption with help of in-built voltage sensors inside the battery and its discharge behavior. Another approach of CMOS power short circuit dissipation is presented in [11] that helps to model the short circuit power dissipation for the configurations when it represents the significant amount of power consumption.

27.2.3 Power Management Techniques

27.2.3.1 CPU Power Management

The main focus of academia and industry has been to target the power consumption of microprocessors. Various methods have been proposed to control the power consumption in microprocessors through logical and system-level techniques.

1. Dynamic voltage and frequency scaling (DVFS): In this method, the voltage across the chip and clock frequency of the transistor is varied (increased or decreased) to control the power consumption and maintain the processing speed at the same time. This method is helpful in preventing computer systems from overheating that can result in system crash. However, the applied voltage should be kept at the level suggested by the manufacturer to keep the system stable for safe operation. DVFS reduces processor idle time by lowering the voltage or frequency, while continuing to process assigned tasks in a permissible amount of time with minimum power consumption. This reduces the dynamic power loss.
2. Dynamic power switching (DPS): In contrast to DVFS, DPS tries to maximize the system idle time that in turn forces the processor to transition to idle or low power mode to reducing power consumption. The only concern is keeping track of the wakeup latency for the processor. It tries to finish the assigned tasks as quickly as possible so that rest of the time can be considered as idle time of the processor. It only reduces leakage current power consumption while increasing the dynamic power consumption due to excessive mode switching of the processor.
3. Standby leakage management (SLM): This technique is close to the strategy used in DPS by keeping the system in low power mode. However, this strategy comes into effect when there is no application running and the system just needs to take care of its responsiveness toward user-related wake-up events (e.g., GUI interaction, key press, or mouse click).

27.2.3.2 RAM Power Management

In general, the CPU is considered to be the dominant component for power consumption in a computing system. However, recent research [17,41] shows that random access memory can also be a significant contributor to system power consumption. Therefore, it should be a target for managing the power consumption specially in case of small-size computers. Currently, memory chips with multiple power modes (e.g., active, standby, nap, and power down) are available in the market and can be used for designing an efficient memory power management technique. The primary idea behind multiple modes of memory operation is that different amounts of power are consumed inside a memory in different states. Memory can execute a transaction only in an active state but can store the data in all of the states. The only concern for utilizing multiple modes is to consider the latency in terms of time and power consumption while switching the modes. Similar to CPU power management, there are primarily two approaches for memory power management through utilizing mode: static and dynamic. In the case of static method, memory is kept at a low power mode for the duration of system operation, while in the dynamic method, memory is placed in a low power mode when its idle time is more than the threshold time. Another approach to memory power management is described in [59] that combines multiple hardware components on a single chip to create smaller and power-efficient components.

A memory power consumption management technique is described in [15] that concentrates on DRAM modules for energy optimization by putting them in low power operating modes during idle periods without paying very high penalties. Another approach for power consumption management in memory is shown in [17] that changes the power state of memory with the change in load on the memory subsystem modules. A feedback control theory–based approach for managing memory power consumption is presented in [41] that puts memory chips in low power mode while maintaining the desired latency and improves memory power efficiency by 19.3%. An operating system scheduler–based approach is presented in [30] where the OS directs the memory module to be in low power mode by keeping track of all processes running on the system. A comprehensive approach of DRAM power management is shown in [24] that combines the benefit of low power mode of modern DRAMs, history-based adaptive memory schedulers, and proposes delaying the issuance of memory commands. A faster and accurate framework for analyzing and optimizing microprocessor power dissipation at the architectural level is described in [12] that validates the power requirement in the design stage of the circuit.

Another application level power management technique is described in [50] that solves the cost-aware application placement problem with an algorithm design that minimizes the cost and migration costs while meeting the performance requirements. Another two-step approach of power-aware processor scheduling is presented in [54] that first performs load balancing among the multiple processors and then applies DVFS to control the speed of the processors to minimize the power consumption. A proactive thermal management approach in data centers is described in [34] that optimizes the air compressor duty cycle and fan speed to prevent heat imbalance in cooling the data center while minimizing the cooling cost. Additionally, it reduces the risk of damage to the data center due to excessive heating. An approach to decrease power consumption in servers is introduced in [48] where NAND flash-based disk caches are extended to replace PCRAM.

27.2.3.3 Miscellaneous Power Consumption

According to [20], miscellaneous components are responsible for large fraction (30%–40%) of power consumption inside a computing system that consists of disk, I/O peripherals, network, and power supplies. The primary contributors in power consumption are disk and power supplies. Device vendors have started implementing their power management protocols to ensure that these devices can run in a low-performance mode. For example, hard disks typically have a timer that measures the time of inactivity and spins the drive down, if possible, to save power.

27.2.4 Control-Based Management of Computing Systems

Control theory offers a promising methodology to automate key system management tasks in distributed computing systems with some important advantages over heuristic or rule-based approaches. It allows us to systematically solve a general class of dynamic resource provisioning problems using the same basic control concepts, and to verify the feasibility of a control scheme before deployment on the actual system. Control theory provides a systematic and well-founded way to design an automated and efficient management technique through continuous observation of system state as well as variation in the environment input, and apply control inputs so as to maintain the system within a desired stable state region corresponding to the desired QoS requirements.

Research in real-time computing has focused on using classical feedback control as the theoretical basis for performance management of single-CPU applications [22]. These reactive techniques observe the current application state and take corrective action to achieve a specified performance metric, and have been successfully applied to problems such as task scheduling [13], CPU provisioning [13,36,38], bandwidth allocation and QoS adaptation in Web servers [8], multitier Websites [37], load balancing in e-mail and file servers [37,46], and CPU power management [39,49]. In more complex control problems, a pre-specified plan, the feedback map, is inflexible and does not adapt well to constantly changing operating conditions. Moreover, classical control is not a suitable option for managing computing systems exhibiting hybrid and nonlinear behavior, and where control options must be chosen from a finite set. Therefore, researchers (including the authors) have studied the use of more advanced state-space methods adapted from model predictive control [40] and limited lookahead supervisory control [14] to manage such applications [6,7,29]. These methods offers a natural framework to accommodate the system characteristics described earlier, and take into account multi-objective nonlinear cost functions, finite control input sets, and dynamic operating constraints while optimizing application performance. The autonomic approach proposed in [29] describes a hierarchical control-based framework

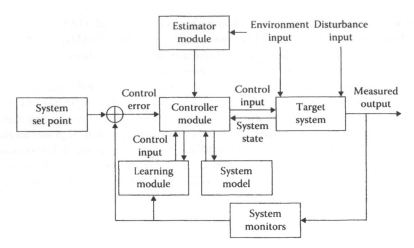

FIGURE 27.1   Elements of a general control system.

to manage the high-level goals for a distributed computing system by continuous observation of the underlying system performance. Additionally, the proposed approach is scalable and highly adaptive even in the case of time-varying dynamic workload patterns.

A typical control system consists of the components shown in Figure 27.1. *System set point* is the desired state of the system in consideration that a system tries to achieve during its operation. *Control error* is the difference between the desired system set point and the measured output during system operation. *Control inputs* are the system parameters, which are applied to the system dynamically for changing the performance level of the system. The *controller module* takes observation of the measured output and provides the optimal combination of different control inputs to achieve the desired set point. The *estimator module* estimates the unknown parameters for the system based upon the previous history using statistical methods. *Disturbance input* can be considered as the environmental input that affects the system performance. The *target system* is the system in consideration, while the *system model* is the mathematical model of the system, which defines the relation between its input and output variables. A *learning module* takes measured output through the monitor and extracts information based on the statistical methods. *System state* typically defines the relationship between control or input variables and performance parameters of the system.

27.2.5 Queuing Models for Multitier Systems

Multitier enterprise systems are composed of various components that typically include Web (http) servers, application servers, and database servers, subjected to a stream of Web requests. Each component performs its function by servicing requests from upper tiers and may submit subrequests to lower tiers. In order to avoid overloading, system administrators usually limit the number of concurrent requests served at each tier that is called "concurrency limit" for that particular tier [16].

In general, a Web request has to wait in a queue for computational resources before it can enter a tier. For example, if the number of maximum threads allowed in the application (app) server is capped to limit concurrency, a new request will wait until an existing running request releases a thread. Clearly, the total service time of the enterprise system is directly affected by the queuing policy at each tier. Therefore, an approximate queuing model can be used to capture the behavior of such systems. Thereafter, it can be used to measure the average number of Web requests in the queue and the average time spent there. During the progression of this work, we have considered four different queuing models. For a detailed description, readers are referred to [32]. These are as follows:

M/M/1: This is the most basic queuing model where both service time and interarrival time are exponentially distributed.

M/G/1 FCFS: The M/G/1 queue assumes that the interarrival time is exponentially distributed, but the service time has a general distribution. This is a more realistic model of Web service behavior since Web requests exhibit a wide variance in their service requirements. The scheduling discipline is *first come first served (FCFS)*. Thus, this model assumes that only one request is serviced at a time, hence restricting the concurrency level to one.

M/G/1 PS: This is an M/G/1 queue with *processor sharing (PS)* scheduling. In such a model, as a Web request arrives, it is able to share the server concurrently with all preceding, existing requests. The computational resource is shared equally by all the requests in a round-robin manner, though with an extremely short slice. When not using the server, a request waits in the queue for its turn to get a slice of the server. In other words, the concurrency level is basically unlimited. The mean analysis of this system is rather simple. In fact, the closed form expression for the mean response time is similar to that of the M/M/1 queue. The only catch is that this model can only be used to study a Web server realistically if we can ensure that the total number of requests in the system do not increase above the maximum concurrency limit. *Thus, if the bottleneck resource utilization is light to moderate, we can use this queue to model Web servers.*

M/G/1 LPS(k): This is an M/G/1 queue with *limited processor sharing* (LPS) scheduling discipline. The parameter k models the concurrency level. In such a model, the first k requests in the queue share the server and the rest of the requests wait in the queue. As requests complete their service and depart from the system, awaiting requests are admitted to be among the k concurrently sharing the server in a processor sharing manner. When $k = 1$ the queue becomes a FCFS queue, and when $k = \infty$ it becomes a PS queue. The LPS(k) is a realistic queuing model for systems that have a limit on the maximum number of concurrently executing jobs. This limit is typically enforced on all Web servers and databases. Even in an operating system, the maximum number of processes that can execute concurrently is capped (e.g., 32,000 for Linux kernel 2.6). The analysis of the LPS queue is rather difficult [55,56] and online prediction for response time and other variables can become intractable.

27.2.6 Kalman Filters

The Kalman filter [26] is an *optimal recursive data processing algorithm*, which estimates the future states of linear stochastic process in the presence of measurement noises. This filter is optimal in the sense that it minimizes the mean of squared error between the predicted and actual value of the state. It is typically used in a predict-and-update loop where knowledge of the system, measurement device dynamics, statistical description of system noise, and the current state of the system is used to predict the next state estimate. Then the available measurements and statistical description of measurement noise is used to update the state estimate.

Two assumptions are made before applying the Kalman filter for state estimation; first, the system in consideration is described by a linear model, or if the system is nonlinear, the system model is linearized at the current state (extended Kalman filter), and second, the measurement and system noise are Gaussian and white, respectively. *Whiteness* indicates that noise is not correlated with time and it has equal impact on all operating modes. Due to its simple approach with optimal results, the Kalman filter has been applied in wide areas of engineering application including motion tracking, radar vision, and navigation systems.

Woodside et al. [51] applied extended Kalman (EK) filtering techniques to the estimation of parameters of a simple closed queueing network model. The population size was approximated by a continuous variable in order to fit the EK mathematical framework. A demonstration of the use of the EK filter to track parameters such as think time and processing time parameters of a time-varying layered queueing system is provided in [58]. The EK filter uses observations of average response time as well as utilization of three resources: a Web processor, a database processor, and a disk. In [52], the technique of using an

EK filter in conjunction with layered queueing models is employed to control the number of allocated servers so as to maintain the average response time within a given range.

The work on applying EK filtering to estimating performance model parameters is extended in [57] where an estimation methodology is sought. Models such as separable closed queueing network models, open queueing network models, and layered models are considered. Parameters may have lower and upper bounds. Tracking periodic deterministic as well as random perturbation of one or more parameters is demonstrated experimentally.

Recently, the Kalman filter has been applied to provision CPU resources in case of virtual machines hosting server applications [27]. In [27], the feedback controllers based on the Kalman filter continuously detect CPU utilization and update the allocation correspondingly for estimated future workloads. Overall, an average of 3% performance improvement in highly dynamic workload conditions over a three-tier Rubis benchmark Web site deployed on a virtual Xen cluster was observed.

In this chapter, we describe our implementation of an *exponential Kalman filter* that we used to predict the computational nature of the incident requests over a Web server by predicting the *service time S* and *delay D* of a request by observing the current average response time of the incident request and request arrival rate on the Web server. This filter uses an $M/G/1$ PS approximate queuing model as the system state equation and considers variation in S and D at previous approximation to estimate the S and D at next sample time. This filter is exponential because it operates on the exponential transformation of the system state variables. This transformation allows us to enforce the ≥ 0 constraint on the state variables. Such a constraint is not possible in typical Kalman filter implementations. Further details are provided in Section 27.6.3.

27.3 Our Approach

A great amount of work has been done in the past for enabling power awareness within computing systems as described earlier at both hardware and software levels. However, it has been difficult to implement those in real systems either due to their implementation complexity or inefficiency to capture the system dynamics completely with multidimensional QoS objectives in a fluctuating environment. Our current work addresses both the modeling and management problems. First we identify the system dynamics accurately with respect to controllable parameters in an off-line as well as on-line manner. Second we utilize the system model within a *model predictive controller* to achieve the desired objectives. The developed predictive controller can accommodate multidimensional QoS objectives easily just by learning the variation of that objective with respect to the system state, control inputs to the system, and environmental changes. The novelty of our approach lies in the collection of state-of-the-art techniques to monitor and model the system performance with the scalable nature of the predictive controller with respect to new QoS objectives for performance optimization.

27.4 Case Study: A Multitier Enterprise System

Multitier enterprise systems are composed of various components that typically include Web (http) servers, application servers, and database servers. Each component performs its function with respect to Web requests and forwards the result to the next component (tier). Generally, system administrators limit the number of concurrent requests served at each tier, which is called "concurrency limit" for that particular tier [16]. In order to experiment and validate our work, we used the following system setup.

System setup: Our system consists of four physical nodes: *Nop01*, *Nop02*, *Nop03*, and *Nop10*. Names *Nop04–Nop09* are reserved for the virtual machines. Table 27.1 summarizes configuration of physical

TABLE 27.1   Physical Machine Configuration

Name	Cores	Description	RAM	DVFS	VMs
Nop01	8	2 Quad core 1.9 GHz AMD opteron 2347 HE	8 GB	No	Nop04, Nop07 (monitoring server)
Nop02	4	2.0 GHz Intel Xeon E5405 processor	4 GB	No	Nop05, Nop08 (client machines)
Nop03	8	2 Quad core 1.9 GHz AMD opteron 2350	8 GB	Yes	Nop06, Nop09 (application server)
Nop10	8	2 Quad core 1.9 GHz AMD opteron 2350z	8 GB	Yes	Nop11, Nop12 (database server)

machines. It also shows the virtual machines (VMs) running on all physical machines and the roles played by those VMs. All VMs run same version of Linux (2.6.18−92.*el5xen*). Client machines are used to generate request load. Application servers run the open source version of IBM's J2EE middleware, *Web Sphere Application Server Community Edition* (WASCE). Database machines run MySQL. *Nop03* and *Nop10* both have dynamic voltage and frequency scaling (DVFS) capability that allows the administrator to tune the complete physical node or its individual cores for the desired performance level. Xen Hypervisor (http://www.xen.org/) was used to create and manage physical resources (CPU and RAM) for the cluster of virtual machines (VMs) on these physical servers.

We used *Daytrader* [3] as our representative application. Daytrader comes with a client that can drive a trade scenario that allows users to monitor their stock portfolio, inquire about stock quotes, buy or sell stock shares, as well as measure the response time for benchmarking. Out of the box, this application puts most of the load on the database server. To emulate business enterprise loads in a highly dynamic environment, we modified the main trade scenario servlet to allow us to shift the processing load of a request from the database node to the computing node. In the modified Daytrader application, the Web server generates a random symbol for each request from the symbol set of available stock names. The Web server performs a database query based on that symbol and returns the result of the query to the client. Also, it computes the integral sum of first N integers, where N is supplied through the client workload.

Due to limitation of workload pattern allowed (only uniform) in Daytrader client, we used the Httperf [5] benchmarking application client tool in all of our experiments. It provides flexibility to generate various workload patterns (Poisson, deterministic, and uniform) with numerous command line options for benchmarking. We modified Httperf to print the performance measurements of our interest periodically while running the experiment. At the end of each sample period, a modified version of Httperf prints out the detailed performance statistics of the experiment in terms of *total numbers of requests sent, minimum response time, maximum response time, average response time, total number of errors with types, and response time* for each request.

Monitoring: Specially developed Python scripts and Xenmon [21] were used as monitoring sensors on all virtual and physical machines. These sensors monitor CPU, disk, and RAM utilization of the nodes (physical and virtual) throughout the system execution and report data after each sampling interval as well as at the end of the experiment. System time was synchronized using NTP. The jitter in monitoring sensors across all servers was controlled using a PID controller as described in [18]. Modifications to the Web server code allowed us to monitor Web server performance in terms of *max threads active* in the Web server, *response time measured* at first tier and at the database tier for each incident request, and *average queue size* in the Web server after each sampling interval. The client returns the measured maximum, minimum, and average response time during the sampling period. We specified 100 s *as the timeout value* for a request response for all of the experiments as it provides enough time for the Web server to serve most of the Web requests even at the maximum utilization of bottleneck resource. Any outstanding request after the timeout was logged as an error. It also returns the number of errors with their type (client timeout, connection reset, etc.) in each sample. Measurement of power consumption of a physical node was done with the help of a real-time wattmeter.

27.5 Model Identification

As a result of different experiments described in Section 27.6, an extensive list of system parameters have been identified. This list is shown in Table 27.2. It contains three different types of parameters: *control variables, state variables,* and *performance variables.* Control variables are those which can be controlled at runtime to tune the system toward the desired performance objectives. State variables describe the current state of the system under observation. Performance variables are used to quantify the QoS level of the system. Additionally, state variables are divided into two different categories: *observable* and *unobservable*. Observable variables can be measured directly through sensors, system calls, or application-related API, while unobservable variables cannot be measured directly; instead they are estimated within a certain accuracy using existing measurements through various techniques at runtime. During our experiments, we used specially written sensors or different tools to measure observable variables, while unobservable variables (e.g., service time and delay) are estimated through the exponential Kalman filter described in Section 27.6.3.

27.6 System Modeling

An accurate system model derivation is necessary to run a computing system efficiently in a power saving environment. The derived model will depict the exact system behavior in terms of various performance objectives with changes in operating environment and controllable parameters. For identifying an accurate system model of our representative distributed multitier system (see Section 27.4), extensive experiments have been performed and results have been analyzed. During these experiments, we analyzed the multitier system performance with respect to system utilization, various workload profiles, bottleneck resource utilization, and its impact on system performance. Additionally, we calculated the work factor of our client requests with the help of linear regression techniques described in [45]. Details of the modeling efforts are described next.

27.6.1 Power Consumption

As a first step toward system model identification, the mutual relationship among physical CPU core utilization, CPU frequency, and power consumption of the physical server was identified. This work is an extension of the power modeling effort described in [19] to model the system power consumption with greater accuracy that can be utilized effectively in real-time physical server deployment. Details of the experiment can be found in our technical report [43]. Figure 27.2 shows the power consumed on one of the physical servers *Nop03* with respect to the aggregate CPU core usage and CPU frequency. An extensive experiment was performed over physical server *Nop03* with the help of a specially written

TABLE 27.2   System Parameters

Control Variables	State Variables	Performance Variables
CPU frequency	CPU utilization (observable)	Average response time
Cap on virtual machine resources	Memory utilization (observable)	Power consumption in watts
Load distribution percentage (in a cluster)	Service time (unobservable)	Percentage of errors
Number of service threads	Queue waiting time (unobservable)	
Number of virtual machines in cluster	Queue size on each server (unobservable)	
	Number of live threads (observable)	
	Peak threads available in a Java VM (observable)	

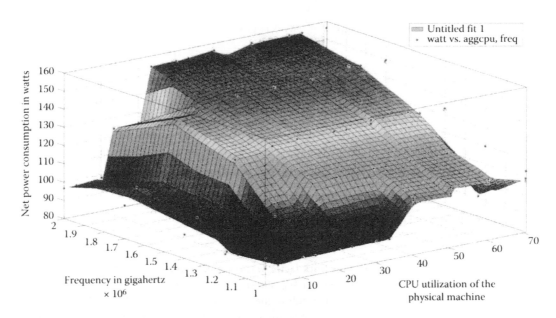

FIGURE 27.2 Power consumption on *Nop03* vs. CPU frequency and aggregate CPU core utilization. This plot uses linear interpolation to create the surface. Note: Power consumption decreases as the frequency decreases.

script, which exhausted a physical CPU core through floating point operations in increments of 10% utilization independent of the current CPU frequency. With multiple instances of this utility, all eight physical CPU cores of the *Nop03* server were loaded in incremental manner for different discrete values of CPU frequencies. CPU frequency across all of the physical cores (1–8) was kept the same during each step. The consumed power was measured with the help of a real-time wattmeter. Based on this experiment, we created a regression model for power consumption at the physical machine with respect to CPU core frequency and aggregate CPU utilization. After analysis of the results (and reconfirmation with several other experiments across other nodes), it was observed that the power consumption model of a physical machine is nonlinear because power consumption in these machines depends not only on the CPU core frequency and utilization, but also depends nonlinearly on other power-consuming devices, e.g., memory, hard drive, CPU cooling fan, etc. As a result, a *lookup table with near neighbor interpolation* was found to be the best fit for aggregating the power consumption model of the physical machine. A combination of CPU frequency and aggregate CPU core usage of the physical machine is used as a key of the lookup table to access the corresponding power consumption value. This aggregate power model was utilized further for the controlled experiments described in Section 27.7 for predicting the estimated power consumption by the physical server at a particular setting of CPU core frequency and aggregate physical CPU utilization.

27.6.2 Request Characteristics

An Httperf benchmark application code was modified to allow the generation of client requests to the Web server, *Nop06*, at a prespecified rate provided from a trace file. At *Nop06*, each request performed a certain fixed floating point computation on the Web server and then performed a random select query on the database machine. To better evaluate the nature of requests, we identified the number of CPU cycles needed to process the request using linear regression [45].

During any sampling interval T, if ρ is virtual CPU utilization, f is CPU frequency, W_{fc} is the work factor of the request (defined in terms of CPU clock cycles), λ is request rate, and ψ is system noise,

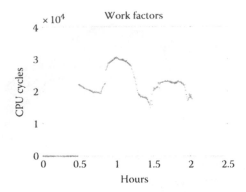

FIGURE 27.3   Work factor plot for request characteristic.

then, $\rho * f = \lambda * W_{fc} + \psi$. The average work factor was computed to be 2.5×10^4 CPU cycles with a coefficient of variation = 0.5. The variation in W_{fc} shows the variation in the nature of a final request based on the chosen symbol for query. The result of the experiment is shown in Figure 27.3. Due to the similar computational nature of all the request incidents on the Web server in a given sample time, we can approximate the total computation time of all the requests, which in turn gives the average response time of the requests in a given sample time. We can use this average response time information to check the status of QoS objective (response time) in the Web server.

27.6.3 Web Server Characteristics

We aimed to identify the Web server bottlenecks and estimate the uncontrolled performance to compare later with the performance under the predictive controller. In this experiment, *Nop09* was the virtual machine (physical machine *Nop03*) running the first tier of the Daytrader application. Virtual CPU of *Nop09* was pinned to a single physical core, and 50% of the physical core was assigned to *Nop09* as the maximum available computational resource. Physical memory was also limited to 1000 MB for *Nop09*. *Nop11* was configured as a database using similar CPU and memory-related operating settings over physical server *Nop10*. To simulate a real-time load scenario, all CPU cores of physical server *Nop03* (except the CPU core hosting *Nop09*) were loaded approximately 50% with the help of utility scripts described in Section 27.6.1. MAX_JAVA_threads, a parameter that sets the maximum concurrency limit was configured as 600 in the Web server application. All CPU cores in physical server *Nop03* were operating at their maximum frequency, 2.0 GHz. The request trace (see Figure 27.4) in this experiment was based on the user request traces from the 1998 World Cup Soccer (WCS-98) Web site [9]. Figure 27.14 shows the response time and power consumption as measured from this experiment. It also shows the CPU utilization at Web server and aggregate CPU utilization of the physical machine. Notably, we saw that the CPU utilization at the Web server (*Nop09* CPU without controller) and aggregate CPU utilization of the physical machine (*Nop03* CPU without controller if we zoom in the line) follow a trend that is similar to the rate of requests made to the first tier. The power consumption curve was almost flat. We also noticed that the Web server response time is correlated to the resident requests (system queue size) in the Web server system (not shown here, but described in [43]).

27.6.3.1 Estimating Bottleneck Resource

To determine the bottleneck resource, we used a queuing approximation for a two-tier system as shown in Figure 27.5. λ is the incoming throughput of requests to an application. ρ is the utilization of the bottleneck resource. S is the average service time on the bottleneck resource. D is the average delay. T is

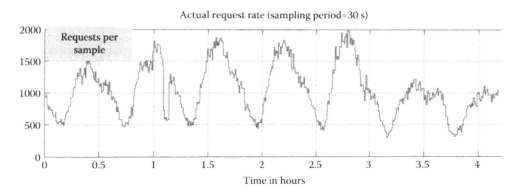

FIGURE 27.4 Http workload based on World Cup Soccer (WCS-98) applied to the Web server.

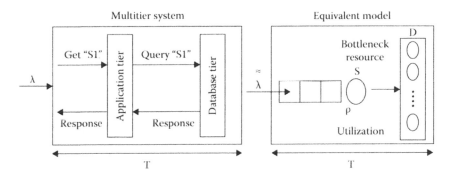

FIGURE 27.5 A queuing model for the two-tier system.

the average response time of a request. The average waiting time for a request is $W = T - S - D$. We define a queue model with state vector $[S; D]$ and observation vector as $[T]$.

An exponential Kalman filter (KF) was used to estimate the system state as mentioned earlier. It is important to note that we can approximate the system as a M/G/1/∞ PS queue if the system has no bottleneck. In the presence of bottleneck, the system utilization (not necessarily CPU) will approach unity. At that time, the system will change to the LPS queue model. However, as mentioned earlier, it is difficult to build a tractable model for LPS queuing systems. Hence, we just identify the operating regions where the system changes the mode between two queuing models and analyze the system in the infinite PS queue region only.

The KF equations, written in terms of exponentially transformed variables, $[x1 \in \Re; x2 \in \Re]$ s.t. $S = exp(x1)$ and $D = exp(x2)$ are as follows. Note that this transformation ensures $S, D \in \Re^+$: For a given timed index of observation, k, the equations $\begin{pmatrix} exp(\hat{x1}_k) \\ exp(\hat{x2}_k) \end{pmatrix} = \begin{pmatrix} exp(x1_{k-1}) \\ exp(x2_{k-1}) \end{pmatrix} + N(0, Q)$ and $T = exp(x1_k) * (1/(1 - \lambda_k * exp(x1_k))) + exp(x2_k) + V(0, R)$ define the state update dynamics and observation. N and V are Gaussian process and measurement noises with mean zero and covariances Q and R, respectively. One can verify that these equations described the behavior of a M/G/1 PS queue. Here, predicted bottleneck utilization is given by $\hat{\rho}_k = \lambda_k^* exp(x1_k)$. Additionally, the KF does not update its state when the predicted bottleneck resource utilization becomes more than 1.

Figure 27.6 shows results of the off-line analysis of data generated through experiment discussed in Section 27.6.3 with the help of KF described in current section. According to Figure 27.6a, the developed exponential KF tracks service time S and delay D at the Web server perfectly with low error variance as the experiment (Section 27.6.3) progresses. Additionally, as per Figure 27.6b, KF tracks bottleneck utilization

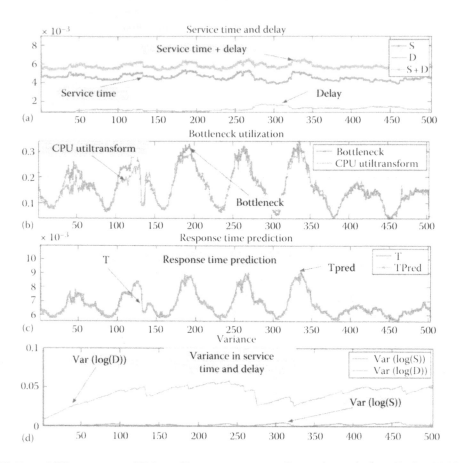

FIGURE 27.6 Off-line exponential Kalman filter output corresponding to the results from Section 27.6.3. Service time and delay are in millisecond range. Response time is specified in seconds.

similar to CPU utilization of the system. However, we noticed that sometimes, the bottleneck utilization might saturate at 1 without the CPU utilization reaching that value. In those cases, we discovered that the number of available system threads acted as the bottleneck. According to Figure 27.6c, predicted response time from the KF T_{pred} and actual response time T are also very close to each other, which indicates efficiency of the KF. The efficiency of the KF indicates that the developed filter is able to capture the response time dynamics of the Web server system perfectly that will be used in an online manner with the predictive control framework in the next subsection.

27.6.4 Impact of Maximum Usage of Bottleneck Resource

We aimed then to observe the effect of high bottleneck resource usage on system performance. Our test setup contains the Daytrader application, which is a multithreaded Java-based enterprise application hosted on WASCE that listens on port number 8080 for incoming http requests. This daytrader application serves the incoming requests through creating a new child Java thread or through an existing Java thread if available in pool of free threads. A newly arrived http request for the Daytrader application is handed over to the newly created or free child thread for further processing. This child thread rejoins the pool of free threads once it finishes the processing. A limit on maximum number of created Java threads can be imposed in the Web server that can also be considered as the maximum concurrency limit of the system. In case of unavailability of a free thread due to already achieved maximum thread limit and an empty pool of free threads, a newly arrived http request has to wait for thread availability that impacts the application performance severely in terms of response time. Therefore, we chose the settings that made the number of available threads as the system bottleneck.

We performed various experiments with different settings for max Java threads. This parameter sets the maximum number of threads that can be used for request processing. Based on our observation, there are typically 90 more system threads that are not accounted under this cap. Figure 27.7 shows the results for one experiment with max threads set to 500. This figure shows that at maximum utilization of the bottleneck resource (Figure 27.7c) system performances decreases significantly and response time

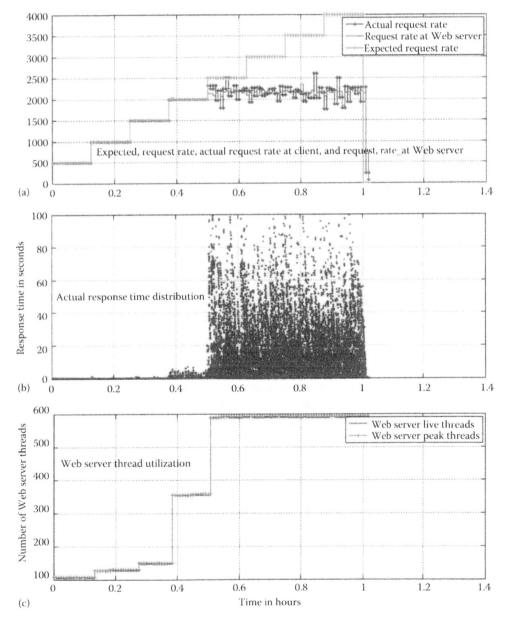

FIGURE 27.7 Impact of maximum utilization of bottleneck resource on performance from Section 27.6.4. Max thread = 500.

(Figure 27.7b) from the Web server becomes unpredictable. Furthermore, this is the region where the system transitions from a PS queue to a LPS queue system.

Once the system reaches the max utilization of the bottleneck resources, it restricts entry for more requests (Figure 27.7a) into the system resulting in maximum utilization of the incoming system queue, which in turn results in rejection of the incoming client requests from the server. Therefore, to achieve predefined QoS specifications, the system should never be allowed to reach the maximum utilization of bottleneck resources. Additionally, this boundary related to max usage of bottleneck resource can also be considered as the "safe limit" of system operation.

27.6.5 Impact of Limited Usage of Bottleneck Resource

Next we observed the Web server performance when the bottleneck resource utilization varied from minimum to maximum and back to minimum. This type of study provides knowledge regarding Web server performance if bottleneck resource utilization is lowered from the maximum limit through a controller that maintains the QoS objective of the multitier system.

The configuration settings for this experiment were same as described in Section 27.6.1. The max number of Java threads for the experiment is 600. The client request-trace profile used for this experiment is shown in Figure 27.8e. According to the results shown in the same figure (Figure 27.8), system utilization (Figure 27.8a) and performance in terms of response time (Figure 27.8d) follow the trend of the applied client request profile (Figure 27.8e). We can also see the sudden jump in size of server queue (Figure 27.8c), which indicates contention of computational resources among all of the pending requests inside the system. The sudden increase in RAM utilization (Figure 27.8a) is due to the increase in thread utilization (Figure 27.8b) of the system. Additionally, from the comparison of request rate and response time plot in Figure 27.8, it is apparent that by lowering the system utilization and client load on the Web server, the Web server can be brought back to state, where it can restore QoS objective of the system that can consist of minimizing the system queue size (Figure 27.8c) and server response time (Figure 27.8d).

27.6.5.1 Kalman Filter Analysis

Results of the experiment were analyzed with the help of the KF described in Section 27.6.3, and results of the analysis are shown in Figure 27.9. According to Figure 27.9, the defined exponential KF tracks service time and delay at the Web server quite well with low variance as the experiment progresses. One can notice the regions (Figure 27.9a) where the bottleneck resource utilization approaches unity but the CPU utilization is less than one. Upon further investigation of those time samples, the bottleneck resource was found to be the maximum number of Java threads available in the Web server. This limit can be changed by the "MAX JAVA Threads" configuration setting. The goal of any successful controller design for performance optimization of the system will be to drive the system to work in the stable region (where the bottleneck resource utilization is less than unity). During the experiment, when predicted utilization of the bottleneck resource is more than one, KF does not update its states.

27.7 Power Management Using Predictive Control

This section describes the implementation of an online predictive controller that uses the KF and the queuing model identified in the previous section (see Figure 27.10). This controller is similar to the *L0 controller* described in [29] and predicts the aggregate response time of the incident requests and the estimated power consumption during the next sample time (look-ahead horizon N) of the system based on different possible combinations of control inputs (CPU core frequency). It optimizes the system behavior in terms of QoS objectives by continuous observation of the underlying system and choosing the best control input for the system in next sample interval.

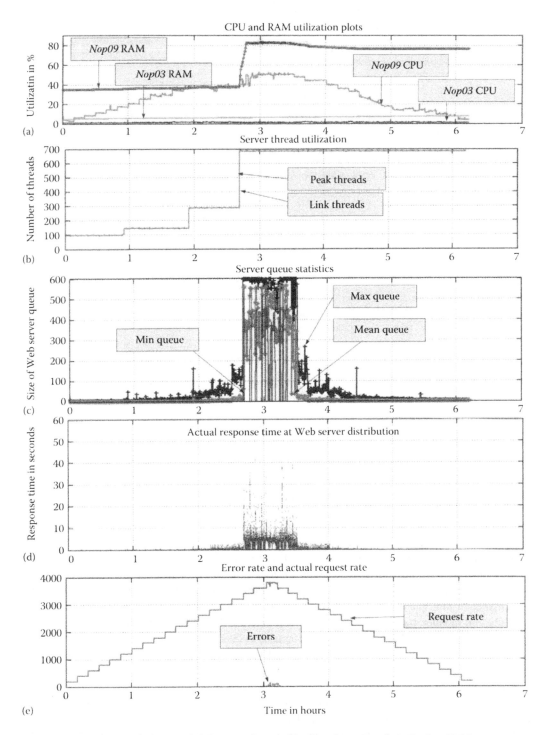

FIGURE 27.8 Web server behavior while limiting the use of bottleneck resource from Section 27.6.5.

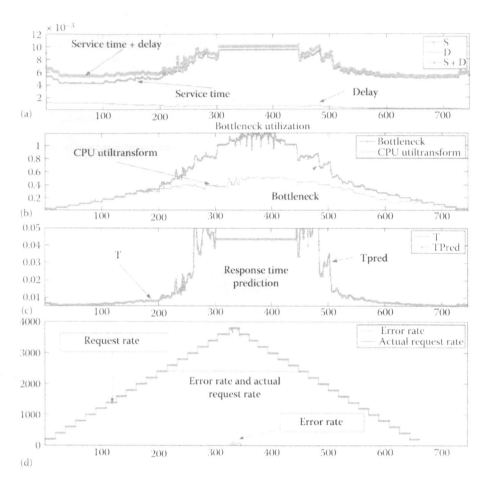

FIGURE 27.9 Off-line KF analysis of the results from Figure 27.8 of Section 27.6.5. Service time and delay are in milliseconds. Response time is specified in seconds.

System variables: Although there is a large number of system parameters listed in Section 27.6, we have chosen a small set of the most important parameters for our predictive controller to show the performance of our modeling approach. The chosen control input is the *CPU core frequency*, due to its impact on system performance in multiple dimensions for response time of the system and power consumption. *System queue size* and *response time* and *power consumption* were the chosen state variables as they are the typical performance variables in the Web service industry used to define a typical multidimensional service level agreement (SLA). Other experiments (reported in detailed report [43]) indicate that the higher value of application queue represents contention in computational resources of the application, and the total response time value indicates the system's capability to process the requests lying in the system queue in a timely manner. Therefore, we try to minimize the application queue size and total response time as one of the components in cost function *J* (described later in this section).

Plant model: The queuing model identified in the previous section was used to estimate the state of the managed system.

Controller model: In order to combine the power consumption QoS and the predicted response time, the controller uses a different internal system model. The controller model uses the estimated system state, predicted response time, and predicted power consumption to make the system decisions. The system state for this experiment, $x(t)$, at time t can be defined as the set of system queue $q(t)$ and

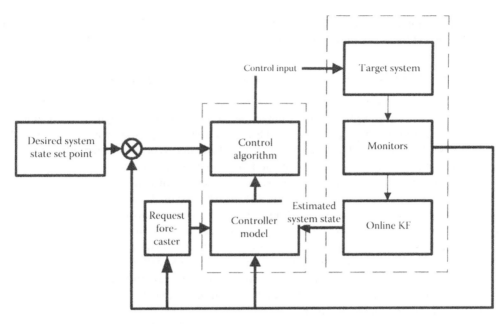

FIGURE 27.10 Elements of the applied predictive control framework.

response time $r(t)$, that is, $x(t) = [q(t); r(t)]$. The queuing system model is given by the equation: $\hat{q}(t+1) = max\{q(t) + (\hat{\lambda}(t+1) - \frac{\alpha(t+1)}{\hat{W}_f}) * T, 0\}$ and $\hat{r}(t+1) = (1 + \hat{q}(t+1)) * \frac{\hat{W}_f}{\alpha(t+1)}$, where at time t, $q(t)$ is the queue level of the system, $\lambda(t)$ is the arrival rate of requests, $r(t)$ is the response time of the system, $\alpha(t)$ is a scaling factor defined as $u(t)/u_{max}$, where $u(t) \in U$ is the frequency at time t (U is the finite set of all possible frequencies that the system can take), u_{max} is the maximum supported frequency in the system, and $\hat{W}_f$ is the predicted average service time (work factor in units of time) required per request at the maximum frequency. An online KF estimates the service time $\hat{S}_t$ of the incident request at current frequency $u(t)$, which is scaled against the maximum supported frequency of the system to calculate the work factor $\left(\hat{W}_f = \hat{S}_t * \frac{u(t)}{u_{max}}\right)$. $E(t)$ is the system power consumption measured in watts at time t.

Estimating environment inputs: The estimation of future environmental input and corresponding output of the system is crucial. In this experiment, an *autoregressive moving average* model was used as estimator of the environmental input as per the following equation. $\lambda(t+1) = \beta * \lambda(t) + \gamma * \lambda(t-1) + (1 - (\beta + \gamma)) * \lambda(t-2)$, where β and γ determine the weight on the current and previous arrival rates for prediction.

Control algorithm and performance specification: In this work we use a limited lookahead controller algorithm, which is a type of model predictive control. Starting from a time t_0, the controller solves an optimization problem defined over a predefined horizon ($t = 1 \ldots N$) and chooses the first input $u(t_0)$ that minimizes the total cost of operating the system J in the future prediction horizon. Formally, the chosen control input $u(t_0) = \arg\min_{u(t) \in U} \left(\sum_{t=t_0+1}^{t=t_0+N} J(x(t), u(t))\right)$. During this work, we limited the horizon to $N = 2$ as there is a significant computation cost associated with a longer horizon.

The cost function, J, at time t, is the weighted conjunction of drift of system state $x(t)$, ($x(t) = [q(t); r(t)]$) from the desired set point x_s, of the system state ($x_s = [q*, r*]$ where $q* =$ desired maximum queue size, $r* =$ desired maximum response time) and power consumption $E(t)$ (desired power consumption is 0). Formally, $J(t) = Q * \|x(t) - x_s\| + R * \|E(t)\|$, where Q and R are user-specified relative weights for the drift from the optimal system state x_s and power consumption $E(t)$, respectively.

The power consumption $E(t)$ is predicted with the help of the *look-up table* generated in Section 27.6.1 based on the current frequency of the CPU core and aggregate system utilization of the physical server.

27.7.1 Power Consumption and Response Time Management

This section uses the concepts introduced in the earlier sections for developing a control structure to manage server power consumption while maintaining the predefined QoS requirement of minimum response time under a time-varying dynamic workload for a daytrader application hosted in a virtualized environment (see Section 27.4). The following sections will give details of these experiments.

Experiment settings: Experimental settings and incoming request profiles were kept similar to Section 27.6.3 for direct comparison between the Web server performance with and without the controller implementation. A local monitor running on the VM (*Nop09*) hosting Web server collected, processed, and reported performance data after every SAMPLE_TIME (30 s) to the controller running on the physical host machine (*Nop03*). These performance data include average service time at Web server (computation time at application tier as well as query time over database tier), average queue size (average resident request into the system) of the system during the time interval of SAMPLE_TIME, and request arrival rate. The average queue size of the system is measured based on the total resident request in the system at previous sample, (plus) total incident request into the system, and (minus) total completed requests from the system in the current sample duration.

System state for predictive control: We used the exponential KF described in Section 27.6.3 to track the system state online. The two main parameters received from the filter are the current service time S and predicted response time T_{pred}. These values are then plugged into the model described in the previous section. The power model described in Section 27.6.1 was used to estimate the system (physical node of Web server) power consumption. With the help of these system and power models, the predictive controller provides the optimal configuration of the system in terms of CPU core frequency. The performance of the online controller directly depends on the accuracy of Kalman filter estimation of the parameters of the Web server application model and the power consumption model of the physical system.

For this experiment, we chose optimal system state set point to be $x_s = [q*, r*]$ where $q* = 0$ and $r* = 0$, which shows our inclination toward keeping system queue and response time both minimum. Q and R (user-specified relative weights for cost function) were chosen as 10,000 and 1, respectively, to penalize the multitier system a lot more for increment in queue size and response time compared to the increment in power consumption. Additionally, look-ahead horizon value N is 2 for the current experiment. Request forecasting parameters β and γ were equal to 0.8 and 0.15, respectively, to put maximum weight on the current arrival rate to calculate future arrival request rate.

Results from the experiment are shown in Figures 27.11 and 27.14, while plots of the estimates from the online KF are shown in Figure 27.13. Additionally, the direct comparison of the response time statistics and power consumption from Sections 27.6.3 and 27.7.1 are shown in Figure 27.14. Validation of our power consumption model defined in Section 27.6.1 is done by comparing the power consumption estimated by predictive controller against actual power consumption reported by a real-time wattmeter (see Figure 27.12).

Observations from Figures 27.11 and 27.14: The aggregate CPU utilization and memory utilization (Figure 27.11a) of the Web server and database tier are shown in Figure 27.11. *Nop03* CPU core frequencies during the experiment are shown in Figure 27.11b, and the Java thread utilization of the Web server is shown in Figure 27.11c. Figure 27.11d shows the queue size of the Web server through the method described in Section 27.7. The most interesting plot in Figure 27.11 is Figure 27.11b, which shows the change in frequency of the CPU core from the controller to achieve predefined QoS requirements based on the control steps taken by observing the system state and estimating the future environmental inputs. After direct comparison of Figure 27.11a and b, we can see that *Nop03* CPU core frequency is changed as the incident request rate at the Web server changes. Additionally, the controller chose a 1.2 GHz frequency for the CPU core until there was some sudden increase or decrease in the incident request rate.

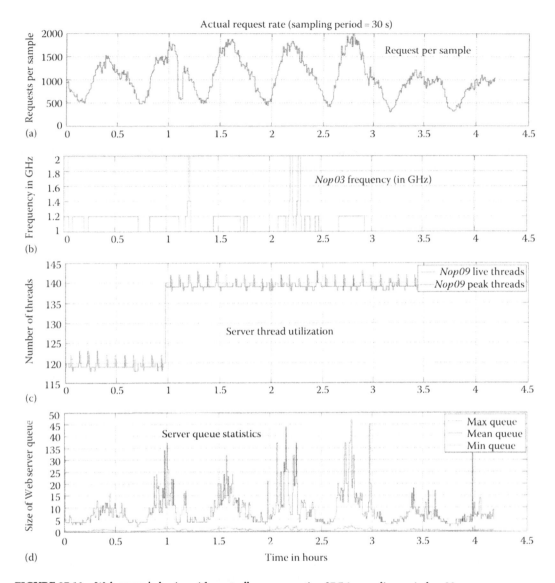

FIGURE 27.11 Web server behavior with controller as per section 27.7.1: sampling period = 30 s.

Furthermore, the controller does not change the frequency of the core too often, even when the incident request rate is changing continuously, which shows the minimal disturbance in the system operation due to predictive controller. The power consumption plot for *No03* is shown in Figure 27.14c, while statistics (max and min) of observed response time at Web server are shown in Figure 27.14a and b.

27.7.2 Performance Analysis

According to Figure 27.13, the online KF tracks average response time (Figure 27.13c) of the incident requests and bottleneck utilization (Figure 27.13b) with high accuracy. The estimated service time of the incident requests by the KF shows minimal variation. According to Figure 27.13c, predicted response time from the KF Tpred and actual response time T are also very close to each other, which indicates the efficiency of the KF. The controlled version runs at a lower frequency most of the time, which results in

FIGURE 27.12 Comparison of power consumption for actual vs. predicted through predictive controller in Section 27.7.1. (a) Predicted and actual power consumption in Web server from Section 27.7.1 (b) Error in predicting power consumption compared to actual in Section 27.7.1

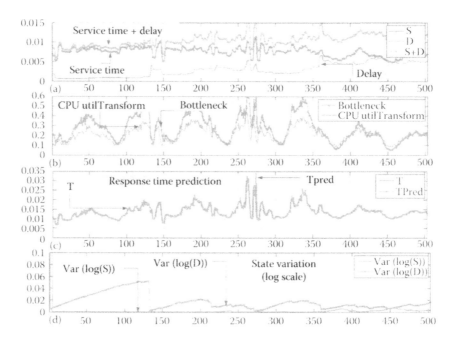

FIGURE 27.13 Online exponential KF output corresponding to the experiment from Section 27.7.1 (Figures 27.11 and 27.14). Service time and delay are in millisecond range. Response time is specified in seconds.

considerable amount of power savings (18%) over a period of 4 h of experiment (Figure 27.14c) compared to the baseline experiment shown in Section 27.6.3. The controller changes the frequency of the CPU core at very few occasions, but it is able to identify the sudden increase in the incident request rate which reflects the adaptive nature of the controller in case of dynamic load conditions.

According to Figure 27.14a and b, even after the presence of a local controller and slow running system (lower frequency), response times at the Web server are in the similar range in both the cases. It shows that the controller, while managing to decrease power consumption, does not affect QoS objectives of the system negatively. Furthermore, it is visible that there is negligible memory and CPU overhead due to the controller (Figure 27.14d). The overhead in virtual CPU utilization over the Web server *Nop09* can mostly be attributed to the lower physical core frequency. According to Figure 27.12, the power model described in Section 27.6.1 estimates the power consumption in the physical machine *Nop03* quite well with only 5% average error in prediction that indicates its effectiveness. Additionally, Java thread utilization is less

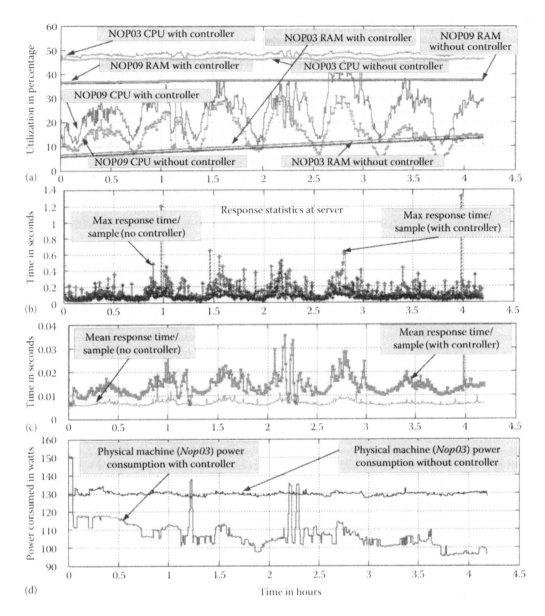

FIGURE 27.14 Comparison of results with controller (Section 27.7.1) and without controller (Section 27.6.3): Sampling period = 30 s. Standard deviation for all response measurements = 0.02 s (without controller), 0.019 s (with controller).

in the case of the controller, which indicates that even after slowing down the system, incident requests are getting served in time without much contention of computational resources. Furthermore, we found out that the mean server queue statistics are also in the same range for both the cases (details about mean queue statistics in the no controller case are available in [43]).

Observations in the previous paragraph indicate that the current system model captures the dynamics of the multitier Web server (daytrader) well. Additionally, the system model uses typical control inputs, state variables, and performance measurements of the multitier Web service domain for achieving QoS objectives that make the proposed framework suitable for any multitier Web service system.

27.8 Conclusion and Discussions

We have presented a simple and novel approach to develop models with low variance for multitier enterprise systems. We showed that the developed model can be integrated with a predictive control framework for dynamically changing the system tuning parameters based on the estimated time-varying workload. According to the results shown in Section 27.7.1, the developed system model in terms of a KF tracks the system performance online with high accuracy. Additionally, the proposed power consumption model of the system used by the controller predicts the overall physical server power consumption well (95% accurate). Using this model we showed that we can optimize system performance and achieve an 18% reduction of power consumption in 4 h of experiment in a single server without affecting the response time too much. Furthermore, the experimental results (CPU and RAM consumption with and without the controller) indicate that the proposed approach has low run-time overhead in terms of computational and memory resources. We further plan to extend this framework and validate its performance over a cluster of multitier computing systems in hierarchical fashion as described in [29].

Enormous research has been done by industry and academia to make computing system infrastructures power aware by applying power management policies. These policies include hardware redesign, application level control, efficient cooling systems, and shutting down the unused servers in a data center. Furthermore, current virtualization technologies are also used for saving the infrastructure and operating costs of the computing infrastructure. Virtualization enables usage of idle CPU cycles by multiple servers while sharing the hardware resources at the same time to reduce overall power consumption of the system, hardware cost, and hosting space. Due to increased power consumption in data centers, there is tremendous need for profiling of data centers with respect to hot spots, air conditioning, and active server usage. Gradually, data center vendors are adopting best practices to increase the efficiency of their infrastructure. These best practices include identifying power consumption in components with respect to impact on performance, enabling the power awareness features, appropriately sized server farms, shutting down unused servers, and removing the unused old servers from the infrastructure [4]. Shutting down the unused servers and installation of an efficient cooling system is easy to accomplish, while other methods, which need careful observation of system behavior with changes in environment, are complex and need a skilled architect to design the power-efficient policies that can be applicable to various scenarios. In conclusion, it is not necessary to replace the existing hardware with a new power-aware hardware; instead a significant amount of power saving can be achieved easily by employing power-aware application (system) controllers and policies on the existing systems.

References

1. B. Tudor and C. Petty. Gartner says data center power, cooling and space issues are set to increase rapidly as a result of new high-density infrastructure deployments. http://www.gartner.com/it/page.jsp?id=1368614 (accessed Nov 2010).
2. C. Petty and B. Tudor. Gartner says energy-related costs account for approximately 12 percent of overall data center expenditures. http://www.gartner.com/it/page.jsp?id=1442113 (accessed Nov 2010).
3. Daytrader. http://cwiki.apache.org/GMOxDOC20/daytrader.html (accessed Nov 2010).
4. M. Blackburn. Five ways to reduce data center server power consumption. http://www.thegreengrid.org/Global/Content/white-papers/Five-Ways-to-Save-Power (accessed Nov 2010).
5. Httperf documentation. Technical report, HP, 2007.
6. S. Abdelwahed, N. Kandasamy, and S. Neema. Online control for self-management in computing systems. In *Proceedings of the 10th IEEE Real-Time and Embedded Technology and Applications Symposium*, Washington, DC, pp. 365–375, 2004.

7. S. Abdelwahed, S. Neema, J. Loyall, and R. Shapiro. A hybrid control design for QoS management. In *The 24th IEEE International Real-Time Systems Symposium*, pp. 366–369, Cancun, Mexico, 2003.

8. T.F. Abdelzaher, K.G. Shin, and N. Bhatti. Performance guarantees for web server end-systems: A control-theoretical approach. *IEEE Transactions on Parallel and Distributed Systems*, 13(1):80–96, January 2002.

9. M. Arlitt and T. Jin. Workload characterization of the 1998 World Cup Web site. Technical Report HPL-99-35R1, Hewlett-Packard Labs, September 1999.

10. M. Bhardwaj, R. Min, and A. Chandrakasan. *Power-Aware Systems*, Cambridge, MA, 2000.

11. L. Bisdounis, S. Nikolaidis, O. Koufolavlou, and C.E. Goutis. Modeling the CMOS short-circuit power dissipation. In *Circuits and Systems, 1996. ISCAS '96., "Connecting the World"., 1996 IEEE International Symposium*, vol. 4, pp. 469–472, Atlanta, GA, May 1996.

12. D. Brooks, V. Tiwari, and M. Martonosi. Wattch: A framework for architectural-level power analysis and optimizations. In *Computer Architecture, 2000. Proceedings of the 27th International Symposium*, pp. 83–94, Vancouver, BC, Canada, 2000.

13. A. Cervin, J. Eker, B. Bernhardsson, and K.-E. Arzen. Feedback-feedforward scheduling of control tasks. *Journal of Real Time Systems*, 23(1–2):25–53, 2002.

14. S. L. Chung, S. Lafortune, and F. Lin. Limited lookahead policies in supervisory control of discrete event systems. *IEEE Transactions on Automatic Control*, 37(12):1921–1935, December 1992.

15. V. Delaluz, N. Vijaykrishnan, A. Sivasubramaniam, and M.J. Irwin. Memory energy management using software and hardware directed power mode control. Technical report, 2000.

16. Y. Diao, J. L. Hellerstein, S. Parekh, H. Shaikh, M. Surendra, and A. Tantawi. Modeling differentiated services of multi-tier web applications. *MASCOTS*, 0:314–326, 2006.

17. B. Diniz, D. Guedes, W. Meira, Jr., and R. Bianchini. Limiting the power consumption of main memory. In *Proceedings of the 34th Annual International Symposium on Computer Architecture*, ISCA '07, pp. 290–301, New York, ACM, 2007.

18. A. Dubey et al. Compensating for timing jitter in computing systems with general-purpose operating systems. In *ISROC*, Tokyo, Japan, 2009.

19. A. Dubey, R. Mehrotra, S. Abdelwahed, and A. Tantawi. Performance modeling of distributed multi-tier enterprise systems. *SIGMETRICS Performance Evaluation Review*, 37(2):9–11, 2009.

20. D. Economou, S. Rivoire, and C. Kozyrakis. Full-system power analysis and modeling for server environments. In *Workshop on Modeling Benchmarking and Simulation (MOBS)*, 2006.

21. D. Gupta, R. Gardner, and L. Cherkasova. Xenmon: QoS monitoring and performance profiling tool. Technical report, HP Labs, 2005.

22. J. Hellerstein, Y. Diao, S. Parekh, and D.M. Tilbury. *Feedback Control of Computing Systems*. Wiley-IEEE Press, Hawthorne, New York, 2004.

23. Y. Hu, Q. Li, and C.-C. J. Kuo. Run-time power consumption modeling for embedded multimedia systems. In *Proceedings of the 11th IEEE International Conference on Embedded and Real-Time Computing Systems and Applications*, pp. 353–356, Hong Kong, China, 2005.

24. I. Hur and C. Lin. A comprehensive approach to dram power management. In *IEEE 14th International Symposium on High Performance Computer Architecture, HPCA*, pp. 305–316, Salt Lake City, UT, 2008.

25. R. Joseph, M. Martonosi. Run-time power estimation in high performance microprocessors. In *International Symposium on Low Power Electronics and Design*, pp. 135–140, Huntington Beach, CA, 2001.

26. R. E. Kalman. A new approach to linear filtering and prediction problems. *Transactions of the ASME Journal of Basic Engineering*, 82(Series D):35–45, 1960.

27. E. Kalyvianaki, T. Charalambous, and S. Hand. Self-adaptive and self-configured cpu resource provisioning for virtualized servers using kalman filters. In *ICAC '09: Proceedings of the 6th International Conference on Autonomic Computing*, pp. 117–126, New York, ACM, 2009.

28. S. Kamil, J. Shalf, and E. Strohmaier. Power efficiency in high performance computing. In *IEEE International Symposium on Parallel and Distributed Processing, IPDPS*, pp. 1–8, Miami, FL, 2008.

29. N. Kandasamy, S. Abdelwahed, and M. Khandekar. A hierarchical optimization framework for autonomic performance management of distributed computing systems. In *Proceedings of 26th IEEE International Conference on Distributed Computing Systems (ICDCS)*, Lisboa, Portugal, 2006.

30. D. S. Kandemir, V. Delaluz, A. Sivasubramaniam, N. Vijaykrishnan, and M. J. Irwin. Scheduler-based DRAM energy management. In *Proceedings of the 39th Conference on Design Automation*, pp. 697–702. ACM Press, New Orleans, LA, 2002.

31. J. O. Kephart and D. M. Chess. The vision of autonomic computing. *Computer*, 36:41–50, January 2003.

32. L. Kleinrock. *Theory, Volume 1, Queueing Systems*. Wiley-Interscience, Richmond, TX, 1975.

33. Z. Qian, Z. Wang, R. P. Dick, Z. Mao, L. Zhang, B. Tiwana, and L. Yang. Accurate online power estimation and automatic battery behavior based power model generation for smartphones. In *Proceedings of International Conference on Hardware/Software Codesign and System Synthesis*, Scottsdale, AZ, 2010.

34. E. Lee, I. Kulkarni, D. Pompili, and M. Parashar. Proactive thermal management in green datacenters. *The Journal of Supercomputing*, Vol. 38, pp. 1–31, 2010. 10.1007/s11227-010-0453-8.

35. W. Liao, L. He, and K. M. Lepak. Temperature and supply voltage aware performance and power modeling at microarchitecture level. *IEEE Transactions on Computer-Aided Design of Integrated Circuits and Systems*, 24(7):1042–1053, 2005.

36. X. Liu, X. Zhu, S. Singhal, and M. Arlitt. Adaptive entitlement control of resource containers on shared servers. In *Proceedings of 9th IFIP/IEEE International Symposium on Integrated Network Management (IM)*, Nice, France, 2005.

37. C. Lu, Guillermo A. Alvarez, and J. Wilkes. Aqueduct: Online data migration with performance guarantees. In *FAST '02: Proceedings of the 1st USENIX Conference on File and Storage Technologies*, p. 21, Berkeley, CA, USENIX Association, 2002.

38. C. Lu et al. Feedback control real-time scheduling: Framework, modeling and algorithms. *Journal of Real-Time Systems*, 23:85–126, 2002.

39. Z. Lu et al. Control-theoretic dynamic frequency and voltage scaling for multimedia workloads. In *International Conference on Compilers, Architectures, and Synthesis Embedded Systems (CASES)*, pp. 156–163, Grenoble, France, 2002.

40. J. M. Maciejowski. *Predictive Control with Constraints*. Prentice Hall, London, 2002.

41. X. Wang, M. Eiblmaier, and R. Mao. Power management for main memory with access latency control. *Febid '09*, San Francisco, CA, ACM, 2009.

42. R. Mehrotra, A. Dubey, S. Abdelwahed, and A. Tantawi. Integrated monitoring and control for performance management of distributed enterprise systems. *International Symposium on Modeling, Analysis, and Simulation of Computer Systems*, 0:424–426, Miami, FL, 2010.

43. R. Mehrotra, A. Dubey, S. Abdelwahed, and A. Tantawi. Model identification for performance management of distributed enterprise systems. Technical Report ISIS-10-104, Institute for Software Integrated Systems, Vanderbilt University, April 2010.

44. M. Rupp, Mostafa E. A. Ibrahim, and Hossam A. H. Fahmy. A precise high-level power consumption model for embedded systems software. *EURASIP Journal on Embedded Systems*, 2011, pp. 1:1–1:14 2010.

45. G. Pacifici, W. Segmuller, M. Spreitzer, and A. Tantawi. CPU demand for web serving: Measurement analysis and dynamic estimation. *Performance Evaluation*, 65(6–7):531–553, June 2008.

46. S. Parekh, N. Gandhi, J. Hellerstein, D. Tilbury, T. Jayram, and J. Bigus. Using control theory to achieve service level objectives in performance management. In *Proceedings of IFIP/IEEE International Symposium on Integrated Network Management*, Seattle, WA, 2001.

47. David A. Patterson and John L. Hennessy. *Computer Organization and Design, The Hardware/Software Interface*, 4th Edn. Morgan Kaufmann, Burlington, MA, 2008.

48. D. Roberts, T. Kgil, and T. Mudge. Using non-volatile memory to save energy in servers. In *Design, Automation Test in Europe Conference Exhibition, 2009. DATE '09.*, pp. 743–748, Nice, France, 2009.

49. T. Simunic and S. Boyd. Managing power consumption in networks on chips. In *Proceedings of Design, Automation, and Test Europe (DATE)*, pp. 110–116, Paris, France, 2002.

50. A. Verma, P. Ahuja, and A. Neogi. Power-aware dynamic placement of HPC applications. In *Proceedings of the 22nd Annual International Conference on Supercomputing*, ICS '08, pp. 175–184, New York, ACM, 2008.

51. M. Woodside, T. Zheng, and M. Litoiu. The use of optimal filters to track parameters of performance models. In *QEST '05: Proceedings of the Second International Conference on the Quantitative Evaluation of Systems*, p. 74, Torino, Italy, September 2005.

52. M. Woodside, T. Zheng, and M. Litoiu. Service system resource management based on a tracked layered performance model. In *ICAC '06: Proceedings of the Third International Conference on Autonomic Computing*, pp. 175–184. Dublin, Ireland, IEEE Press, June 2006.

53. J. Zedlewski, S. Sobti, N. Garg, F. Zheng, A. Krishnamurthy, and R. Wang. Modeling hard-disk power consumption. In *Proceedings of the Second USENIX Conference on File and Storage Technologies*, pp. 217–230, Berkeley, CA, USENIX Association, 2003.

54. F. Zhang and Samuel T. Chanson. Power-aware processor scheduling under average delay constraints. In *Proceedings of the 11th IEEE Real Time on Embedded Technology and Applications Symposium*, pp. 202–212, Washington, DC, IEEE Computer Society, 2005.

55. J. Zhang, J. G. Dai, and B. Zwart. Law of large number limits of limited processor-sharing queues. *Mathematics of Operations Research*, 34:937–970, November 2009.

56. J. Zhang and B. Zwart. Steady state approximations of limited processor sharing queues in heavy traffic. *Queueing Systems*, 60:227–246, 2008.

57. T. Zheng, M. Woodside, and M. Litoiu. Performance model estimation and tracking using optimal filters. *IEEE Transactions on Software Engineering*, 34(3):391–406, May–June 2008.

58. T. Zheng, J. Yang, M. Woodside, M. Litoiu, and G. Iszlai. Tracking time-varying parameters in software systems with extended Kalman filters. In *CASCON '05: Proceedings of the 2005 Conference of the Centre for Advanced Studies on Collaborative Research*, pp. 334–345, IBM Press, Toronto, Ontario, Canada, October 2005.

59. Victor V. Zyuban and Peter M. Kogge. Inherently lower-power high-performance superscalar architectures. *IEEE Transactions on Computers*, 50:268–285, March 2001.

28

Power Measuring and Profiling: State of the Art

Hui Chen
Wayne State University

Weisong Shi
Wayne State University

28.1 Introduction

In the late 1990s, power, energy consumption, and power density had become the limiting factors not only for the system design of portable and mobile devices, but also for high-end systems [40]. The design of the computer system had changed from the performance-centric stage to the power-aware stage.

Accompanied by the increasing of the integration of CMOS (complementary metal-oxide semiconductor) circuits and the clock frequency, the power density of the hardware circuits increased quickly. To cool down the processor with a cost-effective method may be challenging if the power density continues to increase. This problem makes power one of the first-class system design considerations, although power is not a new problem for computer system design, and performance is not the only objective any more. Making performance/power trade-offs becomes very important when we design the hardware architecture of computer systems.

The trend of personal computers, portable computers and mobile devices will require long-lasting batteries, otherwise the user experience will be bad. To extend battery lifetime, we not only need to develop new battery techniques, but to improve the energy efficiency of these devices. Improving energy efficiency of mobile devices requires both hardware and software methods. The hardware methods aim to design low-power circuits and support dynamic power management strategies, and software methods try to make power/performance trade-offs with the APIs supplied by hardware.

Finally, accompanied by the emerging of the Internet, many large data centers are built by big companies to supply stable, good quality service for customers. Even when the workload is low, these servers still have very large power dissipation, because they are designed for the peak workload. These servers always need to be kept "on," even though most of the time the workload is trivial. Furthermore, a similar amount of money, as used for energy consumption, needs to be spent on cooling down the data center. These reasons cause unproportional energy problems in data centers [4].

28.1.1 Terminologies

Before describing more details about power measuring and profiling, we first introduce some basic terminologies.

- *Energy*: In computer systems, energy, in joules, is the electricity resource that can power hardware devices to do computation. More energy is used for the research of mobile platform and data centers. For mobile devices, energy is strongly related to battery lifetime. For data centers, which consume a large amount of energy, energy is used as the concern of electricity costs. Usually, researches in these areas use energy efficiency, such as PUE (power usage effectiveness), as the metric to evaluate their work.
- *Power*: Power is the dissipation rate of energy. The unit of power is watt (W) (or joule per second). Originally, this metric is used to reflect the current delivery and voltage regulator of the circuits. In system research, power may also be used for the abstract concepts, such as process and operating system. For example, we may say that the power of a process is 1 W. This means that the execution of this process causes the hardware circuits to dissipate 1 W of power.
- *Power density*: Power density is the amount of power dissipated per unit area. This metric represents the heat problem of the processor die, because power has a direct relationship with heat.
- *Static power*: Static power, which is mainly leakage power, is the power caused by the incomplete turning off of transistors. This terminology denotes the basic power that a hardware device needs when it is not active.
- *Dynamic power*: Dynamic power is the power caused by the switching of the capacitance voltage states. This terminology denotes the extra part of power needed to make the hardware device work.
- *Idle power*: Idle power is the power of the full system when the system is in the idle state. This terminology is used for system-level research, because hardware devices are not really inactive even when the system is idle. Idle power includes the dynamic power of the system hardware and a part of dynamic power caused by some system processes.
- *Active power*: Active power is the extra power dissipated by the system when it is doing the computation.

28.1.2 Power-Aware System Design

In the early age, researchers tried to decrease the power and eliminate waste of energy during the architecture design stage, because the power problem obstructs the development of computer systems. These techniques include multicore on-chip processor (CMP), core-independent functional units, dynamic frequency and voltage scaling and clock gating. Most of these techniques are designed to decrease the energy consumption of the processor, which is the dominant energy consumer in a typical computer system. In addition, some other techniques, such as the phase-change memory and solid-state disk driver, are also proposed to decrease the power dissipation of other devices.

Although hardware-based power management techniques have been proven useful for reducing the energy consumption of the computer system, and they will be of continuing importance in the future, researchers point out that these techniques alone are not enough, and higher-level strategies for reducing energy consumption will be increasingly critical [10]. Current operating systems are not designed by considering the power problem; thus, even if the system is idle it still consumes a large amount of energy, which is seen as waste and not used for computing. Vahdat et al. argue that traditional operating system design should be revisited for energy efficiency [47]. H. Zeng et al. present EcoSystem [54], which tries to manage power as one kind of system resource. Except for designing new operating systems, some other high-level power-aware strategies, such as power-aware scheduling [38], are also globally researched. Moreover, power analysis is equally important when designing software. Some scholars even proposed energy complexity [26] as one of the metrics for algorithm design.

28.1.3 Power Measuring and Profiling

Power measuring and profiling is a new area that is arising in parallel with the power problem. As energy consumption becomes one of the foremost considerations when designing computer systems, power profiling becomes a key issue in the community of computer systems. As the basis of the power-aware system research, power profiling is not only needed to evaluate power optimization techniques and to make power/performance trade-offs it is also important to supply critical power information for operating systems and power-aware softwares.

The early stage publications [10,34,49,52] are mainly based on simulation because it was used during the hardware design stage to make power/performance trade-offs. Most of them estimate the power consumption of the hardware circuits based on the classical power model shown in Equations 28.1 and 28.2. Here, C is the load capacitance, V_{dd} is the supply voltage, α is the activity ratio of the circuits, f is the clock frequency, and $I_{leakage}$ is the leakage current of the circuits. The content of simulation methods is beyond the scope of this chapter.

$$P_{dynamic} = CV_{dd}^2 \alpha f \qquad (28.1)$$

$$P_{static} = I_{leakage} V_{dd}^2 \qquad (28.2)$$

Even though these circuit-level power models are good to estimate the circuits' power, they cannot be used to estimate the power of the real products. To understand the power dissipation of the real systems becomes critical; thus many researchers use instruments to measure the power of these devices directly. The measured power is more accurate than estimated power. Accordingly it is used by many to analyze the power problems of the computer system [27,30,35], to validate the effect of the power-aware strategies, and to build software power models with the linear regression method [5,23,28]. While the measured power is accurate, these methods also have two limitations. First, low-level power can hardly be measured. Second, the measured power cannot be used by power-aware strategies or the cost is too high.

To cope with the limitations of direct measurement, the following research is trying to use software methods to estimate the power on the online system. These works are done from different levels: component level [8], core level [6], CPU functional unit level [23], process level [15], and virtual machine level [31]. These articles either use operating system events or hardware performance counter events to build their power model.

Basically, we can classify power measuring and profiling techniques into three categories: simulation approach, hardware measurement, and software power profiling. In the following sections, we mainly describe the last two approaches.

28.2 Hardware-Based Power Measurement

Hardware-based power measurement uses instruments to measure the current or voltage of the hardware devices and further uses these measured values to compute the power of the measured object. The instruments used to do the measurement includes different types of meters, special hardware devices that can be embedded into the hardware platforms, and power sensors that are designed within the hardware. Normally, these methods can only measure the component level power, because the high integration of the hardware circuits makes the lower-level functional units become difficult to measure. Most publications [23,27] make use of microbenchmarks, which stress one or several special functional units, to isolate lower-level power.

28.2.1 Power Measurement with Meters

Direct power measurement with meters is a straightforward method to understand the power dissipation of devices and the full system. Many articles [9,23] rely on power meters to measure the real power and use it to validate their research work or to do analysis. Moreover, some works [18] measure the hardware components' power and break it down into lower levels based on some indicators that could reflect the activity of these lower-level units. The differences in these methods are which type of meter is used to do the measurement and at which place the measurement is done.

One type of the globally used meters is the digital multimeter, which is easy to use. Generally, these meters can sample the measured object each second. The result can be collected by the serial port that is connected to the data collection system. Using this method we need to disconnect the wires that we want to measure and connect a small resistor (normally less than 0.5 Ω). Finally, we measure the voltage of the resistor and compute the power on this wire. Figure 28.1 shows an example of this method. Joseph et al. use multimeters to measure the power while executing different benchmarks and make power/performance trade-offs [27]. Using this method, Feng et al. measured the node level and component level power for a node of the distributed system [16].

Another kind of meter that many people use is the clamp meter, which can measure current without disconnecting the wire. Usually a clamp meter has a larger measurement range than multimeter; thus, it can be used to measure the power of systems in which the current is much higher. The connection of the measurement platform of the clamp meter is similar to the multimeter. In [30], the authors adopt the direct power measurement method with clamp meters to measure the power on a Cray XT4 supercomputer under several HPC (high performance computer) workloads. Their results show that computation-intensive

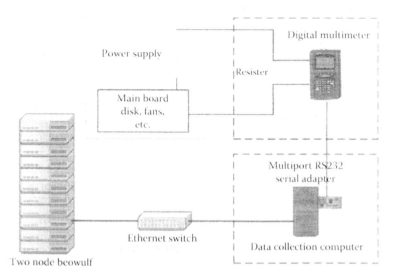

FIGURE 28.1 An example of using a multimeter to measure the power. (From Feng, X. et al., Power and energy profiling of scientific applications on distributed systems, in *Proceedings of the 19th IEEE International Parallel and Distributed Processing Symposium (IPDPS'05)–Papers–Volume 01*, IPDPS '05, p. 34, IEEE Computer Society, Washington, DC, 2005.)

benchmarks generate the highest power dissipation while memory-intensive benchmarks yield the lowest power usage.

Multimeter and clamp meters are mainly used to measure the DC power by connecting them between the power supply and the measured component. In addition, we also use one kind of power meter, such as "Watts UP" [2], to measure the AC power. This kind of meter can only measure the system-level power, because only power supplies are powered by AC. AC power is good for understanding the total system power characters, but improper for lower-level analysis, because the transform efficiency is not constant during the measurement.

28.2.2 Power Measurement with Specially Designed Devices

While direct measuring with meters is simple, it does not supply methods to control the measurement process, for example, to synchronize the measured power to the monitoring of the performance metrics. Thus, some specially designed power measurement devices are presented to measure the power in these circumstances. One of the early methods is by Itsy [48], which is used to measure the power of mobile devices. The platform is integrated between the power supply and the measured mobile device as shown in Figure 28.2. This framework could not only measure the power but could also trigger the measured devices. (Section 28.3.2 will describe an example of using this platform.)

PLEB [45] is a single board on computer that is designed with a set of current sensors on board. Furthermore, the microcontroller of this device is integrated with an analog-to-digital converter to read the sensors. This platform can be used to isolate the power of processor, memory, flash driver, and I/O devices. Figure 28.3 shows the structure of PLEB.

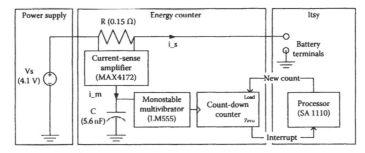

FIGURE 28.2 The Itsy energy-driven statistical sampling prototype. (From Feng, X. et al., Power and energy profiling of scientific applications on distributed systems, in *Proceedings of the 19th IEEE International Parallel and Distributed Processing Symposium (IPDPS '05)–Papers–Volume 01, IPDPS '05*, p. 34, IEEE Computer Society, Washington, DC, 2005.)

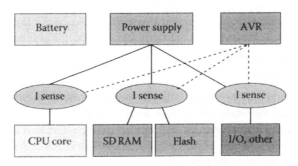

FIGURE 28.3 The structure of PLEB. (From Snowdon, D.C. et al., Power measurement as the basis for power management, In *2005 Workshop on Operating System Platforms for Embedded Real-Time Applications*, 2005.)

28.2.3 Integrating Sensors into Hardware

The last type of approach is mainly used by high-performance servers. In the past several years, the majority of designed servers contains a service processor [19,21], which is a hardware- and software-integrated platform that works independently of the main processor and the server's operating system. The hardware of the service processor may either be embedded on the motherboard or on a plug-in card. Most of the hardware of the service processor include power sensors to monitor the power supplied to the administrator for power management. For example, Intel's service processor Intelligent Platform Management Interface (IPMI) [22] supports APIs to read the power information monitored by the sensors.

Other techniques [1,46] that are designed to improve the energy efficiency of data centers that are used on servers integrate power sensors at a deeper level. IBM BladeCenter and System *x* servers supply PowerExecutive solutions, which enable customers to monitor actual power draw and temperature loading information [1].

Though online power-aware applications can use this method, it is difficult to gain lower-level power information, in which case hardware circuits are too complicated to distinguish the originality of power dissipation. In addition, power monitoring circuits also dissipate a large amount of power as well.

28.3 Software-Based Power Profiling

Although the hardware approach can supply very accurate power information, the problems we listed in the last section limit its application range. The software-based approach, however, can be used to supply more fine-grained online power information, which could be used by power-aware strategies. The live power information of systems is highly needed for designing high-level energy-efficient strategies. For example, Ecosystem [36,54], which proposes the concept of managing system energy as a type of resource, requires the support of real-time power information at different levels. As part of the energy-centric operating system, energy profiles are also needed by new power-aware scheduling algorithms [3,32]. Furthermore, compared with hardware measurement, the software-based method is much more flexible. Usually, we can apply it to different platforms without changing the hardware.

28.3.1 Power Model

The software-based approach usually builds power models to estimate the power dissipation of different levels: instruction level, program block level, process level, hardware component level, system level, and so forth. These methods first try to find the power indicators that could reflect the power of these software or hardware units. Then they build the power model with these power indicators and fine-tune the parameters of the power model. Finally, they verify the accuracy of the power models by comparing with the result measured by hardware or applying the power information into a power-aware strategy to test its usability. Based on the difference of the power indicators, we categorize these methods into two categories: system profile-based method and hardware performance counter (PMC)-based method.

The metrics of the power model may vary for each type of usage. Here, we list several important metrics we use to evaluate a power model.

- *Accuracy:* The accuracy of the power model defines how accurate the estimated power is relative to the measured power. For many applications, accuracy is the foremost requirement. For example, power/performance trade-off relies on the accurate power estimation to analyze, or the result is meaningless.
- *Simplicity:* The simplicity of the power model is also important in some circumstances, such as supplying real-time power information on mobile platforms. In these circumstances, if the power model is too complicated, the overhead of the power estimation will be too high to be able to be used.

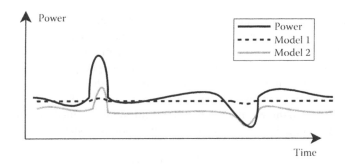

FIGURE 28.4 An example shows the responsiveness of the power model. (From Bertran, R. et al., Decomposable and responsive power models for multicore processors using performance counters, In *Proceedings of the 24th ACM International Conference on Supercomputing*, pp. 147–158, Ibaraki, Japan, ACM Press, 2010.)

This is usually because the power monitor needs to collect the events that will be used by the power model. Based on our experiment, accuracy is usually contradicted to simplicity, because the accuracy of the power model is closely related to the complexity and the sampling rate.

- *Responsiveness:* The responsiveness of the power model means whether it can reflect the variation of the power, as shown in Figure 28.4. In this example, the responsiveness of Model 2 is better than Model 1 even though Model 1 seems more accurate. However, if we use this model for the power-aware policies, the chance of triggering these policies will be lost.
- *Granularity:* Granularity means to which level the power model can estimate. For example, a power model that can estimate the main functional units of CPU has higher granularity than another power model that can only estimate the CPU power. Usually, high granularity means high accuracy and poor simplicity.

Different from the power models that are used for simulation, the power models described in this section are used to supply online (realtime) power information. Furthermore, these power models mainly use system profile or hardware performance counters. These publications are more concentrated on estimating the power of hardware components or software modules, for which the energy efficiency could be improved by using some energy-aware strategies with the power information estimated. In the following sections, we summarize these power models based on the information that the power model uses.

28.3.1.1 System Profile-Based Power Model

System profile or system events are a set of performance statistical information supplied by the operating system. These events reflect the current state of the hardware and software, including the operating system. For example, CPU utilization is the performance metric that can reflect the current workload of the processor. Nearly all operating systems support these system events: Linux saves these values under the "proc" directory, and Windows supplies a set of APIs, called the performance counter, to access these values. Here the performance counter is different from the hardware performance counters that will be talked about in the next section. Some works use them to build power models for the hardware component, process, and even program blocks, because these system events are directly related to the usage of the hardware and software.

The operating system, which is constituted of a set of system processes, consumes a large amount of power even when the system is idle, and it is the main cause of unproportional system usage and power dissipation. Several articles [36,47,54] review traditional operating system design with energy as one of the foremost important considerations. Thus, understanding the power dissipation of these system processes is very important for the energy proportional operating system design. In [33], Li et al. estimate the power dissipation caused by the operating system. First, they find the power behaviors of three types of OS routines: interrupts, process and interprocess control, and file system and miscellaneous services. These OS routines have different power behaviors. Some of them generate constant power dissipation and some others, such as process scheduling and file I/O, show higher standard deviation because they are largely dependent on system status. However, they find that the power of these OS routines has a linear relationship with instructions executed per cycle (IPC). They build the power model based on IPC as shown in Equation 28.3. In this equation k_1 and k_0 are constants obtained from linear regression step. Finally, they define the routine-level-based operating system power model as shown in Equation 28.4.

$$P = k_1 \times IPC + k_0 \tag{28.3}$$

$$E_{OS} = \sum_i P_{OS\_routine,i} \times T_{OS\_routine,i} \tag{28.4}$$

In [15], Do et al. build process-level energy models. First, they build energy models for three main components: CPU, memory, and wireless network interface card (WNIC). The CPU energy model is based on system events such as the active time of the CPU, the time that the CPU worked on each frequency, and frequency transition times. They assume a linear relationship between CPU frequency and power. This equation is shown in Equation 28.5, in which P_j denotes the power of the CPU when the frequency is j, n_k denotes the amount of times transition k occurs, and E_k is the corresponding energy consumption. The energy consumed by disk and WNIC is computed with the amount of data operated by these devices. These two energy models are shown in Equations 28.6 and 28.7. The disk read power (P_{read}), disk write power (P_{write}), WNIC send power (P_{send}), and WNIC receive power (P_{recv}) are assumed as constant values. The energy of a process is calculated with the hardware resource used by each device in the last time interval, shown in 28.8. Here, U_{ij} is the usage of the application i on resource j, $E_{resourcej}$ is the amount of energy consumed by resource j, and $E_{interaction}$ is the amount of energy consumed by the application because of the interaction among system resources, in the time interval t. The energy profiling module is running at kernel space. This tool supplies several APIs that other user space software can use, which returns the energy assumption of a process in a specified time interval.

$$E_{CPU} = \sum_i P_j t_j + \sum_i n_k T_k \tag{28.5}$$

$$E_{DISKi} = t_{readi} P_{read} + t_{writei} P_{write} \tag{28.6}$$

$$E_{NETi} = t_{sendi} P_{send} + t_{recvi} P_{recv} \tag{28.7}$$

$$E_{APPi} = \sum U_{ij} E_{resourcej} + E_{interaction} \tag{28.8}$$

Kansal et al. built the energy model for three main components, CPU, memory, and disk; then they broke down the energy into the virtual machine level based on the utilization of each component [31]. The CPU energy model they proposed is based on CPU utilization; the memory energy model uses the number of LLC (last level cache) misses; and the disk energy model relies on the bytes of data that the disk reads and writes. These three energy models are shown as follows:

$$E_{cpu} = \alpha_{cpu} \mu_{cpu} + \gamma_{cpu} \tag{28.9}$$

$$E_{mem}(T) = \alpha_{mem} N_{LLCM}(T) + \gamma_{mem} \tag{28.10}$$

$$E_{disk} = \alpha_{rb} b_R + \alpha_{wb} b_W + \gamma_{disk} \tag{28.11}$$

In these three equations, μ_{cpu}, $N_{LLCM}(T)$, b_W, and b_R denote CPU utilization, the number of LLC misses in time T, the amount of bytes write into the disk, and the amount of bytes read from disk. α and γ parameters are constants that they get when training the energy model. The last two methods, even though they build simple power models, only generate very low overhead, and thus can be used on real systems to provide online energy information. Similar to Kansal et al., Dhiman et al. also propose an online power prediction system for virtualized environments [14]. Instead of using linear regression, they utilize a Gaussian mixture vector quantization–based training and classification to find the correlation of measured events and the power.

SoftWatt, which models the CPU, memory hierarchy, and the low-power disk subsystem, is described in [20]. This tool is able to identify the power hot spots in the system components as well as the power-hungry operating system services. Zedlewski et al. present Dempsey, a disk simulation environment that includes accurate modeling of the disk power dissipation [53]. Dempsey attempts to accurately estimate the power of a specific disk stage, which includes seeking, rotation, reading, writing, and idle-periods, with a fine-grained model. Molaro et al. also analyze the possibility to create a disk driver power model based on the disk status stages [39].

28.3.1.2 PMC-Based Power Model

Hardware performance counters are a group of special registers that are designed to store the counts of hardware-related activities within computer systems. Compared with the system events supplied by the operating system or special software, hardware performance counters provide low-overhead access to a wealth of detailed performance information related to the CPU's functional units, main memory, and other components. These hardware performance events, such as L1/L2 cache miss times, could reflect the hardware activities. Thus, it is suitable for building the power models of these hardware components.

Even though many works use PMCs to establish the power model, the basic process of these methods is about the same. Here are the general steps for generating the power model with PMC events.

1. Classify the device into several main functional units.
2. Choose a group of hardware performance events that may be related to the power dissipation of the device for which we want to build the power model.
3. Collect the power of the device while different microbenchmarks are executed. These microbenchmarks stress one or several functional units of the device. In parallel with this process, we monitor all the performance counters we choose at the first step. Some early-stage processes only supply a limited number of PMCs; thus, we may need to repeat this process by running the same microbenchmark monitor in different PMC events.
4. Analyze the relationship of each PMC event we choose and the measured power with linear regression. This step finds correlation of PMC events and the power of each functional unit.
5. Select the PMC events that are most related to the functional unit's power to construct the unit-level power model. Usually, the power model of each unit is the product of a constant parameter and a value that reflects the active ratio of this unit, which is converted by the value of the PMC event. Some works omit the transfer and use the value of the PMC event directly.
6. Construct the device's power model by summarizing each functional unit's power model and the static power of this device.
7. Find the value of the parameters in the model and exercise it with some microbenchmarks.

To our knowledge, Bellosa et al. first proposed the idea of using PMCs to estimate power [5]. They ran specific calibration software, which stress one or several functional units, and use performance counters to monitor four types of hardware events, which are integer operations, floating-point operations, second-level address strobes, and memory transactions, individually. Then they analyze the correlation of these events with the power of the component and find that these four events are tightly related to the functional units that they stressed with the calibration software. Because the platform, a Pentium II 350 PC they used only supports two performance counters, they can only use a sampling method to estimate the power of each thread and the system. Similar to Bellosa, Joseph et al. verified the correlation of more than 10 CPU functional units and used several performance events that are most related to the CPU power to model the power of a Pentium Pro processor [28].

In [13], Contreras et al. build power models for the Intel PXA255 processor and memory. The processor power model uses four hardware events, instructions executed, data dependencies, instruction cache miss, and TLB (translation lookaside buffer) misses, and assume these counter values have the linear relationship with the processor power. They build the memory power model with three counters: instruction cache miss, data cache miss, and number of data dependencies. Bircher et al. find that the events that have higher correlation with power are all IPC related [7].

Different from the previous work that ignores the power dissipated by the small subunits of the processor, or consider it as static power, Isci et al. build a more complicated power model for a Pentium 4 processor by considering many more CPU functional units [23]. We cannot neglect this part of energy because it may account for nearly about 24% of the total CPU power [28]. Another difference is that they choose 22 fine granularity physical components that can be identifiable on a P4 die photo. For each of these physical components, they use an event or a group of events that can reflect the access amount of this physical component. For example, they use IOQ Allocation, which counts all bus transactions (reads, writes, and prefetches) that are allocated in the IO Queue, and FSB (front side bus) data activity, which counts all DataReaDY and DataBuSY events on the front side bus, to compute the bus control access rates. In [24], they list the equations that compute all the 22 physical components' access rate. Then they estimate each physical component power used in Equation 28.12 and estimate the total power of CPU with Equation 28.13.

$$Power(P_i) = AccessRate(C_i)$$
$$.Architectual\ Scaling(C_i)$$
$$.Max\ Power(C_i)$$
$$+ NonGatedClock\ Power(C_i) \tag{28.12}$$

$$Total\ Power = \sum_{i=1}^{22} Power(C_i) + Idle\ Power \tag{28.13}$$

In Equation 28.12, $AccessRateC_i$, $Architectual\ Scaling(C_i)$, $Max\ Power(C_i)$, and $NonGatedClock\ Power(C_i)$ denote the access rate, a scaling strategy that is based on micro-architectural and structural properties, the maximum power, and the conditional clock power of physical component C_i. Here the maximum power and the conditional clock power are estimated based on empirical value. For example, the initial maximum power is estimated based on area of this unit on the die. This estimation is reasonable because the area is related to the number of CMOS used by this unit. Then, the summation of all the physical components' power plus the idle power of the processor is the total power. Finally, they fine-tune the parameters in the equations by running a group of training benchmarks, which exercise CPU differently, and compare the estimated power with power meters' measured result.

Although previous efforts have already been able to accurately estimate the CPU power, they are not suitable for new on-chip multicore processors (CMP) because part of the CPU functional unit, such as the last level cache, is shared between these cores. Also, new processors may supply different kinds of hardware performance events since the architecture of the processor changed rapidly. Following the previous works, Bertran et al. proposed a core-level power model [6]. They categorized the processor components into three categories: the in-order engine, the memory subsystem, and the out-of-order engine, based on the extent that the PMCs can monitor their activities. For example, there are no PMC events to monitor the activity of some components in the in-order engine category. Then they select eight functional units, which include the front-end component, the integer component, floating point component, the simple instruction and multiple data component, the branch prediction unit, the L1 cache, the L2 cache, and the front side bus, to build the power model for a core. For the sharing components, such as L2 cache, the usage ratio will be computed individually. After choosing the components, they designed a set of software microbenchmarks that train different components and collect the maximum power information.

Finally, they build the per core power model as Equation 28.14 shows. In this equation, AR_i denotes the activity ratio of this component. The total CPU power is the summation of all the cores' power.

$$Power_{core} = \left(\sum_{i=1}^{i=compas} AR_i \times P_i \right) + Power_{static} \qquad (28.14)$$

The previous works show that performance events can be directly used to build power models for CPU and memory; however, current performance counters do not directly supply useful events that reflect the activity of other devices. In [8], Bircher et al. show that processor-related performance events are highly related to the power of other devices, such as memory, chipset, I/O, and disk. They define chipset subsystem as processor interface chips that are not within other subsystems, and define I/O subsystem as PCI (peripheral component interconnect) buses and all devices attached to them. In order to establish the power models for these subsystems with processor-related performance events, first they need to understand the correlation or explain how these events are propagated in the subsystem. Figure 28.5 shows the propagation of these performance events in all the subsystems they defined.

They select nine performance events based on the average error rate and comparison of the estimated and measured power traces. These events include cycles, halted cycles, fetched UOPs, level 3 cache misses, TLB misses, DMA (dynamic memory access) accesses, processor memory bus transactions, uncacheable accesses, and interrupts. A more detailed description can be found in [8]. Equation 28.15 shows the CPU power model they proposed. In this equation, the numbers 35.7, 9.25, and 4.31 are the maximum power dissipation of one CPU, the minimum power dissipation of one CPU, and a constant value that reflects the relationship of the performance events and the real power, respectively.

$$\sum_{i=1}^{NumCPUs} \left(9.25 + (35.7 - 9.25) \times Percent\ Active_i \right.$$

$$\left. + 4.31 \times \frac{FetchedUOPS_i}{Cycle} \right) \qquad (28.15)$$

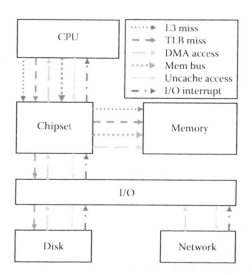

FIGURE 28.5 Propagation of performance events in the subsystems. (From Bircher, W. L. and John, L. K., Complete system power estimation: A trickle-down approach based on performance events, in *IEEE International Symposium on Performance Analysis of Systems and Software* 0, 158–168, San Jose, CA, 2007.)

Usually, it is difficult to model the memory power, because memory accesses may happen both from the CPU-memory side and MEMORY-I/O side. The memory power model they proposed includes two parts: power generated by CPU access and power generated by DMA. Power generated by CPU, as shown in Equation 28.16, is only related to the L3 cache misses and the cycle. Equation 28.17, in which *MCycle* is computed with *Cycle*, represents the other part of memory power. As we can see, it is only related with bus transactions and *Cycle*.

$$\sum_{i=1}^{NumCPUs} \left(28 + \frac{L3LoadMisses_i}{Cycle} \times 3.43 \right.$$
$$\left. + \frac{L3LoadMisses_i^2}{Cycle} \times 7.66 \right) \tag{28.16}$$

$$\sum_{i=1}^{NumCPUs} \left(29.2 - \frac{BusTransitions_i}{MCycle} \times 50.1 \times 10^{-4} \right.$$
$$\left. + \frac{BusTransitions_i^2}{MCycle} \times 813 \times 10^{-8} \right) \tag{28.17}$$

The challenge of modeling disk power with performance events is that disk is logically far from CPU. Finally, they find that interrupt number and DMA accesses are most related to disk power. Equation 28.18 gives this power model. In this equation, the number 21.6 is the idle power of the disk.

$$\sum_{i=1}^{NumCPUs} \left(21.6 + \frac{Interrupts_i}{Cycle} \times 10.6 \times 10^7 \right.$$
$$+ \frac{Interrupts_i^2}{Cycle} \times 11.1 \times 10^{15}$$
$$+ \frac{DMA\,Access_i}{Cycle} \times 9.18$$
$$\left. + \frac{Interrupts_i^2}{Cycle} \times 45.4 \right) \tag{28.18}$$

The I/O subsystem is connected with many types of I/O devices; thus, several performance events, such as DMA accesses, uncacheable accesses, and interrupts, are related to the power of the I/O subsystem. Although the majority of I/O operations are caused by DMA access, they find that interrupts are more related to the power of the I/O subsystem. Equation 28.19 shows the power model of the I/O subsystem.

$$\sum_{i=1}^{NumCPUs} \left(21.6 + \frac{Interrupts_i}{Cycle} \times 1.08 \times 10^8 \right.$$
$$\left. + \frac{Interrupts_i^2}{Cycle} \times 11.1 \times 1.12^9 \right) \tag{28.19}$$

28.3.2 Program Power Analysis

As we have mentioned before, power-aware hardware design is not enough to solve the power problem, and higher-level power-aware strategies are even more important. In the future, both the operating system and normal user space applications should be designed considering energy efficiency. Several articles [26,55] even proposed the concept of energy complexity, similar to time complexity and space complexity, as one of the metrics that evaluates the quality of the algorithms. To understand code-level power dissipation,

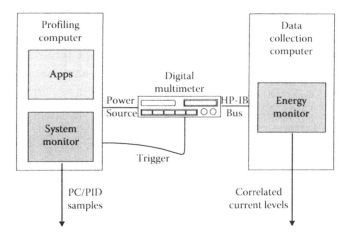

FIGURE 28.6 The architecture of PowerScope. (From Flinn, J. and Satyanarayanan, M., Powerscope: A tool for profiling the energy usage of mobile applications, in *WMCSA '99: Proceedings of the Second IEEE Workshop on Mobile Computer Systems and Applications*, p. 2, IEEE Computer Society, Washington, DC, 1999.)

we need to map the power to corresponding code blocks. With lower-level program power characters, the program designer could find power-hungry areas or hot spots in the code. Furthermore, this information can be used as the guidance for power-aware strategies. This is different from the power/performance trade-offs, which only need to synchronize the measured or estimated power to the counted performance metrics.

PowerScope [17], proposed by Flinn et al., is one of the first tools on the mobile platform that can profile the energy consumption of applications. It can not only determine which fraction of the total system power is caused by a process, but also determine the energy consumption of the procedures within a process. Figure 28.6 shows the architecture of PowerScope. As we can see, it includes three modules: energy monitor, system monitor, and energy analyzer. The system monitor collects information such as the value of PC register, process identifier (pid), and other information, for example, whether the system is currently processing an interrupt. The collected data is used to identify the executed program at a time point. To synchronize the system monitor and the energy monitor, they connect the multimeter's external trigger input and output to the parallel port of the profiling computer. By controlling the parallel port pin, they synchronize the data collection between the profiling machine and the data collection machine. The system monitor module on the profiling machine triggers the digital meter to sample the power; in this way they can synchronize the process events and measured power. Finally, the energy analyzer analyzes the raw data and generates the energy profile.

Different from PowerScope, which uses time-driven method to sample, Chang et al. present an energy-driven method to help programmers evaluate the energy impact of their software design [11]. Energy-driven method refers to sampling of the data for each energy quota, other than for each time interval. They use the Itsy platform [48] to measure the power of the system. If the consumed energy comes to the energy quota, the Itsy platform will generate an interrupt to the profiling system. Then, the interrupt service will collect information that could identify the program block. Similar to PowerScope, they also analyze the program power profile off-line. The architecture of this method is shown in Figure 28.7.

As further research of their earlier work [16], Ge et al. propose a program power analysis method called PowerPack [18] that is different from the last two methods [11,17]. PowerPack is designed for the desktop PC. The architecture of PowerPack is shown in Figure 28.8. They run the profiled application, the system status profiler, a thread that controls the meters and a group of power reader threads on the same platform. PowerPack uses a special method that is different from the last two methods to synchronize measured power to process information. Their method is used to find the power profile of predefined program code blocks. First, they implement a set of functions, as shown in Table 28.1, that will be called

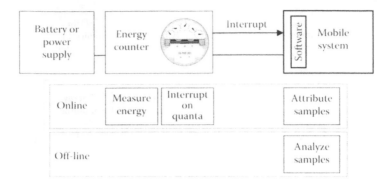

FIGURE 28.7   The architecture of event-driven method. (From Chang, F. et al., Energy driven statistical sampling: Detecting software hotspots, in Falsafi, B. and Vijaykumar, T. (eds.), *Power-Aware Computer Systems*, vol. 2325 of Lecture Notes in Computer Science, pp. 105–108, Springer, Berlin, Heidelberg, Germany, 2003.)

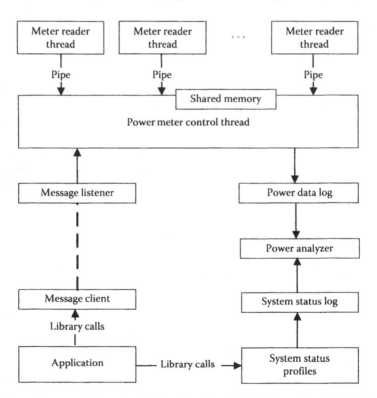

FIGURE 28.8   The architecture of PowerPack. (From Ge, R. et al., *IEEE Trans. Parallel Distrib. Syst.*, 21(5), 658, 2010.)

by the profiled applications before and after some critical code blocks. The executing of these functions then triggers the system status profiler and the meter control thread to sample the data. After this, the power analyzer can analyze the collected data simultaneously. Finally, they propose a method to map the measured power into the application code and analyze the energy efficiency in a multicore system.

Isci et al. developed an experimental framework to compare the control-flow-based with the performance-monitoring-based power-phase detection techniques [25]. Their results show that both the control-flow and the performance statistics provide useful hints of the power phase behaviors.

Table 28.2 summarizes the classification of these previous publications.

TABLE 28.1 The PowerPack Power Meter Profile API

Function	Description
Pmeter_init	Connect to meter control thread
Pmeter_log	Set power profile log file and options
Pmeter_start_session	Start a new profile session and label it
Pmeter_end_session	Stop current profile session
Pmeter_finalize	Disconnect from meter control thread

Source: Ge, R. et al., *IEEE Trans. Parallel Distrib. Syst.*, 21(5), 658, 2010.

TABLE 28.2 Classification of Power Profiling Articles

Simulation	Hardware Measurement	Software Power Profiling
Wattch [10]	Itsy00 [48]	PowerScope [17]
SimplePower [52]	Joseph0106 [27]	Isci03 [23]
SoftWatt [20]	PowerPack [16,18]	Li03 [33]
Orion [49]	PowerExecutive [1]	Chang03 [11]
SimWattch [12]	IMPI [22]	Lorch98 [35]
Dake94 [34]	Kamil08 [30]	Bellosa00 [5]
	DCEF [46]	Dempsey [53]
	Snowdon05 [45]	vEC [29]
		Powell09 [41]
		Bertran10 [6]
		Joseph0108 [28]
		Kansal10 [31]
		pTop [15]
		Dhiman10 [14]
		SoftWatt [20]
		Tempo [39]
		Contreras05 [13]
		Bircher [7,8]
		SPAN [50]

28.4 Case Studies

Power measuring and profiling are globally used in many areas of power/energy-aware system designs: making power/performance analysis, supplying directive power information for power-aware strategies, and analyzing the power behaviors of programs, especially the hot spots, and so forth. In this section, we describe several case studies using power measuring and profiling techniques.

28.4.1 Energy-Aware Scheduling

Energy-aware scheduling not only refers to scheduling algorithms that try to manage the system energy consumption, but also includes other scheduling algorithms that try to manage power dissipation and component temperature. More specifically, energy-aware scheduling includes scheduling methods that use system power characters to meet the following aims: conserving the energy consumption without substantially decreasing the performance [51], managing the heap power or temperature within a specified value [38,42,43], managing the energy consumption within a budget to extend the battery lifetime [54], and so on.

Based on task power profiles, Merkel et al. propose a method for scheduling tasks on a multiple processor system to maintain the power balance of CPUs and to reduce the need for throttling [38]. In this section, we describe their method in detail.

28.4.1.1 Task Energy Estimation

First, we need to estimate the energy consumption of an individual task, because the energy-aware scheduling method makes decisions at the task level. This includes estimating how much energy a task spent in the last time unit and predicting the energy consumption in the next time unit. Although the accuracy of the power model is important, a task-level power profiler must choose simple power models to decrease the overhead generated by the data collection module.

To estimate the energy consumption of a processor, they use the performance counter to build a simple power model, as described in the last section. Then they determine the energy consumption of a task in the last time unit by reading these chosen event counters before and after this time unit. Except for the energy consumption of the last time unit, it is also critical to predict the underlining energy consumption in the next time interval. However, estimating this is nearly inevitable because many factors may influence the power of the task. Merkel et al. assume that the energy consumption of the next time interval equals that of the last time interval, since they do a group of experiments and find that the energy consumption during two consecutive time intervals only changes by a very small percentage, normally below 6%. Table 28.3 shows their experiment results. We could see that although the difference may be very high in some circumstances, the average difference is very low. Thus, it is reasonable to use the last time interval's energy consumption as the prediction and make the scheduling decision.

To eliminate the task migration caused by misprediction, they use an exponential average function, shown in Equation 28.20, to make the prediction by using the task's past energy consumption. The weight p they used is based on the length of the time interval.

$$\overline{x_i} = p \cdot x_i + (1 - p) \cdot \overline{x_{i-1}} \tag{28.20}$$

28.4.1.2 Design of the Scheduling Algorithm

Although task migration between CPUs can balance the temperature and workload of these CPUs, it may decrease the performance of the system because this strategy contradicts with processor affinity. Processor affinity is a kind of scheduling strategy that schedules a task to run on the CPU it has just run before; in this way the system may not need to reload the data and instructions of this task into the cache. Thus, task migration should be limited when designing an energy-aware scheduling.

They design two energy-aware scheduling strategies: energy balancing and hot task migration. Energy balancing is used in circumstances when multiple tasks can be executed on one CPU. By rescheduling the execution of hot tasks and cool tasks, this strategy could balance the energy consumption on one CPU. Figure 28.9 shows the flow of this scheduling algorithm. In this way, further task migration may not be needed. However, this may not always be the case. In circumstances when only one task can be executed on the CPU, they use hot task migration, which tries to migrate a hot task to another cool CPU if the current CPU's temperature exceeds the limitation. Figure 28.10 shows the flow of this scheduling strategy.

TABLE 28.3 Change in Power Consumption during Successive Timeslices

Program	Maximum (%)	Average (%)
Bash	19.0	2.05
Bzip2	88.8	5.45
Grep	84.3	1.06
Sshd	18.3	1.38
Openssl	63.2	2.48

Source: Merkel, A. and Bellosa, F., *SIGOPS Oper. Syst. Rev.*, 40, 403, April 2006.

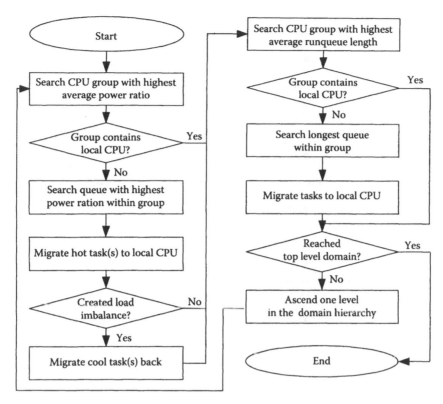

FIGURE 28.9 Energy and load balancing algorithm. (From Merkel, A. and Bellosa, F., *SIGOPS Oper. Syst. Rev.*, 40, 403, April 2006.)

28.4.1.3 Implementation and Evaluation

Merkel et al. implemented these two power-aware scheduling strategies on a Linux 2.6.10 kernel. Furthermore, they implemented the CPU energy estimation module and task energy profiling module on the same platform. The CPU's event counters are read on every task switch and at the end of each time interval. The values are transferred to energy value and saved in a data structure.

To test the load balancing and temperature control, they run a group of programs on the system. First, they run these programs without using the power-aware strategies. The result is shown in Figure 28.11. Then they run the same group of programs with energy-aware strategies available on the same system. Figure 28.12 shows this result.

28.4.2 Software Power Phase Characterizing

Characterizing the power phase behavior of application has become more and more important because of the increase in the complexity of the architecture and the global use of dynamic power-management strategies. For example, power phase characterizing can help the system optimize performance/power trade-offs, and help the programmer identify critical execution areas that generate abnormal power dissipation. The basic aim of power phase characterizing is classifying software power behaviors into self-similar operation regions. Isci et al. present a method for characterizing the running application's power phase in [25]. In this section, we describe their method in detail as a case study.

They use Pin [37] to collect synchronous information, such as system events, PMC events and measured power. Pin is a platform-independent tool that is designed for software instrumentation, which is essential

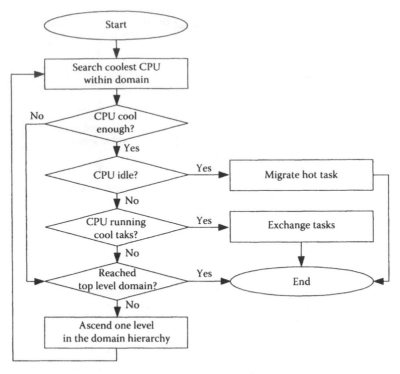

FIGURE 28.10 Hot task migration algorithm. (From Merkel, A. and Bellosa, F., *SIGOPS Oper. Syst. Rev.*, 40, 403, April 2006.)

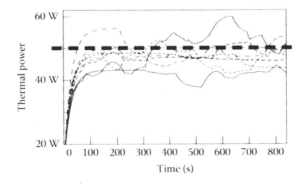

FIGURE 28.11 Thermal power of the eight CPUs with energy balancing disabled. (From Merkel, A. and Bellosa, F., *SIGOPS Oper. Syst. Rev.*, 40, 403, April 2006.)

for applications to evaluate performance and to detect a bug. With the Pin API we can observe all the architectural state of a process, such as the contents of registers, memory, and control flow. The power information is required to measure the current flow into the processor directly. The architecture of their method is shown in Figure 28.13.

The control-flow-based application phase is tracked with the basic block vector (BBV) method [44], which maps an executed PC address to the basic blocks of an application binary. Each item of BBV is corresponded to a specific block and saves a value that denotes how often this block is executed during the sampling period. Based on their past work [23], they select 15 PMC events that are more related to the CPU power to track the power phase. After that, the measured PMC events are converted into per-cycle

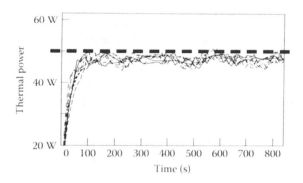

FIGURE 28.12 Thermal power of the eight CPUs with energy balancing enabled. (Merkel, A. and Bellosa, F., Balancing power consumption in multiprocessor systems, *SIGOPS Oper. Syst. Rev.*, 40, 403–414, April 2006.)

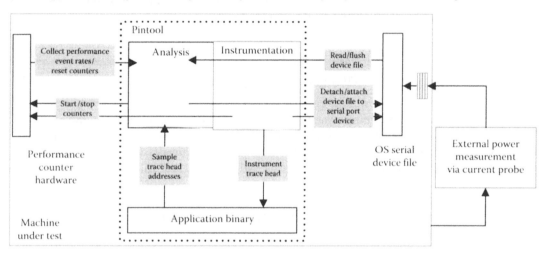

FIGURE 28.13 Experimental setup for power phase analysis with pin. (From Isci, C. and Martonosi, M., Phase characterization for power: evaluating control-flow-based and event-counter-based techniques, in *The 12th International Symposium on High-Performance Computer Architecture*, 2006, pp. 121–132, Austin, TX, February 2006.)

rates, and saved in a data structure similar to BBV as a 15-dimensional vector. Finally, the analyzer clusters the sampling result saved in the BBV and the PMC vector into power phases with different clustering algorithms, such as first pivot clustering and agglomerative clustering.

28.4.3 SPAN: A Realtime Software Power Analyzer

Section 28.3.2 presents several techniques for analyzing software power behavior. These methods use different strategies to synchronize the hardware-measured power with the software execution information. In this section, we describe a real-time power analyzer, called SPAN [50], which can find the power dissipation at the function-block level. Different from other articles, SPAN uses PMC events to estimate the power of the processor. After that, it analyzes how this power dissipation is related to the current program's function blocks.

28.4.3.1 Power Estimation

Similar to other PMC-based power models, SPAN also uses the same strategy to estimate the CPU power. To get high adaptability, SPAN only uses IPC and CPU frequency to build the power model. Previous articles [7,33] have proven high correlation between IPC-related PMC events and power.

Assuming that a CPU supports n frequencies, f_i, $i = 1, 2, 3 \ldots n$, $P(f_i)$ is the power of CPU for each frequency f_i. $P(t_j, f_i)$ is the CPU power when executing benchmark t_j at frequency f_i and $IPC(t_i, f_i)$ is the corresponding IPC. After running a group of microbenchmarks, they calculate $P(f_i)$ and $IPC(f_i)$ as the median of $P(t_j, f_i)$ and $IPC(t_i, f_i)$ individually. After that, they compute $\Delta P(t_j, f_i)$ as the difference between $P(f_i)$ and $P(t_i, f_i)$ for each training benchmark, shown in Equation 28.21. Similarly, they calculate ΔIPC (t_j, f_i) as the IPC difference of training benchmark t_i to the median value, shown in Equation 28.22.

$$\Delta P(t_j, f_i) = P(t_i, f_i) - P(f_i) \tag{28.21}$$

$$\Delta IPC(t_j, f_i) = IPC(t_i, f_i) - IPC(f_i) \tag{28.22}$$

Targeting on predicting $\Delta P(t_j, f_i)$, they use $\Delta IPC(t_j, f_i)$ as model input to derive linear regression parameters $P_{inct}(f_i)$ and $P_\Delta(f_i)$ with Equation 28.23. The final predicted power dissipation is shown in Equation 28.24, in which $P(f_i)$ dominates the calculated value.

$$\Delta P(t_j, f_i)_{pret} = P_{inct}(f_i) + P_\Delta(f_i) * \Delta IPC(t_i, f_i) \tag{28.23}$$

$$P(t_j, f_i)_{pret} = \Delta P(t_j, f_i)_{pret} + P(f_i) \tag{28.24}$$

The modeling of multiple cores is based on the assumption that each core has similar power behavior. Therefore, they apply the single core model to each core in the system. The equation for the total power dissipation is shown in Equation 28.25, in which a_j is the target benchmark, $\Delta P(a_j, f_i, k)_{pret}$ is generated at per core level because different cores might have the varied value of $\Delta IPC(t_i, f_i, k)$.

$$P(a_j, f_i)_{pret\_total} = \sum_{k=1}^{k=cores} (\Delta P(a_j, f_i, k)_{pret} + P(f_i)) \tag{28.25}$$

28.4.3.2 Design and Implementation of SPAN

In order to synchronize program executing information to estimated power, they design a group of APIs for applications. Table 28.4 lists some of these APIs. Developers need to call these APIs at some critical

TABLE 28.4   SPAN APIs

APIs	Description
`span_create()`	Prepare a power model profile which records basic parameters
`span_open()`	Initialize a SPAN control thread and targeting PMCs
`span_start(char* func, char* log)`	Record the targeting application function and specify the log file name
`span_stop(char* func, char* log)`	Stop the power estimation for a specified application function
`span_pause()`	Temporally stop reading PMCs
`span_continue()`	Resume reading PMCs
`span_change_rate(int freq)`	Shift the estimation rate, basically this methods control the PMC sampling rate
`span_change_model(float* model, File* model)`	Modify the model parameters in the model file according to the platform
`span_close()`	Close the opened PMCs and SPAN control thread
`span_output(char* log, FILE* power)`	Invoke SPAN analyzer thread and produce the detailed power estimation information with respective to the profiled functions to the destination file

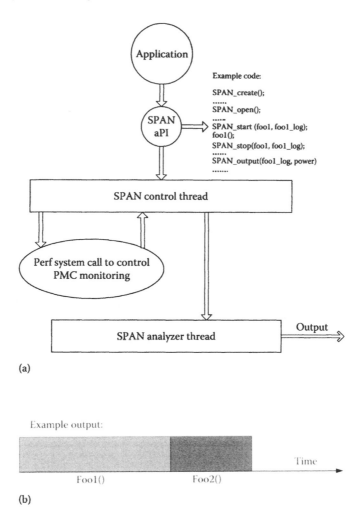

FIGURE 28.14 Design of SPAN. (a) The flowchart of SPAN. (b) The example output of SPAN.

area of their code. The basic flow of the SPAN tool is illustrated in Figure 28.14a. The two inputs of SPAN are the application information and PMC counter values. At the application level, the application information and the estimation control APIs are passed to the control thread through the designed SPAN APIs. Finally, the SPAN outputs a figure of estimated power dissipation represented by different shades distinguishing different application functions, such as shown in Figure 28.14b.

28.4.3.3 Evaluation

In this article, they mainly evaluate two aspects of the SPAN, overhead and responsiveness. They use two benchmarks to do the evaluation. One is the FT benchmark from the NAS Parallel Benchmark suite. Another is a synthetic benchmark that we designed with a combination of integer operation, *PI* calculation, prime calculation, and bubble sort. First, they measured the execution with and without the SPAN instrumentation for 10 times each. The differences of execution time are within 1% on average.

To evaluate the responsiveness of the power model, they compare the continuous measured and estimated values. From Figure 28.15, it is easy to observe that the estimated power is closely related to the measured power dissipation at the overall shape. Furthermore, the figure marks the corresponding benchmark functions. The first iteration of benchmark FT mainly consists of two functions, *compute_initial_conditions*() and *fft*(). The rest iterations follow the same procedure, which can be clearly

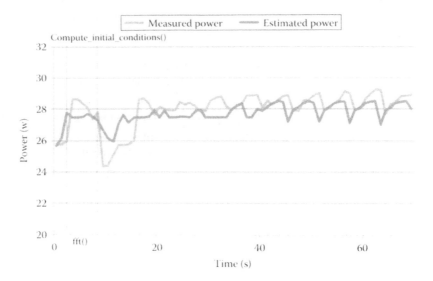

FIGURE 28.15   The FT benchmark with SPAN instrumentation. (Wang, S. et al., Span: A software power analyzer for multicore computer systems, in *Sustainable Computing: Informatics and Systems*, Elsevier, Amsterdam, the Netherlands, 2011.)

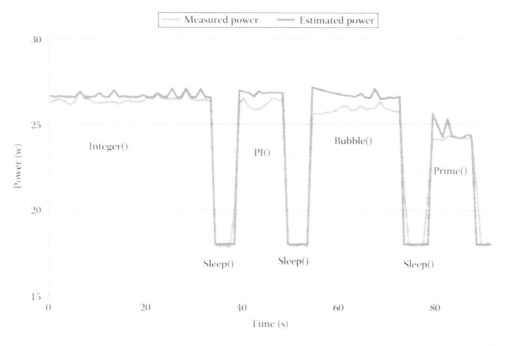

FIGURE 28.16   The synthetic benchmark with SPAN instrumentation. (From Wang, S. et al., Span: A software power analyzer for multicore computer systems, in *Sustainable Computing: Informatics and Systems*, Elsevier, Amsterdam, the Netherlands, 2011.)

observed from Figure 28.15, but the estimations present a certain level of delay due to the rapid function changes in the source code. Moreover, in Figure 28.16, they deliberately insert *sleep()* function between each sub-benchmark in the synthetic workload in order to distinguish each one of them easily. The achieved error rate is as low as 2.34% for the two benchmarks on average.

28.5 Summary and Future Work

This chapter has provided an introduction to the power measuring and profiling techniques that are globally used in power-aware system design. By summarizing the direct power measurement method and software-based power estimation techniques and describing several case studies we show researchers how these methods can be used in their future research.

Power measuring and profiling has already been studied extensively. However, more investigations are needed in the following two areas: improving power profiling techniques and using these strategies in power-aware system design. As we have mentioned earlier, accuracy is not the only requirement, simplicity and adaptability are equally important. In the future, the operating system should supply a module to do the power profiling and supply configurable accuracy. In addition, because of the fast-changing nature of hardware architectures, new methods should adapt to these new architectures. For example, as the number of cores on a single chip keeps increasing, on-chip network fabrics become one of the main power dissipation resources. Thus, future research needs to consider this unit and reevaluate the power indicators that are currently used.

Currently, power measurements and profiling has been used for making power/performance trade-offs during hardware designs, supplying basic information for power-aware strategies, and so on. Exploiting how to make use of power profiling approaches to design energy-efficient software is also an interesting direction, which is highly related to the research activities in the operating systems and software engineering field.

References

1. IBM PowerExecutive. http://www-03.ibm.com/systems/management/director/about/director52/extensions/powerexec.html
2. Watts up. https://www.wattsupmeters.com
3. I. Ahmad, S. Ranka, and S. U. Khan. Using game theory for scheduling tasks on multi-core processors for simultaneous optimization of performance and energy. In *Parallel and Distributed Processing, 2008. IPDPS 2008. IEEE International Symposium on*, Miami, FL, pp. 1–6, April 2008.
4. L. A. Barroso and U. Hölzle. The case for energy-proportional computing. *Computer*, 40:33–37, December 2007.
5. F. Bellosa. The benefits of event: Driven energy accounting in power-sensitive systems. In *Proceedings of the Ninth Workshop on ACM SIGOPS European Workshop: Beyond the PC: New Challenges for the Operating System*, EW 9, Kolding, Denmark, pp. 37–42, ACM, New York, 2000.
6. R. Bertran, M. Gonzalez, X. Martorell, N. Navarro, and E. Ayguade. Decomposable and responsive power models for multicore processors using performance counters. In *Proceedings of the 24th ACM International Conference on Supercomputing*, Tsukuba, Ibaraki, Japan, pp. 147–158, ACM Press, New York, 2010.
7. W. L. Bircher, M. Valluri, J. Law, and L. K. John. Runtime identification of microprocessor energy saving opportunities. In *ISLPED 05: Proceedings of the 2005 International Symposium on Low Power Electronics and Design*, San Diego, CA, pp. 275–280, ACM Press, New York, 2005.
8. W. Lloyd Bircher and Lizy K. John. Complete system power estimation: A trickle-down approach based on performance events. In *IEEE International Symposium on Performance Analysis of Systems and Software*, San Jose, CA, pp. 158–168, 2007.
9. P. Bohrer, E. N. Elnozahy, T. Keller, M. Kistler, C. Lefurgy, C. McDowell, and R. Rajamony. *The Case for Power Management in Web Servers*, pp. 261–289, Kluwer Academic Publishers, Norwell, MA, 2002.

10. D. Brooks, V. Tiwari, and M. Martonosi. Wattch: A framework for architectural-level power analysis and optimizations. In *ISCA '00: Proceedings of the 27th Annual International Symposium on Computer Architecture*, Vancouver, British Columbia, Canada, pp. 83–94, ACM, New York, 2000.

11. F. Chang, K. Farkas, and P. Ranganathan. Energy-driven statistical sampling: Detecting software hotspots. In B. Falsafi and T. Vijaykumar, (eds.), *Power-Aware Computer Systems*, vol. 2325 of Lecture Notes in Computer Science, San Deigo, CA, pp. 105–108, Springer, Berlin, Heidelberg, Germany, 2003.

12. J. Chen, M. Dubois, and P. Stenström. Simwattch: Integrating complete-system and user-level performance and power simulators. *IEEE Micro*, 27(4):34–48, 2007.

13. G. Contreras and M. Martonosi. Power prediction for intel xscale processors using performance monitoring unit events. In *Proceedings of IEEE/ACM International Symposium on Low Power Electronics and Design*, San Diego, CA, pp. 221–226, 2005.

14. G. Dhiman, K. Mihic, and T. Rosing. A system for online power prediction in virtualized environments using Gaussian mixture models. In *Proceedings of the 47th ACM IEEE Design Automation Conference*, Anaheim, CA, pp. 807–812, ACM Press, New York, 2010.

15. T. Do, S. Rawshdeh, and W. Shi. PTOP: A process-level power profiling tool. In *Proceedings of the Second Workshop on Power Aware Computing and Systems (HotPower'09)*, Big Sky, MT, October 2009.

16. X. Feng, R. Ge, and K. W. Cameron. Power and energy profiling of scientific applications on distributed systems. In *Proceedings of the 19th IEEE International Parallel and Distributed Processing Symposium (IPDPS'05)–Papers*–Volume 01, IPDPS '05, p. 34, IEEE Computer Society, Washington, DC, 2005.

17. J. Flinn and M. Satyanarayanan. Powerscope: A tool for profiling the energy usage of mobile applications. In *WMCSA '99: Proceedings of the Second IEEE Workshop on Mobile Computer Systems and Applications*, New Orleans, LA, p. 2, IEEE Computer Society, Washington, DC, 1999.

18. R. Ge, X. Feng, S. Song, H. C. Chang, D. Li, and K. W. Cameron. Powerpack: Energy profiling and analysis of high-performance systems and applications. *IEEE Transactions on Parallel and Distributed Systems*, 21(5):658–671, 2010.

19. R. A. Giri and A. Vanchi. Increasing data center efficiency with server power measurements. Technical report, Intel Information Technology, 2010.

20. S. Gurumurthi, A. Sivasubramaniam, M. J. Irwin, N. Vijaykrishnan, M. Kandemir, T. Li, and L. K. John. Using complete machine simulation for software power estimation: The softwatt approach. In *HPCA '02: Proceedings of the 8th International Symposium on High-Performance Computer Architecture*, Boston, MA, p. 141, IEEE Computer Society, Washington, DC, 2002.

21. L.P. Hewlett-Packard Development Company. Service processor users guide. Technical report, Hewlett-Packard Development Company, L.P., 2004.

22. Intelligent platform management interface. http://www.intel.com/design/servers/ipmi/index.htm.

23. C. Isci and M. Martonosi. Runtime power monitoring in high-end processors: Methodology and empirical data. In *MICRO 36: Proceedings of the 36th Annual IEEE/ACM International Symposium on Microarchitecture*, San Diego, CA, p. 93, IEEE Computer Society, Washington, DC, 2003.

24. C. Isci and M. Martonosi. Runtime power monitoring in high-end processors: Methodology and empirical data. Technical report, Princeton University Electrical Engineering Department, September 2003.

25. C. Isci and M. Martonosi. Phase characterization for power: Evaluating control-flow-based and event-counter-based techniques. In *the Twelfth International Symposium on, High-Performance Computer Architecture, 2006*, pp. 121–132, Austin, TX, February 2006.

26. R. Jain, D. Molnar, and Z. Ramzan. Towards a model of energy complexity for algorithms [mobile wireless applications]. In *Wireless Communications and Networking Conference*, New Orleans, LA, *2005 IEEE*, pp. 1884–1890, Vol. 3, March 2005.

27. R. Joseph, D. Brooks, and M. Martonosi. Live, runtime power measurements as a foundation for evaluating power/performance tradeoffs. In *Workshop on Complexity Effective Design WCED, Held in Conjunction with ISCA-28*, Goteberg, Sweden, June 2001.

28. R. Joseph and M. Martonosi. Run-time power estimation in high performance microprocessors. In *ISLPED '01: Proceedings of the 2001 International Symposium on Low Power Electronics and Design*, pp. 135–140, Huntington, Beach, CA, ACM, New York, 2001.

29. I. Kadayif, T. Chinoda, M. Kandemir, N. Vijaykirsnan, M. J. Irwin, and A. Sivasubramaniam. VEC: Virtual energy counters. In *PASTE '01: Proceedings of the 2001 ACM SIGPLAN-SIGSOFT Workshop on Program Analysis for Software Tools and Engineering*, pp. 28–31, Snowbird, UT, ACM, New York, 2001.

30. S. Kamil, J. Shalf, and E. Strohmaier. Power efficiency in high performance computing. In *IEEE International Symposium on Parallel and Distributed Processing, 2008. IPDPS 2008*. Miami, FL, pp. 1–8, April 2008.

31. A. Kansal, F. Zhao, J. Liu, N. Kothari, and A. A. Bhattacharya. Virtual machine power metering and provisioning. In *Proceedings of the First ACM Symposium on Cloud Computing*, SoCC '10, Indianapolis, IN, pp. 39–50, ACM, New York, 2010.

32. S. U. Khan and I. Ahmad. A cooperative game theoretical technique for joint optimization of energy consumption and response time in computational grids. *Parallel and Distributed Systems, IEEE Transactions on*, 20(3):346–360, March 2009.

33. T. Li and L. K. John. Run-time modeling and estimation of operating system power consumption. *SIGMETRICS Performance Evaluation Review*, 31(1):160–171, 2003.

34. D. Liu and C. Svensson. Power consumption estimation in cmos vlsi chips. *Solid-State Circuits, IEEE Journal of*, 29(6):663–670, June 1994.

35. J. R. Lorch and A. J. Smith. Apple macintosh's energy consumption. *IEEE Micro*, 18(6):54–63, 1998.

36. Y. H. Lu, L. Benini, and G. D. Micheli. Power-aware operating systems for interactive systems. *IEEE Transactions on Very Large Scale Integration Systems*, 10(2):119–134, 2002.

37. C. K. Luk, R. Cohn, R. Muth, H. Patil, A. Klauser, G. Lowney, S. Wallace, V. J. Reddi, and K. Hazelwood. Pin: Building customized program analysis tools with dynamic instrumentation. *SIGPLAN Notices*, 40:190–200, June 2005.

38. A. Merkel and F. Bellosa. Balancing power consumption in multiprocessor systems. *SIGOPS Operating Systems Review*, 40:403–414, April 2006.

39. D. Molaro, H. Payer, and D. L. Moal. Tempo: Disk drive power consumption characterization and modeling. In *IEEE Thirteenth International Symposium on Consumer Electronics, 2009. ISCE '09*. Kyoto, Japan, pp. 246–250, May 2009.

40. T. Mudge. Power: A first-class architectural design constraint. *Computer*, 34:52–58, 2001.

41. M. D. Powell, A. Biswas, J. S. Emer, S. S. Mukherjee, B. R. Sheikh, and S. Yardi. Camp: A technique to estimate per-structure power at run-time using a few simple parameters. In *IEEE Fifteenth International Symposium on High Performance Computer Architecture*, Raleigh, NC, 2009. HPCA, pp. 289–300, 2009.

42. M. D. Powell, M. Gomaa, and T. N. Vijaykumar. Heat-and-run: Leveraging smt and cmp to manage power density through the operating system. In *Proceedings of the 11th International Conference on Architectural Support for Programming Languages and Operating Systems*, Boston, MA, pp. 260–270, 2004.

43. E. Rohou and M. D. Smith. Dynamically managing processor temperature and power. In *Second Workshop on Feedback-Directed Optimization*, Haifa, Israel, 1999.

44. T. Sherwood, E. Perelman, G. Hamerly, and B. Calder. Automatically characterizing large scale program behavior. In *ASPLOS-X: Proceedings of the Tenth International Conference on Architectural Support for Programming Languages and Operating Systems*, San Jose, CA, pp. 45–57, ACM, New York, 2002.

45. D. C. Snowdon, S. M. Petters, and G. Heiser. Power measurement as the basis for power management. In *2005 Workshop Operating System Platforms for Embedded Real-Time Applications*, 2005.

46. Intel Information Technology. Data center energy efficiency with intel power management technologies. Technical report, Intel Information Technology, February 2010.

47. A. Vahdat, A. Lebeck, and C. S. Ellis. Every joule is precious: The case for revisiting operating system design for energy efficiency. In *Proceedings of the Ninth Workshop on ACM SIGOPS European Workshop: Beyond the PC: New Challenges for the Operating System*, EW 9, pp. 31–36, Kolding, Denmark, ACM, New York, 2000.

48. M. A. Viredaz, M. A. Viredaz, D. A. Wallach, and D. A. Wallach. Power evaluation of Itsy version 2.3. Technical report, Compaq, Western Research Laboratory, 2000.

49. H. S. Wang, X. Zhu, L. S. Peh, and S. Malik. Orion: A power-performance simulator for interconnection networks. In *MICRO 35: Proceedings of the 35th Annual ACM/IEEE International Symposium on Microarchitecture*, Istanbul, Turkey, pp. 294–305, IEEE Computer Society Press, Los Alamitos, CA, 2002.

50. S. Wang, H. Chen, and W. Shi. Span: A software power analyzer for multicore computer systems. *Elsevier Sustainable Computing: Informatics and Systems*, 1:23–34, 2011.

51. A. Weissel and F. Bellosa. Process cruise control: Event-driven clock scaling for dynamic power management. In *Proceedings of the 2002 International Conference on Compilers, Architecture, and Synthesis for Embedded Systems*, CASES '02, Grenoble, France, pp. 238–246, ACM, New York, 2002.

52. W. Ye, N. Vijaykrishnan, M. Kandemir, and M. J. Irwin. The design and use of Simplepower: a cycle-accurate energy estimation tool. In *DAC '00: Proceedings of the 37th Annual Design Automation Conference*, Los Angels, CA, pp. 340–345, ACM, New York, 2000.

53. J. Zedlewski, S. Sobti, N. Garg, F. Zheng, A. Krishnamurthy, and R. Wang. Modeling hard-disk power consumption. In *FAST '03: Proceedings of the Second USENIX Conference on File and Storage Technologies*, San Fransisco, CA, pp. 217–230, USENIX Association, Berkeley, CA, 2003.

54. H. Zeng, C. S. Ellis, A. R. Lebeck, and A. Vahdat. Ecosystem: Managing energy as a first class operating system resource. *SIGPLAN Notices*, 37(10):123–132, 2002.

55. K. Zotos, A. Litke, E. Chatzigeorgiou, S. Nikolaidis, and G. Stephanides. Energy complexity of software in embedded systems. *ACIT—Automation, Control, and Applications*, Novosibirsk, Russia, 2005.

29

Modeling the Energy Consumption of Distributed Applications

Georges Da Costa
University of Toulouse

Helmut Hlavacs
University of Vienna

Karin Anna
Hummel
University of Vienna

Jean-Marc Pierson
University of Toulouse

29.1 Problem Presentation and Motivation

The demand in research in energy efficiency in large-scale systems is supported by several incentives [12,31,35], including financial incentives by government or institutions for energy-efficient industries/companies [29]. Indeed, studies like [9] report that IT consumption accounts for between 5% and 10% of the global growing electricity demand, and for a mere 2% of the energy while data centers alone account for 14% of the ICT footprint. It is projected that by 2020 the energy demand of data centers will represent 18% of the ICT footprint, the carbon footprint rising at an annual 7% pace, doubling between 2007 and 2020 [39]. The last 5 years have witnessed the increase of research focusing especially in energy reduction. While being a major concern in embedded systems for decades, the problem is quite new in large-scale infrastructures where performances have been for long the sole parameters to optimize. The motivation comes from two complementary concerns: First, the electrical cost of running such infrastructure is equivalent nowadays with the purchase costs of the equipment during a 4-year usage [27]. Second, electricity providers are not always able to deliver the needed power to run the machines, capping the amount of electricity delivered to one particular client.

Energy concerns have been integrated in many works at hardware, network, middleware, and software levels in large-scale distributed systems. The eEnergy Conference Proceedings [26] and the COST IC0804 Proceedings [22] are providing good insights.

Most of the works base their approaches on models for power consumption. Indeed, it is often needed that the decision system carefully monitors the current status of the system and derives some energy reduction opportunities by predicting the impact of any decision on the system. These predictions are based on models fed by actual values of the system parameters, but must not distort the observed system. These models hopefully derive an accurate view of the system (i.e., the observation is actually accurate) and allow for a predicted behavior of the system in the future. The diverse granularities of the predictions allow for short-, middle-, or long-term decisions on the system.

As we will see in the upcoming sections, many use more or less advanced models. For these models to be widely accepted, they have to be evaluated against real-world measurements of applications and infrastructures, ideally on benchmarks. Direct measurement of the energy consumed by the infrastructure is used to validate the models, either component by component or directly at the plug.

This chapter gathers most of the literature on power and energy modeling, from component to machine to distributed system level. The approaches differ on the measurement platform, the system observations, the granularity of the models (at the component, machine, and distributed systems levels), and on their evaluation.

The chapter is organized as follows: Section 29.2 sets the context of our study. Section 29.3 focuses on measurement platforms and benchmarks, while Section 29.4 investigates the literature on energy consumption modeling. Section 29.5 outlines some of the limits of the current approaches and discusses the difficulty to obtain widely accepted models. Finally, Section 29.6 concludes the chapter.

29.2 Context of Our Study

The focus and context of our study is not on embedded systems, nor on hardware-specific optimization for energy saving, but on energy-efficient computation in distributed, large-scale systems. The components of these systems consist of thousands of heterogeneous nodes that communicate via heterogeneous networks and provide different memory and storage as well as different processing capabilities. Examples for very large-scale distributed systems are computational and data grids, data centers, clouds, etc.

In such large-scale distributed systems, power consumption monitoring is not only of local interest to accurately measure power consumed by the CPU, memory, communication interface, but it relies on distributed monitoring including network components like routers. In the case of migrating tasks, for instance, by utilizing virtualization and switching hosts to low-power modes, a decision concerning the whole large-scale distributed system has to be made for energy-efficient operation. Hence, modeling energy consumption and energy-efficient operation requires a distributed system model.

29.3 Measuring the Energy Consumption

29.3.1 Hardware Measurements

Direct measurement of the energy consumed by the infrastructure is used to validate the models, but these measures suffer from two flaws: If one wants to evaluate very precisely one component of one host, sensitive power meters and oscilloscopes are used. These precise measures are coming at a high cost in maintenance and shipping. Other less-expensive and easy-to-use solutions measure the power directly at the electrical plug.

When used on large scale, hardware monitoring requires a large hardware and software infrastructure. In Grid5000 (a French scientific grid of several thousands of nodes), an experiment shows some of those difficulties [15]. Figure 29.1 exhibits a sample of the live energy consumption measurements on this platform with 17 nodes and 1 switch: As we can see, the measurement is at the machine level, monitoring the energy consumption at the plug.

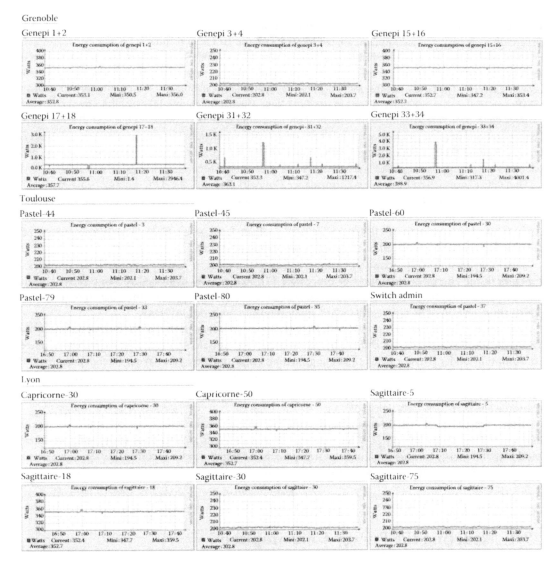

FIGURE 29.1 Live energy measurements on the Grid5000 platform.

Hardware measurement will be simplified as more and more hardware components start to include some measurement system (mainly processors) such as temperature, energy consumption, etc. But heterogeneity of such hardware (vendors do not always stick to the same interface to query those captors) and the size of large distributed systems put back the complexity on the measurement, interoperability, and management software infrastructure.

29.3.2 Benchmarks

Several standards have been proposed for evaluating the energy consumption of nodes in clusters, of multitier servers and of supported applications, the latter being evaluated on specific hardware: The benchmark evaluates, then, both the application and the infrastructure. The reader can refer to [33] for a more comprehensive study on energy-related benchmarks.

The Green500 [14] initiative is challenging the most powerful machines in terms of flops/watts. In the same manner as the Top500, Linpack is used to compute the performance. The first in the ranking achieves 1684 MFlops/W, with a total consumption of the corresponding machine of 38.80 kW. The Green500 is exploring at the moment a new list, based on the HPCC benchmark.

SpecPower [34] is an industry-standard benchmark that evaluates the power and performance of servers and multinode computers. The initial benchmark addresses only the performance of server-side Java. It exercises the CPUs, caches, memory hierarchy, and the scalability of shared memory processors (SMPs), as well as the implementations of the JVM (Java Virtual Machine), JIT (Just-In-Time) compiler, garbage collection, threads, and some aspects of the operating system. It computes the overall server-side-Java operations per watt, including the idle time on specific workloads. The comparison list includes 172 servers. Among these, a Fujitsu server with 76 quadcores (304 cores) reaches a maximum value of 2927 ssj_ops/W.

The TPC (Transaction Processing Performance Council) proposes the TPC-Energy benchmark [17] for transactional applications: Web/application services, decision support, on-line transaction processing. TPC measures watts/operations on the TPC benchmarks (for instance, transaction per seconds). Only few servers from HP have now been evaluated. As an example, 5.84 W/transaction per seconds is given for a typical online transaction processing workload.

The Embedded Microprocessor Benchmark Consortium has a similar approach. It provides a benchmark for energy consumption of processors [7]. It is mainly dedicated for embedded systems and computes number of operations per joule linked to the over performance benchmarks of the consortium, measured on different standard applications for embedded systems.

Additionally, manufacturers provide information about the consumption of their components, using average loads. For instance, AMD describes the average CPU power (ACP) [1] that characterizes power consumption under average loads (including floating point, integer, Java, Web, memory bandwidth, and transactional workloads, subset of TPC and SPEC benchmarks). Interestingly, this work shows, for instance, that cores can consume between 61% and 80% of the processor power and that processors consume less than 25%–35% when idle.

In the best cases, all these benchmarks provide information about standard applications on specific hardware. They limit their purpose on ranking between different architectures for specific workload. Modeling the energy consumption must go a step further: From the raw experimental data, it consists of deriving a mathematical relationship between the consumed energy and the utilization of the system.

29.4 Modeling the Energy Consumption

29.4.1 Modeling at the Component Level

Most studies try to model each component individually as it is far easier not to take into account the multiple interactions between subsystems.

The processor is the main focus of most studies [5,21,23,24] as it is one of the main dynamic part of a computer's energy consumption. As energy consumption of processors usually cannot be measured directly using hardware sensors, models use indirect measures such as performance counters. Performance counters monitor specific internal elements of processors, such as number of memory access, use of floating point unit, and branch misses. Usually processors have a large number of possible performance counters, but can use only a subset of them at a specific time.

Accuracy of such a system is linked to a trade-off between precision and performance. In [23], authors use two methods, either only performance counters or a processor simulator. Simulation is done for a simple alpha processor and gives a precision of about 2%. For performance counters, error is way higher, about 15%. These numbers are obtained using simplistic workload (Spec95 Int and Float). Simulation gives a far better precision but cannot be used in real time: for instance, the alpha processor cannot monitor more than two performance counters at the same time.

In the same way, there is a trade-off between precision and the specificity of the model. Isci and Martonosi [21] propose a model of Intel P4 processors based on linear equation of performance counters. This model is obtained using specific information of P4 architecture. Based on the knowledge of what type of process units are present and using microbenchmarks to evaluate energy consumption of each of those units, the model is built using a simple aggregation (an addition) of small models. This system allows obtaining a rather precise model (in the order of a millisecond timescale on the Spec2000 benchmark). Unfortunately it cannot be reused for another processor.

In order to avoid specificity and provide high efficiency, current approaches are limited to a few percent. This level of accuracy already allows usage of this technique. In [5], a performance counter model is used to provide feedback to prevent a processor from overheating. The performance counter model is precise at about 4%. When temperature rises too much, the scheduler reduces the resource allocation based on the performance counter model. This system has been tested with a variety of workloads such as benchmarks (JVM98, MiBench, Caffeine, Perl) and real applications (kernel build with GCC, Web browsing).

The processor is not the only element that is modeled using performance counters. Kadayif et al. [24] provide a model of memory based on a linear combination of performance counters (cache misses, memory accesses, etc.). This model is precise (about 2.4% error when doing unrolled matrix multiplication) as memory is quite simpler than intrinsic processors. Constants of the model were obtained for a particular system using circuit-level simulation (for an UltraSparc architecture).

Recently, hybrid computing started to obtain a wide audience, and hybrid-related models began to be built. Since this field is quite recent and hardware components are quite complex, current approaches are not precise enough. The methodology is the same as with processors: these computing elements (GPU, FPGA, accelerators, CELL) usually have internal performance counters. In [28], a GPU performance events model is described (in this case based on NVIDIA GPUs). Microbenchmarks are used to correlate performance counters with power consumption. Current results are not stable, as error is between 3% and 30% depending on benchmarks.

Network power consumption has also been widely studied [13,20,30]. In [20], for instance, the authors model the power consumption of residential and professional switches as a function of several elements: bandwidth, packet-rate, packet size, protocols, and so on. As an example, Figure 29.2 displays the power consumption of one HP switch as a function of the bandwidth for TCP connections. Their main findings in this work is that basically the major power cost is due to the static energy (i.e., switching on the equipment) and that the dynamic part has little influence on the power consumption.

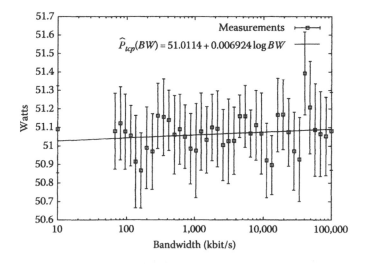

FIGURE 29.2 Energy consumption of a HP switch as a function of the bandwidth, TCP protocol.

29.4.1.1 Interaction of Components

Models of components are already complex. It becomes worse when trying to create a model of the interaction of several components. In [25], authors propose a methodology to produce a model combining processor and disk. They use performance counters to take into account the processor activity, and Windows events to obtain information on the disk activity. The resulting model combines linear counters and events. This work shows one of the limits to modeling energy consumption of a complex system only as a sum of the consumption of each subsystem.

29.4.2 Modeling at Machine Level

Rising from the complexity of only a small number of elements, the final step is to be able to model whole computers. Such models use usually only a small subset of available information such as performance counters in order to derive energy consumption. In [38], authors use performance counters and OS utilization metrics. With such a coarse-grained monitoring system, the energy consumption precision is around 10%. In order to obtain more precise measures, plenty of captors are needed. In [16] authors measure 165 different captors (performance counters, process-related captors, OS wide captors) to derive regression models for workload types: CPU, memory, network, tar, and I/O. Figure 29.3 shows the empirical cumulative distribution functions of the measured power consumptions for the different workload types, showing vast differences in shape. Regressing this energy consumption to the 165 measured OS variables, the prediction error of the respective regression models becomes as low as 3%.

Virtual machines (VMs) allow flexible migration of tasks and can be used to reach a consolidation of the resources required in a large-scale distributed system. For instance, data centers are and will be even more composed of multicore systems running a considerable large number of virtual machines. Thus, models are required on a machine level, which allows estimation of energy consumption on virtual machine level.

In [10], a power modeling methodology dedicated to virtualized systems is described. The energy consumption of a virtual machine is based on counting the number of cycles it executes. The resulting energy consumption is estimated with an error below 5%. In principle, the modeling has to take into consideration requests from users to start a virtual machine, management components (centralized or decentralized) coordinating the allocation of VMs on the host processors, migration cost estimates of VMs and data together with resizing and resulting communication costs, and actual migration

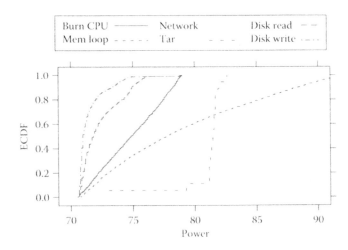

FIGURE 29.3 Empirical cumulative distribution functions of the power consumption of a PC for different workload types.

actions. In [6], global managers are introduced to take resource allocation and VM migration decision based on policies. The considered policies are reallocated according to (1) current utilization of the resources, (2) optimized virtual network topologies (between VMs), and (3) the thermal status of the resources. These heuristics could improve energy consumption of a considered use case cloud system significantly by including live migration of VMs. In [8], a machine learning approach was taken to optimize schedule in terms of energy efficiency based on the MP5 method (based on decision trees).

29.4.3 Modeling at the Distributed System Level

At the distributed system level, energy models take into account a plethora of relevant components directly or indirectly related to computing or networking. These components include computing nodes, networking nodes, storage and backup facilities, devices for power transformation, distribution and storage, computer room air conditioners (CRACs), devices for heat and air flow, humidifiers, lights, etc. The influence of the computing load may be incorporated as an abstract utilization number, or models may explicitly capture the influence of distributed algorithms and application-level protocols on the amount of energy spent to accomplish a task.

In a data center only a fraction of the delivered power is actually spent for IT equipment, the rest is considered to be overhead. A measure for efficiency of a data center is thus how large this overhead is compared to the power consumed for IT [37]:

$$\text{Power usage effectiveness (PUE)} = \frac{\text{Total data center input power}}{\text{IT equipment power}},$$

its reciprocal is called data center efficiency (DCE) or data center infrastructure efficiency (DCiE) [4]. Typical PUE values are around 2, but green data centers can reach PUEs of as little as 1.2.*

Rasmussen [37] reports three popular misconceptions that may result in wrong data center power models: First the efficiency of power and cooling components is not fixed but depends on the load. This is especially true for CRAC and UPS units. The efficiency of the latter drops significantly below a load of 15%. Second, power and cooling units usually do not operate near their maximum load but at much lower loads. Third, the heat output of power and cooling units must also be taken into consideration.

In [32], the authors propose a model capturing the total power consumption of a whole data center. Two external factors driving a data center power consumption are mentioned, i.e., the computing load offered to the servers, and the outside climate, including temperature, humidity, etc. The total power draw is then modeled as a function of this load, load consolidation, and temperature. Further components of the power model include the consumption of power distribution units (PDUs) responsible for transforming high voltages as delivered to the data center to low voltages as needed by servers, and uninterruptible power supplies (UPSs), which guarantee power even in the event of a power outage. The model further describes the power used by the computer room air handler (CRAH), the chiller plant using a compressor to move heat in water or other liquids to a cooling tower.

Models for distributed systems may also focus on the algorithmic side, capturing performance and power consumption of distributed algorithms or application-level protocols. Since currently hundreds of millions of PCs are interconnected by the Internet worldwide, it makes sense to look for applications or generally large-scale distributed systems that consume a lot of power. One such application is given by peer-to-peer file sharing systems, since they require many long running PCs. Some authors have thus asked how to develop especially green versions of P2P file sharing protocols, or how to generally increase the efficiency of such systems.

---

* http://www.google.com/corporate/green/datacenters/measuring.html

Anastasi et al. [2] describe an architecture using proxies that should carry out long-running downloads for PCs. Similar to server consolidation, the idea is to increase the utilization of a proxy and switching on PCs only when downloads have arrived. The authors use mathematical modeling to describe the power savings, which are mainly due to downloads being carried out in parallel.

Whereas [2] is a centralized approach, in [18] a similar, but decentralized approach is described. The authors use Markov models to evaluate the energy savings of a P2P file sharing scenario where each peer may also temporarily act as a download proxy for other peers. Again the main idea is to consolidate hardware and increase the efficiency of the whole system.

Examples for modeling and improving the energy efficiency of BitTorrent can be found in [11,19]. Blackburn and Christensen [11] consider remotely hibernating and waking up BitTorrent nodes after they have downloaded the whole file (so-called seeders), as a modeling methodology they use simulation. On the other hand, the work of [19] is based on a fluid model of BitTorrent published by [36] and explicitly derives the optimal time a seeder should stay online in order to minimize the system power consumption.

29.5 Limits of the Current Approaches and Research Opportunities

As we have seen so far, the modeling of the energy consumption of applications is still in its infancy when considering large-scale communicating systems. Obstacles include the precision of the measurements and of the model, especially when interconnecting individual components. We have seen that even at the single-component level, models do not encompass every workload and are often limited to one kind of hardware component, leading to nongeneric solutions. Most studies focus on the CPU since it consumes the major part of the system energy. It should be noted that (1) multi- and many-core systems are now widely adopted, therefore the models should work at the core level; and (2) the share of the CPU in energy consumption on a server is continuously decreasing [3]: models should take into account other components increasingly.

Interaction of components cannot be modeled simply by adding component-level models since the interaction is in general nonlinear: Indeed the effect of a workload on the processor will not have the same effect on the disk or the memory for instance. When virtual machines share memories (spatial sharing) and CPU (timesharing), it is obvious that the effect of a typical workload on the energy consumption will not have the same impact on both. Difficulty of merging and combining individual models is still to be investigated: Approaches such as metamodeling and loose and/or competing combinations of individual models will provide interesting opportunities (e.g., distributed agents and game theory).

At the machine level, such aggregations of individual models will have to be compared with global models deriving the energy consumption from the machine configuration and runtime. This aspect is particularly important when applications are embedded in virtual machines and the actual access to the physical components is abstracted.

In distributed systems, communications are a key factor. Unfortunately, communications are not easily taken into account so far. Being nonintrusive, it is still difficult to assess the runtime communication patterns of applications. While it is easy to know how much an application is communicating in terms of bandwidth, it is difficult (without access to the application itself) to derive with which other applications (and therefore machines) it is actually communicating. The resulting model of energy consumption for a distributed application is then limited to those cases where access to communicating applications is possible.

Finally, one should consider taking into account not only the distributed system but also the environment where it is running, enhancing the metrics beyond the traditional PUE toward holistic metrics: The energy production means, the manufacturing and provisioning of the infrastructure, the heat reuse, and the building could also be include into energy models.

29.6 Conclusion

In this chapter we described the current state of the art on the energy consumption of applications in distributed systems. After describing the modeling at the individual interacting components, we showed the machine and distributed system levels.

We finished the chapter highlighting the current difficulties in modeling the energy consumption of distributed applications and we provided some research trends and opportunities.

Acknowledgment

This work was partially supported by the COST (European Cooperation in Science and Technology) framework, under Action IC0804 (www.cost804.org).

References

1. AMD-ACP. www.amd.com/us/documents/43761c_acp_wp_ee.pdf
2. G. Anastasi, I. Giannetti, and A. Passarella. A BitTorrent proxy for green internet file sharing: Design and experimental evaluation. *Computer Communications*, 33:794–802, 2010.
3. L. A. Barroso and U. Hölzle. The case for energy-proportional computing. *IEEE Computer*, 40(12):33–37, 2007.
4. C. Belady, A. Rawson, J. Pflueger, and T. Cader. The green grid data center power efficiency metrics: PUE and DCiE. White Paper 6, The Green Grid, 2007.
5. F. Bellosa, A. Weissel, M. Waitz, and S. Kellner. Event-driven energy accounting for dynamic thermal management. In *Proceedings of the Workshop on Compilers and Operating Systems for Low Power (COLP'03)*, New Orleans, LA, 2003.
6. A. Beloglazov and R. Buyya. Energy efficient resource management in virtualized cloud data centers. In *CCGrid'10: 10th IEEE/ACM International Conference on Cluster, Cloud and Grid Computing (CCGrid)*, Melbourne, Australia, 2010.
7. Embedded Microprocessor Benchmark Consortium. Energy benchmarks. www.eembc.org/benchmark/power_sl.php
8. J. L. Berral, Í. Goiri, R. Nou, F. Julià, J. Guitart, R. Gavaldà, and J. Torres. Towards energy-aware scheduling in data centers using machine learning. In *e-Energy'10: First International Conference on Energy-Efficient Computing and Networking*, Passaw, Germany, pp. 215–224. ACM, 2010.
9. P. Bertoldi and B. Atanasiu. Electricity consumption and efficiency trends in the enlarged European Union. European Commission, Joint Research Centre, Institute for Energy, http://ie.jrc.ec.europa.eu/press/electronic_press_pack/Status%20Report%202009.pdf, 2006.
10. R. Bertran, M. Gonzélez, Y. Becerra, D. Carrera, V. Beltran, X. Martorell, J. Torres, and E. Ayguadé. Accurate energy accounting for shared virtualized environments using PMC-based power modeling techniques. In *the 11th ACM/IEEE International Conference on Grid Computing (Grid 2010)*, Brussels, Belgium, 25–29 October 2010.
11. J. Blackburn and K. Christensen. A simulation study of a new green BitTorrent. In *Proceedings of the First International Workshop on Green Communications (GreenComm 2009)*, Dresden, Germany, June 2009.
12. K. W. Cameron, K. Pruhs, S. Irani, P. Ranganathan, and D. Brooks. Report of the science of power management workshop. Available at scipm.cs.vt.edu/scipm-reporttonsf-web.pdf, April 2009.
13. J. Chabarek, J. Sommers, P. Barford, C. Estan, D. Tsiang, and S. Wright. Power awareness in network design and routing. In *IEEE Infocom*, Phoenix, AZ, April 2008.

14. W. C. Feng and T. Scogland. The Green500 list: Year one. In *23rd IEEE International Parallel and Distributed Processing Symposium (IPDPS)—Workshop on High-Performance, Power-Aware Computing (HP-PAC)*, Rome, Italy, May 2009.

15. G. Da-Costa, J.-P. Gelas, Y. Georgiou, L. Lefèvre, A.-C. Orgerie, J.-M. Pierson, O. Richard, and K. Sharma. The green-net framework: Energy efficiency in large scale distributed systems. In *HPPAC 2009: High Performance Power Aware Computing Workshop in Conjunction with IPDPS 2009*, Rome, Italy, May 2009.

16. G. Da Costa and H. Hlavacs. Methodology of measurement for energy consumption of applications (regular paper). In *Energy Efficient Grids, Clouds and Clusters Workshop (co-located with Grid) (E2GC2)*, Brussels, 25/10/2010–29/10/2010, page (electronic medium), http://www.ieee.org/, October 2010. IEEE.

17. Transaction Processing Performance Council. TPC-Energy. http://www.tpc.org/tpc_energy/

18. H. Hlavacs, K. A. Hummel, R. Weidlich, A. Houyou, and H. de Meer. Modelling energy efficiency in distributed home environments. *International Journal of Communication Networks and Distributed Systems*, 4(2):161–182, 2010.

19. H. Hlavacs. Modeling energy efficiency of file sharing. *e&i Elektrotechnik und Informationstechnik*, 11, 2010.

20. H. Hlavacs, G. Da Costa, and J.-M. Pierson. Energy consumption of residential and professional switches. *IEEE International Conference on Computational Science and Engineering*, 1:240–246, 2009.

21. C. Isci and M. Martonosi. Runtime power monitoring in high-end processors: Methodology and empirical data. In *Proceedings of the 36th Annual IEEE/ACM International Symposium on Microarchitecture*, San Diego, CA, IEEE Computer Society, 2003.

22. P. Jean-Marc and H. Hlavacs, eds. *Proceedings of the COST Action IC804 on Energy Efficiency in Large Scale Distributed Systems—1st Year*. Available online on www.cost804.org. IRIT, July 2010.

23. R. Joseph and M. Martonosi. Run-time power estimation in high performance microprocessors. In *International Symposium on Low Power Electronics and Design*, Huntington Beach, CA, 2001.

24. I. Kadayif, T. Chinoda, M. Kandemir, N. Vijaykirsnan, M. J. Irwin, and A. Sivasubramaniam. VEC: Virtual energy counters. In *Proceedings of the 2001 ACM SIGPLAN-SIGSOFT Workshop on Program Analysis for Software Tools and Engineering*, PASTE'01, Snowbind, UT, ACM, 2001.

25. A. Kansal and F. Zhao. Fine-grained energy profiling for power-aware application design. *SIGMETRICS Performance Evaluation Review*, 36, August 2008.

26. R. Katz and D. Hutchison, eds. *ACM/IEEE International Conference on Energy-Efficient Computing and Networking (e-Energy)*, Passau, Germany, 13/04/2010–15/04/2010. ACM, April 2010.

27. J. G. Koomey, C. Belady, M. Patterson, A. Santos, and K.-D. Lange. Accurate energy accounting for shared virtualized environments using PMC-based power modeling techniques. Microsoft, Intel and Hewlett-Packard Corporation Technical Report, August 2009.

28. X. Ma, M. Dong, L. Zhong, and Z. Deng. Statistical power consumption analysis and modeling for GPU-based computing. In *Workshop on Power Aware Computing and Systems (HotPower'09)*, Big sby, MT, 2009.

29. B. Naegel. Energy efficiency: The new SLA. GreenBiz.com http://www.greenbiz.com/news/2009/01/12/energy-efficiency-new-sla Jan. 12, 2009.

30. T. T. T. Nguyen and A. Black. Preliminary study on power consumption of typical home network devices. Technical Report 07011A, Centre for Advanced Internet Architectures (CAIA), October 2007.

31. COST action IC0804 on energy efficiency in large scale distributed systems. www.cost804.org

32. S. Pelley, D. Meisner, T. F. Wenisch, and J. W. VanGilder. Understanding and abstracting total data center power. In *Proceedings of the 2009 Workshop on Energy Efficient Design (WEED)*, Austin, TX, 2009.

33. M. Poess, R. O. Nambiar, K. Vaid, J. M. Stephens, K. Huppler, and E. Haines. Energy benchmarks: A detailed analysis. In *ACM/IEEE International Conference on Energy-Efficient Computing and Networking (e-Energy)*, Passau, Germany, 13/04/2010–15/04/2010, pp. 131–140. ACM, 2010.

34. Standard Performance Evaluation Corporation. Power and performance. www.spec.org/power_ssj2008/

35. U.S. Environmental Protection Agency ENERGY STAR Program. Report to Congress on server and data center energy efficiency, available online www.energystar.gov/ia/partners/prod_development/downloads/epa_datacenter_report_congress_final1.pdf, August 2007.

36. D. Qiu and R. Srikant. Modeling and performance analysis of BitTorrent-like peer-to-peer networks. In *SIGCOMM 2004*, Portland, OR, 2004.

37. N. Rasmussen. Electrical efficiency modeling for data centers. White Paper 113, American Power Conversion, 2006–2007.

38. S. Rivoire, P. Ranganathan, and C. Kozyrakis. A comparison of high-level full-system power models. In *Workshop on Power Aware Computing and Systems (HotPower)*, San Diego, CA, 2008.

39. The Climate Group. SMART 2020: Enabling the low carbon economy in the information age. Technical report, 2008.

30

Comparative Study of Runtime Systems for Energy-Aware High-Performance Computing

Chung-Hsing Hsu
Oak Ridge National Laboratory

Stephen W. Poole
Oak Ridge National Laboratory

30.1 Introduction

Energy consumption has become the primary cost of running a high-performance computing (HPC) system [1,3–5,9,12,54]. In 2007, the annual power and cooling budget matched the annual budget for new servers for the first time [31]. Today, the leading petaflop supercomputers consume a range of 1.4–7 MW of electrical power, with 3.5 MW on average.* The 3.5 MW can be easily translated into 3.5 millions of dollars per year in electricity bill. This is why Google locates its data center in rural Oregon in order to take advantage of the cheap hydroelectric power generated by the nearby Grand Coulee Dam.

The megawatt power consumption required by an HPC system also poses a reliability threat. As more and more heat is generated, system temperature increases and the current thermal solution is closer to running at its full capacity. If the thermal behavior cannot be well controlled, the HPC system will become very unreliable to use. As a rule of thumb, Arrhenius' equation as applied to microelectronics

---

* A petaflop is a measure of a computer's processing speed and can be expressed as a thousand trillion (10^{15}) floating point operations per second. The calculation of average power is based on the 36th edition of the TOP500 list of the world's most powerful supercomputers announced on November 2010. There are seven petaflop systems on that list.

notes that for every 10°C (18°F) increase in temperature, the failure rate of a system doubles. To many companies, the unavailability of an HPC system is equivalent to losing business in the amount of millions of dollars [11]. Because of the preceding two reasons, an HPC system needs to be energy-aware.

The energy awareness of HPC systems has been approached in many ways, from electrical materials to circuit design to systems integration to systems software. These techniques may be different, but the goal is the same: substantially reduce overall system power consumption without a corresponding negative impact on delivered performance. A strategy for achieving the goal is to direct power *only* to useful work. This chapter presents a study of a systems-software approach based on this strategy.

From a systems-software perspective, the strategy of supplying minimal power to useless work can be accomplished through powering down system components when they are underutilized. One example is dynamic voltage and frequency scaling (DVFS). DVFS allows the systems software to reduce a CPU's voltage and frequency when an application is running but the CPU is not fully utilized. Since the power consumption of the CPU is proportional to the frequency and the square of the voltage [41], DVFS allows the systems software to supply the high power only when needed. DVFS is supported by every modern CPU including Intel Xeon, AMD Opteron, and IBM Power7.

Although DVFS is highly promising, the systems software still needs to determine when to use which voltage and frequency during application execution. Otherwise, in the worst case, not only performance will be degraded, more energy will be consumed [13,18,27,45]. This is because the reduced frequency leads to potentially longer running time which in turns increases the energy consumption of the non-CPU components in the system such as memory and disks. There are also time and energy overheads associated with changing the voltage and frequency. Finding which voltage and frequency are the most appropriate to use at each time point is not an easy task. As a result, there is a huge body of work on DVFS scheduling algorithms in the literature since the introduction of DVFS in 1994. These algorithms are generally implemented as runtime systems. In this chapter we discuss some representative DVFS-based runtime systems in terms of their algorithmic features.

The current state of the art in DVFS scheduling algorithms is mainly for consumer electronics and laptop markets. For HPC, the notion of energy awareness is relatively new [11]. Why the distinction? First, the computational characteristics in two platforms differ markedly. The workload on the mobile computing platform has a lot of interactivity with the end user, but the workload on the HPC platform does not. Second, tasks executed in a mobile device tend to share more system resources. In contrast, each job in a HPC system runs with dedicated resources. Third, a HPC system is at a much larger scale than a mobile system, which makes it more challenging to collect information, coordinate decisions, and execute the global decisions. Therefore, it is worthwhile to know whether a DVFS scheduling algorithm, which works well for mobile computing, remains effective for HPC.

This chapter presents a comparative study of DVFS scheduling algorithms for energy-aware HPC. To facilitate an in-depth study, we only investigate ubiquitous algorithms. A DVFS scheduling algorithm is *ubiquitous* if it merely looks into the performance status of the system every time it makes a decision on which voltage and frequency to use. Thus, we do not consider any application- or language-dependent approach such as compiler-directed or MPI-based algorithms. Although ubiquitous algorithms may not be as effective as application-directed and language-directed algorithms, they form a first line of defense against the power and energy issues of current and future HPC systems. This is particularly important as the exact specs of future exaflop systems are still unknown.

In this chapter, we will present a framework to compare different DVFS scheduling algorithms. The goal is to be able to gain insight on which feature of each algorithm leads to effectiveness for energy-aware HPC. In the framework we consider each feature addressing one of the following three design issues in a typical DVFS scheduling algorithm: the abstraction of CPU utilization, the prediction of the trend in CPU utilization, and the association of the voltage and frequency values with CPU utilization. In this way we can understand better the effectiveness of each feature.

The rest of the chapter is organized as follows. Section 30.2 gives a brief history of DVFS in terms of its technology advancement and use scenarios. Then in Section 30.3 we detail a framework for comparing different DVFS scheduling algorithms. We identify three different abstractions of CPU utilization and discuss the associated DVFS scheduling algorithms. These three classes of algorithms are presented in Sections 30.4 through 30.6, respectively. We conclude the chapter in Section 30.7.

30.2 Brief History of DVFS and Its Use

We mention that DVFS is a mature technology which we can leverage for energy-aware HPC. In this section, we present a brief history of the technology and its use. We refer the reader to Keating et al.'s book [33] for more details on the technology itself. For other types of power-reduction techniques, Venkatachalam and Franz [62] have published a good survey on that.

The technology of DVFS on CPU can be traced back as early as 1994 [44]. It was motivated by the following two relations. First, the dynamic power of a CPU ($P_{dynamic}$) depends on the CPU's supply voltage (V) and operating frequency (f):

$$P_{dynamic} \propto V^2 \cdot f. \tag{30.1}$$

The dynamic power accounts for the switchings between 0 and 1 in the CPU when the CPU is in use. According to this relation, lowering voltage or frequency will reduce the CPU power. Second, the operating frequency of the CPU is dependent on its supply voltage:

$$f \propto V. \tag{30.2}$$

This equation implies that we can lower both voltage and frequency at the same time to reduce the CPU power even further. Given that the CPU performance is only affected linearly, we get a cubic reduction in CPU power and a squared reduction in CPU energy when running the CPU at lower voltage and frequency. Thus, DVFS provides a good energy-time trade-off and is applicable to execution scenarios which do not require the highest performance level.

One execution scenario in which DVFS is applicable is when the CPU is underutilized. Weiser et al. [63] presented one of the earliest DVFS scheduling algorithms on general-purpose operating system. In their 1994 seminal paper, Weiser et al. present a collection of interval-based algorithms that take advantage of DVFS. They divide time into fixed-length intervals and determine the CPU voltage and frequency at the beginning of each interval so that most work can be completed by the interval's end. To detect when the CPU is underutilized, they use CPU utilization ratio, i.e., the fraction of time that the CPU spends non-idle in an interval. Low CPU utilization ratio implies that low CPU voltage and frequency are required.

Around the same time, Yao et al. [67] presented one of the first task-based DVFS scheduling algorithms for real-time systems. Both this work and Weiser et al.'s work motivated a huge body of work on DVFS scheduling theory and practice. Over 100 papers have been published in this area of research; for example, [14,15,21–23,38,42,47–49,52,58,59,63,65,68] just to name a few. Most work targets embedded and mobile computing platforms.

Commodity CPUs that support DVFS did not appear on the market until early 2000, when they were first seen in mobile devices. It was not until 2003 that DVFS made its way into desktop CPUs. By late 2004, DVFS started to gain its support in server-class CPUs such as AMD Opteron and Intel Xeon. Today, DVFS is supported by almost every modern CPU.

Along with the wide support of DVFS is the introduction of an open standard called ACPI (Advanced Configuration and Power Interface) [25]. The intent is to bring power management into operating system's control. First released in 1999, ACPI defines the so-called performance states for CPU. From the ACPI viewpoint, a CPU has several performance states. When it is active, the CPU must be in one of the performance states. In addition, a low (or high) power state goes hand in hand with a low (or high) performance state. The introduction of ACPI and performance states allow DVFS scheduling algorithms to be portable across a range of CPU products.

Also introduced are the so-called implementable algorithms. The early work such as Weiser et al.'s requires knowledge of the future; for example, a knowledge of how much workload left to be completed. This knowledge is generally unknown on a general-purpose system. As a result, Pering et al. [46] and Grunwald et al. [23] modified the original algorithms so that the revised versions are implementable.

In early 2000, another execution scenario in which DVFS is applicable was identified. In this scenario, the CPU is stalled, and thus underutilized, due to off-chip memory accesses. For an application that often runs in this scenario, its performance will be affected sublinearly, not linearly. In other words, the added time when reducing the CPU frequency in half will be vastly less than double. This scenario, called sublinear performance scaling, is commonly seen in the execution of HPC applications. It is a direct consequence of the "memory wall" problem [40,64] currently faced by HPC, meaning that the CPU is stalled most of the time waiting for data retrieval because the rate of data retrieval is much slower than the rate of data processing.

There are some DVFS scheduling algorithms designed to exploit sublinear performance scaling. For example, Hsu and Kremer [29,30] used offline profiling to help the compiler identify program regions whose execution time is insensitive to CPU frequency changes and thus can be run at low CPU frequency without significant impact to the overall program performance. Other algorithms, such as [16,17,19,37, 45,56], constantly monitor the memory access rate to get a hint on when a memory-intensive program region is entered and exited. In this case, we say that memory access rate is used to detect when the CPU is underutilized.

DVFS for energy-aware HPC started to gain more attention by 2005. The first workshop in this area of research, called HPPAC (Workshop on High-Performance, Power-Aware Computing), was held. New algorithms were proposed to apply DVFS to communication-bound execution phases. In addition, researchers found out that a CPU may also be underutilized whenever it gets to an early finish in a synchronization task. Examples of DVFS scheduling algorithm exploiting these HPC-specific execution scenarios include [2,10,34,37,55,56,66]. These algorithms are generally not ubiquitous.

30.3 Comparison Framework

In this section we present a framework to compare different DVFS scheduling algorithms. We characterize each algorithm with respect to three design issues: the abstraction of CPU utilization, the prediction of the trend in CPU utilization, and the association of performance states with CPU utilization. The goal is to be able to understand how the three design issues are addressed by each algorithm at high level.

Specifically, we assume that a CPU provides n performance states $f_1, f_2, \ldots, f_n$ (in GHz) where $f_{min} = f_1 < f_2 < \cdots < f_n = f_{max}$. A DVFS scheduling algorithm conceptually divides the time into a sequence of intervals, $I_1, I_2, I_3, \ldots$, each of which has length t_i (in seconds). These intervals do not have to be of the same length. At the beginning of the interval I_i, the algorithm will determine a desired performance state s_i to use for the interval. The *desired* performance state can be one of the n directly supported performance states f_i or a combination of them.

The desired performance state is computed in two steps: first *utilization prediction* and then *state selection*. The DVFS scheduling algorithm first predicts the utilization of CPU based on the historical values of a set of observable events. It then selects the most appropriate performance state to use. We use u to denote the utilization metric. We also distinguish between the predicted and observed utilizations for interval I_i as u_i^p and u_i^o separately. Note that some DVFS scheduling algorithms treat $\{u_i\}$ as a time series. But not all algorithms are time series-based.

We view the state-selection step as a process in which the algorithm solves a minimization problem: find the performance state that minimizes energy consumption while completing the predicted workload within some performance bound. We denote the workload as ω and assume that the algorithm uses a performance model $T(\omega, s)$ to estimate the running time of executing workload ω by performance state s. The minimization problem that the algorithm solves can be formulated as follows:

$$\min_{s} \ P(s) \cdot \max(t, T(\omega^p, s))$$

such that

$$\max(t, T(\omega^p, s)) \leq \Delta(\delta, t, T(\omega^p, f_{max})).$$

where $P(s)$ represents the rate of average energy consumption per second (in watts) of the performance state s, and $\Delta(\delta, t, T(\omega, s))$ defines the performance bound with a user-defined parameter δ as the maximum allowable performance slowdown, e.g., 5%. Note that, in this view the performance model $T(\omega, s)$ is algorithm specific.

To illustrate the use of the comparison framework, let us consider the following simple DVFS scheduling algorithm. The algorithm measures the number of CPU non-idle cycles per second in the previous interval and run the CPU at that speed, i.e.,

$$s_i = \frac{\left(t_{i-1} - t_{i-1}^{idle}\right) \cdot s_{i-1}}{t_i} \tag{30.3}$$

where t_i^{idle} denotes the CPU idle time in interval I_i. We can view that the algorithm characterizes the workload as the number of non-idle cycles, predicts that the current interval will have the same amount of the workload as the previous interval, and selects the lowest performance state that can complete these cycles by the end of the interval, i.e.,

1. $\omega = (t - t^{idle}) * s$
2. $\omega_i^p = \omega_{i-1}^o$
3. $T(\omega, s) = \frac{\omega}{s}$
4. $\Delta(\delta, t, T(\omega, s)) = t$
5. $P(s)$ is decreasing
6. $t_i = t_{i-1} = 1$

Figure 30.1a outlines the algorithm. We can easily convert the algorithm into utilization-based, as shown in Figure 30.1b. This algorithm is similar to the one used in the LongRun technology for the Transmeta CPU [38].

As another example, we distinguish between two definitions of the CPU utilization ratio. The CPU utilization ratio is generally defined as the fraction of time that the CPU spends non-idle in an interval.

Assumption:
$t_i = t_{i-1}.$
Algorithm:
$\omega_{i-1}^o = (t_{i-1} - t_{i-1}^{idle}) \cdot s_{i-1}.$
$\omega_i^p = \omega_{i-1}^o.$
$s_i = \omega_i^p / t_i.$

Assumption:
$t_i = t_{i-1}.$
Algorithm:
$u_{i-1}^o = 1 - t_{i-1}^{idle} / t_{i-1}.$
$u_i^p = u_{i-1}^o.$
$s_i = u_i^p \cdot s_{i-1}.$

(a)   (b)

FIGURE 30.1 A simple DVFS scheduling algorithm (a) workload-based (b) utilization-based.

However, the unit of the time is not specified in the definition. Thus, the unit can be either in seconds or in cycles. A different choice of unit results in a different state-selection formula. Specifically, the two definitions can be expressed as $u = (\omega/s)/t$ and $u = \omega/(t/f_{max})$, respectively, where ω is the number of CPU non-idle cycles. An appropriate state-selection formula for the first definition is $s = u \cdot s$ whereas it is $s = u \cdot f_{max}$ for the second definition. Most modern DVFS scheduling algorithms calculate the CPU utilization ratio from the data stored in /proc/stat and fall into the first definition.

The comparison framework presented earlier is similar to the framework presented by Govil et al. [21] in 1995. In that work, the two steps, utilization prediction and state selection, are also identified. The new framework extends the old one by standing out the performance model which was implicitly assumed in a DVFS scheduling algorithm. The performance model not only relates state selection with CPU utilization in a more integrated manner, it also introduces the dualism between the utilization view and the workload view of the algorithm as illustrated in Figure 30.1.

More specifically, energy-aware HPC requires approaches that can tightly control application performance for every application. This is different from approaches that put the energy saving as the first priority. On one hand, having high-performance penalty is unacceptable. On the other hand, having no performance penalty misses opportunities for energy savings. By highlighting the performance model used in a DVFS scheduling algorithm, we ensure that the DVFS-induced performance slowdown is well-bound.

To recap, we characterize each ubiquitous DVFS scheduling algorithm with respect to three design issues:

- The choice (and definition) of CPU utilization u
- The prediction of future CPU utilization, typically through smoothing of $\{u_i\}$ as a time series
- The selection of s for u via a performance model

In the next several sections, we will present common DVFS scheduling algorithms with respect to these design issues so that we can better understand the inventions from each algorithm at high level.

30.4 Utilization as CPU Utilization Ratio

In this section we present a collection of DVFS scheduling algorithms that define utilization u as the CPU utilization ratio. The CPU utilization ratio is the fraction of time that the CPU spends non-idle in an interval, and thus its value is between 0 and 1. We start with algorithms that estimate future values of utilization as a linear function of previous utilization samples. We present these algorithms in terms of the techniques used for utilization prediction and state selection.

For utilization prediction, two broad classes of technique are used. One class is to treat $\{u_i\}$ as a time series and use smoothing techniques (e.g., moving average) to predict future events. The other class is to view a DVFS scheduling algorithm as a control system and apply control-theoretical techniques for utilization prediction. We start with techniques based on moving average.

1. *SMA$_N$ (simple moving average)*: A SMA$_N$ is the unweighted average of the previous N data points. When calculating successive values, a new value u^o_{i-1} comes in and an old value u^o_{i-N-1} drops out, meaning that a full summation each time is unnecessary. However, a queue is needed to hold the most recent N values.

$$u^p_i = \frac{1}{N} \sum_{j=i-N}^{i-1} u^o_j = u^p_{i-1} + \frac{u^o_{i-1} - u^o_{i-N-1}}{N} \tag{30.4}$$

2. *LongShort* [21]: LongShort is a type of weighted moving average. It averages the 12 most recent intervals' utilizations, weighting the three most recent utilizations three times more than the others.

$$u_i^p = \frac{1}{12} \left(3 \sum_{j=i-3}^{i-1} u_j^o + \sum_{j=i-12}^{i-4} u_j^o \right) \tag{30.5}$$

3. *CMA (cumulative moving average)*: A CMA is the unweighted average of all the data points up to the most current one.

$$u_i^p = \frac{1}{i-1} \sum_{j=1}^{i-1} u_j^o = \left(\frac{i-2}{i-1} \right) u_{i-1}^p + \left(\frac{1}{i-1} \right) u_{i-1}^o \tag{30.6}$$

4. *EMA$_\lambda$ (exponential moving average)* [38,39]: An EMA$_\lambda$ is a weighted moving average with weights decreasing exponentially over time. The parameter λ, $0 \leq \lambda \leq 1$, represents the degree of weight decrease, and a lower λ discounts older observations faster. It is called aged-*a* in [38].

$$u_i^p = (1 - \lambda) \sum_{j=1}^{i-1} \left(\lambda^{(i-1)-j} \right) u_j^o = \lambda \cdot u_{i-1}^p + (1 - \lambda) \cdot u_{i-1}^o \tag{30.7}$$

5. *AVG$_N$* [23,46]: This is an EMA$_\lambda$ with $\lambda = N/(N+1)$.

$$u_i^p = \frac{1}{N+1} \sum_{j=1}^{i-1} \left(\frac{N}{N+1} \right)^{(i-1)-j} u_j^o = \left(\frac{N}{N+1} \right) u_{i-1}^p + \left(\frac{1}{N+1} \right) u_{i-1}^o \tag{30.8}$$

6. *PAST* [63]: It predicts that the upcoming interval's utilization will be the same as the past interval's. Mathematically, it is equivalent to AVG$_0$ and SMA$_1$.

$$u_i^p = u_{i-1}^o \tag{30.9}$$

One nice property of moving average is that all the weights sum up to one. As a result, the predicted utilization u^p will lie between 0 and 1. The normalized value of utilization allows us to make a one-to-one correspondence with the desired performance state. Thus, if the utilization is 0.5 on a 2-GHz machine, for example, the performance state with 1-GHz CPU frequency is desirable since it is predicted to eliminate all CPU idle time. This correspondence certainly simplifies state selection, but not all utilization-prediction techniques hold this property.

Another class of utilization-prediction technique is to view a DVFS scheduling algorithm as a control system and apply control-theoretical techniques for the prediction. The following lists a few techniques.

1. *nqPID$_N$ (not quite PID)* [61]: PID (proportional-integral-derivative) is a classical control-systems technique that is applied to find an appropriate value that represents the past, the present, and the changing workload of the system. The nqPID simplifies the standard PID algorithm by removing feedback and replacing the continuous-time integral and derivative by their discrete-time counterparts (summation and difference).

$$u_i^p = K_P \cdot u_{i-1}^o + K_I \cdot \left(\frac{1}{N} \sum_{j=i-N}^{i-1} u_j^o \right) + K_D \cdot \left(u_{i-1}^o - u_{i-2}^o \right) \tag{30.10}$$

where K_P, K_I, and K_D are the weights of the corresponding terms. The proportional term acts similarly to PAST, and the integral term acts as SMA$_N$. The derivative term expects the coming change in utilization to be the same as the difference between the previous two utilization samples.

2. *PD (proportional differential)* [43]: PD is one of the traditional control theories and estimates the change concerning the direction of a slope.

$$u_i^p = u_{i-1}^o + K_P \cdot \left(1 - u_{i-1}^o\right) + K_D \cdot \left(\left(1 - u_{i-1}^o\right) - \left(1 - u_{i-2}^o\right)\right) \qquad (30.11)$$

where K_P and K_D are the weights of the corresponding terms.

30.4.1 Some Common Algorithms

In this section we present three DVFS scheduling algorithms commonly seen in the Linux computing environment. They are powernowd, CPUSpeed, and ondemand. As we will see, these algorithms use the simple PAST method for utilization prediction. They differ in the formula used for state selection. We start with powernowd.

30.4.1.1 powernowd

The powernowd algorithm, developed by John Clemens, is a very simple algorithm that will adjust the CPU performance state based on the CPU utilization ratio. Although obsolete now, this algorithm represents some of the early ideas for state selection. Figure 30.2 outlines the algorithm.

For state selection, powernowd establishes the high and low thresholds (τ_H and τ_L) to trigger a step either from a low-performance state to a high performance or in the other direction. As a result, the selection of the threshold values becomes important. Unfortunately, there is no clear guideline on how to choose appropriate threshold values. Typically an end user resorts to tune these values empirically with respect to his workload.

Figure 30.2 shows a new function called **down**(s). This function is introduced to deal with CPUs that only support a limited set of performance states. The following list is a set of techniques that addresses this limitation.

1. **down**(s): This technique selects the closest lower performance state that a CPU directly supports. Mathematically, it can be defined as

$$\mathbf{down}(s) = \max\{f_i | f_i \leq s\}. \qquad (30.12)$$

We can define **up**(s) in a similar way.

powernowd
Version: 1.00 (final)
Parameters:
    τ_H, τ_L (default: 0.8, 0.2)
Algorithm:

$$u_{i-1}^o = 1 - \frac{t_{i-1}^{idle}}{t_{i-1}}.$$

$$u_i^p = u_{i-1}^o.$$

$$s_i = \begin{cases} f_{max} & \text{if } u_i^p \in |\tau_H, 1| \\ s_{i-1} & \text{if } u_i^p \in (\tau_L, \tau_H) \\ \mathbf{down}(s_{i-1}) & \text{if } u_i^p \in |0, \tau_L| \end{cases}$$

FIGURE 30.2   The powernowd algorithm.

2. **round$_r$(s)**: This technique selects the closest performance state that a CPU directly supports. The typical half-half split is implemented by setting r to 1/2.

$$\textbf{round}_r(s) = \begin{cases} \textbf{up}(s) & \text{if } s \in [f_L + r \cdot (f_H - f_L), f_H] \\ \textbf{down}(s) & \text{otherwise} \end{cases} \quad (30.13)$$

3. **emulate(s)**: This technique emulates a desired performance state s with two supported performance states f_L and f_H such that $f_L < s < f_H$. For an interval of length t, the technique executes f_L for a duration of t_L and executes f_H for the duration of $t - t_L$:

$$t_L = t \cdot \frac{(f_H - s)}{(f_H - f_L)}. \quad (30.14)$$

Many DVFS scheduling algorithms set $f_L = \textbf{down}(s)$ and $f_H = \textbf{up}(s)$, i.e., using the two neighboring performance states to emulate.

30.4.1.2 CPUSpeed

The CPUSpeed algorithm, developed by Carl Thompson, can be found in many Linux distributions such as SUSE Linux (9.3 or later) and Fedora Core 4. In fact, it has been activated on every Dell PowerEdge server with Red Hat Enterprise Linux, and Dell calls it "demand-based switching." The state-selection formula is similar to powernowd's. It is considered to be more "conservative" because it uses high-performance states more often. The interval length is set to 2 s by default. Figure 30.3 outlines the subsystem.

30.4.1.3 ondemand

The ondemand algorithm is part of the Linux kernel starting from 2.6.9 (October 2004). In earlier versions, the algorithm reduces CPU frequency at minimum steps of 5% of the peak frequency f_{max}, i.e., $s = s - 5\% \cdot f_{max}$. In recent versions, the formula is changed. The algorithm looks for the lowest frequency that can sustain the load while keeping idle time over 30%. If such a frequency exists,

---

CPUSpeed
Version: 1.5 (latest)
Parameters:
  τ_H and τ_L (default: 0.9 and 0.75)
Algorithm:

$$u^o_{i-1} = 1 - \frac{t^{idle}_{i-1}}{t_{i-1}}.$$

$$u^p_i = u^o_{i-1}.$$

$$s_i = \begin{cases} f_{max} & \text{if } u^p_i \in [\tau_H, 1] \\ \textbf{up}(s_{i-1}) & \text{if } u^p_i \in (\tau_L, \tau_H) \\ s_{i-1} & \text{if } u^p_i = \tau_L \\ \textbf{down}(s_{i-1}) & \text{if } u^p_i \in [0, \tau_L) \end{cases}$$

FIGURE 30.3 The CPUSpeed algorithm.

ondemand
Version: 2.6.24.3
Parameters:
τ_H (default: 0.8)
Algorithm:

$$u_{i-1}^o = 1 - \frac{t_{i-1}^{idle}}{t_{i-1}}.$$

$$u_i^p = u_{i-1}^o.$$

$$s_i = \begin{cases} f_{max} & \text{if } u_i^p \in (\tau_H, 1] \\ s_{i-1} & \text{if } u_i^p \in |\tau_H - 0.1, \tau_H| \\ \frac{u_i^p}{\tau_H - 0.1} \cdot s_{i-1} & \text{if } u_i^p \in |0, \tau_H - 0.1) \end{cases}$$

FIGURE 30.4   The ondemand algorithm.

the algorithm tries to decrease to this frequency. In other words, the optimal frequency is the lowest frequency that can support the current CPU usage without triggering the up policy. To be safe, the algorithm focuses 10% under the threshold. Figure 30.4 outlines the algorithm.

30.5 Utilization as Memory Access Rate

Some DVFS scheduling algorithms use memory access rate (MAR) to define utilization. The rationale is that, while modern processors can hide certain memory latencies, many memory accesses cannot be hidden. For these outstanding memory accesses, the CPU must wait until the data is returned. Therefore, MAR is a good indicator for when the CPU is underutilized due to memory accesses. In this section we will examine several DVFS scheduling algorithms based on MAR.

MAR comes in different forms. The two most common ones are the rate of cache misses and the ratio of data cache misses (m) to instruction executed (II). The latter metric, misses per instruction (MPI), has been used in [6,24,57] to study the memory behavior of HPC benchmarks. According to [57], there are several reasons to use this particular metric. First, MPI is a direct indication of the amount of memory bandwidth that must be supported for each instruction. Moreover, given the average memory cycles per cache miss, it is straightforward to compute the memory component of the cycles per instruction. In addition, MPI has the advantage of being independent of the current CPU frequency.

30.5.1 Load Aware

Poellabauer et al. [50,51,53,60] proposed a DVFS scheduling algorithm, load aware (LD), based on MPI. In LD, the number of cache misses and the number of executed instructions are first predicted, and then MAR is calculated. For state selection, the algorithm relies on a precomputed table that maps MARs to desired performance states and consults the table at runtime. Figure 30.5 outlines the core of the algorithm. Note that the algorithm is originally designed for a real-time system to maximize the system utilization. Here we present a variant that fits in our comparison framework.

There are three distinct features of LD. The first feature is how LD derives the desired performance state. Poellabauer et al. [51] assumed that a program's execution time can be broken down into two parts, the time spent in computations and the time spent in memory accesses:

$$T_f(u) = c_0 + c_1 \cdot u \tag{30.15}$$

LD
Parameters:
$T_f(u)$.
Algorithm:

$$m^p_{i-1} = \left(\frac{m^p_{i-2}}{2}\right) + m^o_{i-1}.$$

$$I^p_{i-1} = \left(\frac{I^p_{i-2}}{2}\right) + I^o_{i-1}.$$

$$u^p_i = \frac{m^p_{i-1}}{I^p_{i-1}}.$$

$$r_i = \min_{f_j}\left\{\frac{T_{f_j}\left(u^p_i\right)}{T_{f_{max}}\left(u^p_i\right)} \leq 1 + \delta\right\}.$$

$$s_i = \frac{f_{max}}{r_i}.$$

emulate(s_i).

FIGURE 30.5 The LD algorithm.

where c_0 and c_1 are two constants whose values will be determined empirically through regression analysis. The empirical determination of the two constants, especially c_1, ensures that processor techniques to hide memory latencies are considered. The performance models, one for each supported performance state, allow LD to estimate the impact of running the application at a different performance state with respect to a specific memory access rate. The performance models are constructed off line.

The second feature is how LD executes the desired performance state. Many DVFS scheduling algorithms use the neighboring performance states to emulate the desired state. Poellabauer et al. [51] found out that this particular combination does not necessarily yield the highest energy savings. As a result, LD computes the combination that can yield the best possible energy savings for a given taskset and workload.

The third feature is how LD controls the frequency of changes in a CPU performance state, thereby avoiding the large overheads associated with these changes. LD uses a Schmitt-trigger-style function to avoid frequent changes. The function takes in an arbitrary MAR value and returns 1 of 12 bins. Fixing the number of bins also helps to reduce the overheads associated with making DVFS scheduling decisions. The scaling factors r_i can be precomputed for different MAR bins and performance states.

30.5.2 Memory Aware

Liang et al. [35,36] proposed a DVFS scheduling algorithm, memory aware (MA), that defines utilization as MPI. Figure 30.6 outlines the essence of the algorithm. The most distinct feature of MA is in its state selection. Liang et al. found out that an approximation equation based on the correlation of MAR and the desired performance state can be established as a low-order polynomial. For example, the equation for an Intel PXA270 Xscale platform is $S(u) = -61385u^3 + 12328u^2 - 102.82u + 0.8345$. Each platform has a different approximation equation.

The desired performance state in MA is the state in which a system has the minimum energy consumption. Researches [13,18,27,32,45,69] have shown that the minimum energy consumption may not appear at the lowest performance state, and different applications will have different desired performance states. Clearly, the performance penalty due to low-performance states is not bound in MA.

$$
\begin{array}{|l|}
\hline
\textbf{MA} \\
\text{Parameters:} \\
\quad S(u). \\
\text{Algorithm:} \\
\quad u_i^p = \dfrac{m_{i-1}^o}{T_{i-1}^o}. \\[2mm]
\quad s_i = S\left(u_i^p\right) \cdot f_{max}. \\[2mm]
\quad \textbf{emulate}(s_i). \\
\hline
\end{array}
$$

FIGURE 30.6   The MA algorithm.

30.6 Utilization as CPU Nonstall Time

In the previous section, we have discussed DVFS scheduling algorithms that use memory access rate (MAR) to define utilization. In this section we examine DVFS scheduling algorithms that define utilization as CPU nonstall time. Defining utilization this way allows us to account for other causes of CPU stalls (e.g., I/O) that MAR cannot account for.

Similar to MAR, utilization based on CPU nonstall time can be defined in many ways. For the four algorithms we examine here, two different definitions are used. They are all based on the decomposition of the execution time $T(s)$ into two parts: a frequency-dependent part $T_{on}(s)$ representing the time during which the CPU is not stalled, and a frequency-independent part T_{off} representing the time during which the CPU is stalled, i.e.,

$$
T(s) = T_{on}(s) + T_{off} \tag{30.16}
$$

where we interpret T_{off} as a constant in this section for simplicity. The two possible definitions of utilization are $u = T_{off}/T_{on}(s)$ and $u = T_{on}(f_{max})/T(f_{max})$. One algorithm uses the first definition and the other three algorithms use the second definition. Note that other forms of utilization are possible; for example, $u = T_{on}(s)/T(s)$.

An immediate difference between the two definitions is that the first definition captures CPU stalls whereas the second definition captures CPU nonstalls. In addition, the second definition bounds the metric value between zero and one whereas the first definition does not. In fact, the second definition leads to a normalized metric with respect to the highest performance state f_{max}, and thus its value is invariant to CPU frequency changes. It represents the fraction of the program workload that scales with the CPU frequency, which is similar to the parallelizable part in the Amdahl's Law except that the number of processors is replaced by the CPU frequency [26].

From a workload perspective, consider the number of CPU clock cycles ω split into two parts ω_{on} and ω_{off}:

$$
\omega = \omega_{on} + \omega_{off} \tag{30.17}
$$

where
   ω_{on} is the number of cycles that the CPU is busy executing the on-chip workload
   ω_{off} is the number of CPU stall cycles

Now we can model $T_{on}(s)$ and T_{off} as ω_{on}/s and ω_{off}/s, respectively. If we convert both definitions of utilization into functions of workload, we can derive

$$
u = \frac{T_{off}}{T_{on}(s)} = \frac{w_{off}}{w_{on}} \tag{30.18}
$$

and

$$u = \frac{T_{on}(f_{max})}{T(f_{max})} = \frac{w_{on}}{w_{off}(f_{max}/s)}. \tag{30.19}$$

We notice that the second definition has an additional factor (f_{max}/s) for the off-chip workload w_{off}. This factor is used to normalize the cycle count because the cycle count cannot be interpreted as a constant. It is actually a function of performance state s, i.e., $w_{off}(s) = s \cdot T_{off}$. For different performance states, w_{off} are has different values.

30.6.1 Choi, Soma, and Pedram

Choi, Soma, and Pedram (CSP) [7,8] proposed a DVFS scheduling algorithm which we call CSP. The algorithm defines utilization using the first definition, i.e., $u = T_{off}/T_{on}(s)$. The algorithm assumes that both terms, T_{off} and $T_{on}(s)$, are unknown a priori, and it calculates their values at runtime. Choi et al. found out that utilization can be calculated by the following equation:

$$u = \frac{CPI}{CPI_{on}} - 1 \tag{30.20}$$

where CPI is the number of CPU cycles per instruction for the total workload, and CPI_{on} is the number of CPU cycles per instruction for the on-chip workload; namely, $CPI = w/I$ and $CPI_{on} = w_{on}/I$ where I is the number of instructions. CSP uses the formula

$$s = \frac{f_{max}}{1 + \delta\left(1 + u\left(\frac{f_{max}}{s}\right)\right)} \tag{30.21}$$

to derive the desired performance state s. If the desired performance state is not directly supported, the closet higher supported state is used instead. Figure 30.7 outlines the CSP algorithm.

A distinct feature of CSP is to derive CPI_{on} (and thus utilization u) through online regression analysis. Choi et al. observed the linear relationship between CPI and MPI as follows:

$$CPI = a_s \cdot MPI + CPI_{on}, \tag{30.22}$$

> **Algorithm: Choi et al.**
> **Parameters:**
> $\quad N$ and δ (default: 25 and 5%)
> **Algorithm:**
> $\quad x_{i-1} = \frac{m^o_{i-1}}{I^o_{i-1}}.$
> $\quad y_{i-1} = \frac{w^o_{i-1}}{I^o_{i-1}}.$
>
> $\quad$ fit $\{(x_j, y_j)\}^{i-1}_{j=i-N}$ to
> $\qquad y = a \cdot x + b.$
>
> $\quad u^p_i = \left(\frac{y_{i-1}}{b}\right) - 1.$
>
> $\quad s_i = \frac{f_{max}}{1 + \delta\left(1 + u^p_i(f_{max}/s_{i-1})\right)}.$
>
> $\quad s_i = \mathbf{up}(s_i).$

FIGURE 30.7 The CSP algorithm.

one for each performance state s. Here we provide a formalism to justify its existence. We start with the performance model:

$$T(s) = \frac{I \cdot (w_{on}/I)}{s} + m \cdot l_m \tag{30.23}$$

where

m is the number of memory accesses
l_m is the (average) memory latency

After some algebraic manipulation, we derive the following equation:

$$\left(\frac{w}{I}\right) = (s \cdot l_m)\left(\frac{m}{I}\right) + \left(\frac{w_{on}}{I}\right). \tag{30.24}$$

which is exactly the same as Equation 30.22. The equation also allows us to interpret the coefficient a_s as $(s \cdot l_m)$ which is the number of CPU clock cycles for the memory latency. Since the memory latency l_m is considered as a constant, the value of a_s varies with each performance state s.

Using the same formalism, we can also justify why Equation 30.20 holds:

$$u = \frac{T_{off}}{T_{on}(s)} = \frac{T(s)}{T_{on}(s)} - 1 = \frac{I \cdot (w/I)/s}{I \cdot (w_{on}/I)/s} = \frac{(w/I)}{(w_{on}/I)} - 1.$$

We can also justify the state-selection formula Equation 30.21. Here we provide an alternative formula that selects the same performance state:

$$s = \frac{1 - u \cdot \delta}{1 + \delta} \cdot f_{max}. \tag{30.25}$$

Both formulas select the lowest performance state, which still satisfies the performance requirement $T(s)/T(f_{max}) \leq 1 + \delta$. In addition, we can correlate the two definitions of utilization using this formalism:

$$1 + \left(\frac{T_{off}}{T_{on}(s)}\right)\left(\frac{f_{max}}{s}\right) = \frac{1}{(T_{on}(f_{max})/T(f_{max}))}. \tag{30.26}$$

The main point is that, through the performance model, we can relate utilization prediction with state selection and link them in a more formal manner.

Choi et al. noticed the high variation of u_i^p over time, especially for memory-intensive applications. They proposed a complex method based on prediction errors to handle that. Prediction errors are the slack time (in seconds) between the allocated time for a task and the task's real execution time. The method is more applicable to task-based systems in which multiple tasks share the same computing resources. For HPC systems, the method is less applicable, and thus we omit its study here.

CSP uses the common least-squared method to fit the past N reports of MPI (x_j) and CPI (y_j) into Equation 30.22. For completeness, we list the formula for calculating CPI_{on} (b):

$$b = \frac{\left(\sum x_j^2\right)\left(\sum y_j\right) - \left(\sum x_j y_j\right)\left(\sum x_j\right)}{N\left(\sum x_j^2\right) - \left(\sum x_j\right)^2} \tag{30.27}$$

30.6.2 β-Adaptation

Hsu and Feng [26,28] proposed a DVFS scheduling algorithm called β-adaptation (BA). The algorithm defines utilization using the second definition, i.e., $u = T_{on}(f_{max})/T(f_{max})$. Given that $T_{on}(s)$ and $T(s)$

$$
\boxed{
\begin{aligned}
&\textit{Algorithm: } \textbf{BA} \\
&\textit{Parameters:} \\
&\quad \delta \text{ (default: 5\%)} \\
&\textit{Algorithm:} \\
&\quad I^o[f_k] = \sum_{j=1}^{i-1} I_j^o[f_k]. \\
&\quad t^o[f_k] = \sum_{j=1}^{i-1} t_j[f_k]. \\[6pt]
&\quad x_k = \left(\frac{f_{max}}{f_k}\right) - 1. \\[6pt]
&\quad y_k = \left(\frac{I^o[f_k]}{t^o[f_k]}\right) - 1. \\[6pt]
&\quad \text{fit } \{(x_k, y_k)\}_{k=1}^{n} \text{ to} \\
&\quad\quad y = a * x. \\[4pt]
&\quad u_i^p = a. \\[6pt]
&\quad s_i = \frac{f_{max}}{\left(1 + \delta / u_i^p\right)}. \\[6pt]
&\quad \textbf{emulate}(s_i).
\end{aligned}
}
$$

FIGURE 30.8 The β-adaptation (BA) algorithm.

are unknown a priori, they proposed a method to calculate their values indirectly. Hsu and Feng calculate utilization u using the equation:

$$
u = \frac{\text{IPS}(f_{max})/\text{IPS}(s) - 1}{f_{max}/s - 1} \tag{30.28}
$$

where IPS(s) is the number of instructions per second for performance state s, i.e., $\text{IPS}(s) = I/T(s)$. For state selection, BA uses the following formula:

$$
s = \frac{f_{max}}{1 + \delta/u}. \tag{30.29}
$$

If the desired performance state is not directly supported, BA emulates the state. Figure 30.8 outlines the BA algorithm.

Using the same formalism as we did for CSP, we can justify Equation (30.28) and the state-selection formula for BA. We leave the proof to interested reader. Here we focus on the comparison between BA and CSP.

A major difference between the two algorithms is that BA regresses the workload over the performance states whereas CSP regresses it over time [20]. In this sense, BA introduces a new dimension on the information needed to make effective DVFS scheduling decisions. New algorithms may base their decisions on information from both dimensions.

Another difference between BA and CSP is the complexity of the implementation. Both algorithms use online regression, but CSP's equation is more complex to solve than BA's. BA also uses the common least-squared fit. For CSP, $N \cdot n$ data points need to be preserved at all times. In contrast, BA only requires to store n data points. As we mentioned earlier, there is a high variation in CSP's utilization prediction. BA addresses the problem by using cumulative average to reduce the variation. The downside of the approach is slow response to phase changes.

30.6.3 CPU MISER

Ge et al. [20] proposed a DVFS scheduling algorithm called CPU MISER. The algorithm defines utilization using the second definition, i.e., $u = T_{on}(f_{max})/T(f_{max})$. A novelty of the algorithm is that it estimates

Algorithm: **CPU MISER**
Assumption:
    $l_j, j = \text{CPU}, \ldots, \text{L2}.$
Parameters:
    λ and δ (default: 0.5 and 5%).
Algorithm:

$$t_{on}(s) = \sum_j a_j \cdot \left(\frac{l_j \cdot f_{max}}{s} \right).$$

$$t_{off} = m \cdot l_m.$$

$$\alpha = \frac{(t_{i-1} - t_{off})}{t_{on}(s)}.$$

$$u_{i-1}^o = \frac{\alpha \cdot t_{on}(f_{max})}{\alpha \cdot t_{on}(f_{max}) + t_{off}}.$$

$$u_i^p = \lambda \cdot u_{i-1}^p + (1 - \lambda) u_{i-1}^o.$$

$$s_i = \frac{f_{max}}{\left(1 + \delta / u_i^p\right)}.$$

$$s_i = \textbf{round}_{1/3}(s_i).$$

adjust t_i.

FIGURE 30.9   The CPU MISER algorithm.

$T_{on}(s)$ and $T(s)$ directly. Recall that both CSP and BA estimate the two execution times indirectly. CPU MISER uses EMA for utilization prediction. For state selection, it uses the same formula as BA. Figure 30.9 outlines CPU MISER.

CPU MISER uses the following model to calculate $T_{on}(s)$ directly:

$$T_{on}(s) = \alpha \cdot \sum_{j=\text{CPU}}^{\text{L2}} a_j \cdot \left(\frac{l_j \cdot f_{max}}{s} \right) \tag{30.30}$$

where α is an overlapping factor such that $0 \leq \alpha \leq 1$, a_j is the number of accesses to cache level j (with $a_{\text{CPU}} = I$), and l_j is the access latency (in seconds) for cache level j at f_{max} (with $l_{\text{CPU}} = 1$). Since the access counts a_j and m can be measured directly and T_{off} can be modeled as $(m \cdot l_m)$, the overall time $T(s)$ can be estimated directly. This "direct computation" distinguishes CPU MISER from CSP and BA.

Another novelty of CPU MISER is that it adjusts the length of the interval based on the selected performance state. CPU MISER decreases the interval length when a low-performance state is selected and increases the length for a high-performance state. The rationale is that misprediction may be more severe at low-performance states. The interval is shortened so as to reduce the negative effect of misprediction. The following shows one implementation of variable-length time interval:

$$t_i = \begin{cases} 250 \text{ ms} & \text{if } s_i \geq f_H \\ 50 \text{ ms} & \text{otherwise} \end{cases}$$

30.6.4 ECO

Huang and Feng [31] proposed a DVFS scheduling algorithm called ECO. Similar to BA and CPU MISER, ECO defines utilization using the second definition, i.e., $u = T_{on}(f_{max})/T(f_{max})$. Similar to CPU MISER, ECO estimates $T_{on}(s)$ and $T(s)$ directly. However, ECO uses a different method to estimate them. In

Algorithm: **ECO**
Assumption:
l_m.
Parameters:
N and δ (default: 3 and 5%).
Algorithm:

$$w_{off}^{on} = w^s - w_b^s - w_r^s.$$

$$w_{off}^{off} = m \cdot (l_m \cdot s).$$

$$w_{off} = \min(w_{off}^{on}, w_{off}^{off}).$$
$$w_{on} = w_{i-1} - w_{off}.$$

$$u_{i-1}^o = \frac{w_{on}}{w_{on} + w_{off}(f_{max}/s)}.$$

$$u_i^p = \frac{\left(\sum_{j=i-N}^{i-1} u_j^o\right)}{N}.$$

emulate(s_i).

FIGURE 30.10 The ECO algorithm.

contrast to CPU MISER, ECO uses SMA for utilization prediction. For state selection, it uses the same formula. Figure 30.10 outlines ECO.

Similar to CPU MISER, ECO assumes that $w = t \cdot s = w_{on} + w_{off}$ and focuses on getting a more precise estimation of w_{off}. The novelty of the algorithm is that ECO uses two different measurements to estimate w_{off}, i.e.,

$$w_{off} = \min\left(w_{off}^{on}, w_{off}^{off}\right) \tag{30.31}$$

where w_{off}^{on} represents the on-chip measurement of CPU stall cycles due to off-chip workload, and w_{off}^{off} represents the off-chip measurement. Both measurements overestimate the number of CPU stall cycles. But each is more precise in a different scenario. Specifically, w_{off}^{on} is more precise for non-CPU-bound applications whereas w_{off}^{off} is more precise for CPU-bound applications.

The novel measurement method of ECO for w_{off} is to address a limitation of modern CPUs that there exists no such hardware event that can measure CPU stall cycles due to off-chip activities directly. The on-chip measurement w_{off}^{on} subtracts the stall cycles due to on-chip activities of branch misprediction w_b^s and full reorder buffer w_r^s from the total stall cycles w^s. Since not all on-chip activities are subtracted, the measurement is an overestimate. The off-chip measurement w_{off}^{off}, on the other hand, relies on the count of L2 cache misses m which overestimates the number of memory accesses since not every cache miss results in a memory access. Multiple cache misses to a memory block may result in a single transfer of the memory block.

30.7 Conclusions

We have presented, in this chapter, a comparison of several representative DVFS scheduling algorithms for energy-aware HPC. These algorithms were characterized with three design issues: the abstraction of CPU utilization, the prediction of the trend in CPU utilization, and the association of the voltage and frequency values with CPU utilization. We identified three types of CPU utilization based on CPU utilization ratio, memory access rate, and CPU nonstall time. For each type of CPU utilization, we detailed several algorithms and discussed their major differences in terms of high-level ideas. This comparison provides us a general view of DVFS scheduling algorithms for energy-aware HPC.

In the future, we believe that DVFS scheduling algorithms will become more adaptive. We have seen how CPU MISER adapts the length of an interval based on how confident it is with the prediction result from the state-selection formula. We have also seen that ECO takes advantages of hardware-specific features to improve the precision of its utilization calculation. We consider these as a first step toward more adaptive DVFS scheduling algorithms. The next logical step would be to identify the execution scenario in which each technique is most effective. With this knowledge, an adaptive DVFS scheduling algorithm will be able to select the best technique to use once it detects the current execution scenario. The selection can range from different state-selection formulas to different abstractions of CPU utilization.

Acknowledgments

This work was partially supported by the Extreme Scale Systems Center at Oak Ridge National Laboratory. The submitted manuscript has been authored by a contractor of the U.S. government under Contract No. DE-AC05-00OR22725. Accordingly, the U.S. government retains a nonexclusive, royalty-free license to publish or reproduce the published form of this contribution, or allow others to do so, for U.S. government purposes.

References

1. A. Beloglazov, R. Buyya, Y.C. Lee, and A. Zomaya. In M.V. Zelkowitz, ed., A taxonomy and survey of energy-efficient data centers and cloud computing systems, *Advances in Computers*, 82: 47–111, 2011.
2. M. Bhadauria, V. Weaver, and S.A. McKee. Accomodating diversity in CMPs with heterogeneous frequencies. In *International Conference on High Performance Embedded Architectures and Compilers*, Paphos, Cyprus, January 2009.
3. D.J. Brown and C. Reams. Toward energy-efficient computing. *Communications of the ACM*, 53(3):50–58, March 2010.
4. K.W. Cameron. The challenges of energy-proportional computing. *IEEE Computer*, 43(5):82–83, May 2010.
5. L. Chang, D.J. Frank, R.K. Montoye, S.J. Koester, B.L. Ji, P.W. Coteus, R.H. Dennard, and W. Haensch. Practical strategies for power-efficient computing technologies. *Proceedings of the IEEE*, 98(2): 215–236, February 2010.
6. M. Charney and T. Puzak. Prefetching and memory system behavior of the SPEC95 benchmark suite. *IBM Journal of Research and Development*, 41(3):265–286, May 1997.
7. K. Choi, R. Soma, and M. Pedram. Fine-grained dynamic voltage and frequency scaling for precise energy and performance trade-off based on the ration of off-chip access to on-chip computation time. In *Design, Automation and Test in Europe Conference*, Paris, France, February 2004.
8. K. Choi, R. Soma, and M. Pedram. Fine-grained dynamic voltage and frequency scaling for precise energy and performance trade-off based on the ratio of off-chip access to on-chip computation times. *IEEE Transactions on Computer-Aided Design of Integrated Circuits and Systems*, 24(1):18–28, January 2005.
9. D. Donofrio, L. Oliker, J. Shalf, M.F. Wehner, C. Rowen, J. Krueger, S. Kamil, and M. Mohiyuddin. Energy-efficient computing for extreme-scale science. *IEEE Computer*, 42(11):62–71, November 2009.
10. M. Etinski, J. Corbalan, J. Labarta, M. Valero, and A. Veidenbaum. Power-aware load balancing of large scale MPI applications. In *Workshop on High-Performance, Power-Aware Computing*, Rome, Italy, May 2009.

11. W. Feng. Making a case for efficient supercomputing. *ACM Queue*, 1(7):54–64, October 2003.

12. W. Feng and K.W. Cameron. The Green500 list: Encouraging sustainable supercomputing. *IEEE Computer*, 40(12):50–55, December 2007.

13. X. Feng, R. Ge, and K.W. Cameron. Power and energy profiling of scientific applications on distributed systems. In *19th IEEE International Symposium on Parallel and Distributed Processing*, Denver, CO, April 2005.

14. K. Flautner, S. Reinhardt, and T. Mudge. Automatic performance-setting for dynamic voltage scaling. In *International Conference on Mobile Computing and Networking*, Rome, Italy, July 2001.

15. J. Flinn and M. Satyabarayanan. Energy-aware adaptation for mobile applications. In *ACM Symposium on Operating Systems Principles*, Charleston, SC, December 1999.

16. V.W. Freeh, N. Kappiah, D.K. Lowenthal, and T.K. Bletsch. Just-in-time dynamic voltage scaling: Exploiting inter-node slack to save energy in MPI programs. *Journal of Parallel and Distributed Computing*, 68(9):1175–1185, September 2008.

17. V.W. Freeh, D.K. Lowenthal, F. Pan, and N. Kappiah. Using multiple energy gears in MPI programs on a power-scalable cluster. In *ACM SIGPLAN Symposium on Principles and Practice of Parallel Programming*, Raleigh, NC, June 2005.

18. V.W. Freeh, D.K. Lowenthal, F. Pan, N. Kappiah, and R. Springer. Exploring the energy-time tradeoff in MPI programs on a power-scalable cluster. In *IEEE International Symposium on Parallel and Distributed Processing*, Denver, CO, April 2005.

19. V.W. Freeh, F. Pan, D.K. Lowenthal, N. Kappiah, R. Springer, B.L. Rountree, and M.E. Femal. Analyzing the energy-time tradeoff in high-performance computing applications. *IEEE Transactions on Parallel and Distributed Systems*, 18(6):835–848, June 2007.

20. R. Ge, X. Feng, W. Feng, and K.W. Cameron. CPU MISER: A performance-directed, run-time system for power-aware clusters. In *International Conference on Parallel Processing*, Xi'an, China, September 2007.

21. K. Govil, E. Chan, and H. Wasserman. Comparing algorithms for dynamic speed-setting of a low-power CPU. In *International Conference on Mobile Computing and Networking*, Berkeley, CA, November 1995.

22. F. Gruian. Hard real-time scheduling for low-energy using stochastic data and DVS processors. In *International Symposium on Low Power Electronics and Design*, Huntington Beach, CA, August 2001.

23. D. Grunwald, P. Levis, K. Farkas, C. Morrey III, and M. Neufeld. Policies for dynamic clock scheduling. In *USENIX Symposium on Operating System Design and Implementation*, San Diego, CA, October 2000.

24. J. Hennessy and D. Patterson. *Computer Architecture: A Quantitative Approach*. 4th edn. Morgan Kaufmann, San Mateo, CA, 2006.

25. Hewlett Packard, Intel, Microsoft, Phoenix, and Toshiba. *Advanced Configuration and Power Interface*. http://www.acpi.info

26. C. Hsu and W. Feng. Effective dynamic voltage scaling through CPU-boundedness detection. In *Workshop on Power-Aware Computer Systems*, Portland, OR, December 2004.

27. C. Hsu and W. Feng. A feasibility analysis of power awareness in commodity-based high-performance clusters. In *IEEE International Conference on Cluster Computing*, Boston, MA, September 2005.

28. C. Hsu and W. Feng. A power-aware run-time system for high-performance computing. In *International Conference for High Performance Computing, Networking, Storage and Analysis*, Seattle, WA, November 2005.

29. C. Hsu and U. Kremer. The design, implementation, and evaluation of a compiler algorithm for CPU energy reduction. In *ACM SIGPLAN Conference on Programming Languages Design and Implementation*, San Diego, CA, June 2003.

30. C. Hsu and U. Kremer. Compiler support for dynamic frequency and voltage scaling. In J. Henkel and S. Parameswaran, eds., *Designing Embedded Processors: A Low Power Perspective*. Kluwer Academic Press, Boston, MA, 2007.

31. S. Huang and W. Feng. Energy-efficient cluster computing via accurate workload characterization. In *International Symposium on Cluster, Cloud and Grid Computing*, Melbourne, Victoria, Australia, May 2009.

32. R. Jejurikar and R.K. Gupta. Dynamic voltage scaling for system-wide energy minimization in real-time embedded systems. In *International Symposium on Low Power Electronics and Design*, Newport Beach, CA, August 2004.

33. M. Keating, D. Flynn, R. Aitken, A. Gibbons, and K. Shi. *Low Power Methodology Manual: For System-on-Chip Design*. Springer Publishing Company, Incorporated, Dordrecht, the Netherlands, 2007.

34. D. Li, B.R. de Supinski, M. Schulz, K.W. Cameron, and D.S. Nikolopoulos. Hybrid MPI/OpenMP power-aware computing. In *International Symposium on Parallel and Distributed Processing*, Taipei, Taiwan, April 2010.

35. W.-Y. Liang, S.-C. Chen, Y.-L. Chang, and J.-P. Fang. Memory-aware dynamic voltage and frequency prediction for portable devices. In *International Conference on Embedded and Real-Time Computing Systems*, Kaohsiung, Taiwan, August 2008.

36. W.Y. Liang and P.-T. Lai. Design and implementation of a critical speed-based DVFS mechanism for the Android operating system. In *International Conference on Embedded and Multimedia Computing*, Cebu, Philippines, August 2010.

37. M.Y. Lim, V.W. Freeh, and D.K. Lowenthal. Adaptive, transparent frequency and voltage scaling of communication phases in MPI programs. In *International Conference for High Performance Computing, Networking, Storage and Analysis*, Tampa, FL, November 2006.

38. J. Lorch and A. Smith. Improving dynamic voltage algorithms with PACE. In *International Conference on Measurement and Modeling of Computer Systems*, Cambridge, MA, June 2001.

39. Y.-H. Lu, L. Benini, and G. de Micheli. Power-aware operating systems for interactive systems. *IEEE Transactions on Very Large Scale Integration Systems*, 10(2):119–134, April 2002.

40. S. McKee. Reflections on the memory wall. In *International Conference on Computing Frontiers*, Ischia, Italy, April 2004.

41. T. Mudge. Power: A first class design constraint for future architectures. *IEEE Computer*, 34(4):52–58, April 2001.

42. B. Mochocki, X. Hu, and G. Quan. A unified approach to variable voltage scheduling for nonideal DVS processors. *IEEE Transactions on Computer-Aided Design of Integrated Circuits and Systems*, 23(9):1370–1377, September 2004.

43. K.-Y. Mun, D.-W. Kim, D.-H. Kim, and C.-I. Park. dDVS: An efficient dynamic voltage scaling algorithm based on the differential of CPU utilization. In *The Asia-Pacific Computer Systems Architecture Conference*, Beijing, China, September 2004.

44. L. Nielsen, C. Niessen, J. Sparsø, and C. Van Berkel. Low-power operation using self-timed circuits and adaptive scaling of the supply voltage. *IEEE Transactions on Very Large Scale Integration Systems*, 2(4):391–397, December 1994.

45. F. Pan, V.W. Freeh, and D.M. Smith. Exploring the energy-time tradeoff in high-performance computing. In *Workshop on High-Performance, Power-Aware Computing*, Denver, CO, April 2005.

46. T. Pering, T. Burd, and R. Brodersen. The simulation and evaluation of dynamic voltage scaling algorithms. In *International Symposium on Low Power Electronics and Design*, Monterey, CA, August 1998.

47. T. Pering, T. Burd, and R. Brodersen. Voltage scheduling in the lpARM microprocessor system. In *International Symposium on Low Power Electronics and Design*, Rapallo, Italy, July 2000.

48. N. Pettis, L. Cai, and Y.-H. Lu. Dynamic power management for streaming data. In *International Symposium on Low Power Electronics and Design*, Newport Beach, CA, August 2004.

49. P. Pillai and K. Shin. Real-time dynamic voltage scaling for low-power embedded operating systems. In *ACM Symposium on Operating Systems Principles*, Chateau Lake Louis, Banff, Canada, October 2001.

50. C. Poellabauer, D. Rajan, and R. Zuck. LD-DVS: Load-aware dual-speed dynamic voltage scaling. *International Journal of Embedded Systems*, 4(2):112–126, 2009.

51. C. Poellabauer, L. Singleton, and K. Schwan. Feedback-based dynamic voltage and frequency scaling for memory-bound real-time applications. In *IEEE Real-Time and Embedded Technology and Applications Symposium*, San Francisco, CA, March 2005.

52. J. Pouwelse, K. Langendoen, and H. Sips. Dynamic voltage scaling on a low-power microprocessor. In *International Conference on Mobile Computing and Networking*, Rome, Italy, July 2001.

53. D. Rajan, R. Zuck, and C. Poellabauer. Workload-aware dual-speed dynamic voltage scaling. In *International Conference on Embedded and Real-Time Computing Systems*, Toyama, Japan, August 2006.

54. P. Ranganathan. Recipe for efficiency: Principles of power-aware computing. *Communications of the ACM*, 53(4):60–67, April 2010.

55. I. Rodero, S. Chandra, M. Parashar, R. Muralidhar, H. Seshadri, and S. Poole. Investigating the potential of application-centric aggressive power management for HPC workloads. In *International Conference on High Performance Computing*, Goa, India, December 2010.

56. B. Rountree, D.K. Lowenthal, B.R. de Supinski, M. Schulz, V.W. Freeh, and T. Bletsch. Adagio: Making DVS practical for complex HPC applications. In *International Conference on Supercomputing*, Yorktown Heights, NY, June 2009.

57. S. Sair and M. Charney. Memory behavior of the SPEC2000 benchmark suite. Technical Report RC-21852, IBM Thomas J. Watson Research Center, New York, October 2000.

58. Y. Shin, K. Choi, and T. Sakurai. Power optimization of real-time embedded systems on variable speed processors. In *Proceedings of the 2000 IEEE/ACM International Conference on Computer-Aided Design*, San Jose, CA, November 2000.

59. T. Simunic, L. Benini, and G. de Micheli. Dynamic power management of portable systems. In *International Conference on Mobile Computing and Networking*, Boston, MA, August 2000.

60. L. Singleton, C. Poellabauer, and K. Schwan. Monitoring of cache miss rates for accurate dynamic voltage and frequency scaling. In *Proceedings of the 14th, Annual Multimedia Computing and Networking Conference*, San Jose, CA, January 2005.

61. A. Varma, B. Ganesh, M. Sen, S. Choudhary, L. Srinivasan, and B. Jacob. A control-theoretic approach to dynamic voltage scaling. In *Proceedings of the International Conference on Compilers, Architectures, and Synthesis for Embedded Systems*, San Jose, CA, October 2003.

62. V. Venkatachalam and M. Franz. Power reduction techniques for microprocessor systems. *ACM Computing Surveys*, 37(3):195–237, September 2005.

63. M. Weiser, B. Welch, A. Demers, and S. Shenker. Scheduling for reduced CPU energy. In *USENIX Symposium on Operating System Design and Implementation*, Monterey, CA, November 1994.

64. W. Wulf and S. McKee. Hitting the memory wall: Implications of the obvious. *ACM SIGARCH Computer Architecture News*, 23(1):20–24, March 1995.

65. R. Xu, C. Xi, R. Melhem, and D. Mossé. Practical PACE for embedded systems. In *International Conference on Embedded Software*, Pisa, Italy, September 2004.

66. W. Yang and C. Yang. Exploiting energy saving opportunity of barrier operation in MPI programs. In *Asia International Conference on Modelling and Simulation*, Kuala Lampur, Malaysia, May 2008.

67. F. Yao, A. Demers, and S. Shenker. A scheduling model for reduced CPU energy. In *IEEE Annual Symposium on Foundations of Computer Science*, Milwaukee, WI, October 1995.

68. H. Zeng, X. Fan, C. Ellis, A. Lebeck, and A. Vahdat. ECOSystem: Managing energy as a first class operating system resource. In *International Conference on Architectural Support for Programming Languages and Operating Systems*, San Jose, CA, October 2002.

69. X. Zhong and C.-Z. Xu. Frequency-aware energy optimization for real-time periodic and aperiodic tasks. In *Joint Conference on Languages, Compilers, and Tools for Embedded Systems and Software and Compilers for Embedded Systems*, San Diego, CA, June 2007.

31

Tool Environments to Measure Power Consumption and Computational Performance

Timo Minartz
University of Hamburg

Daniel Molka
Technical University of Dresden

Julian Kunkel
University of Hamburg

Michael Knobloch
Forschungszentrum Jülich

Michael Kuhn
University of Hamburg

Thomas Ludwig
University of Hamburg

Most software utilizes only a low percentage of available hardware resources; either the software does not need the resources to perform the task or the software does not handle the resources in an efficient manner. This is the case for server applications [1] as well as for desktop applications [2]. Low utilization is problematic, because the utilization and power consumption of hardware are not proportional [1]. To deal with unneeded resources, hardware vendors reacted with low-power states, which can be activated when the hardware is idle, meaning zero utilization. But this solves the problem only partially, because infrequently and inefficiently utilized hardware still consumes a high percentage of the maximum power. Unfortunately, there is no energy-proportional hardware available yet [3].

This problem may be disregarded in single personal desktop computers, but as the number of computers vastly increases the energy wastage increases too. This is the case for larger offices using various counts of

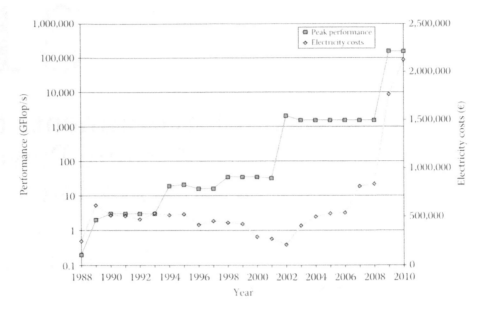

FIGURE 31.1   Increase of performance and annual electricity costs at an example HPC center.

desktop computers, more so for data centers housing larger counts of servers or centers built to operate supercomputers.

Because such centers deal with large energy bills, there are many tools available to measure the performance and power consumption of individual components, which can also be used to investigate the efficiency of server and desktop systems.

Supercomputers combine the performance of hundreds or thousands of office computers to tackle problems that could not be solved on normal PCs in adequate time. High-performance computing (HPC) is an important tool in natural sciences to analyze scientific questions in silico. With the capabilities offered by HPC, scientists no longer have to perform time-consuming and error-prone experiments. Instead modeling and simulation of the law of nature within computer systems provides a well-defined environment. Models from weather systems to protein folding and nanotechnology are simulated and manipulated at the will of the scientists, leading to new observations and understanding of phenomena that are too fast to grasp in vitro [4].

As performance of these supercomputers is the crucial factor, there are several methods to speed up systems. The processing speed of a single processor can be increased, or the number of processors is increased—either by fusing more chips into a single machine or by interconnecting multiple machines into a cluster system.

In former times, the performance of applications was improved by packing more functionality into a single chip and incrementing the clock frequency of the processor; however, the power consumption of the processor is proportional to the square of its clock frequency. Combined with the steady miniaturization, chips would soon have been designed that produce more heat than a nuclear power plant or even the sun [5].

To reach the performance increase implied by Moore's law,* multiple chips are packed on a single chip while trying to keep the power consumption as low as possible [6]. Figure 31.1 shows the performance and operation cost increase of a typical HPC center. While we needed about €400,000 to operate a cluster with a peak performance of 0.2 GFlop/s (200,000,000 floating point operations performed

---

* Moore's law describes the trend of an increasing number of transistors on an integrated circuit. On an average, the number doubles every 18–24 months.

per second)* in 1980, we now need about €2,000,000 to reach a peak performance of 158 TFlop/s (158,000,000,000,000 floating point operations performed per second). These costs only include the electricity costs; further costs like acquisition costs or maintaining costs are not included here, but these costs are included in the total costs of ownership (TCOs). The TCOs are a multiple of the electricity costs.

The main cause for the high energy consumption is that, unfortunately, the resources provided by any hardware are not even utilized by parallel applications [7]. Because the hardware power consumption depends heavily on the utilization and we do not have energy-proportional devices [3], the utilization has to be improved to increase the efficiency. But near-optimal utilization of all resources provided on one single chip (or processor) is already a challenging task to developer, compiler, and middleware. This list includes the operating system as well, which manages the low-level hardware and schedules tasks to the available resources.

The complexity of designing and implementing of parallel applications is even higher than for their sequential versions. Scientific applications intended to run on supercomputers already using multiple programming concepts to get the most performance out of the hardware. Today, in the era of multi-core and multisocket processors, the challenge is almost the same for developer of desktop and server applications because the independent processors/cores must exchange intermediate results by means of communication. This communication process incurs additional latency and might cause idle processors waiting for new data to process. Consequently, careful attention must be given to balance the work evenly among the resources.

There are several programming concepts worth mentioning in this context. These can be roughly separated into concepts for shared and distributed memory. On shared memory architectures parallel programs are usually implemented using threads. POSIX and other standardized interfaces—like Windows or Boost Threads—provide ways to use threads manually. These usually involve programming on a very low level; additionally, more abstract concepts are available. The OpenMP[†] standard provides semiautomatic parallelization using compiler pragmas and library functions. The GNU Compiler Collection provides full OpenMP support as of version 4.4. More advanced approaches include Intel's Threading Building Blocks and Microsoft's Parallel Patterns Library. On distributed memory architectures the de facto standard is to use some kind of message passing, most prominently via the Message Passing Interface (MPI). MPI provides a standardized interface which enables parallel programs to send messages over the network in an efficient manner. Obviously, depending on the network technology used, this introduces even more latency. For this reason, hardware vendors often provide implementations specifically tuned for their architecture. MPICH and OpenMPI provide open-source implementations of the MPI standard. Additionally, version 2 of the MPI standard provides an interface for efficient parallel I/O.

While some numerical algorithms are capable to utilize computing resources on a parallel computer to a high degree using these concepts, most applications exploit only a small percentage of peak performance [8]. To mention a highly optimized and well parallel algorithm that can saturate the theoretical peak performance to about 60% is the LINPACK benchmark [9], which solves a system of linear equations.

Nowadays, our experience shows that people are happy to achieve 10% peak performance of a given system for real applications. Therefore, tuning and optimization of applications to exploit more of the available resources is an important task to improve performance and efficiency of the facility. It is important to optimize from the most promising and performance- or energy-boosting bottleneck to the least.

Often, performance and energy are directly correlated, more efficient and high-performance codes finish earlier, causing less energy consumption. In many cases, using less resources with a higher

---

* Today's quad-core CPUs reach about 10 GFlop/s.
† Open Multi-Processing.

efficiency is more energy efficient; however, a fast execution is mandatory for the scientist to deliver results, and, therefore, the focus is on the execution time. By reducing the executing time by 1 min, you save about 3.80 €.* Because runtimes of hours, days, and even months are regular, huge savings can be reached. Another upcoming task is, of course, to reduce the general operating costs without or with only a small impact on the performance. This optimization process is described in the next section.

In the further sections, tools are described to measure performance and related power consumption. We will describe several metrics and tools to measure performance and energy-related characteristics to localize performance and power issues. We will describe how to integrate these tools into existing HPC tool environments to be able to visualize and analyze the connection between power and performance.

31.1 Optimization Process

In industry the process of system and application tuning is often referred as performance engineering. As scientific programs usually require a huge amount of resources, one could expect they are especially designed for performance. However, in most cases, the performance optimization is performed after the program output is validated. At this late stage a version of the code exists, which is tested to some extent, but a complete redesign is usually out of reach.

A schematic view of the typical iterative optimization process is shown in Figure 31.2. In general, the *closed loop of optimization* is not limited to source code; it could be applied to any system. To measure performance and power consumption, hardware and software configurations must be chosen including the appropriate input, i.e., problem statement. It might happen that optimizations made for a particular configuration degrade performance or increase energy consumption on a different setup. Therefore, multiple experimental setups could be measured together. Often, the measurement itself influences the system by degrading performance, which must be kept in mind. Picking the appropriate measurement tools helps to reveal the real behavior of the system and applications.

In the next step, obtained empirical data is analyzed to identify optimization potential in the source code and on the system. As execution of each instruction requires some resources, the code areas must

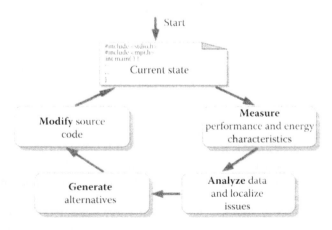

FIGURE 31.2 Closed loop of optimization and tuning.

---

* Assuming annual costs of €2,000,000, which is realistic for electricity costs (see Figure 31.1).

be rated. First, code regions, the execution of which requires a significant portion of runtime (or system resources and resulting energy consumption), are identified. Once the issues are identified, alternative realizations are developed, which tend to mitigate the problems. Then, tuning potential of those hot spots is assessed based on possible performance gains/power reductions and time to modify the solution. Changing a few code lines of the main execution base to improve performance is more efficient than recoding whole processes that might be active for only a small percentage of the total runtime. From the point of the computing facility, optimizing a program by even 1% increases the benefit of the hardware (because it runs 24 h a day for months).

At the end of a loop, the current source code is modified to verify the improvement of the new design. The systems get reevaluated in the next cycle until potential gains are either too low, because the results are already near-optimal, or the time to change the source code outweighs the improvements.

Large systems are often complex; therefore, application logic may be reduced to the core of the algorithm—the application kernel—which is then implemented in a benchmark to show potential of system and modifications. Benefit of such a benchmark is the reduced complexity of the program analysis.

The tools to be mentioned in the course of this chapter focus on the measuring (Sections 31.4 and 31.5) and analyzing (Section 31.7) steps.

31.1.1 Data Sources

In the *closed loop*, data is collected that characterizes the application run and the system. Data can be generated by hardware devices or within software, either the operating system, the application itself, or additional tools.

In some devices hardware sensors are available that measure internal utilization of components. All decent consumer hardware has a built-in mechanism for performance reporting, while power consumption has to be measured using additional measurement devices. Depending on the CPU architecture, a wide range of counters is available that measure efficiency of cache and branch-prediction, or determine the number of instructions run, or the number of floating point operations performed. Newer network interface cards accumulate the number of packets, and the amount of data received and transmitted. The operating system provides a rich set of interesting characteristics of the whole system and for individual applications: memory, network, and CPU usage. These mechanisms are further described in Sections 31.4 and 31.5, respectively.

One way of keeping the information is to store *statistics*, e.g., absolute values like the number of function invocations, the average execution time of a function, or the performed floating point operations. *Statistics* of a program represent the *profile* during the runtime.* In contrast to a *profile*, a *trace* records states and events of a program together with a timestamp, allowing analysis of temporal dependencies and event-specific information, e.g., the communication pattern between processes. Tracing of an application produces much more data, potentially degrading performance and distorting attempts to analyze observations.

Whatever way of aggregation is chosen, the data must be correlated to the behavior of the investigated application.

31.1.2 Collecting Data

There are several approaches to measure the performance and the power consumption of a given application. A *monitor* is a system that collects data about the program execution. Approaches could be classified based on where, when, and how runtime behavior is monitored. Monitors mostly rely on software to measure the state of the system; data from available hardware sensors are queried from the software on demand.

---

* In fact, the user could instruct some systems to collect statistics for individual program phases.

Due to overhead caused by the measurement infrastructure, the user must turn it on explicitly. Usually, changes are made to the program under inspection. The process of changing a program is called *instrumentation*. Popular methods are to either alter source code, relink object files with patched functions, or modify the machine code directly [10].

The altered program invokes functions of the monitoring environment to store performance and power consumption data about the program execution and the system state.

However, the actual storage of traces poses some problems. Saving them to the local hard disk may influence the measurements in several ways. On the one hand, there may be an impact on the performance, because depending on the amount of the traced information, the disk may be busy writing out the traces. On the other hand, writing the traces may inhibit the storage system from using a power-saving idle mode. One way to circumvent this problem is to use volatile-memory-based storage. However, this solution also has drawbacks. For example, traces can use quite a large amount of space, lowering the total amount of main memory for other available purposes. Due to swapping, this may also have effects on performance.

Another way to deal with this is to store the traces on some kind of remote storage system. However, this requires the use of the network. Again, due to the utilization of the network, this may have an impact on both the performance and the power consumption.

Overall, a solution based on the specific environment and problem must be used. For example, if a (possibly slower) service network is available, it may be used for sending the traces to a remote storage system. If most of the main memory is unused, it can be utilized to temporarily store the traces.

31.1.3 Relation between the Data

It is desirable to be able to learn about the relationship of certain events that happen at different levels of the system—for example, in the user's program and in the file system server. So it may be useful to be able to tell which specific disk activity on the server was triggered by which I/O command issued from within a program. This usually is impossible, because of concurrent operations and complex optimizations on each level of the system.

One way to accomplish this is to generate unique request identifiers (IDs), which can be used to correlate different events with each other. For example, these IDs can be associated with exactly one request in the user's program. They get passed down through the levels of the system and are written into the traces alongside the actual event. In this way, it is possible to establish an explicit relationship between them. An implementation that allows this kind of tracing for MPI-I/O applications on top of PVFS* can be found in [11]. Another way is to use timestamps, which may be necessary to integrate data from external devices, e.g., a power meter.

Obviously, both methods require that all involved components trace their respective activities. This may involve retrofitting components with tracing abilities or the addition of support for request IDs. Additionally, this can introduce the need for some kind of post-processing. For example, it may be necessary to merge the different traces into one unified trace, which can then be used by trace analysis tools.

31.1.4 Analyzing Data

Users analyze the data recorded by the monitoring system to localize optimization potential. The data is either recorded during program execution and assessed after the application finished. This approach of postmortem analysis is referred to as *offline* analysis. An advantage of this methodology is that data can be analyzed multiple times and compared with older results.

---

* Parallel Virtual File System, http://www.pvfs.org/

Another approach is to gather and assess data *online*, while the program runs. This way feedback is provided immediately to the user, who could adjust settings to the monitoring environment depending on the results.

Due to the vast amount of data, sophisticated tools are required to localize performance and power issues of the system and correlate them with application behavior and finally source code. Tools operate either manually, i.e., the user must inspect the data himself, it could give hints to the user where abnormalities or inefficiencies are found (semiautomatic tool), or try to assess data automatically.

Tool environments, which localize and tune code automatically, without user interaction, are on the wish list of all programmers. However, because of the system and application complexity automatic tools are only applicable for a very small set of problems.

We will focus on trace-based, offline analysis using the examples of Vampir, Sunshot, and the Scalasca tool set in Section 31.7. Further tools to measure power and performance are introduced after explaining elementary hardware characteristics that influence the measurement in the following section.

31.2 Hardware Characteristics

Because desktop or even server applications do not often fully utilize all cores, modern processors have multiple performance and power states. With dynamic frequency voltage scaling (DVFS) it is possible to reduce the core voltage and the corresponding frequency to reduce power consumption in phases with low utilization. Because the lower frequency of course decreases the performance, these frequency states are named *Performance States* (short: *P-States*) by the ACPI standard [12]. The maximum performance state is labeled *P0*, the next lower one *P1*, and so on. Table 31.1 summarizes the performance states of the Intel® Xeon® 5560 processor series. Besides the performance states, modern hardware makes use of enhanced sleep modes when not utilized. In case of the processor, these states are named *CPU States* (short: *C-States*). For the Xeon 5560 processor [13], the state *C0* indicates the normal operating state of the processor. The *C1* state is a low-power state entered when all threads within a core execute an HLT instruction. The processor will transition to the *C0* state upon occurrence of an interrupt. While in *C1* state, the clock of the core is gated. The ACPI RESET instruction will cause the processor to initialize itself and return to *C0*. The next deeper sleep state in this example is the *C3* state, where the core flushes the contents of its caches and stops all of the clocks. With each generation of processors the set of states increases, because the hardware inside the processor can be better controlled and managed. As an example, Table 31.2 summarizes the power specifications of the Xeon 5500 series for each *C-State*.

For other hardware components (e.g., hard disks) there exist also multiple power states named *Device Power States* (short: *D-States*). For example, a disk can have multiple power states: operating state (reading or writing to/from disk), idle state, and sleep state (disk stops spinning, caches are flushed, . . .).

TABLE 31.1   Example
P-States for X5560 Series

P-State	Frequency (MHz)
P0	2800
P1	2667
P2	2533
P3	2400
P4	2267
P5	2133
P6	2000
P7	1867
P8	1733
P9	1600

TABLE 31.2 Example Power Specifications for the Xeon 5500 Series

Package C-State	Power Consumption (W)				
	130	95	80	60	38
C1E	35	3	30/40	22	16
C3	30	26	26/35	18	12
C6	12	10	10/15	8	8

Source: Intel Corporation, Intel Xeon processor 5500 series datasheet, Volume 1, March 2009.

Besides to the already mentioned states, for the processor some further states exist, the *Thermal States* (short: *T-States*) [14]. The *T-State* is one of the three execution states that processors execute code in. Much like the emergency brake of a car, the *T-State* is used to forcefully reduce the processor's execution speed. The purpose of the *T-State* is to prevent the processor from overheating by lowering its temperature. This is done by introducing idle cycles in the processor, out of this *T-States* do not control the voltage. Because idle cycles do not get into power-saving *C-States*, *T-States* should be avoided if possible (turned off in the BIOS).

In general (for all components), a higher state resulting in a lower power consumption means a higher wake-up time (latency) for the component: The disk needs to accelerate its platters to the needed revolutions per minute to write or read data, the processor needs to populate its cache, and so on. Because of this, sleep and performance states are in general controlled by the operating system using governors (in case of *P-*, *D-States* and *T-States*) or by the hardware itself (in case of *C-States*); the actual state is very important for the power and performance measurement. The interfaces to the operating system are discussed in detail in Section 31.5.2.

31.3 Measurement Metrics

Before explaining the tools for performance and power measuring in the next section, Table 31.3 summarizes the metrics used for inspections. The choice of the right metric is important to have the capabilities to compare different aspects of the code/machine, for example, focusing on performance, power, or a ratio of both. There are many other metrics available with different focuses (see for example [17–19]).

As a side note the LINPACK benchmark is used to measure performance of supercomputers, and these results are the performance numbers (in Flop/s) published in the Top500 lists,* the corresponding power efficiency (in Flop/s/W) in the Green500 lists† to rank the installations.

31.4 Performance Measurement

From the beginning of computing there was a need for measuring the performance of the system. In today's highly complex systems, we have a couple of highly sophisticated tools that help programmers tune their applications in order to achieve the best performance on a certain system.

Performance measurement can mean two different things, first measuring the performance of an application on a certain system and second measuring the performance of the system itself.

It is very important to know both the characteristics of the application as well as the characteristics of the available platforms in order to choose the platform that best fits the application to reduce runtime and energy consumption.

---

* http://www.top500.org/
† http://www.green500.org/

TABLE 31.3   Metrics for Measuring Performance and Power Consumption

Name	Unit	Description
Performance	Flop/s	Floating point operations per second
Energy	J	$1 J = 1 kg\ m^2\ s^{-2}$, one joule is defined as the amount of work done by a force of 1 N moving an object through a distance of one meter [15]
Energy	Wh	$1\ Wh = 3600\ Ws = 3600\ J$ [15]
Power	W	VA, watt, equivalent to energy per second $(J\ s^{-1})$. Meaning the rate at which energy is generated and consumed [15]
Thermal design power (TDP)	W	Represents the average maximum power the cooling system in a computer is required to dissipate
Power efficiency	Flop/s/W	Power efficiency measured in performance per power [16]
Time-to-solution (TTS)	s	Time needed so solve a specific problem
Energy-to-solution (ETS)	J	Amount of energy needed to solve a specific problem
Energy-delay product (EDP)	Js	Product of ETS and TTS [17]

We first focus on the application performance analysis.

The aim of performance analysis is to examine how an application behaves on a certain system. Here we have to differentiate whether the whole application is regarded or just a kernel, e.g., the main loop of the application. When examining the kernel we usually focus on single-node or single-core performance while concentrating on the communication- and I/O-behavior when examining the whole application.

Performance analysis is especially important for scientific codes running on HPC systems, where a shorter runtime usually is directly correlated to reduced energy consumption.

There are several sources a tool can exploit to gain information on resource usage and performance of an application.

The Linux operating system provides the /proc and the /sys file system with very detailed information on hardware and software usage. A more convenient way to access these information is through libgtop.*

Other processors used in computers today have special registers to count specific hardware events like floating point operations or cache misses. The Linux kernel version 2.6.30 and above has the perfcounter interface to provide access to hardware counters. Kernels prior to 2.6.30 had to be patched with either the perfmon2[†] or the perfctr[‡] kernel patch to provide this functionality.

The most commonly used library to read these counters is PAPI, the Performance API [20]. PAPI is a portable library that works with the perfcounter kernel interface as well as with a patched kernel and supports all modern x86 and Power processors. PAPI provides both a high level and a low-level API. The high level API provides the ability to start, stop, and read predefined counters. It is a very lightweight API that allows the developer to quickly insert counter collections into their codes. However, the cost of simplicity is a slightly higher overhead and less functionality compared to the low-level API.

Since version 4 PAPI also supports other components than the CPU to collect device information, included in the release version of PAPI are components for network devices (ethernet and infiniband), ACPI[§], and lm-sensors.**

---

* http://library.gnome.org/devel/libgtop/stable/
† http://perfmon2.sourceforge.net/
‡ http://user.it.uu.se/~mikpe/linux/perfctr/
§ http://www.acpi.info/
** http://www.lm-sensors.org/

On x86 platforms, an alternative to PAPI, which still involves changes of the code (i.e., API calls), is Likwid [21], which provides a very lightweight interface to hardware performance counters. Likwid supports all current x86 processors from Intel and AMD and directly reads the machine-specific registers of those CPUs and thus needs neither the `perfcounter` interface nor a patched kernel. Likwid is, unlike PAPI, a stand-alone tool which starts the counters before the application is started and stops the counters and prepares a report after the application finishes. It also provides a marker API for the user to control which code regions are measured.

However, hardware counter measurements all share one problem. While modern processors provide over 100 different counters, they only have a very limited number of registers, typically 4–6, to store counter values.

Thus, a reasonable subset of counters has to be chosen to get meaningful results. Several tools, e.g., KOJAK [22], the predecessor of Scalasca, or Likwid, provide so-called counter groups and derived metrics thereof to aid the developer, i.e., the developer chooses the metric he is interested in and the tools choose the right counters. Sometimes multiple application runs are necessary in order to collect all the desired information.

Example 31.1 Likwid

This example shows the output of `Likwid` for the Flop/s measurement of a simple calculation of π with OpenMP on an Intel Nehalem processor. This processor has six registers to count events, two of these are fixed (`INSTR_RETIRED_ANY` and `CPU_CLK_UNHALTED_CORE`). `Likwid` prints for each counter and derived metric the values per core and some statistics.

```
likwid-perfCtr -c 0-3 -g FLOPS_DP likwid-pin -c 0-3 ./pi
------------------------------------------------------------
CPU type:       Intel Core Bloomfield processor
CPU clock:      2.67 GHz
Measuring group FLOPS_DP
------------------------------------------------------------
likwid-pin -c 0-3 ./pi
computed pi =          3.14159265358967
CPU time =        0.19 sec
```

Event	core 0	core 1	core 2	core 3
INSTR_RETIRED_ANY	5.29443e+07	5.2507e+07	5.25049e+07	5.25066e+07
CPU_CLK_UNHALTED_CORE	1.39793e+08	1.32563e+08	1.32529e+08	1.32555e+08
FP_COMP_OPS_EXE_SSE_FP_PACKED	0	0	0	0
FP_COMP_OPS_EXE_SSE_FP_SCALAR	1.79474e+07	1.75005e+07	1.75002e+07	1.75006e+07
FP_COMP_OPS_EXE_SSE_SINGLE_PRECISION	4	0	0	0
FP_COMP_OPS_EXE_SSE_DOUBLE_PRECISION	1.79474e+07	1.75005e+07	1.75002e+07	1.75006e+07

Event	Sum	Max	Min	Avg
INSTR_RETIRED_ANY	2.10463e+08	5.29443e+07	5.25049e+07	5.26157e+07
CPU_CLK_UNHALTED_CORE	5.3744e+08	1.39793e+08	1.32529e+08	1.3436e+08
FP_COMP_OPS_EXE_SSE_FP_PACKED	0	0	0	0
FP_COMP_OPS_EXE_SSE_FP_SCALAR	7.04487e+07	1.79474e+07	1.75002e+07	1.76122e+07
FP_COMP_OPS_EXE_SSE_SINGLE_PRECISION	4	4	0	1
FP_COMP_OPS_EXE_SSE_DOUBLE_PRECISION	7.04487e+07	1.79474e+07	1.75002e+07	1.76122e+07

Metric	core 0	core 1	core 2	core 3
Runtime [s]	0.0524201	0.0497094	0.0496964	0.0497063
CPI	2.64037	2.52468	2.52413	2.52455

DP MFlops/s (DP assumed)	306.664	299.028	299.022	299.03
Packed MUOPS/s	0	0	0	0
Scalar MUOPS/s	306.664	299.028	299.022	299.03
SP MUOPS/s	6.83472e-05	0	0	0
DP MUOPS/s	306.663	299.028	299.022	299.03

Metric	Sum	Max	Min	Avg
Runtime [s]	0.201532	0.0524201	0.0496964	0.0503831
CPI	10.2137	2.64037	2.52413	2.55343
DP MFlops/s (DP assumed)	1203.74	306.664	299.022	300.936
Packed MUOPS/s	0	0	0	0
Scalar MUOPS/s	1203.74	306.664	299.022	300.936
SP MUOPS/s	6.83472e-05	6.83472e-05	0	1.70868e-05
DP MUOPS/s	1203.74	306.663	299.022	300.936

Another approach for the performance analysis of an application is profiling using debug information of the application. There are several tools doing this, from the basic yet quite powerful gprof to more advanced tools like the `Intel VTune Amplifier XE`,* which combines profiling and hardware counter analysis.

Example 31.2 gprof

This example illustrates the call-tree output of gprof for a simulation of Conway's Game of Life. Every section represents one element of the call tree along with its callers and callees, the number of calls to each of those, and the time spent, respectively.

```
granularity: each sample hit covers 4 B for 2.50% of 0.40 s

index % time    self  children    called     name
                0.02    0.15      12/12           main [2]
[1]     42.5    0.02    0.15      12          Life::update(void) [1]
                0.15    0.00   48000/48000          Life::neighbor_count(int, int) [4]
-----------------------------------------------
                0.00    0.17       1/1            _start [3]
[2]     42.5    0.00    0.17       1          main [2]
                0.02    0.15      12/12             Life::update(void) [1]
                0.00    0.00      12/12             Life::print(void) [13]
                0.00    0.00      12/12             to_continue(void) [14]
                0.00    0.00       1/1              instructions(void) [16]
                0.00    0.00       1/1              Life::initialize(void) [15]
-----------------------------------------------

[3]     42.5    0.00    0.17                   _start [3]
                0.00    0.17       1/1            main [2]
-----------------------------------------------
                0.15    0.00   48000/48000          Life::update(void) [1]
[4]     37.5    0.15    0.00   48000          Life::neighbor_count(int, int) [4]
-----------------------------------------------
```

When analyzing the performance of an application on a larger scale, more sophisticated tools are needed. This is the domain of tools like Vampir, Sunshot, or Scalasca, see Section 31.7 for a more detailed description of these tools.

---

* http://software.intel.com/en-us/articles/intel-vtune-amplifier-xe/

To use the information obtained by performance analysis tools, the developer needs to know characteristics of the machine the application is running on—that is the domain of benchmarking.

There are several benchmarks used to rate the performance of a computer system. The most prominent among them is certainly the (High Performance) LINPACK benchmark.

However, while the LINPACK benchmark is used to compare different computer systems because it results in a single number, i.e., the Flop/s the machine achieved solving a system of linear equations, the significance of the result is matter to vivid discussions in the (HPC-)community, since the LINPACK results are usually not representative for real-world applications.

That led to the development of several other benchmark suites which usually consist of a number of kernels to cover most standard (HPC-)applications. These are, among others, the HPC Challenge (HPCC),[*] the NAS Parallel Benchmark[†], and the DEISA Benchmark Suite.[‡]

Another approach is taken by the Standard Performance Evaluation Corporation (SPEC).[§] This nonprofit organization publishes a number of performance benchmarks for computer systems, with some of those benchmarks aiming at HPC systems, particularly the SPEC MPI and the SPEC OMP benchmark, which evaluate the MPI and OpenMP performance, respectively. Also noteworthy for performance analysis is the SPEC CPU benchmark, which rates the combined performance of CPU, memory, and compiler. Most other benchmarks measure Java performance for certain tasks.

A special case, and very interesting in the context of this book, is the Java benchmark SPECpower, which rates the energy efficiency of a computer system using different workload levels. However, the relevance of this benchmark is unclear, since several problems occur when measuring the power consumption of a machine. This topic is discussed in Section 31.5.

In summary, as performance tuning is always platform dependent, the first step is to choose the right system for the application, e.g., no memory-bound application should run on a BlueGene system. This information can be obtained by benchmarking. The next step is to investigate single-core performance by profiling (e.g., with gprof) and hardware-counter analysis with either PAPI or Likwid (depending on architecture and whether source code modification is possible or not). Multithreaded programs can be analyzed, for example, with the Intel VTune Amplifier, which is also a very good tool for sequential analysis, or with parallel analysis tools like the timeline-based Vampir or Sunshot or the automatic analysis tool Scalasca. However, the main focus of the latter tools is the analysis of the communication behavior of MPI or hybrid MPI/OpenMP programs.

31.5 Power Measurement

Because of performance driven development, the energy consumption of hardware (especially processors) has become a real problem in recent years. This is mainly caused by the increasing power density of more advanced manufacturing processes which rises as the amount of transistors per mm^2 increases faster than the power consumption per transistor decreases. For example, Intel's Pentium-I processor (P5, 0.8 μm, 1993) has a maximal power consumption of 14 W at a chip size of 292 mm^2 and a resulting power density of about 4.8 W/cm^2. Intel's Itanium2 processor (Fanwood, 130 nm, 2004) has a maximal power consumption of 130 W, a chip size of 374 mm^2 and a resulting power density of about 34.8 W/cm^2 (about seven times higher). The TDP has not increased further since then as it is constrained by the cooling system. Thus, a current generation high-end server processor like the Intel Xeon X5680 (Westmere-EP, 32 nm, 2010) still has the same TDP of 130 W. However, the power density increased to 54.1 W/cm^2 (more than a factor of 10 compared to the Pentium) as the die size of that processor is only 240 mm^2. This heat has to be dissipated, so further energy is needed to cool down the chips, e.g., air cooling (using fans)

---

[*] http://icl.cs.utk.edu/hpcc/
[†] http://www.nas.nasa.gov/Resources/Software/npb.html/
[‡] http://www.deisa.eu/science/benchmarking/
[§] http://www.spec.org/

or water cooling (using pipes and hydraulic pumps). In this section, we want to limit our examination to the node power measurement, so further cooling, of e.g., the machine room, is disregarded in this context. We split our further examinations into two subsections dealing with external sources, which are additional hardware to gather hardware information, and internal sources, which use existing hardware or software mechanisms to monitor the system state.

31.5.1 External Sources

To measure the power consumption, we first focus on primary measurement, meaning how to measure the power consumption of nodes or servers as a whole. For this purpose, usually power measurement devices such as power or watt meters are looped through between the power distribution unit and the power supply of the node. There exist a bunch of different power meters, which mainly differ in the count of measurement channels, the accuracy of the measurement, the interface to read the measurement values, and, of course, the price. Table 31.4 gives an overview about some device types.

The cost per measurement channel for each device type scales with the accuracy and the available interface types. While the external devices in general have a higher accuracy, in most cases another hardware API is needed to extract the measurement values from the serial/Firewire/USB/LAN interface. If the device supports some higher-level protocols, e.g., SNMP,* the fetching of the measurement data is in most cases easier, but it is possible that the fetching interval and the measurement interval are different. If using integrated measurement devices in the PDU or the power supply, the advantage is that no additional space (e.g., in the computing racks) is needed for measurement devices and the power cable connection is as regular as without power measurement.

From a computer scientist's point of view, power meters all work the same way: At every fixed timestep a measurement takes place, which results in discrete measurement values (in general, an interpolation of device internal measurements). These measurement values can be read from the device and can be further analyzed. The general problem when interpreting the data is the problem of buffers and conversion losses. The power supply has capacitive buffers to compensate for short-term variation. Thus a short increase of the power consumption of the node devices may be unrecognized by the measurement device. Further the measurement device itself may have a buffer for the measurement values, so the correlation of the measurement data and the resources utilization has to be validated. In most cases this is possible with timestamps provided by the measurement device on the one hand and the node on the other hand, potential communication delays have to be deducted. However, the breakdown to the component power consumption in most cases is difficult, because of conversion losses inside the power supply. An example breakdown with primary measurement can be found in [26].

With secondary measurement it is possible to determine the power consumption of the node internal devices as each outlet of the power supply can be measured separately. The simplest methods are power

TABLE 31.4   Example Measurement Devices by Type, Internal Accuracy, Measurement Interval, Measurement Channels, and Data Interface

Device	ZES LMG [23]	Watts up [24]	PX-5528 [25]
Type	External	External	Integrated
Accuracy	High	Low	Mid-range
	$(+/-0.1\%)$	$(+/-1.5\%)$	$(+/-1.0\%)$
Interval	10 ms	1 s	1 s
Channels	1–8	1	24
Interface	Serial/Firewire	USB/LAN	Serial/LAN/SNMP

---

* Simple Network Management Protocol.

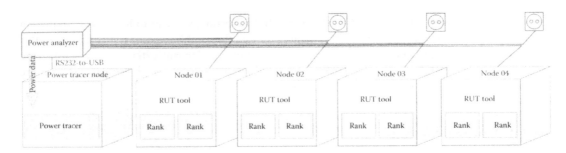

FIGURE 31.3 Tracing power consumption and node statistics of processes by using an accurate power meter with four measurement channels. On each node run two MPI processes and one further process to collect further data, e.g., the utilization.

supplies with integrated measurement devices which distribute the data, for example, via PMBus* or the IPMI† interface. Unfortunately, few power supply vendors integrate such capabilities.

Another approach is to use direct current sensor clamps and connect them to each outlet of the power supply. These sensors use the Hall effect to measure the power consumption contact-less. Each of these clamps has to be connected to other devices, e.g., a power meter or oscilloscope. There are two main problems with using the clamps: On one hand, the general acceptable accuracy of about 2% cannot be guaranteed for low currents. On the other hand, this results in a large number of measurement devices, which is not really practicable for production machines. However, it is comparatively simple to measure the power consumption of a disk, because each disk has its own power outlet. For the processors it is difficult to determine the power distribution, as some outlets could be shared with other devices (e.g., the ATX connector) and each processor could be supplied by multiple outlets. A possible solution is special measurement system boards that provide interfaces to measure the power consumption of each device. However, such boards are custom products mainly used for verification purposes during product development. They are usually not easily available for outsiders.

A more feasible approach is to build a model to estimate or approximate the power consumption. This can be done, for example, by using the devices utilization [3] or the usage of performance counters [27].

Figure 31.3 visualizes the tracing of the power consumption with an external power analyzer with four channels, each of the channels connected to one node. The Power Tracer Node reads the measurement values via the serial port from the measurement device, as would be the case for the LMG 450. Besides the power consumption, it is of course possible to trace further data simultaneously on each node, for example, the resources utilization or device states as discussed in the following subsection.

31.5.2 Internal Sources

This subsection contains methods to record hardware characteristics influencing the power consumption as described in Section 31.2. For the Linux operating system the /proc and /sys interfaces, respectively, can monitor and manage hardware states. To manage, for example, the *P-States* of the processor, the corresponding kernel module has to be loaded. In case of Intel processors, this is the cpufreq module, for AMD processors the powernow-k8 module is needed. With these modules the operating system can manage the states using multiple so-called governors.‡ Each governor implements a different strategy for switching between the *P-States*. The standard ones are the *Conservative, Ondemand, Performance, Powersave,* and *Userspace* governors. The cpufreq governor decides (dynamically or

---

* Power System Management Protocol, http://pmbus.org/
† Intelligent Platform Management Interface, http://www.intel.com/design/servers/ipmi/
‡ http://www.mjmwired.net/kernel/Documentation/cpu-freq/governors.txt

statically) which target frequency to set within the limits of each policy. For example, the *Ondemand* governor sets the CPU depending on the current usage, the *Performance* governor instead sets the processor statically to the highest available frequency. With the Linux commands `cpufreq-info` and `cpufreq-set` the governor and the frequencies can be monitored and changed, respectively, for each logical processor.

The additional `cpufreq-stats` module* provides information about the usage of different *P-States*, *C-States*, and transitions between them. For example, the `time_in_state` file located under `/sys/devices/system/cpu` includes the amount of time spent in each of the frequencies supported by the processor. The output has a frequency and time pair in each line, which means this processor spent time (in 10 ms) at the corresponding frequency. The output has one line for each of the supported frequencies.

The main problem when monitoring *P-* and *C-States* are the fast transitions between the states. The time it takes, for example, for the Intel Xeon X5560 processor to switch between two frequencies is about 10,000 ns, for the *C-States* switches the duration ranges from 3 to 245 ns. A feasible method for monitoring is collecting the data per timestep; the smaller the timestep, the greater the accuracy and the amount of data. This is exactly what PowerTop[†] does. The user gets feedback about the usage of the different idle and performance states. The software further lists the processes polling the devices and preventing idle states. So it is possible to analyze the system behavior (e.g., unnecessary frequent disk polling for journal writing) and further to analyze own software to avoid, for example, active waiting for devices that possibly have a positive impact on the performance, but for sure a negative impact on the power consumption.

One other software to analyze software in terms of energy is the Intel Energy Checker.[‡] The software faces the problem that all too often activity is measured by how busy a server is while running an application rather than by how much work that application completes. The Intel Energy Checker API provides the functions required for exporting and importing counters from an application to measure the real work done.

For HPC environments there already exist multiple frameworks to access power consumption values and to manage the device hardware state based on policies. PowerPack[§] is one example for such frameworks. It is a tool to isolate the power consumption of devices including disks, memory, NICs, and processors in a high-performance cluster and to correlate these measurements to application functions. Further frameworks are available from hardware vendors, for example, the Active Energy Manager Plugin[**] for the IBM Systems Director. The plugin measures, monitors, and manages the energy and thermal components of IBM systems and allows to control power and limit power consumption (power capping).

When using frequency reduction or idle states, this naturally has a high potential for wrong decisions in terms of performance. Goal of the *eeClust*[††] project is to get knowledge about the execution behavior of the applications currently executing to reduce the energy consumption without impairing program performance. This can be done, for example, using statistical hardware usage patterns (e.g., based on performance counters or utilization) or using code annotations to communicate the future hardware utilization.

As described in this section, many aspects have to be concerned when measuring the power consumption. The measurement itself using dedicated measurement devices is straightforward, but the desired measurement granularity has to be chosen in advance regarding costs and data amount. Besides the

---

* http://www.mjmwired.net/kernel/Documentation/cpu-freq/cpufreq-stats.txt

[†] http://www.lesswatts.org/projects/powertop/

[‡] http://software.intel.com/en-us/articles/intel-energy-checker-sdk/

[§] http://scape.cs.vt.edu/index.php/home/

[**] http://www-03.ibm.com/systems/software/director/aem/

[††] http://www.eeclust.de/

interpretation of the measurement data (because of buffers, etc.), the main challenge is the correlation of power consumption with hardware utilization. Further the hardware utilization (and the device power states) must be correlated with the currently executed code. The next sections introduce multiple tool environments that face this challenge with trying to generalize the usage of the tools for measuring power and performance.

31.6 Trace Generation

Most of the mentioned tools to measure performance and power consumption have different focuses and interfaces. Additional tools are necessary to collect all this information and store it in a format that can later be analyzed. If the generated data retains the chronology of events, it is referred to as tracing, whereas profiling only provides aggregate information. In this section we introduce two tracing tools, namely VampirTrace and HDTrace. An exemplary trace evaluation using the corresponding analysis tools is done in Section 31.7.

VampirTrace is traditionally focused on the performance analysis of parallel programs. It is based on instrumentation of the application in order to generate a sequence of events that are recorded at runtime. It collects information from the perspective of the individual processes or threads and can only access internal sources directly. However, as energy consumption is becoming increasingly important, VampirTrace has recently been extended to record energy-related information from external sources as well. HDTrace uses a different approach. It basically records information as to how the running applications affect the hardware in a cluster. Applications do not need to be manipulated.

31.6.1 VampirTrace

VampirTrace [28,29] is an open-source tracing tool developed by the Center for Information Services and High Performance Computing (ZIH). It is used to instrument applications in order to generate trace files in the OTF* format that can be analyzed using Vampir and Scalasca later on (see Sections 31.7.1 and 31.7.3). Depending on the type of instrumentation, the trace files contain events like function entries end exits, MPI messages send between processes, hardware performance counters, etc. All events are stored with a timestamp, thus events of different processes can be visualized with the correct timing behavior. Figure 31.4 shows the possible data sources that can be accessed using VampirTrace. In order to

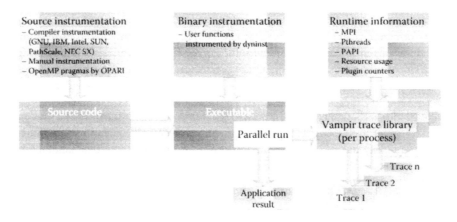

FIGURE 31.4 VampirTrace data sources.

---

* Open Trace Format.

generate a trace file using VampirTrace the application usually has to be recompiled using the provided wrappers:

```
CC=icc              -> CC=vtcc
CXX=icpc            -> CXX=vtcxx
F90=ifort           -> F90=vtf90
MPICC=mpicc         -> MPICC=vtcc -vt:cc mpicc
```

After being compiled using the VampirTrace wrappers, the application generates a trace file when it is run. If the source code is not available, Dyninst* [30] can be used to instrument binaries.

The compiler based instrumentation modifies all functions in order to generate events at every entry and exit. OPARI [31] is used to additionally trace OpenMP directives. MPI functions are instrumented using a replacement library that records additional information at runtime (e.g., sender, receiver, message size) before calling the actual MPI function. Thread creation using the pthread library is detected as well.

The default instrumentation settings record as much information as possible about the program behavior. However, this can result in a huge runtime overhead and large trace files. Filters can be used to reduce the size of the trace files. Events of filtered functions will not be added to the trace file. Unfortunately, this does not reduce the overhead at runtime as the compiler-based instrumentation will generate an event that is then checked against the filter list. In order to prevent the event generation at runtime, manual instrumentation can be used. In that case the compiler-based instrumentation is disabled. Instead the programmer has to add VT_USER_START("eventname"); and VT_USER_END("eventname"); calls to the source code at positions where enter or leave events should be generated. The usual way is to start with automatic instrumentation, then apply more and more filters until the trace file includes just enough information for the analysis, and then instrument the remaining functions manually to reduce the overhead at runtime.

VampirTrace natively supports performance counters via PAPI [20]. Thus all sources for which a PAPI component is available can be recorded as additional information for every event. Tracing of hardware performance counters is configured using the VT_METRICS environment variable that can contain multiple PAPI event names in a colon separated list. If tracing of performance counters is enabled, the selected counters are recorded for every generated event. Another possibility to incorporate information about the hardware is to use VampirTrace's Plugin Counter Interface [32]. This interface can be used to record the power consumption using an external power meter. The power consumption is stored in a database during the program execution [33]. The power consumption plugin accesses the database after the program has finished and adds the information about the power consumption to the trace. Other plugins exist that record *P-State* and *C-State* changes using the Linux Perf Events subsystem included in Linux kernels since version 2.6.31.

31.6.2 HDTrace

The *HDTrace environment* is another tracing environment. It is developed under the GPL-license and consists of the components shown in Figure 31.5. HDTrace concentrates on evaluation of new ideas and therefore has not the maturity level of VampirTrace. The TraceWriting-C library is responsible for storing events in XML trace files and statistics in a binary format with an XML description header. A project file links together all trace and statistic files of multiple sources without conversion. Statistics, which include operation system information about network, CPU, and memory, are recorded periodically. *PIOsim* is an event-based simulator that reads the application event traces and allows them to run in artificial cluster environments. The simulation generates trace files of the run and internal

---

* http://www.dyninst.org/

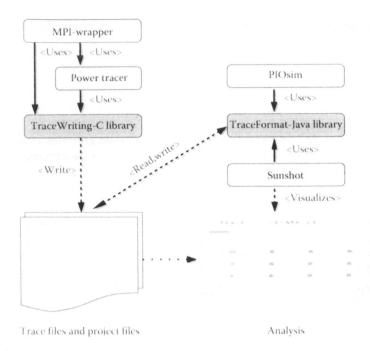

FIGURE 31.5 HDTrace components.

components for further inspection. Several extensions to the trace format are available; with the trace format, it is possible to trace file server activities of PVFS2 servers and visualize them together with client activity to understand causal relations. Trace files of application or simulation runs are visualized by Sunshot, a Java-Swing application in which the original design is based on Jumpshot [34]. The Jumpshot viewer is part of the MPI implementation MPICH2, which allows visualization of the *SLOG2* trace format of the *MPI Parallel Environment* (MPE). *PowerTracer* is an extension to the trace environment, which periodically traces information about power usage from an external power meter in statistic files. Information about the interaction between the PowerTracer and the trace environment is shown in Figure 31.6.

31.7 Trace Evaluation

In this section we introduce tools that can be used to analyze the traces generated by the tools discussed in the previous section. Vampir and Sunshot are used to visualize the tracefiles generated by VampirTrace and HDTrace, respectively. Scalasca is another tool that can be used to analyze the OTF files generated by VampirTrace.

Vampir and Sunshot are timeline-based tools that allow navigation through the whole sequence of events. This provides as much information as possible. However, with the ever increasing number of cores in current HPC systems it becomes difficult to identify potential problems. In contrast to that, Scalasca automatically searches the trace for known issues and presents only these. Another feature of Scalasca is the higher scalability compared to the other tools. If the trace can be kept in memory all the time, Scalasca can analyze traces with hundreds of thousands of processes. In this case the file system bottleneck can be avoided by not creating intermediate files for each process but only storing the result of the automatic analysis.

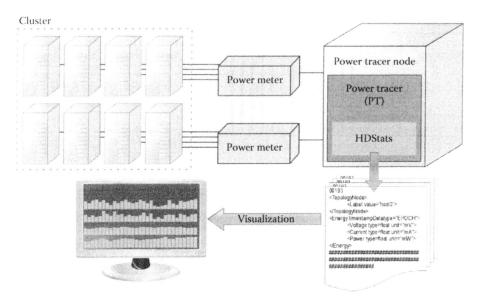

FIGURE 31.6   Tracing and analysis workflow from multiple nodes.

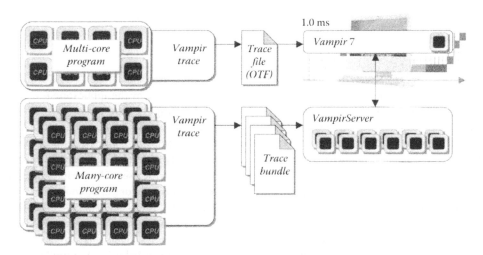

FIGURE 31.7   Usage of VampirTrace and Vampir.

31.7.1 Vampir

Vampir [28,29] is a performance analysis and optimization tool developed by the Center for Information Services and High Performance Computing (ZIH). It visualizes trace files that have been generated with VampirTrace (see Section 31.6.1). Long running applications on systems with many cores can produce large trace files that cannot be analyzed sufficiently fast using a single workstation. In order to analyze such large trace files, Vampir can connect to a VampirServer that performs the resource consuming operations on an HPC system. Figure 31.7 shows how the tools work together.

Vampir allows the programmer to analyze the chronology of activities in parallel processes or threads. Figure 31.8 shows a typical configuration of Vampir displays to analyze MPI applications. All displays can be enabled and disabled independently. When zooming within on display, all displays adapt to the selected interval.

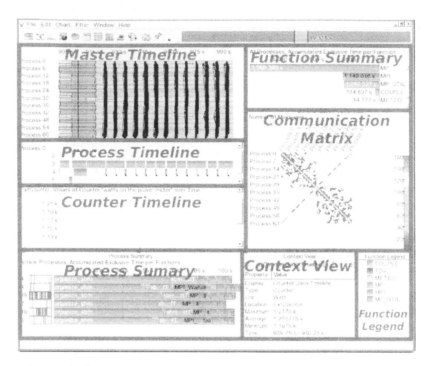

FIGURE 31.8   Vampir displays.

The *Master Timeline* shows the sequence of events for all processes and threads. Time spent in MPI functions is displayed in red by default whereas the CPU time consumed by the application is represented with green bars. If function groups were defined during the tracing they are represented by different colors to quickly identify program phases. Messages between processes are represented by black lines. A well-balanced application usually shows a regular pattern of calculation and communication phases. Load imbalances as well as abnormally behaving processes can be quickly identified in the master timeline.

The *Process* and *Counter Timelines* allow analysis of individual processes. The performance of individual functions can be determined based on the associated hardware performance counters. Combined with the *Function* or *Process Summary*, which reveals how much time is spent in each function, this allows identification of code regions with a high optimization potential (i.e., a significant portion of the runtime combined with low utilization of the hardware). The *Counter Timeline* is also used to display power consumption, *P-State*, and *C-State* changes that have been recorded using VampirTrace's plugin interface for external counters. The example in Figure 31.8 shows a higher power consumption during the communication phases compared to the calculation phases.

The *Context View* shows additional information about anything that is selected in another display. In the example it shows the power consumption at a certain timestamp. Another use case is to identify the source code file and line number of a function. The *Communication Matrix* shows the communication pattern between the processes. There are some more displays available that are not shown in the aforementioned example: a call tree for the selected process, message statistics, and a summary of I/O events.

The rest of this section gives some examples how Vampir can be used to investigate the impact of the processor and main memory utilization on the total energy consumption.

The example shown in Figure 31.9 reveals the power consumption of an MPI application during idle (sleep) and compute phases as well as during a MPI barrier. The power consumption within the MPI barrier is very high as it uses busy waiting in the tested implementation.

The power consumption of ALU and FPU operations as well as data transfers between registers and memory was measured using Microbenchmarks [35]. Data transfers are examined by performing the same

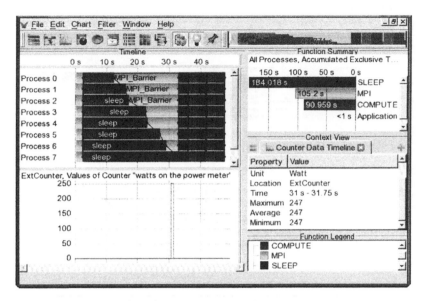

FIGURE 31.9   Power consumption of MPI barrier (OpenMPI 1.3).

operation on data from different levels in the memory hierarchy. The ALU and FPU load is controlled by substituting the performed operation on the same chunk of data. This generates varying pressure for the execution units while keeping the utilization of other units (decoders, branch prediction, load-store unit, etc.) constant.

The benchmarks use the manual instrumentation provided by VampirTrace in order to record only necessary information. Instead of function entries and exits, the threads record the current phase of the benchmark. Furthermore the benchmarks measure performance counters internally only for the measurement routine. This avoids counting of hardware events outside the actual measurement. In principle the measurement works as follows:

```
start_counters()
vt_user_start(REGION)
read timestamp counter
perform operation on data set
read timestamp counter
vt_user_end(REGION)
stop_counters()
```

A REGION can be "L1," "L2," "L3," or "RAM" depending on the current data set size. The additional regions "IDLE," "WAIT," and "COM" visible in Figure 31.10 are used during the synchronization phases between the individual measurements for different data set sizes. The performance counters (start_counters(), stop_counters()) are accessed via PAPI or the perfmon2 interface and are added to the trace file using the user-defined counter feature. User-defined counters are defined as follows:

```
int gid,*cid;
cid = (int*) malloc(num_events*sizeof(int));
gid = VT_COUNT_GROUP_DEF("performance counter");
for (i=0;i<num_events;i++)
  cid[i]=VT_COUNT_DEF(event[i],"value",VT_COUNT_TYPE_DOUBLE,gid);
```

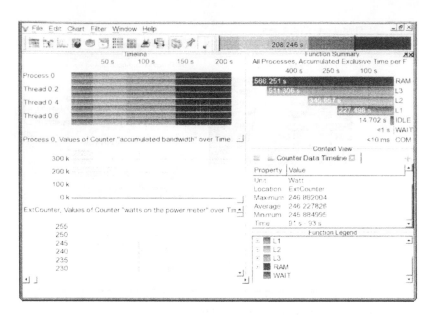

FIGURE 31.10   Power consumption of data transfers from different cache levels and main memory into registers.

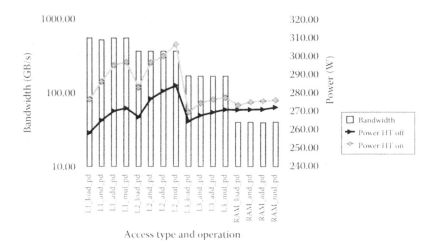

FIGURE 31.11   Power consumption of different operations on a dual socket Intel X5670 system.

The read counter values (stop_values - start_values) are stored in the trace using

```
for (i=0;i<num_events;i++)
  VT_COUNT_DOUBLE_VAL(cid[i],counter_values[i]);
```

Some results are shown in Figures 31.10 and 31.11. Figure 31.10 shows the power consumption of data transfers from different cache levels. It can be seen that the individual cache levels do not only have decreasing performance the farther away they are from the core, but also as the power consumption rises as more and more parts of the chip are used.

Figure 31.11 shows the different power consumption of arithmetic operations that were obtained using Vampir. It shows significant differences in power consumption between the different arithmetic operations. Furthermore, it shows that hyperthreading has a substantial overhead in terms of power consumption in situations where it does not improve performance.

31.7.2 Sunshot

In the following, traces of the *HPC Challenge Benchmark* (HPCC)* demonstrate the features of *Sunshot* in detail. The goal is to understand how energy and performance could be related with the visualizing tool. Additionally, the HPCC results are discussed briefly.

An excerpt of the HPCC trace including energy metrics and client activity is shown in Figure 31.12. In the left a tree view visualizes the mapping of the metrics and traces to nodes—in this case *hpcc* is run on node06 to node09, process 0 and process 4 are run on node06, and the energy metrics (I, P, U) belong to node06. To the right of the tree view, the activity and statistics for each timestamp are drawn—black areas in the process activity correspond to computation on the client processes, the colors encode calls to MPI (communication) functions. The statistic metrics compute the maximum independently for each timeline. Uninteresting timelines or statistics can be removed from the view to dig into the issues.

From this view a slight fluctuation of the power consumption can be observed between different nodes, during the broadcast operation (purple operation on the right), the power consumption is lower.

In Figure 31.13 only the timelines for energy metrics and some metrics for node local activity of node06 are shown. Moving the mouse over an event or statistic adds further information in the bar above the timeline canvas. In this example the bar prints the yellow color and the text "Items" underneath, the average current on node06 is 0.838 A and the value under the mouse pointer is about 0.92 A.

The performance statistics provided by the operating system provide a hint as to how the power consumption is related to node behavior—CPU utilization is almost proportional to the current and power consumption.

However, as the height of each bar is scaled linear between 0 and the maximum value of this timeline, the fluctuation cannot be seen easily in this figure. Therefore, the values can be scaled logarithmically or across multiple timelines of the same category, in Figure 31.14 the energy metric timelines are scaled to the minimum and maximum value of all nodes, therefore the minimal fluctuation between 229 and

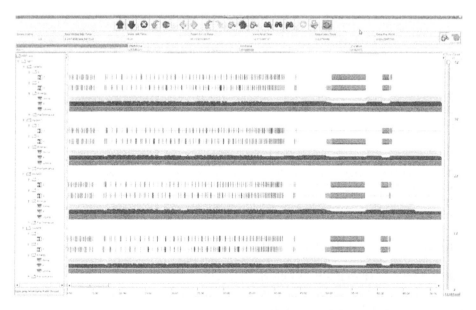

FIGURE 31.12 Sunshot: Timelines of the first phase of the HPCC-run including energy metrics and client side events.

---

* The HPC Challenge Benchmark consists of seven applications: the High Performance LINPACK, DGEMM, Stream, PTRANS, RandomAccess, FFT, and a set to benchmark communication bandwidth and latency [36].

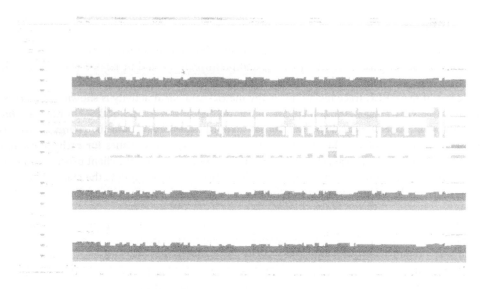

FIGURE 31.13   Sunshot: Energy metric timelines of the HPCC-run.

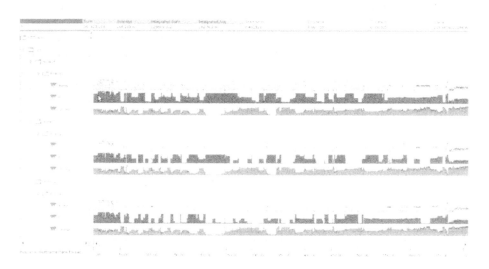

FIGURE 31.14   Sunshot: Energy metric timelines of the HPCC-run—all timelines are scaled with the global minimum and maximum of one category (I, P, or U).

230.4 V becomes visible as well. During the first phase of the benchmark, the observed power utilization between the nodes varies; however, the supply voltage stays the same for all nodes as excepted, because all nodes are connected to the same power distribution unit.

The space to draw a statistic timeline (in this case the three energy timelines) is limited. Therefore, if the sampling frequency of the value is more frequent than it can be drawn in Y-axis due to the amount of pixels, then a shifted color scheme shows minimum and maximum observed for a given time, the drawn value is the average value for the time span covering a particular pixel. This behavior is visualized in Figure 31.15a. Figure 31.15b is an excerpt of a zoomed-in timeline from Figure 31.15a showing the values as they were observed on the measuring device.

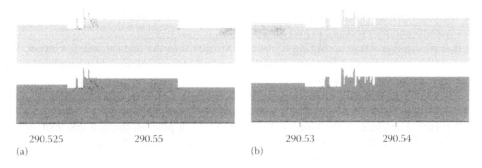

(a)   (b)

FIGURE 31.15 Sunshot: Excerpt of the first process timelines of I and P to demonstrating how minimum, average, and maximum are visualized in the timelines. (a) Overview. (b) Zoomed.

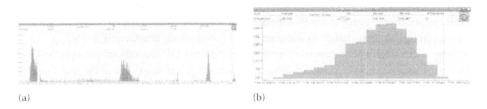

(a)   (b)

FIGURE 31.16 Sunshot: Energy histograms for a node. (a) Power. (b) Voltage.

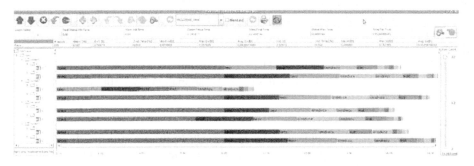

FIGURE 31.17 Sunshot: Trace profile window of the first phase of the HPCC-run showing the aggregated time spend in each MPI call.

The values of a given statistic metric can be visualized in histograms with configurable bin number to show how these values are distributed for a given timeline. Power and voltage for the first node are presented in Figure 31.16, the power fluctuates between 135 and 220 W. Three peaks can be observed at 138, 175, and 208 W, this is due to the CPU activity. During the program run, there are phases in which both CPUs are idle, then sometimes one CPU is completely utilized while the other process sleeps, or both CPUs are idle because the processes wait for input from processes hosted on other nodes.

Profiles of timeline events can be created with Sunshot as well. The profile for the first phase of the HPCC-run is provided in Figure 31.17. In this view all events on each timeline are aggregated by a given metric, for example, by summing up the inclusive time these events required. During this phase broadcast operations account for about 8 s on most processes, but process 2 shows a different behavior. In the given example the other processes wait for process 2 to join the collective broadcast operation, i.e., all other processes have to wait for process 2 to join the operation.

Note that all displays adjust according to the area visualized in the main timeline window; if a user zooms in or moves the clipping area, other displays, e.g., all histograms, update accordingly to present their content only for the visible area.

In the trace environment MPI internal communication was traced as well for this experiment. Due to the communication intense benchmark this leads to big trace files, 800 MB of compressed XML files to record all events. Therefore, in the visualized example for each process, the first 200,000 lines of each XML file (out of 101,768,032 lines) are used to render the event timelines. However, tracing performance and power statistics results in less than 1 MB of data.

31.7.3 Scalasca

Scalasca [37], a joint development of Jülich Supercomputing Centre (JSC) and the German Research School for Simulation Sciences (GRS), is an integrated performance analysis toolset to automatically analyze large-scale parallel applications, using the MPI, OpenMP, or a hybrid MPI/OpenMP programming model. It has proven scalability to up to 294,912 processes [38], i.e., the whole BlueGene/P at JSC.

The novelty of Scalasca is an automatic search for wait-states, i.e., sections where a process has to wait for another process at a synchronization point (either local, e.g., a point-to-point message, or global, e.g., a barrier) due to work imbalance. For manual analysis, e.g., with Vampir, EPILOG files—the Scalasca tracefiles—can be converted to other trace formats like OTF (see Section 31.6.1).

Scalasca contains both a sequential analyzer (EXPERT) and a parallel analyzer (SCOUT) to identify wait-state patterns in EPILOG traces. SCOUT is a parallel program that has to run with the same amount of processes as the original application since it "replays" the communication of the original program, sending trace data instead of the original data. Thus it scales with the original application. However, currently the analysis capabilities, i.e., the number of detectable patterns, of EXPERT are higher than those of SCOUT, so in some cases the sequential analyzer, though only usable on a smaller scale, is needed.

The Scalasca workflow, as shown in Figure 31.18, is a three-phased workflow: program instrumentation, execution measurement and analysis, and analysis report examination. For all three phases the `scalasca` command can be used:

1. `scalasca -instrument`
   Used to instrument the application by adding calls to the Scalasca measurement system into the application's code. Prepends the compiler command. Instrumentation is either automatic (i.e., compiler-based), semiautomatic using source-code transformation, or linked with a pre-instrumented library, e.g., using the PMPI-interface. Additionally there is an API for manual instrumentation, i.e., direct calls to the measurement system by the user.

2. `scalasca -analyze`
   Used to control the measurement environment during the application execution, and to automatically perform a trace analysis after measurement completion (if tracing was enabled).
   The Scalasca measurement system supports both runtime summarization and event trace collection and analysis, optionally including hardware-counter information.

3. `scalasca -examine`
   Used to post-process the analysis report generated by the measurement runtime summarization and/or postmortem trace analysis, and to start Scalasca's analysis report examination browser CUBE3.

The reports generated by Scalasca are visualized by the CUBE3 browser, as shown in Figure 31.19. The view is divided into three columns, the first displays the regarded metric (i.e., which problem), the

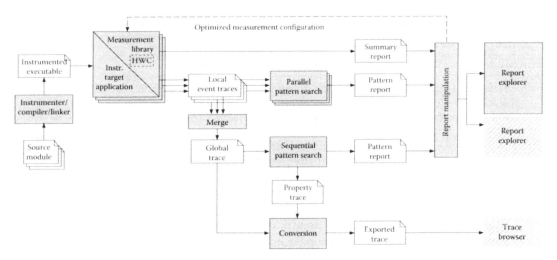

FIGURE 31.18   The Scalasca workflow.

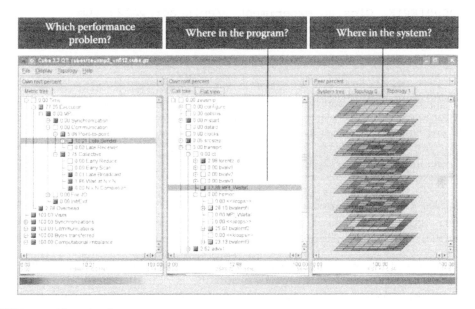

FIGURE 31.19   The Cube3 browser.

second and third column display where in the call tree (i.e., which function) and where on the machine (i.e., which process) this problem occurs.

Example 31.3 Scalasca Full Workflow Example

This example describes the usage of Scalasca on the BlueGene/P machine at Forschungszentrum Jülich with the SOR-Application. SOR consists of the single source file `sor.c`, so instrumentation is straightforward:

```
scalasca -instrument mpixlc -c sor.c
scalasca -instrument mpixlc sor.o -o sor.x
```

The resulting instrumented binary `sor.x` has to be run in the Scalasca measurement and analysis nexus:

```
scalasca -analyze mpirun -mode vn -np 128 ./sor.x
```

The Scalasca analyzer will take care of certain control variables, which assist in the measurement of your application. The default behavior of the Scalasca analyzer is to create a summary file. It is also possible to create a detailed event trace, as indicated by the initial messages from the measurement system (called EPIK).

```
S=C=A=N: Scalasca 1.3 runtime summarization
S=C=A=N: ./epik_sor_vn128_sum experiment archive
S=C=A=N: Collect start
mpirun -mode vn -np 128 ./sor.x
[00000]EPIK: Created new measurement archive ./epik_sor_vn128_sum
[00000]EPIK: Activated ./epik_sor_vn128_sum [NO TRACE]

        [... Application output ...]

[00000]EPIK: Closing experiment ./epik_sor_vn128_sum
...
[00000]EPIK: Closed experiment ./epik_sor_vn128_sum
S=C=A=N: Collect done
S=C=A=N: ./epik_sor_vn128_sum complete.
```

After successful execution of the job, a summary report file is created within a new measurement directory. In this example, the automatically generated name of the measurement directory is `epik_sor_vn128_sum`, indicating that the job was executed in BlueGene's virtual node mode (`-mode vn`) with 128 processes (`-np 128`). The suffix `_sum` refers to a runtime summarization experiment. The summary analysis report can then be post-processed and examined with the Scalasca report browser:

```
scalasca -examine epik_sor_vn128_sum
INFO: Post-processing runtime summarization report ...
INFO: Displaying ./epik_sor_vn128_sum/summary.cube ...
```

A screenshot of the Scalasca report browser CUBE3 with the summary analysis report of SOR opened is shown in Figure 31.20. The examination of the application performance summary may indicate several influences of the measurement on your application behavior. For example, frequently executed, short functions may lead to significant perturbation and would be prohibitive to trace: These need to be eliminated before further investigations using trace analysis are taken into account.

During trace collection, information about the application's execution behavior is recorded in so-called event streams. The number of events in the streams determines the size of the buffer required to hold the stream in memory. To minimize the amount of memory required, and to reduce the time to flush the event buffers to disk, only the most relevant function calls should be instrumented.

When the complete event stream is larger than the memory buffer, it has to be flushed to disk during application runtime. This flush impacts application performance, as flushing is not coordinated between processes, and runtime imbalances are induced into the measurement. The Scalasca measurement system uses a default value of 10 MB per process or thread for the event trace: When this is not adequate, it can be adjusted to minimize or eliminate flushing of the internal buffers. However, if the value specified is too large for the buffers, the application may be left with insufficient memory to run, or run adversely with paging to disk. Larger traces also require more disk space (at least temporarily, until analysis is complete), and are correspondingly slower to write to

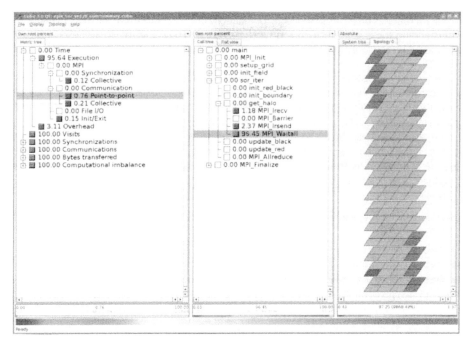

FIGURE 31.20   Viewing a runtime summary in CUBE3.

and read back from disk. Often it is more appropriate to reduce the size of the trace (e.g., by specifying a shorter execution, or more selective instrumentation and measurement) than to increase the buffer size.

To estimate the buffer requirements for a trace measurement, `scalasca -examine -s` will generate a brief overview of the estimated maximal number of bytes required:

```
scalasca -examine -s epik_sor_vn128_sum

[cube3_score epik_sor_vn128_sum/summary.cube]
Reading ./epik_sor_vn128_sum/summary.cube... done.
Estimated aggregate size of event trace (total_tbc): 25698304 bytes
Estimated size of largest process trace (max_tbc):   215168 bytes
(When tracing set ELG_BUFFER_SIZE > max_tbc to avoid intermediate flushes
 or reduce requirements using a file listing USR regions to be filtered.)
```

flt	type	max_tbc	time	%	region	
	ANY	215168	11849.04	100.00	(summary)	ALL
	MPI	195728	147.47	1.24	(summary)	MPI
	COM	9696	465.93	3.93	(summary)	COM
	USR	9744	11235.64	94.82	(summary)	USR

The line at the top of the table referring to `ALL` provides the aggregate information for all executed functions. In this table, the column `max_tbc` refers to the maximum of the trace buffer capacity requirements determined for each process in bytes. If `max_tbc` exceeds the buffer size available for the event stream in memory, intermediate flushes during measurement will occur. To prevent flushing, you can either increase the event buffer size or exclude a given set of functions from measurement.

To aid in setting up an appropriate filter file, this "scoring" functionality also provides a breakdown by different categories, determined for each region according to its type of call path. Type `MPI` refers to function calls to the MPI library and type `OMP` either to OpenMP regions or calls to the OpenMP API.

User-program routines on paths that directly or indirectly call MPI or OpenMP provide valuable context for understanding the communication and synchronization behavior of the parallel execution, and are distinguished with the COM type from other routines that are involved with purely local computation marked USR.

Routines with type USR are typically good candidates for filtering, which will effectively make them invisible to measurement and analysis (as if they were "inlined"). COM routines can also be filtered; however, this is generally undesirable since it eliminates context information. Since MPI and OMP regions are required by Scalasca analyses, these cannot be filtered.

By comparing the trace buffer requirements with the time spent in the routines of a particular group, the initial scoring report will already indicate what benefits can be expected from filtering. However, to actually set up the filter, a more detailed examination is required. This can be achieved by applying the cube3_score utility directly on the post-processed summary report using the additional command-line option -r:

```
cube3_score -r epik_sor_vn128_sum/summary.cube

Reading summary.cube... done.
Estimated aggregate size of event trace (total_tbc): 25698304 bytes
Estimated size of largest process trace (max_tbc):   215168 bytes
(When tracing set ELG_BUFFER_SIZE > max_tbc to avoid intermediate flushes
 or reduce requirements using a file listing USR regions to be filtered.)
```

flt	type	max_tbc	time	%	region	
	ANY	215168	11849.04	100.00	(summary)	ALL
	MPI	195728	147.47	1.24	(summary)	MPI
	COM	9696	465.93	3.93	(summary)	COM
	USR	9744	11235.64	94.82	(summary)	USR
	MPI	80000	2.14	0.02	MPI_Irsend	
	MPI	73600	1.07	0.01	MPI_Irecv	
	MPI	16040	20.77	0.18	MPI_Allreduce	
	MPI	16000	14.32	0.12	MPI_Barrier	
	MPI	9600	87.25	0.74	MPI_Waitall	
	COM	9600	304.28	2.57	get_halo	
	USR	4800	5432.60	45.85	update_red	
	USR	4800	5432.87	45.85	update_black	
	MPI	240	0.54	0.00	MPI_Gather	
	MPI	200	3.63	0.03	MPI_Bcast	
	USR	48	368.66	3.11	TRACING	
	USR	48	0.50	0.00	looplimits	
	MPI	24	0.52	0.00	MPI_Finalize	
	USR	24	0.54	0.00	init_boundary	
	USR	24	0.48	0.00	init_red_black	
	COM	24	2.88	0.02	sor_iter	
	COM	24	156.25	1.32	init_field	
	COM	24	0.82	0.01	setup_grid	
	MPI	24	17.23	0.15	MPI_Init	
	COM	24	1.70	0.01	main	

(The basic form of this command is reported when running scalasca -examine -s.) As the maximum trace buffer required on a single process for the SOR example is approximately 215 KB, there is no need for filtering in this case.

When all options of the Scalasca measurement system are set in a way that measurement overhead and space requirements are minimized, a new run of the instrumented application can be performed, passing the -t option to `scalasca -analyze`. This will enable the tracing mode of the Scalasca measurement system. Additionally, the parallel postmortem trace analyzer searching for patterns of inefficient communication behavior is automatically started after application completion:

```
scalasca -analyze -t mpirun -mode vn -np 128 ./sor.x

S=C=A=N: Scalasca 1.3 trace collection and analysis
S=C=A=N: ./epik_sor_vn128_trace experiment archive
S=C=A=N: Collect start
mpirun -mode vn -np 128 ./sor.x
[00000]EPIK: Created new measurement archive ./epik_sor_vn128_trace
[00000]EPIK: Activated ./epik_sor_vn128_trace [10000000 bytes]

          [... Application output ...]

[00000]EPIK: Closing experiment ./epik_sor_vn128_trace
[00000]EPIK: Flushed file ./epik_sor_vn128_trace/ELG/00000
...
[00013]EPIK: Flushed file ./epik_sor_vn128_trace/ELG/00013
[00000]EPIK: Closed experiment ./epik_sor_vn128_trace
S=C=A=N: Collect done
S=C=A=N: Analysis start
mpirun -mode vn -np 128 scout.mpi ./epik_sor_vn128_trace
          [... SCOUT output ...]
S=C=A=N: Analysis done
S=C=A=N: ./epik_sor_vn128_trace complete.
```

This creates an experiment archive directory epik_sor_vn128_trace, distinguishing it from the previous summary experiment through the suffix _trace. A separate trace file per MPI rank is written directly into a subdirectory when measurement is closed, and the parallel trace analyzer SCOUT is automatically launched to analyze these trace files and produce an analysis report. This analysis report can then be examined using the same commands and tools as the summary experiment:

```
scalasca -examine epik_sor_vn128_trace
INFO: Post-processing trace analysis report ...
INFO: Displaying ./epik_sor_vn128_trace/trace.cube ...
```

The screenshot in Figure 31.21 shows that the trace analysis result at first glance provides the same information as the summary result. However, the trace analysis report is enriched with additional performance metrics that show up as submetrics of the summary properties, such as the fraction of point-to-point communication time potentially wasted due to late sender situations where early receives had to wait for sends to be initiated. That is, the trace analysis can reveal detail of inefficient execution behavior.

The file system requirements for an EPILOG event trace and its analysis are much higher than for a runtime summary. The runtime of a batch job will also increase due to additional file I/O at the end of measurement and for analysis. After successful tracing, the Scalasca measurement has created a directory containing the event trace and its analysis files. The default behavior of Scalasca in tracing mode is to create a runtime summary report (stored in `summary.cube`) as well as a trace analysis report (stored in `trace.cube`).

After successful trace analysis, and before moving the experiment archive, the trace files can be removed by deleting the ELG subdirectory in the experiment archive.

Current developments in Scalasca include the analysis of time-dependent behavior, i.e., different iterations of a main loop of a computational kernel, and a root-cause analysis, i.e., finding the root of

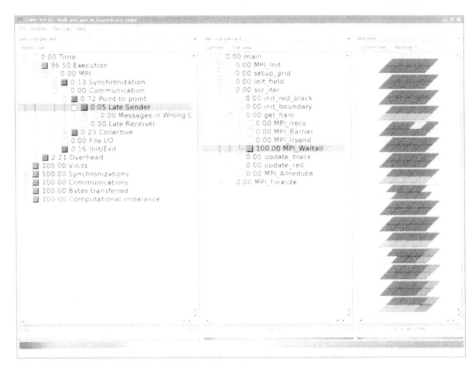

FIGURE 31.21   Determine a late sender in CUBE3.

a wait-state and how it propagates [39]. Within the *eeClust* project Scalasca, is extended to perform an analysis to identify energy saving possibilities in applications.

31.8 Conclusion

In this chapter, tools were introduced that allow to measure performance and power on all layers of abstraction. Most of those are also open source—from the presented tools, only Vampir and Intel's VTune are commercial tools.

To measure pure performance, PAPI and Likwid provide access to embedded performance counters on the CPU. The measurement of power consumption using dedicated measurement devices is straightforward, but the desired measurement granularity has to be chosen in advance regarding device costs and data amount. VampirTrace and HDTrace allow recording of performance and power metrics together with process activity into trace files which can be visualized and analyzed by the programmer. Scalasca replays trace files and automatically detects common issues for degraded performance.

Typically the developer of a parallel program will first ensure correctness of the application and then use these tools to evaluate the scalability and energy efficiency to identify bottlenecks in system and application.

While all these solutions help to analyze power and performance characteristics, there are limitations that are subject to discussion.

First, these tools do not allow adjustment of analysis metrics to match all analysis-scenarios of users. Imagine the user of a tool might want to visualize the average value of a metric (power or performance) over a set of processes, for example, to see the power consumption at rack level. To allow high flexibility, the tools must know the physical arrangements and relation among all hardware and software, which is not the case yet. In addition, flexibility must be provided by the tools to allow arbitrary functions to be applied on groups of timelines.

Even further, aggregation and analysis could be interesting based on conditions of a program's status. A user might want to analyze power consumption of the system based on the program's activity to rate functions on their energy efficiency. This would require the tools to aggregate metrics from multiple timelines if a particular event is observed on another timeline.

Measuring the power consumption of a specific component and assigning that value to a particular event is difficult or sometimes even impossible. A measured value of, e.g., 50 mJ on the I/O-system, could be caused by a write operation that contains data from multiple write operations of one or even multiple processes. In the future, this issue will remain tricky, as optimizations inside any library or middleware might defer and aggregate operations to a later time.

Scalability of the introduced parallel performance tools varies. While the measurement systems are able to trace applications running on many processors, the analysis limits the scaling.

This is especially true for the timeline-based tools, where a large number of timelines cannot be displayed because of screen resolution and usability—if every timeline is just one pixel high, then the user does not see anything at all. While Sunshot is designed to visualize a small number of timelines together with metrics on node level, Vampir scales to a higher number of processes and can display statistics in a separate window. However, both tools allow visualization of any subset of an interesting timeline, at a time.

Scalasca with its event-based parallel replay scales to a much higher number of processes—in principle it scales with the analyzed application—but is limited to the identification of known performance degradation patterns and currently does not support energy-related patterns. However, this usually is enough to get the user started and gives good hints for further in-depth analysis with one of the other presented tools.

References

1. L. A. Barroso and U. Hölzle. The case for energy-proportional computing. *Computer*, 40:33–37, December 2007.
2. M. A. Kazandjieva, B. Heller, P. Levis, and C. Kozyrakis. Energy dumpster diving. In *Workshop on Power Aware Computing and Systems (HotPower)*, Big Sky, MT, 2009.
3. T. Minartz, J. Kunkel, and T. Ludwig. Simulation of power consumption of energy efficient cluster hardware. *Computer Science–Research and Development*, 25:165–175, 2010.
4. W. J. Kaufmann and L. L. Smarr. *Supercomputing and the Transformation of Science*. W. H. Freeman & Co., New York, 1992.
5. R. Ronen, S. Member, A. Mendelson, K. Lai, S. L. Lu, F. Pollack, and J. P. Shen. Coming challenges in microarchitecture and architecture. In *Proceedings of the IEEE*, 89(3):325–340, 2001.
6. S. E. Thompson and S. Parthasarathy. Moore's law: The future of Si microelectronics. *Materials Today*, 9(6):20–25, 2006.
7. W. Schönauer and H. Häfner. Explaining the gap between theoretical peak performance and real performance for supercomputer architectures. *Scientific Programming*, 3(2):157–168, 1994.
8. L. Oliker, A. Canning, J. Carter, C. Iancu, M. Lijewski, S. Kamil, J. Shalf, H. Shan, E. Strohmaier, S. Ethier, and T. Goodale. Scientific application performance on candidate petascale platforms. In *Proceedings of the International Parallel and Distributed Processing Symposium (IPDPS)*, Long Beach, CA, 2007.
9. J. J. Dongarra, P. Luszczek, and A. Petitet. The LINPACK benchmark: Past, present, and future. *Concurrency and Computation: Practice and Experience*, 15:803–820, 2003.
10. S. Shende, A. D. Malony, and R. Ansell-bell. Instrumentation and measurement strategies for flexible and portable empirical performance evaluation. In *International Conference on Parallel and*

Distributed Processing Techniques and Applications (PDPTA 2001), Las Vegas, NV, pp. 1150–1156, 2001.

11. T. Ludwig, S. Krempel, J. Kunkel, F. Panse, and D. Withanage. Tracing the MPI-IO calls' disk accesses. In B. Mohr, J. L. Träff, J. Worringen, and J. Dongarra, eds., *Recent Advances in Parallel Virtual Machine and Message Passing Interface*, Lecture Notes in Computer Science 4192, pp. 322–330. Berlin/Heidelberg, Germany: C&C Research Labs, NEC Europe Ltd., and the Research Centre Jülich, Springer, 2006.

12. Hewlett-Packard Corporation, Intel Corporation, Microsoft Corporation, Phoenix Technologies Ltd. and Toshiba Corporation. Advanced configuration and power interface specification, Revision 3.0a, December 2005.

13. Intel Corporation. Intel Xeon Processor 5500 Series datasheet Volume 1, March 2009.

14. Intel Corporation. LessWattsorg.

15. Physikalisch Technische Bundesanstalt. The legal units in Germany, February 2004.

16. W.-C. Feng and K. W. Cameron. The Green500 list: Encouraging sustainable supercomputing. *Computer*, 40(12):50–55, December 2007.

17. R. Gonzalez and M. Horowitz. Energy dissipation in general purpose microprocessors. In *IEEE Journal of Solid-State Circuits*, 31:1277–1284, September 1996.

18. C. Bekas and A. Curioni. A new energy aware performance metric. *Computer Science–Research and Development*, 25:187–195, 2010. 10.1007/s00450-010-0119-z.

19. C.-H. Hsu, W. C. Feng, and J. S. Archuleta. Towards efficient supercomputing: A quest for the right metric. In *International Symposium on Parallel and Distributed Processing*, Denver, CO, 12, p. 230a, 2005.

20. D. Terpstra, H. Jagode, H. You, and J. Dongarra. Collecting performance data with PAPI-C. In *Proceedings of the Third International Workshop on Parallel Tools for High Performance Computing*, Dresden, Germany, pp. 157–173. Springer, Berlin/Heidelberg, Germany, 2009.

21. J. Treibig, G. Hager, and G. Wellein. LIKWID: A lightweight performance-oriented tool suite for x86 multicore environments. In *39th International Conference on Parallel Processing Workshops (ICPPW 2010)*, San Diego, CA, April 2010.

22. B. J. N. Wylie, B. Mohr, and F. Wolf. Holistic hardware counter performance analysis of parallel programs. In *Proceedings of Parallel Computing 2005 (ParCo 2005)*, Malaga, Spain, pp. 187–194, October 2005.

23. ZES ZIMMER Electronic Systems GmbH. Precision Power Analyzer.

24. Electronic Educational Devices. Watt's up internet enabled power meters.

25. Raritan Inc. Dominion PX-5528 Technical Specifications, October 2010.

26. T. Minartz, J. Kunkel, and T. Ludwig. Model and simulation of power consumption and power saving potential of energy efficient cluster hardware. Master's thesis, Institute of Computer Science, University of Heidelberg, Heidelberg, Germany, 2009.

27. R. Zamani and A. Afsahi. Adaptive estimation and prediction of power and performance in high performance computing. *Computer Science—Research and Development*, 25:177–186, 2010. 10.1007/s00450-010-0125-1.

28. A. Knüpfer, H. Brunst, J. Doleschal, M. Jurenz, M. Lieber, H. Mickler, M. S. Müller, and W. E. Nagel. The Vampir performance analysis tool-set. In *Proceedings of the Second International Workshop on Parallel Tools for High Performance Computing*, Stuttgart, Germany, pp. 139–155. Springer, Berlin/Heidelberg, Germany, 2008.

29. M. S. Muller, A. Knupfer, M. Jurenz, M. Lieber, H. Brunst, H. Mix, and W. E. Nagel. Developing scalable applications with Vampir, VampirServer and VampirTrace. In *Parallel Computing: Architectures, Algorithms and Applications*, Advances in Parallel Computing 15, pp. 637–644. Amsterdam, the Netherlands: IOS Press, 2007.

30. B. Buck and J. K. Hollingsworth. An API for runtime code patching. *International Journal of High Performance Computer Applications*, 14:317–329, November 2000.

31. B. Mohr, A. D. Malony, S. Shende, and F. Wolf. Design and prototype of a performance tool interface for OpenMP. *The Journal of Supercomputing*, 23:105–128, August 2002.

32. R. Schöne, R. Tschüter, D. Hackenberg, and T. Ilsche. The VampirTrace plugin counter interface: Introduction and examples. In *Third Workshop on Productivity and Performance (PROPER 2010)*, Naples, Italy, August 2010.

33. S. Krempel, J. Kunkel, and T. Ludwig. Design and implementation of a profiling environment for trace based analysis of energy efficiency benchmarks in high performance computing. Master's thesis, Institute of Computer Science, University of Heidelberg, Heidelberg, Germany, 2009.

34. O. Zaki, E. Lusk, W. Gropp, and D. Swider. Toward scalable performance visualization with jumpshot. *International Journal of High Performance Computing Applications*, 13(2):277–288, Fall 1999.

35. D. Molka, D. Hackenberg, R. Schöne, and M. S. Müller. Characterizing the energy consumption of data transfers and Arithmetic Operations on x86-64 Processors. In *Proceedings of the First International Green Computing Conference*, Chicago, IL, pp. 123–133. IEEE Computer Society, Washington, DC, 2010.

36. P. Luszczek, J. J. Dongarra, D. Koester, R. Rabenseifner, B. Lucas, J. Kepner, J. Mccalpin, D. Bailey, and D. Takahashi. Introduction to the HPC Challenge Benchmark Suite. Technical report, Knoxville, TN: University of Tennessee, 2005.

37. M. Geimer, F. Wolf, B. J. N. Wylie, E. Abraham, D. Becker, and B. Mohr. The Scalasca performance toolset architecture. *Concurrency and Computation: Practice and Experience*, 22(6):277–288, April 2010.

38. B. J. N. Wylie, D. Böhme, B. Mohr, Z. Szebenyi, and F. Wolf. Performance analysis of Sweep3D on Blue Gene/P with the Scalasca toolset. In *Proceedings 24th International Parallel and Distributed Processing Symposium and Workshops (IPDPS)*, Atlanta, GA. IEEE Computer Society, Washington, DC, April 2010.

39. M. Geimer, F. Wolf, B. J. N. Wylie, D. Becker, D. Böhme, W. Frings, M.-A. Hermanns, B. Mohr, and Z. Szebenyi. Recent developments in the Scalasca toolset. In *Proceedings of the Third International Workshop on Parallel Tools for High Performance Computing*, Dresden, Germany, Springer, Berlin/Heidelberg, Germany, 2009.

32

BlueTool: Using a Computing Systems Research Infrastructure Tool to Design and Test Green and Sustainable Data Centers

Sandeep Gupta
Arizona State University

Georgios Varsamopoulos
Arizona State University

Anna Haywood
Arizona State University

Patrick Phelan
Arizona State University

Tridib Mukherjee
Xerox Corporation

32.1 Growth and Technological Trends in Data Centers

With the boom of Internet-based mass services of this decade, from massive multiplayer games to Web-based e-mail and personal pages, and additionally with the increase of corporate server farms and data storage facilities, there has been a great growth of data centers [4,20]. As an effect, energy used in data centers has been increasing: An IDC IT Experts Survey reported that in 2000, data centers consumed 1% of the total power budget of the United States; in 2005 they reached 2% of that. In 2006, data centers in the United States used 59 billion KWh of electricity per year, costing US $4.1 billion and generating 864 million metric tons of carbon dioxide (CO_2) in emissions; this is expected to double in every subsequent year [20]. Moreover, the power density increases: the energy efficiency is doubling every 2 years, the performance of integrated circuits, however, triples in the same period. Hence, hardware power efficiency is offset by the increasing miniaturization and density of equipment; consequently, power draw at-the-plug will only increase [4]. Systems will keep increasing their number of cores and their power density; a recent trend in data center provisioning is the move from individual rack-mounted servers to blade-based servers, where a set of blades is mounted in chassis with large shared power supplies. Altogether, this means that we will see hotter data centers in the near future.

Moreover, the cooling cost is about half of the total energy cost [4], which is an indication of low energy efficiency. Many existing data centers are doing very little to be energy efficient: power management software is disabled [27], cooling is overprovisioned, and some still use low-efficiency power supplies. Toward enforcing energy efficiency, the U.S. Congress passed a bill in 2006 to ask government agencies to investigate the energy efficiency of data centers as well as industry's efforts of developing energy-efficient technologies.

32.1.1 Related Research Work on Data Center Management

The two aforementioned trends on (1) the increased density and (2) the increased energy-saving programmability of equipment have sprouted research on how to use these new capabilities to save energy. There are many emerging technologies that make "greener" data centers a realistic possibility. In addition to simple systemic improvements such as higher-efficiency power supplies, finer-grained control over cooling systems and processor speed offer hope for dynamic, software-controlled improvements in overall power consumption. To that extent, it is expected that more server boards will support the advanced configuration and power interface (ACPI) (http://www.acpi.info/) and dynamic frequency and voltage scaling (DVFS) as a reactive protection mechanism against overheating, while cooling systems will support network connectivity allowing remote configuration and control.

System level power state transition and frequency scaling: Managing power state transition for the CPU as well as other components such as hard drive, display, and memory has been broadly studied. The idea is to save energy through transition to the lower power states of system components when the workload is low. The DVFS of the CPU is also of interest in some works [8,21,23] in which the idea is to scale the frequency of the hardware components to the offered workload in order to save energy. Some works argue that the *power proportionality*, i.e., scaling the power of the data center in proportion to the incoming workload rate, can be achieved through these system level component management [15].

Dynamic server (and resource) provisioning: The idea in this approach is to adjust the number of active servers in a data center to the offered workload and suspend others. Related works [6,7,15] show that this approach saves significant energy under Web-oriented workloads.

Thermal-aware workload scheduling: The effectiveness of thermal-aware scheduling is based on the nonuniformity of the heat dissipated and recirculated in an air-cooled data center room; therefore, selecting the servers/workload scheduling that have the least thermal impact saves energy [18,19] help to save energy. Thermal-aware scheduling has been proposed in some works [18,19,26]. Moore et al. [16,18], and Bash and Forman [3] show that thermal-aware workload placement can save energy. Mukherjee et al. [19] and Tang et al. [26] model the heat that is inefficiently recirculated among the servers; using this model, they propose spatiotemporal, thermal-aware job scheduling algorithms for high-performance computing (HPC) batch job data centers. The effectiveness of thermal-aware scheduling for Internet data centers (IDCs) through analytical and simulation-based study is shown in [1]. Although the nonlinearity of the cooling system behavior adds to the complexity of the problem, they include thermal awareness in the workload management to maximize energy efficiency and provide a holistic approach for saving energy.

32.1.2 Shortcomings of Current Research and the Lack of a Testing Infrastructure

Current research in data centers is facing several shortcomings:

Resolving system-model assumptions: Most of the related research work relies on several simplifying system-model assumptions such as *equipment homogeneity*, *task homogeneity*, the *task arrival process*, *queuing*, or the *CRAC behavior*. Although there exists research that deals with energy conservation in heterogeneous data centers [12,28], there is little research toward resolving the choice of such system-modeling assumptions. The greatest challenge in addressing and resolving these assumptions is to maintain a low complexity and good scalability of the system identification models. The derived models will need to be used for online decision making and they have to efficiently describe the data center system.

Proactivity and coordination of management: Power density trends show that data centers can easily reach 500 W/m^2 ($\sim$50 W/ft^2), or about 1.5 KW per chassis. In such densities, it can take only a minute or two before the systems reach critically high temperatures when the cooling equipment fails, while in higher densities, cooling failure can cause the equipment reach those temperatures in only seconds [25]. At the same time, it takes at least a few minutes before a CRAC unit reaches a desired output temperature. This is the main cause of the *cooling delay problem* (see Figure 32.1b, thick line shows the cooling need while the thin line shows the cooling supply) that will be more dominant as the power density of data centers increases, enabling *proactivity* and *coordination* between the *task management* software, the *power management* software and the *cooling control*. These controls are either completely reactive in behavior, or even when proactive they have a limited view of the parameters. Lack of coordination and (predictive) proactivity can cause energy-inefficient operation or delays.

Scalable and efficient derivation of thermal maps: Another issue is enabling an adaptive, scalable, low-cost process to dynamically derive a thermal model of a data center. One dominant reason for the low applicability of thermal-aware algorithms is the high cost to setup and configure the thermal-aware software system. Many existing thermal-aware scheduling algorithms rely on a thermal map or profile of the data center to predict the impacts of possible task placements. Conventionally, creating a thermal map requires exclusive access to a data center in order to test, document, and calibrate for all possible task placement configurations. This process can be extremely time intensive, taking up to weeks at higher

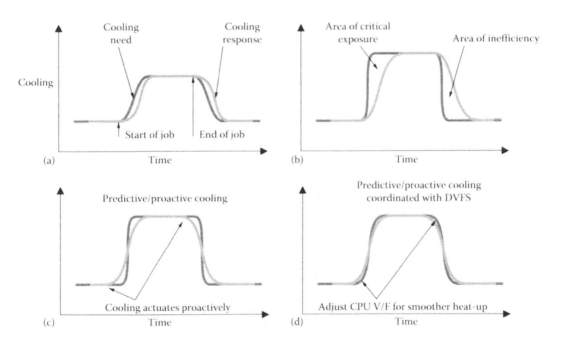

FIGURE 32.1 Conceptual plots of the cooling delay problem and need for proactivity and coordination. (a) A reference plot for medium density data center: The thick line shows the need for cooling while the thin line shows the cooling supply; the supply closely follows the need. (b) In highly dense data centers, the equipment heats up faster and higher; however, the cooling technology is still as slow to respond, creating windows of critical heat exposure and of inefficiency. (c) Predictive, proactive cooling can partly answer the cooling delay problem, while (d) coordinating that with adapting the voltage/frequency at the equipment can yield a closer match up.

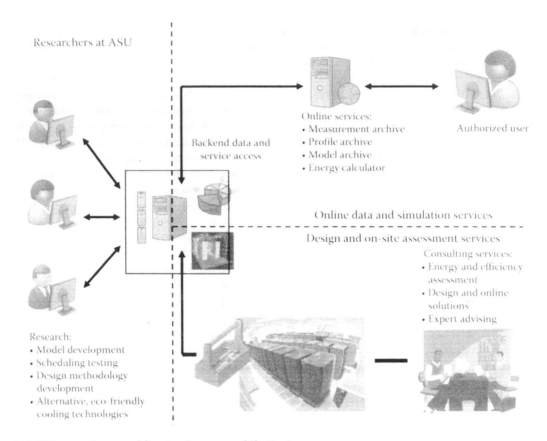

FIGURE 32.2   Conceptual functional structure of BlueTool.

granularity, and the data center is unavailable for use during these periods. Obviously this is an expensive and undesirable property that can prevent potentially advantageous research from seeing use.

A turn toward sustainability: Despite the efforts for reducing the energy wastage in data centers, the research is not focused toward ensuring the sustainability of these facilities. Only recent works [5,9,10] have raised the question of sustainability; sustainability of data centers is still an open topic of research.

The prohibitive cost of algorithm and model testing: Most of the models and solutions developed are at best examined and tested using computational fluid dynamics (CFD), numerical or analytical methods. This is because there is a prohibitive cost of testing any ideas on an actual and real data center. This cost is twofold: (a) paying the bills of running the tests and (b) paying the cost, in any form this may manifest, of delaying or canceling all other useful work that the data center would have done if it were not for the tests.

32.1.2.1 Need for a Testing Infrastructure

All the aforementioned shortcomings are rooted on the lack of testing infrastructure. BlueTool's call is to answer this lack and develop an affordable and open data center for research. The rest of the chapter presents BlueTool's structure and example research tasks to utilize it (Figure 32.2).

32.2 Description of BlueTool

In order to achieve the aforementioned objectives, the BlueTool infrastructure is designed to offer the following facilities:

BlueCenter: A small-scale experimental data center which is the heart of the BlueTool project. It will consist of several racks of servers, enclosed in a insulated room, cooled by a combination of heat-activated and conventional cooling equipment.

BlueSense: A data center assessment toolkit. This toolkit will consist of wireless portable sensor nodes to be used for on-site thermal assessment of data centers.

BlueSim: A simulation toolkit with a workstation with the enhanced CFD simulation software will provide the means to use the measurement readings to verify models and management algorithms through detailed CFD simulation. Part of BlueSim is to offer a Web-accessible energy calculator tool where a user can provide the required parameters and view the savings of the thermal-aware techniques.

BlueWeb: The infrastructure will provide a means to collect, archive, and classify measurement data and simulation results. This "raw" material can be used in the related research projects. This repository shall be available to the research and engineering community.

Figure 32.3 depicts the architecture of functionality to be supported by BlueTool.

32.2.1 BlueCenter

BlueCenter is the mini data center being developed as part of the BlueTool project. It is located on the ASU campus and its purpose is to serve as a testbed for developing and testing

- Quantitative and analytical models of heat recirculation and of power usage
- Management algorithms and software
- The sustainability of a data center

and evaluate various techniques, e.g., thermal-aware server provisioning or heat-capturing schemes.

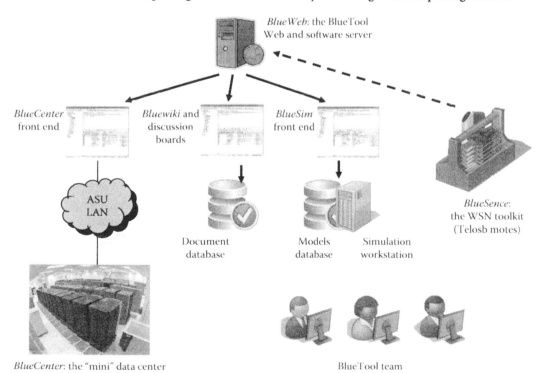

FIGURE 32.3 Architectural structure of the BlueTool project. It consists of several subprojects to create and offer certain facilities.

Computing equipment: The larger objective is to equip the data center with diverse servers to test for heterogeneity scenarios. Currently, the project has several Dell PowerEdge 1855 and IBM xSeries 336 servers.

Cooling devices: The energy consumption of CRAC units can be reduced by improved control strategies, but an approach that has not been widely attempted is to use some of the heat generated by the electronic equipment to drive a heat-activated cooling unit, which can supplement the cooling provided by the conventional CRACs and thus reduce their energy consumption.

Measuring and monitoring equipment: For measuring purposes, the built-in sensors of the computing and cooling equipment will be used, as well as a set of extra sensors, including wireless sensors from the BlueSense subproject.

Room enclosure: BlueCenter is planned to be housed in an insulated enclosure. The enclosure will feature panels and heats will provide internal contraptions to test for various levels of heat recirculation.

32.2.2 BlueSense

Current measuring equipment includes several MICA and TelosB wireless motes and a power meter. The motes will be used in measuring temperature and other conditions at locations not covered by built-in sensors in other equipment.

32.2.3 BlueSim

BlueSim is an integrated chain of various software tools that allows users to model physical attributes of a data center with a GUI interface. The purpose of BlueSim is to provide a high-level and user-friendly way of specifying a data center to be simulated. This is done by use of a GUI interface and a high-level specification language called Ciela. The BlueSim tool chain then translates the high-level specification to a "CFD-ready" file for use by OpenFOAM.

BlueSim's tool chain: As it can be seen in Figure 32.4, based on a user's input an XML file is generated, this XML file is parsed to generate a 3D data center model using Gmsh software. The output of Gmsh

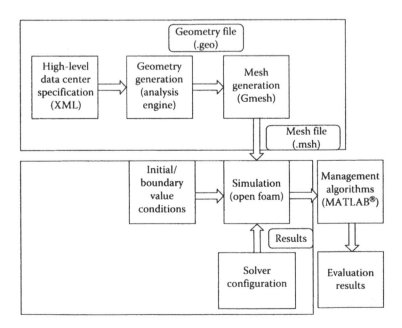

FIGURE 32.4 Block diagram of BlueSim's software tool chain. Data centers are specified in a high-level form. This high-level specification is then translated to a geometry and then to a mesh file, ready to be input to a CFD simulator. After the simulation stage is over, various scripts can process the output and return results to the user.

software along with simulation parameters are fed to OpenFOAM simulator. Figure 32.5 shows the architecture of BlueSim.

Ciela, BlueSim's specification language: Ciela, an acronym for "Computer Infrastructure Engineering LAnguage," is a high-level, XML-based specification language for data center rooms. It is developed on several common properties of data center rooms. One of these properties is the hierarchical structure of computing equipment, i.e., the equipment is organized into rows which consist of racks, which in turn consist of servers or chassis, which in turn consist of servers. The language also allows for specifying multiplicity of items, thus reducing the overall size of the specification script.

The purpose of BlueSim is to allow for a quick description of data centers and reduce the time to develop a complete CFD model. Students took only a few minutes to specify data centers in Ciela given a sample template. This is a desirable feature, especially when engineers want to explore what-if scenarios or several alternative layouts.

32.2.4 BlueWeb

All the services and subprojects will have a front end at what is called BlueWeb, a Web server with all the necessary software to interface to the functionality of each subproject. For example, BlueWeb will offer an interface to BlueCenter so that researchers can test their techniques.

BlueWiki is a repository of articles that relate to research on data center management. It serves as a online hub for engineers that either want to get updated on data center research or want to contribute new knowledge.

32.3 Enabling Research on Data Center Design and Management with BlueTool

This section includes several examples of how BlueTool can be used to assist the research in data center design, modeling, and management.

32.3.1 Developing and Testing Recirculation Models

Previous work proposed a linear, low-complexity heat recirculation model, called *abstract heat recirculation model* (AHRM), and a methodology to assess the model. According to this model, the air recirculation inside a data center can be characterized as cross interference among server nodes. Figure 32.6 demonstrates the AHRM and the correlation between distributed nodes. Node i has inlet temperature T_{in}^i, which is a mix of supplied cold air with temperature T_{sup} and recirculated exhaust warm air from other nodes. The outlet warm air of node i will partially return to the air conditioner, and partially recirculate into other nodes with constant rate. It also draws in exhaust hot air from the outlet of other nodes due to recirculation. The rate, or the percentage of recirculated heat a_{ij}, is defined as cross-interference coefficients (Figure 32.7).

Why modeling heat recirculation is important: Despite considerable progress in data center cooling technologies, substantially many data centers remain to be air cooled without aisle separation. This means that hot air from the equipment's air outlets can indeed recirculate into air inlets instead of reaching the CRAC. Heat recirculation is a dominant cause of energy inefficiency in data centers. Modeling it, i.e., quantifying it and predicting it, can be useful in both assessing the effect it has on the energy efficiency of the data center, as well as devising methods of avoiding it, e.g., using thermal-aware workload scheduling.

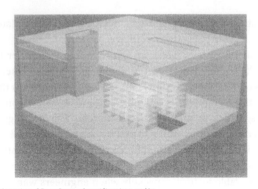

```
<DataCenter xsi:noNamespaceSchemaLocation="DataCenterSpecification.xsd">
  <RoomArchitecture>
    <Shape>
      <Wall Height="8.5" Length="16" Width="" Units="ft" name="Wall1" orientation="0.0"/>
      <Wall Height="8.5" Length="16" Width="" Units="ft" name="Wall2" orientation="90.0"/>
      <Wall Height="8.5" Length="16" Width="" Units="ft" name="Wall3" orientation="90.0"/>
      <Wall Height="8.5" Length="16" Width="" Units="ft" name="Wall4" orientation="90.0"/>
    </Shape>

    <RaisedFloor Height="1.167" Units="ft">
      <Tiles>
        <Collection MultiplicityOfRows="1" Offset="10" Reference="Wall1" x="1" z="92" UnitsOfOffset="in">
          <TileModel>
            <BlockOfTiles MultiplicityOfTiles="7" ModelOfTile="tileTest1" Offset="0" Orientation="0"/>
          </TileModel>
        </Collection>
      </Tiles>
      <Vents>
        <Collection MultiplicityOfRows="2" Offset="76" Reference="Wall1" x="1" z="42" UnitsOfOffset="in">
          <TileModel>
            <BlockOfTiles MultiplicityOfTiles="7" ModelOfTile="ventTest1" Offset="0" Orientation="0"/>
          </TileModel>
        </Collection>
      </Vents>
      <Ceilings Height="1.167" Units="ft"> </Ceilings>
    </RaisedFloor>
  </RoomArchitecture>

  <CRAC CFM="12" Capacity="0" UnitsOfCapacity="S.1" Gradient="-130720">
    <Location>
      <Inside Distance="3" Reference="Wall1" UnitsOfDistance="m" x="1" Depth="1">
        <Inside>Inside</Inside>
      </Inside>
    </Location>
  </CRAC>

  <Equipment>
    <HorizontalRow>
      <Collection MultiplicityOfRows="1" Offset="10" Reference="Wall1" x="1" z="67" Orientation="0" UnitsOfOffset="in">
        <RackModel>
          <BlockOfRacks MultiplicityOfRacks="4" ModelOfRack="rackTest1" Offset="0" RackOpening="1" UnitsOfOffset="in">
            <Chasis NumberOfBladeServers="10" MultiplicityOfChasis="5" FlowRate="0.25" Gradient="3268">
              <BladeServer Powerdissipation="100" Model="Dell" UnitsOfPowerDissipation="kw"/>
            </Chasis>
          </BlockOfRacks>
        </RackModel>
      </Collection>

      <Collection MultiplicityOfRows="1" Offset="10" Reference="Wall1" x="1" z="117" Orientation="0" UnitsOfOffset="in">
        <RackModel>
          <BlockOfRacks MultiplicityOfRacks="4" ModelOfRack="rackTest1" Offset="0" RackOpening="0" UnitsOfOffset="in">
            <Chasis NumberOfBladeServers="10" MultiplicityOfChasis="5" FlowRate="0.25" Gradient="3268">
              <BladeServer Powerdissipation="100" Model="Dell" UnitsOfPowerDissipation="kw"/>
            </Chasis>
          </BlockOfRacks>
        </RackModel>
      </Collection>
    </HorizontalRow>
  </Equipment>
</DataCenter>
```

FIGURE 32.5   A rendering of the generated data center room on OpenFOAM/ParaView, along with the Ciela source code that specifies it. The specification relies on implicit assumptions to maintain concision, e.g., walls are assumed to succeed one another.

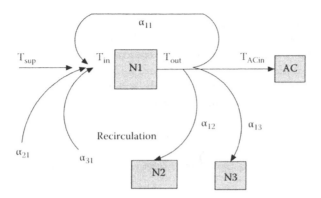

FIGURE 32.6 The recirculation among server nodes is modeled as coefficients α_{ij} of "cross-interference."

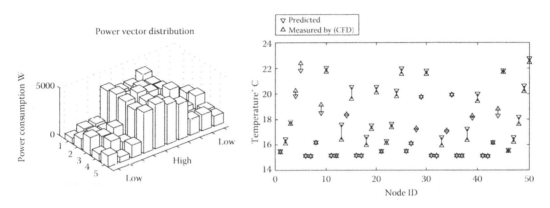

FIGURE 32.7 Early simulations showed that the model can be made quite accurate in predicting the inlet temperatures given the power distribution at the computing nodes. BlueTool allows to test under various configurations and allows for a better evaluation of this modeling method.

How BlueTool can help validate recirculation models: BlueCenter will contain certain air-handling contraptions that will allow adjusting the amount of air that recirculates and how it recirculates. Conventionally, an AHRM is built using CFD-based simulations. Using the AHRM of the BlueCenter for any given configuration of the air-handling contraptions, we can contrast the predicted temperature values for the AHRM versus the observed temperature values in order to reach a conclusive validation of the AHRM. The methodology would involve the following steps:

1. Specify a configuration using Ciela.
2. Feed the specification to the BlueSim tool to yield the AHRM.
3. Run the data center at various workload levels to emulate various conditions.
4. Contrast the gathered data (e.g., temperatures, pressure, air speed) to the predicted values from the AHRM and the CFD simulation.

A further step would be to investigate the possibility of autonomous model generation [13,14], which can greatly reduce need to take data centers offline to perform modeling tests.

32.3.2 Thermal-Aware Resource Management and Workload Scheduling

Thermal-aware job scheduling describes any scheduling effort that has some degree of knowledge of the thermal impact (i.e., the effect on the heat and temperature distribution in the data center room) of a schedule to the data center. Thermal-aware job scheduling is a cyber-physical approach to scheduling, which aims to produce schedules that yield minimal cooling needs and therefore are energy efficient. It is cyber-physical because it takes into consideration both the cyber (computing) performance and the physical (energy, cooling) performance. Most of the previous work on thermal-aware job scheduling has focused on the spatial aspect of job scheduling, i.e., the server assignment of jobs [11,17,18,26]. Spatial thermal-aware job scheduling tries to address the effects of heat recirculation, a common phenomenon which is responsible for much of the energy inefficiency in contemporary data centers. Algorithms such as MinHR [18] and XInt [26] perform thermal-aware spatial job scheduling and manage to reduce the heat recirculation; they can effectively reduce the supply heat index [24], a metric that describes the energy efficiency of a data center, by more than 15% for job sizes of about half the data center's capacity (Figure 32.8) [26].

The performance of XInt has been tested using a simulated small-scale data center of 7U blade-server equipment. XInt was compared to simple thermal-aware approaches, such as the uniform outlet profile (UoP), which tries to equalize the temperatures at the air outlets of the nodes, minimal computing energy (MCE), which assigns the task to as few nodes as possible, and uniform placement of tasks around (UT), which equally distributes the task among all the available nodes. The comparison was made simulating a small-scale data center of 1000 processors [26].

The work in [19] leverages the equipment heterogeneity and slack in execution estimates, extends the work in [26], and shows that thermal-aware *spatiotemporal* job scheduling can yield a synergistic effect of energy efficiency benefits. For example, a result shows that thermally optimized spatiotemporal job scheduling may yield double the savings over the spatial-only scheduling by XInt with first-come, first-serve (FCFS) temporal scheduling. The synergy derives from the temporal smoothing of the workload, yielding a large pool of available servers at any time for thermally efficient spatial placement of the jobs.

The work in [2] proposes the *highest thermostat setting (HTS)* algorithm, which performs *cooling-aware* and *thermal-aware* spatial job scheduling, and integrates such scheduling with cooling management (dynamic variation of CRAC thermostat setting to meet the cooling demands). Cooling-aware approaches can foresee the cooling delay problem and avoid its consequences.

Why thermal-aware scheduling can be a valuable tool: Figure 32.9, obtained from simulations using AHRM, shows that XInt consistently has the minimal cooling cost. At 50% utilization rate, XInt can save 24%–35% energy cost compared with UT and UoP. In addition, the performance of MCE has the worst energy efficiency for most of the utilization rates. Figure 32.9 also shows the ideal optimal lower bound for

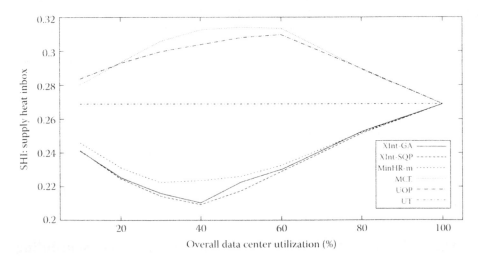

FIGURE 32.8 Supply heat index (SHI) for the compared placement algorithms. According to this evaluation metric, XInt has the minimal SHI.

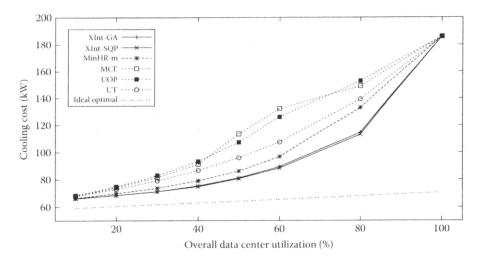

FIGURE 32.9 An overview of the cooling cost induced by each placement policy. The XInt algorithm achieves the lowest cost among the compared schemes.

the cooling cost, which assumes nonexistence of heat recirculation, with the supplied cold air and all inlet temperatures being equal to a redline temperature of 25°C. This means that thermal-aware scheduling is a very promising way of reducing waste energy when compared to the conventional workload assignment methods.

How BlueTool can help validate thermal-aware scheduling: Similarly to the validation of heat recirculation models, BlueCenter's power-measuring equipment can be used to show the effectiveness of the thermal-aware management approaches in saving energy. At the same time it will be possible to demonstrate, quantify, and model the cooling delay problem. This can be achieved by adjusting the cool-air flow from the cooling unit(s) to emulate conditions of high power density. An example methodology for demonstrating the cooling delay model is as follows:

1. Run the BlueCenter servers at idle and let BlueCenter reach a steady temperature profile.
2. Suddenly increase the server utilization to a much higher level and record the temperature rise over time.
3. Analyze the delay using the observed flow values of air and cooling liquid.

32.3.3 Developing and Testing Heat-Activated Cooling

It is well known that data centers generate a great deal of heat. All of the electric power input to a data center eventually converts to heat, and at present that heat is generally discharged to the environment without attempting to use it for some other purpose. In the electric power and other industries, it has long been recognized that combined heat and power, also called cogeneration, leads to very efficient use of energy resources (e.g., see Ref. [22]). That same concept, however, has not yet penetrated the world of data center design and management, at least not to any significant extent. One of the few organizations attempting to recover and reuse data center heat is IBM Zurich [5], which reports that by coupling microchannel heat sinks directly to the CPUs enables as much as 85% of the waste heat to be recovered, at a temperature of 60°C. Their intended application for the heat is largely district heating of nearby buildings, which is a relatively straightforward use of heat at this temperature.

Given the relatively low upper limit of the temperature of the available heat (95°C), it is prudent to minimize the temperature difference between the CPU and wherever the heat is to be used. That is, the heat exchange between the CPU and the application needs to be as efficient as possible. Even though we are potentially working with a large quantity of waste heat, its usefulness (in thermodynamic terms, its exergy) is severely limited by its temperature. The higher the temperature of the recovered heat, the greater is the type and number of applications to which it can be put to use. A preliminary heat removal design focused on processes involving the phase change of a fluid, akin to the heat pipes commonly found in laptops today. In such systems, the heat generating sources (here the CPUs) cause a working fluid to boil, and the resulting vapor to rise until it reaches a condenser, where the heat is rejected to its ultimate application and the vapor condenses back to liquid. The liquid returns to the heat generating source, either through gravity or through capillary action. Although such phase-change systems can transport heat very efficiently with minimal temperature drop, flow management of the vapor and liquid return is not trivial, especially for a complex system involving perhaps hundreds (even thousands!) of CPUs. This approach was therefore abandoned in favor of a simpler, single-phase convection system, similar to that employed by IBM [5]. We are in the process now of testing alternative heat sink designs, with the goal being not only to make the heat sink as efficient as possible, but also to minimize its cost given the large number that will be required in any realistic data center.

Applications for the captured heat in BlueTool: It has already been mentioned earlier that using the heat for buildings via a district heating scheme is one application. This makes sense in areas that require significant heating for building ventilation and/or domestic water use, but probably not for warm areas like the southwestern United States. In such places, cooling is in far greater demand than heating. We are thus proposing to recover data center heat and use it as the energy source for heat-driven cooling [10]. Our initial idea is to make use of a commercially available LiBr/H_2O absorption chiller, in which the recovered data center heat is transferred to the generator of the chiller where the refrigerant mixture boils to separate the H_2O refrigerant from the aqueous LiBr solution. Our calculations indicate that such a chiller can operate with generator temperatures as low as 65°C, although the recommended generator temperature is near our maximum CPU temperature of 95°C. We are faced with the considerable challenge of delivering the heat to the chiller generator at as high a temperature as possible, with very low tolerance for leaks and other maintenance issues, while minimizing the system cost. Although many hurdles must be overcome to meet this challenge, the potential reward of cost-effective data center heat recovery provides terrific motivation for continuing work.

32.3.4 Other Uses of BlueTool

In addition to the previous examples, BlueTool can be used for other purposes such as

- Evaluating the thermal efficiency of a data center by submitting a specification of it in Ciela.
- Testing the efficacy of management software by running it on BlueCenter.
- Announcing a new technology or explaining a new concept through BlueWiki.

32.4 Summary

This chapter introduced the BlueTool research infrastructure project, an effort to address various research shortcomings related to data centers. BlueTool consists of several subprojects and their corresponding tools to achieve its objectives. The flagship subproject is BlueCenter, a small-scale data center enclosed in an isolated room; the main objective of BlueTool is to provide the BlueCenter to researchers and engineers to test and develop models and methods for the engineering, design, and management of data centers.

In addition to the BlueCenter, BlueTool consists of the BlueSim, BlueWeb, and BlueSense subprojects. BlueSim provides simulation-based assessments of data center or management designs using a data center specification language called Ciela. BlueWeb provides a Web front end to BlueTool's facilities. BlueSense provides a kit of wireless sensor nodes to be used in profiling other data centers.

After the description of the project's tools, the chapter presented a series of research tasks (e.g., validating heat recirculation and power consumption models, and testing scheduling algorithms), and described how BlueTool can be used to carry out these tasks.

BlueTool is an ongoing and evolving effort, and is seeking feedback and collaboration from the academic and the industrial community in order to become a useful research tool and infrastructure. The BlueTool's Web site can visited at http://impact.asu.edu/BlueTool/.

Acknowledgments

The authors would like to thank the editors of the handbook for the opportunity to present this work. The BlueTool project is funded by NSF grant #0855277. Part of the work presented in this chapter has been funded by NSF grant #0834797.

References

1. Z. Abbasi, G. Varsamopoulos, and S. K. S. Gupta. Thermal aware server provisioning and workload distribution for internet data centers. In *ACM International Symposium on High Performance Distributed Computing (HPDC '10)*, Chicago, IL, June 2010.

2. A. Banerjee, T. Mukherjee, G. Varsamopoulos, and S. K. S. Gupta. Cooling-aware and thermal-aware workload placement for green HPC data centers. In *International Conference on Green Computing Conference (IGCC2010)*, Chicago, IL, August 2010.

3. C. Bash and G. Forman. HPL-2007-62 cool job allocation: Measuring the power savings of placing jobs at cooling-efficient locations in the data center. Technical report HPL-2007-62, HP Laboratories, Palo Alto, CA, August 2007.

4. K. G. Brill. The invisible crisis in the data center: The economic meltdown of Moore's law. Technical report, Uptime Institute, Santa Fe, NM, July 2007.

5. T. Brunschwiler, B. Smith, E. Ruetscheo, and B. Michel. Toward zero-emission data centers through direct reuse of thermal energy. *IBM Journal of Research and Development*, 53(3):11:1–11:13, 2009.

6. J. Chase, D. Anderson, P. Thakar, A. Vahdat, and R. Doyle. Managing energy and server resources in hosting centers. In *SOSP '01: Proceedings of the Eighteenth ACM Symposium on Operating Systems Principles*, pp. 103–116, New York, 2001. ACM.

7. G. Chen, W. He, J. Liu, S. Nath, L. Rigas, L. Xiao, and F. Zhao. Energy-aware server provisioning and load dispatching for connection-intensive internet services. In *NSDI'08: Proceedings of the Fifth USENIX Symposium on Networked Systems Design and Implementation*, pp. 337–350, Berkeley, CA, 2008. USENIX Association.

8. M. Elnozahy, M. Kistler, and R. Rajamony. Energy conservation policies for web servers. In *Proceedings of the Fourth Conference on USENIX Symposium on Internet Technologies and Systems (USITS'03)*, Berkeley, CA, 2003. USENIX Association.

9. S. K. S. Gupta, T. Mukherjee, G. Varsamopoulos, and A. Banerjee. Research directions in energy-sustainable cyber-physical systems. *Sustainable Computing: Informatics and Systems*, 1(1):57–74, March 2011.

10. A. Haywood, J. Sherbeck, P. E. Phelan, G. Varsamopoulos, and S. K. S. Gupta. A sustainable data center with heat-activated cooling. In *Proceedings of ITHERM 2010*, San Diego, CA, June 2010.

11. T. Heath, A. P. Centeno, P. George, L. Ramos, and Y. Jaluria. Mercury and freon: Temperature emulation and management for server systems. In *ASPLOS-XII: Proceedings of the 12th International Conference on Architectural Support for Programming Languages and Operating Systems*, pp. 106–116, New York, 2006. ACM Press.

12. T. Heath, B. Diniz, E. V. Carrera, W. Meira Jr., and R. Bianchini. Energy conservation in heterogeneous data centers. In *ACM SIGPLAN Symposium on Principles and Practice of Parallel Programming (PPoPP)*, pp. 186–195, Chicago, IL, June 2005.

13. M. Jonas, G. Varsamopoulos, and S. K. S. Gupta. On developing a fast, cost-effective and non-invasive method to derive data center thermal maps. In *GreenCom'07 Workshop, Proceedings of the IEEE International Conference on Cluster Computing*, Austin, TX, September 2007.

14. M. Jonas, G. Varsamopoulos, and S. K. S. Gupta. Non-invasive thermal modeling techniques using ambient sensors for greening data centers. In *ICPPW, 39th International Conference on Parallel Processing*, pp. 453–460, San Diego, CA, 2010.

15. D. Meisner, B. T. Gold, and T. F. Wenisch. Powernap: Eliminating server idle power. *SIGPLAN Notices*, 44:205–216, March 2009.

16. J. Moore, J. Chase, and P. Ranganathan. Weatherman: Automated, online, and predictive thermal mapping and management for data centers. In *IEEE International Conference on Autonomic Computing (ICAC)*, pp. 155–164, Dublin, Ireland, June 2006.

17. J. Moore, J. Chase, and P. Ranganathan. Weatherman: Automated, online, and predictive thermal mapping and management for data centers. In *Third IEEE International Conference on Autonomic Computing*, Dublin, Ireland, June 2006.

18. J. Moore, J. Chase, P. Ranganathan, and R. Sharma. Making scheduling "cool": Temperature-aware workload placement in data centers. In *ATEC '05: Proceedings of the Annual Conference on USENIX Annual Technical Conference*, pp. 5–5, Berkeley, CA, 2005. USENIX Association.

19. T. Mukherjee, A. Banerjee, G. Varsamopoulos, S. K. S. Gupta, and S. Rungta. Spatio-temporal thermal-aware job scheduling to minimize energy consumption in virtualized heterogeneous data centers. *Computer Networks*, 53(17):2888–2904, 2009. Virtualized Data Centers.

20. R. Mullins. HP service helps keep data centers cool. Technical report, IDG News Service, July 2007.

21. S. Nedevschi, L. Popa, G. Iannaccone, S. Ratnasamy, and D. Wetherall. Reducing network energy consumption via sleeping and rate-adaptation. In *USENIX Symposium on Networked Systems Design and Implementation, NSDI08*, pp. 323–336, San Francisco, CA, April 2008.

22. N. Petchers. *Combined Heating, Cooling & Power Handbook: Technologies & Applications: An Integrated Approach to Energy Resource Optimization*. Fairmont Press, New York, 2003.

23. P. Ranganathan, P. Leech, D. Irwin, and J. Chase. Ensemble-level power management for dense blade servers. In *ISCA'06 33rd International Symposium on Computer Architecture*, pp. 66–77, Boston, MA, 2006.

24. R. K. Sharma, C. E. Bash, and C. D. Patel. Dimensionless parameters for evaluation of thermal design and performance of large scale data centers. In *Proceedings of the American Institute of Aeronautics and Astronautics (AIAA)*, p. 3091, Chicago, IL, 2002.

25. R. Sullivan and K. G. Brill. Cooling techniques that meet "24 by forever" demands of your data center. Technical report, Uptime Institute, Inc., January 2006.

26. Q. Tang et al. Energy-efficient thermal-aware task scheduling for homogeneous high-performance computing data centers: A cyber-physical approach. *IEEE Transactions on Parallel and Distributed Systems*, 19(11):1458–1472, 2008.

27. D. Tennant (ed.). The greening of IT. *Computerworld*, July 2007.

28. X. Wang and M. Chen. Adaptive power control for server clusters. In *NSF Next Generation Software Workshop at International Parallel & Distributed Processing Symposium*, Miami, FL, April 2008.

VII

Software Systems

VII

33

Optimizing Computing and Energy Performances in Heterogeneous Clusters of CPUs and GPUs

Stephane Vialle
L'École Supérieure d'Électricité (Supélec)

Sylvain Contassot-Vivier
Henri Poincaré University (Nancy 1)

Thomas Jost
Henri Poincaré University (Nancy 1)

33.1 Introduction

Today multicore CPU clusters and GPU clusters are cheap and extensible parallel architectures, achieving high performances with a wide range of scientific applications. However, depending on the parallel algorithm used to solve the addressed problem and on the available features of the hardware, relative computing and energy performances of the clusters may vary. In fact, modern clusters cumulate several levels of parallelism. Current cluster nodes commonly have several CPU cores, each core supplying SSE units (streaming SIMD extension: small vector computing units sharing the CPU memory), and it is easy to install one or several GPU cards in each node (graphics processing unit: large vector computing units with their own memory). So, different kinds of computing *kernels* can be developed to achieve computations on a same node, some for the CPU cores, some for the SSE units, and some others for the GPUs. Also several combinations of those kernels can be used considering:

1. A cluster of multicore CPUs
2. A cluster of GPUs
3. A cluster of multicore CPUs with SSE units
4. A hybrid cluster of both GPUs and multicore CPUs
5. A hybrid cluster of both GPUs and multicore CPUs with SSE units

Each solution exploits a specific hardware configuration and requires a specific programming, and the different solutions lead to different execution times and energy consumptions. Moreover, the optimal combination of kernel and hardware configuration also depends on the problem and its data size.

Another aspect that impacts the performances lies in the communications. According to the algorithm used and the chosen implementation, communications and computations of the distributed application can overlap or be serialized. Overlapping communications and computations is a strategy that is not adapted to every parallel algorithm nor to every hardware, but it is a well-known strategy that can sometimes lead to serious performance improvements. In that context, *asynchronous parallel algorithms* are known to be very well suited. Asynchronous schemes present the great advantage over their synchronous counterparts to perform an implicit overlapping of communications by computations, leading to a better robustness to the interconnection network performance fluctuations and, in some contexts, to better performances [4]. Moreover, although a bit more restrictive conditions apply on their use, a wide family of scientific problems support them.

So, some problems can be solved on current distributed architectures using different *computing kernels* (to exploit the different available computing hardware), with *synchronous* or *asynchronous* management of the distributed computations, and with *overlapped* or *serialized* computations and communications. These different solutions lead to various computing and energy performances according to the hardware, the cluster size, and the data size. The optimal solution can change with these parameters, and application users should not have to deal with these parallel computing issues.

The main goal in this field is to develop auto-adaptive multialgorithms and multikernel applications, in order to achieve optimal runs according to a user-defined criterion (minimize the execution time, the energy consumption, or minimize the energy-delay product, etc.). A *multi-target* and *multi-code* program should include several solutions (implementations) and should be able to automatically select the right one, according to a criterion based on execution speed, on energy consumption or on a speed-energy trade-off. In order to implement this kind of auto-selection of a hardware configuration to exploit and code to run, it is necessary to design a computing and energy performance model.

However, the development of this kind of auto-adaptive solution remains a real challenge as it requires an *a priori* knowledge of the behavior of each software solution on each class of hardware. Obviously, it is not possible to collect such information for every possible combination. Nonetheless, the design of a model considering heterogeneous distributed architectures, computing performances and energy performances could fill that gap. Such a design is usually achieved by a theoretical analysis of the behaviors of the main classes of parallel algorithms. Then, the result is commonly a model having several parameters depending on the problem and algorithm (nature and data size) but also on the hardware features. This last information is obtained by the use of generic benchmarks on the target systems.

In the following section, we present the feedback from our previous experiments on clusters of CPUs and GPUs and identify some pertinent benchmarks and optimization rules that can be respectively used to feed a model and to enhance the overall performances. Then, our methodology and metrics for performance evaluation are presented in Section 33.4. Based on the experience gained in our previous works together with a theoretical analysis, a model of computing and energy performance is proposed in Sections 33.5 and 33.6 and is validated in Section 33.8. All the tests are done with a representative example of scientific computing algorithms, which is a partial differential equation (PDE) solver detailed in Section 33.7. Finally, we conclude over the current degree of development of fully auto-adaptive algorithms and the short and middle term achievements that can be expected.

33.2 Related Works

Many research efforts have focused on designing energy performance models of GPUs. They usually focus on modeling one GPU chip, or one GPU card.

Rofouei et al. [21] introduce a new hardware and software monitoring solution (called LEAP-server), to achieve real-time measurement of the energy consumption of the CPU chip, the GPU card, and the PC motherboard. Then the authors run some benchmarks and attempt to link computing performances (execution times and speedup) to energy performances, and they achieve fine measures exhibiting different energy consumptions in function of the GPU memory used. GPUs have different memories, with different performances in term of speed and energy consumption (see [17]). Finally, their measurement solution and performance modeling aims at deciding whether the computations has to be run on the CPU or on the GPU to optimize the performances. Ma et al. [20] use conventional hardware, classical electrical power measurement mechanisms and functions of the NVIDIA toolkit to measure the workload of the different components of the GPU. It leads to fine measurement and complex workload and electrical power profiles of different benchmark applications. Then, the authors run a statistical analysis, with a *support vector regression* model (SVR) trained on the benchmark application profiles. The resulting SVR is used to predict the energy performances of new applications in the function of their workload profiles, in order to automatically select CPU or GPU computing kernels.

SPRAT is a language designed by Takizawa et al. [22] to process some *streams* on a CPU or a GPU. It aims at reducing the energy consumption without degrading the computing performance. A performance model is introduced by the authors to take into account the execution times on CPU and on GPU, the data transfer time between CPU and GPU, and the energy consumption. The authors focus on the cost of data transfers, which can be excessive and then lead to longer execution times on the GPU than on the CPU. Some *credits* are introduced in SPRAT runtime. The number of credits increases when the GPU kernel execution appears to be efficient, and they take the risk of running another GPU kernel (instead of a CPU one). This model and language seems especially suited to applications that require frequent data transfers between CPU and GPU.

All these models and choice strategies between a CPU and a GPU kernel have been designed to optimize the usage of one GPU card and one CPU motherboard. However, our goal is to use a *cluster of PCs with CPU and GPU on each node*. So, we decided to monitor the energy consumed by each node (i.e., by each complete PC), and to optimize the global energy consumption of our clusters (i.e., the energy required by the nodes and the interconnection network). Moreover, we are not only interested in choosing the right kernel (CPU or GPU kernel) run on each node, but also in determining the right operating mode (typically between synchronous and asynchronous versions of a parallel algorithm). The next section introduces our first parallel performance model and experimental measures.

The choice between computing and energy performances can be relevant when the fastest solution is not the less energy-consuming. A global performance criterion can be required to achieve the right choices of computing kernels and operating mode. An interesting global criterion is the *energy delay product* (EDP) introduced in [16] in 1996. It is the product of the energy consumption and the execution time: the two parameters we want to decrease. Looking for the solution that minimizes their product can lead to a good compromise between computing and energy performances. We plan to include the EDP in our future model, in order to track this compromise.

33.3 First Experiments and Basic Model

In this section, we consider three benchmarks: three applications distributed on computing clusters with CPU or GPU nodes. These applications are classical intensive computations: (1) a *European option pricer* corresponding to embarrassingly parallel computations, (2) a *PDE solver* corresponding to an iterative algorithm including a large amount of computations and several communications at each iteration, and (3) a *Jacobi relaxation* corresponding to an iterative algorithm with a huge number of iterations and a small amount of computations and communications at each iteration. Each benchmark implementation has been optimized both on CPUs and GPUs (especially the memory accesses), and has been experimented on a CPU cluster and a GPU cluster. Execution times and consumed energies have been measured and are introduced, analyzed, and modeled in this section.

33.3.1 Testbed Introduction and Measurement Methodology

Our first testbed is a cluster of 16 nodes. Each node is a PC composed of an Intel Nehalem CPU with four hyperthreaded cores at 2.67 GHz, 4 GB of RAM, and a NVIDIA GTX285 GPU with 1 GB of memory. This cluster has a Gigabit Ethernet interconnection network built around a small DELL Power Object 5324 switch (with 24 ports). The energy consumption of each node is monitored by a Raritan DPXS20A-16 device that continuously measures the electric power dissipation (in watts) and can monitor up to 20 nodes. This device hosts a SNMP server that a client can question to get the instantaneous power consumption of each node.

In order to achieve both computing and energy performance measurements of our parallel application, we run a shell script that: (1) runs a Perl SNMP client sending requests to the SNMP server of the Raritan monitor device to sample the electric power consumption, and storing data in a log file, (2) runs the parallel application on a cluster (executing a mpirun command), (3) extracts the right data from the log file and computes the energy consumption when the parallel application has finished. The sampling period of the power dissipation on any node of the cluster is approximately 300 ms, and the consumed energy is computed as a definite integral using the trapezoidal rule. The measurement resolution of our complete system appears to be close to 6 W, i.e., approximately 4% of the measured values.

We only consider the energy consumption of the nodes that are actually used during the computation, as it is easy to remotely switch off unused nodes of our GPU cluster. However, it is not possible to switch off the GPU of one node when using only its CPU, and we have not yet tried to reduce the frequency and the energy consumption of the CPU when using mainly the GPU. We also consider the energy consumption of the cluster interconnection switch, which appears to be very stable, independently of the communications achieved across the cluster.

33.3.2 Observation of Experimental Performances

33.3.2.1 European Option Pricer Benchmark

Pricing is a very classic and frequently run computation in the financial industry. Many banks aim at speeding up and improving their energy consumption using different parallel architectures, including

GPUs. This pricer of European options is based on independent Monte Carlo simulations, and from a pure parallel algorithmic point of view, this is an *embarrassingly parallel* algorithm: Each computing node processes some independent Monte Carlo trajectories, and communications are limited to data distribution at the beginning of the application, and to the gathering of the results of each computing node at the end. However, we have been very careful about the parallelization of the random number generator (RNG), to be able to generate uncorrelated random numbers from thousands of threads spread on different nodes [1] without decreasing the pricing accuracy.

The implementation on multicore CPU clusters has been achieved using both MPI, to create one process per node and insure the few internode communications, and OpenMP to create several threads per core and take advantage of each available core. The OpenMP parallelization has been optimized to create the required threads only once (inside a large parallel region) and balance the work among these threads. Moreover, inside each thread, data storage and data accesses are implemented in order to optimize cache memory usage. The implementation on GPU clusters uses the same MPI-based computation distribution and internode communication, while CUDA is used to send data and Monte Carlo trajectory computations on the GPU of each node. In order to avoid frequent data transfers between CPU and GPU, we have ported our RNG to the GPU and all node computations are executed on the GPU. Moreover, we have optimized the GPU memory accesses, minimizing the usage of the (slow) global memory of the GPU, and using mostly (fast) GPU registers.

Figure 33.1 shows quasi-ideal execution time decreases on CPU and GPU clusters, while the energy consumption remains approximately constant. Finally, the execution on the GPU cluster seems really more efficient than the execution on the multicore CPU cluster.

33.3.2.2 PDE Solver Benchmark

This application performs the resolution of PDEs using the multisplitting-Newton algorithm and an efficient linear solver using the biconjugate gradient algorithm. The solver is applied to the resolution of a 3D transport model, which simulates chemical species in shallow waters. That application is fully detailed in Section 33.7.

Figure 33.2 exhibits very good decrease of the execution times, and using a GPU cluster seems more interesting than using a CPU cluster. However, the consumed energy does not remain constant and increases significantly when using more than four GPUs.

33.3.2.3 Jacobi Relaxation Benchmark

This application solves the Poisson's equation on a 2D grid using a Jacobi algorithm. The values on the border of the grid are fixed. The goal is to compute the values inside the grid. In these implementations, at each iteration, a new grid is computed such that the value at each point becomes the mean value

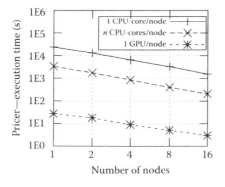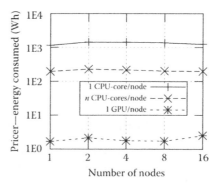

FIGURE 33.1 First benchmark: European option pricer.

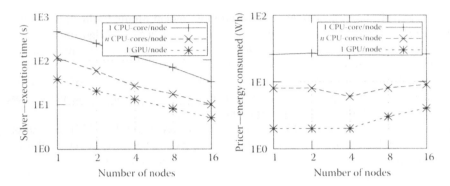

FIGURE 33.2   Second benchmark: PDE solver.

of its four neighbors in the previous grid state. Hence, for each point, only three additions and one division by four are required, and with a naive implementation, five memory accesses would be required (four reads and one write). But on modern architectures, memory accesses are much more expensive than computations [23], and this application is *memory bound*. This is the reason why all our optimizations on CPU and on GPU aim at reducing bandwidth consumption and at using the memory bus efficiently.

The CPU implementation has been designed to access to contiguous data elements in order to use the cache memory efficiently. For the considered grid sizes, several rows fit in the cache memory. Hence data elements will be computed row by row. This way, the number of cache misses is minimized and as a consequence the number of memory accesses is minimized as well. A blocked version computing 8 × 8 blocks has been tried but it showed less interesting performance. The only two effective improvements we have found are loop-unrolling and padding to grid sizes that are multiples of 16 elements, to avoid accesses separated by the *critical stride* as explained in [14]. The GPU implementation has been designed for GPU without generic cache mechanisms. It aims at optimizing the usage of the small *shared* memories available on chip, which can be used as a *software-managed cache memory*. However, the shared memory is too small to contain several rows of the grid. As a consequence, data partitioning techniques have been used inside each computing node to process the Jacobi Grid per blocks. The size of these blocks has been optimized to get *coalesced* memory accesses.

Moreover, as we use clusters, internode communications are required at each iteration. But they exchange data between CPU memories or between GPU memories (on GPU clusters), which can be long compared to the computation speed of each node. So we have optimized these communications, implementing overlapped asynchronous communications, and overlapping with CPU-to-GPU and GPU-to-CPU data transfers when possible.

Figure 33.3 shows better performance when using the GPU cluster instead of the CPU one, but the speedup is poor and the energy consumption clearly increases with the number of used nodes. Finally, the GPU cluster appears always more efficient than the CPU cluster. But the computing and energy performance profiles of these three benchmarks are different, and the superiority of the GPU cluster seems to evolve with the number of used nodes. This issue is investigated in the next section.

33.3.3 Relative Performances of the CPU and GPU Clusters

In order to analyze the interest to use a GPU cluster in place of a CPU cluster, we have computed the relative speedup (SU) and the relative energy gain (EG) of the GPU cluster compared to the CPU one. Figure 33.4 illustrates these computing and energy relative performances for the three benchmarks introduced in the previous section. We can observe that:

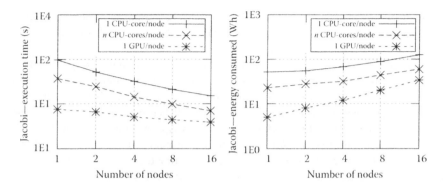

FIGURE 33.3 Third benchmark: Jacobi relaxation.

- The option pricer with embarrassingly parallel computations achieves a speedup and an energy gain close to 100 on a GPU cluster compared to a multicore CPU cluster, while the PDE solver and the Jacobi relaxation, including computations and communications, reach speedup and energy gain in the range from 1.6 to 10.
- Speedup and energy gain decrease when the number of used nodes increases. This phenomena is limited with the option pricer, but is stronger with the PDE solver and very clear with the Jacobi relaxation. On larger clusters, performances could become higher on multicore CPU clusters.
- Speedup and energy gain curves exhibit very similar profiles. They seem to have the same behavior.

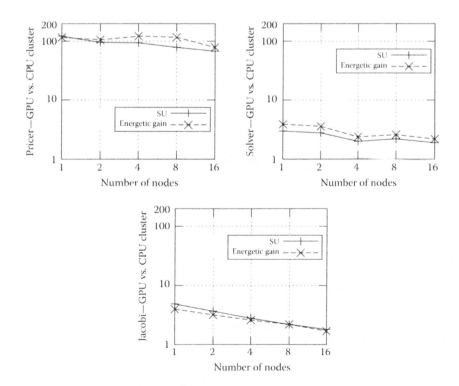

FIGURE 33.4 GPU cluster vs. CPU cluster relative performances.

In fact, when the interconnection network is the same, the communications are a little bit longer on a GPU cluster. As we need to exchange some data located in GPU global memories, data have to be transferred from GPU memories to CPU ones. Then, internode communications must be performed (using MPI for example) and, finally, the received data must be transferred into the global GPU memories. On the opposite, computations are faster on GPUs. So, the ratio of communication time in the total execution time significantly increases when using a GPU cluster. That leads to a smaller performance increase when using more nodes, and finally the multicore CPU cluster can become more efficient.

The next section establishes a first performance and energy consumption model on two different clusters. This model is limited to *scalability areas* of the performance curves and is applied to the cases of a multicore CPU cluster and a GPU cluster. It aims at predicting the most efficient cluster as a function of the number of used nodes, and at helping us to always choose the best solution.

33.3.4 Basic Modeling of CPU and GPU Cluster Performances

33.3.4.1 Execution Time and Energy Consumption on One Cluster

Considering A: a cluster, N: the number of used nodes, $T(A,N)$: the execution time of an application, and $E(A,N)$: the consumed energy, in the ideal case we have: $T(A,N) = T(A,1)/N$, and $E(A,N) = E(A,1)$ is constant. Our first benchmark (European option pricing achieving embarrassingly parallel computations) exhibits performances close to this ideal case (see Figure 33.1). But the other experiments do not exhibit these ideal performances. However, lines appear approximately straight when using logarithmic scales, meaning the parallelization *scales*, so we can write:

$$T(A,N) = T(A,1)/N^{\sigma_T^A}, \quad 0 \leq \sigma_T^A \leq 1 \tag{33.1}$$

$$E(A,N) = E(A,1) \cdot N^{\sigma_E^A}, \quad 0 \leq \sigma_E^A \leq 1 \tag{33.2}$$

where σ_T^A and σ_E^A are the slopes of the straight lines of the execution time and energy consumption on curves drawn with logarithmic scales. In the ideal case, these parameters values would be 1, but in practice they are less than 1 (see Figures 33.2 and 33.3).

So, when the parallelization scales, the speedup and energy gain of cluster A compared to a sequential run on one of its nodes (*level 1* gains) are:

$$SU_1(A,N) = \frac{T(A,1)}{T(A,N)} = N^{+\sigma_T^A} \quad (\geq 1) \tag{33.3}$$

$$EG_1(A,N) = \frac{E(A,1)}{E(A,N)} = N^{-\sigma_E^A} \quad (\leq 1) \tag{33.4}$$

In this basic model we make the following assumptions:

1. The electric power dissipated by one node of cluster A during a sequential computation remains constant and is equal to $P(A,1) = E(A,1)/T(A,1)$.
2. The electric power dissipated by any used node of cluster A during a parallel computation is independent of the number of used nodes, and is equal to $P(A,1)$.
3. The electric power dissipated by the network switch of the cluster remains constant and is equal to $P_{switch}(A)$.

Hypothesis 3 matches all our observations on different clusters. But we will see in Section 33.5 that hypotheses 1 and 2 are approximations. However, assuming these hypotheses, the energy consumed on one node during a sequential run is

$$E(A,1) = P(A,1) \cdot T(A,1) + \frac{P_{switch}(A)}{N_{max}} \cdot T(A,1) \tag{33.5}$$

where N_{max} is the maximal number of available nodes in the cluster. And the energy consumed on N nodes during a parallel run is

$$E(A, N) = P(A, 1) \cdot T(A, N) \cdot N + \frac{P_{switch}(A) \cdot N}{N_{max}} \cdot T(A, N) \tag{33.6}$$

$$E(A, N) = \frac{T(A, N)}{T(A, 1)} \cdot N \cdot \left(P(A, 1) \cdot T(A, 1) + \frac{P_{switch}(A)}{N_{max}} \cdot T(A, 1) \right)$$

Using Equations 33.1 and 33.5, we get

$$E(A, N) = N^{1 - \sigma_T^A} \cdot E(A, 1) \tag{33.7}$$

and using Equation 33.2 we can deduce the relation between the execution time and energy consumption formulas on one cluster, in the scalability area of the performance curves:

$$\sigma_E^A = 1 - \sigma_T^A \tag{33.8}$$

33.3.4.2 Relative Speedup and Energy Gain between Two Clusters

To identify the most efficient parallelization on two different clusters A and B, we compute the respective gains on cluster A compared to the gains on cluster B (level 2 gains): $SU_2^{A/B}(N)$ and $EG_2^{A/B}(N)$. This leads to

$$SU_2^{A/B}(N) = \frac{T(B, N)}{T(A, N)}$$

$$= \frac{T(B, 1)}{T(A, 1)} \cdot N^{\sigma_T^A - \sigma_T^B}$$

$$= SU_2^{A/B}(1) \cdot N^{\sigma_T^A - \sigma_T^B} \tag{33.9}$$

and to

$$EG_2^{A/B}(N) = \frac{E(B, N)}{E(A, N)}$$

$$= \frac{E(B, 1)}{E(A, 1)} \cdot N^{\sigma_E^B - \sigma_E^A}$$

$$= EG_2^{A/B}(1) \cdot N^{\sigma_T^A - \sigma_T^B} \tag{33.10}$$

These two relative gains have different initial values (on 1 node) but similar evolutions (same $\sigma_T^A - \sigma_T^B$ exponent in the gain expressions). This is the reason why the speedup and energy gain curves on Figure 33.4 are nearly parallel.

As an example, let us consider that cluster A has faster nodes than cluster B. Depending on the relative values of σ_T^A and σ_T^B, it is possible that the relative gain $SU_2^{A/B}$ becomes smaller than 1 inside the scalability area (the validity domain of Equation 33.9) beyond a threshold number of nodes $N_{A/B,T}$. Then, cluster B runs faster than cluster A beyond this threshold, and similarly, cluster B can be less energy-consuming beyond a threshold $N_{A/B,E}$. Using Equations 33.9 and 33.10 we can determine these two thresholds:

$$SU_2^{A/B}(N_{A/B,T}) = 1 \iff N_{A/B,T} = \left(SU_2^{A/B}(1) \right)^{\frac{-1}{\sigma_T^A - \sigma_T^B}} \tag{33.11}$$

$$EG_2^{A/B}(N_{A/B,E}) = 1 \iff N_{A/B,E} = \left(EG_2^{A/B}(1) \right)^{\frac{-1}{\sigma_T^A - \sigma_T^B}} \tag{33.12}$$

33.3.4.3 Application of the Model to a CPU and a GPU Cluster

When using one GPU, the computation time is smaller than on one CPU (or we do not use the GPU), and $SU_2^{g/c}(1) > 1$. But communication times are longer on a GPU cluster than on a CPU cluster with the same interconnection network, because they require the same CPU communications plus some CPU–GPU data transfers. So, the scalability is *weaker* on the GPU cluster and the σ_T parameter is smaller: $0 \le \sigma_T^{GPU} \le \sigma_T^{CPU} \le 1$. Moreover, although the dissipated electric power of a single GPU node is larger than the one of a CPU node, the overall energy consumption of the GPU cluster is usually lower than the CPU cluster: $EG_2^{g/c}(1) > 1$. These hypotheses are verified on our three benchmarks (see Figures 33.1 through 33.3). When these hypotheses are true, the two threshold numbers of nodes (Equations 33.11 and 33.12) exist and are greater than 1. Then we define:

$$N_{g/c,min} = \min(N_{g/c,T}, N_{g/c,E}) \tag{33.13}$$

$$N_{g/c,max} = \max(N_{g/c,T}, N_{g/c,E}) \tag{33.14}$$

When $N < N_{g/c,min}$ the GPU cluster solution is faster and less energy-consuming, when $N_{g/c,max} < N$ the multicore GPU cluster is slower and more energy-consuming, and when the number of nodes is in the range $[N_{g/c,min}; N_{g/c,max}]$ the GPU cluster is either faster or less energy-consuming.

So, assuming our different hypotheses are true (mainly expressed by the existence of a scalability area), three different *execution configurations* exist:

- $N < N_{g/c,min}$: the GPU cluster is more interesting.
- $N_{g/c,min} \le N \le N_{g/c,max}$: the GPU cluster is more or less interesting than the CPU cluster, depending on the relative importance of the computation speed and the energy consumption.
- $N_{g/c,max} < N$: the CPU cluster is more interesting.

33.3.5 Need for Predicting the Best Operating Mode

Using a GPU cluster to run an embarrassingly parallel program (without internode communications) and with all computations running on the GPU, such as our European option pricer, can be very efficient (see Section 33.3.2 and Figure 33.1). The speedup and energy gain of one GPU node compared to one multicore CPU node are close to 100 and remain close to 100 when the number of nodes increases. Such a case corresponds to the first category described above.

When all computations are not run on the GPU and frequent data transfer are required between CPU and GPU on each node, and/or frequent internode communications are required, the speedup and energy gain on one node are more limited and decrease when the number of nodes increases. Our PDE solver and Jacobi relaxation benchmarks belong to this second category (see Figures 33.2 and 33.3).

Finally, most parallel programs include internode communications and data transfers between CPUs and GPUs, and do not perform all their computations on the GPUs. So, the three *execution configurations* introduced at the end of Section 33.3.4 exist, and using a GPU cluster with a large number of nodes may lead to an important gain or to an important waste of time and energy.

Usually, the end user is not a computer scientist. Although he is capable of specifying if he wishes to decrease the execution time or the energy consumption, he is not able to choose whether or not to use GPUs, depending on the number of available and used nodes. A heuristic has to be designed and implemented in order to achieve an automatic choice of the right computing kernel to use in execution configurations 1 and 3, and to respect the user objective in execution configuration 2.

33.3.6 Interests and Limits of the Basic Model, and Need for a New One

The heuristic we introduced in the previous section needs a model to predict performances and choose a computing kernel. Our previous model, defined in Section 33.3.4, does not make any assumption on the internal architecture of the clusters *A* and *B*. It is based on the observation of a scalability area and on measurement of computing and energy performances of clusters *A* and *B* when using two different numbers of nodes. This provides at least four benchmarks of the target application, and to establish some kind of experimental reference curves. This approach is acceptable before entering an *exploitation mode*, running the application many times. Then, achieving four extra-runs of the application code in order to improve all others, seems the right solution. Of course it is not always possible to achieve some runs on only one node, depending on the problem size, but the model equations could be adapted to not require experiments on one node.

However, many researchers have to run their parallel application in *experimentation mode*: They achieve only a few runs (but large runs) before upgrading the application. Then, execution of four benchmarks before optimizing the execution of each new version of the application can be prohibitive. To optimize the execution of a parallel application in experimentation mode we need a more accurate model of our architecture, requiring only *elementary benchmarks* to be calibrated for our machine. Such a model will be designed in Sections 33.5 and 33.6.

33.4 Measurement Issues and Alternative Modeling Approach

33.4.1 Measurement Methodology

Figure 33.5 illustrates the hardware and software configuration of our CPU+GPU cluster PC, the energy devices and monitors as well as the benchmarking mechanism. As already mentioned in Section 33.3.1, each PC is electrically connected to an output port of an energy device (a Raritan DPXS20A-16), that measures the instantaneous electrical power dissipation of all its outputs. Those measures are collected via SNMP requests.

In order to measure both computing and energy performances of our application, we offer the following benchmarking methodology:

1. We allocate some nodes on our cluster, through the OAR cluster management environment and run a main shell script on the first node we get (nodes are sorted by alphabetic order).
2. The main shell script starts a SNMP client (Perl script) for each SNMP server (i.e., for each electric device and monitor to use).
3. Each SNMP client sends requests to a SNMP server to sample the electrical power dissipated by each of its outputs. The sampling interval is close to 300 ms in our experiments, but this can be

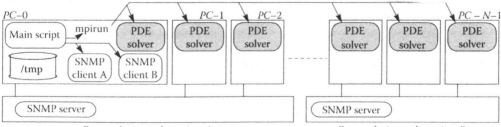

FIGURE 33.5 Hardware and software architecture of the test platform.

tuned. Each SNMP client stores the acquired data in log files on the local disk of the first PC. Each line of a log file stores the sample number, the sample time, and the electrical power dissipation of each output port of the electrical device at the sample time.

4. The main shell script waits for the different log files to be created (meaning the energy monitoring mechanism is running). Then, it reads these files and gets the current sample number from each file. The main shell script runs the parallel application on the allocated nodes, executing a mpirun command, while the electrical power dissipation sampling continues.

5. The execution time of the MPI parallel program is measured internally using gettimeofday.

6. When the mpirun command finishes, the main shell script reads the log files and gets the new current sample numbers. Then it stops the SNMP clients and computes the energy consumed by all the PC nodes involved in the parallel run. It computes a definite integral of the instantaneous power measures between the sample numbers surrounding the mpirun execution, using the trapezoidal rule and considering time measures stored in each line of the log files.

Finally, each benchmark is repeated five times, and then the average values of execution time and energy consumption are computed.

33.4.2 Technical Limitations of the Measures

A first point that may induce small perturbations in the performance measurements is the concurrent execution of several SNMP clients on the first allocated node. These processes run concurrently with the parallel application and might disturb it. Nonetheless, during our benchmarks we did not observe any significant impact on the execution time measures when running the energy consumption measurement. But users must be aware that it could happen on a larger system, using many energy devices and monitors and running many SNMP clients.

The first real limitation of our benchmarking system is about the uncertainties of the measures. For the whole set of experiments involving several nodes of the cluster (with or without GPU, in synchronous or asynchronous mode), the average variations of the execution times are close to 7%, and the average variations of energy consumptions are around 7.5% and are a bit less stable. The good point is that those variations seem to be strongly correlated. Although they may be partially explained by the sensor quality concerning the energy consumption measures, it remains quite difficult to precisely identify their exact sources.

We also observed four other sources of measurement uncertainties. But those ones can be analyzed and taken into account in our model:

1. There are serious variations in the power consumptions of the different PCs in a homogeneous cluster. Current PCs with identical external features have very close computing performances, but their energy consumptions can vary a lot. So, we decided to consider the energy consumption of each node in our model rather than considering a global average energy consumption of the cluster.

2. We observe some steps in electrical power dissipation of each cluster node. When running computations on the GPU, the dissipated electrical power increases, stabilizes and then increases again before stabilizing again. We observe the symmetric phenomena when computations end. This is mostly due to the GPU fans start/stop cycles. They do not start and stop exactly when computations begin and end, but a little bit later when the GPU temperature increases or decreases and crosses thresholds.

3. We do not observe an instantaneous power dissipation decrease when stopping computations to enter an internode communication substep of the application. First, sensors do not detect an immediate decrease of power dissipation of the node when intensive computations end. Second, according to the previous point, the fans continue to run up to 1 min before stopping. As a consequence, entering a communication substep leads to a slow decrease of the energy consumption during this substep. In particular, when the substep is short, the energy decrease is

likely to be unnoticeable. However, for sufficiently long substeps, the decrease should reach the idle power level.

4. Performing communications has a negligible impact over the energy consumption. Hence, overlapping computations and communications, as in our asynchronous algorithms, does not lead to any additional energy consumption.

Our model has been designed to take into account these general features of parallel systems, and then to minimize their associated uncertainties.

33.5 Node Level Model

In this section and the following one, we present a theoretical model linking together the energy and computing performances of two sets of nodes executing a same application. That model is decomposed in two nested levels. The inner one is the node level, described in this section, which consists in a single machine. The second one is the cluster level, described in Section 33.6.

In the scope of this study, dealing with the comparison between CPU and GPU clusters, we focus on intensive computing applications in which other activities (especially disk accesses) are negligible with respect to the computations. Nonetheless, if necessary, such additional activities could be quite easily added to the model presented throughout the two following sections. They would be inserted into the model in a similar way as the computing activities.

Concerning the node level, we link the energy and computing performances of two node configurations. In the following, we do not make any distinction between the denominations *the two node configurations* and *the two nodes*, but the reader must keep in mind that those *two nodes* may correspond to the same physical node used in two different configurations (use of different numbers of CPU cores and/or additional accelerators: SSE, GPU, FPGA: field-programmable gate array, etc.).

33.5.1 Complete Model

The hypotheses made over the hardware configuration of a node are very general as a node contains:

- At least one and possibly several CPU cores
- 0 or more additional accelerator cards (GPU and/or FPGA)

In fact, on one node, the power can be divided into several parts: P_I, the power consumed by the system node in *idle* state, which is assumed to be constant during the execution of the application, and a series of U powers $P_C(u_i, t), 1 \leq i \leq U$, each one corresponding to the additional power (in addition to P_I) used by computing unit u_i at time t if it is active (i.e., making computations).

If we assume that

- The powers of the computing units are cumulative when used together
- For every computing unit u_i, its additional power $P_c(u_i, t)$ is either null (not active) or maximal (active)

then the total power used at each time t on one node is $\sum_{i=1}^{U} P_C(u_i, t) + P_I$, and the total amount of energy consumed during a period T is given by

$$E = \int_0^T \left(\sum_{i=1}^{U} P_C(u_i, t) + P_I \right) dt = T \cdot P_I + \int_0^T \sum_{i=1}^{U} P_C(u_i, t) dt \quad (33.15)$$

Then, if we denote by $\beta(u_i)$ $(0 \leq \beta(u_i) \leq 1)$ the ratio of the total execution time of the application during which unit u_i is active, and by $P_C(u_i)$ the additional power used by unit u_i (assumed constant during u_i

activity), (33.15) can be reformulated as

$$E = T \cdot \left(\sum_{i=1}^{U} \beta(u_i) P_C(u_i) + P_I \right) \qquad (33.16)$$

The transition from (33.15) to (33.16) is mathematically valid because the functions $P_C(u_i)$ are piecewise continuous over the integration interval (duration T).

Although this formulation is very interesting, it requires a very accurate knowledge on the behavior of the studied application, which is not always available in practice. Indeed, it is necessary to know the respective periods of use of every computing unit during the execution of the application. In some simple cases, where just a few different types of units are used, it may be possible to get reasonable estimations of those information. However, in most cases, some benchmarks of the application itself are mandatory to get accurate evaluations in its context of use.

33.5.2 Discussion on the Importance to Avoid Benchmarking the Target Application

An important aspect that has to be taken into account is the way to deduce the different computation ratios. Those ratios can be deduced either by a theoretical analysis or via a series of executions of the target application on the target cluster.

The first solution is not possible when the algorithm and source code of the considered application are not available. Even in the opposite case, it may be a hard task to theoretically deduce the ratios, especially in algorithms where the execution path is irregular. And finally, the time to lead such a detailed study is not always available.

The second solution only depends on the availability of the target cluster, but most of the times this is not an obstacle. Nevertheless, executing the target application several times to deduce its behavior is kind of a nonsense in many situations, especially if we want to minimize the overall energy expenses (which include the benchmarks required for the application setting).

This is why we are concerned with minimizing application-dependent benchmarks by using as much as possible small generic benchmarks. And when benchmarks of the application are mandatory, we suggest using reduced configurations (problem size, etc.) to get minimal execution times and energy expenses. The information retrieved from such executions may not be completely representative of the exploitation case, but they should be sufficiently accurate in most cases to allow the model to produce estimations of acceptable quality. Moreover, the experience of the user may help to deduce more accurate parameters from the measured ones.

33.5.3 Simplified Model

As we focus on the practical usability of the energy model, we propose making some approximations over the additional power used during the computations (as opposed to idle times). Thus, we consider only one $P_C(u, t)$, which corresponds to the maximal additional power used by the entire set of computing units u (typically CPU core(s) and/or GPU(s)) used to perform the computations. However, as one may be confronted with the comparison of nodes with different hardware configurations (different numbers of CPU cores, presence or absence of a GPU, model of the GPU if present, etc.) it is useful to make the distinction between hardware configurations in the model. So, we respectively denote by $P_I(X)$ the idle power of a node with configuration X (e.g., all the nodes of type X in a cluster), and $P_C(X, u, t)$ the additional power consumed during the computations performed on a node of type X, with units u at time t.

So, at each time t, the total power used on one node of type A is $P_C(A, u, t) + P_I(A)$ and the total amount of energy consumed during a period T is

$$\int_0^T (P_C(A, u, t) + P_I(A))dt = T \cdot P_I(A) + \int_0^T P_C(A, u, t)dt \qquad (33.17)$$

Now, let us consider two distinct executions of the same application on two nodes A and B. The computing units used on node A for the execution are u_A (CPU core(s) and/or GPU(s)) and the ones used on node B are u_B (CPU core(s) and/or GPU(s)). Also, we denote by $T_A(u_A)$ the total execution time on node A using computing units u_A, and $T_B(u_B)$ the total execution time on node B using units u_B. Then, we will have a lower energy consumption on node A when

$$T_A(u_A) \cdot P_I(A) + \int_0^{T_A(u_A)} P_C(A, u_A, t)dt \leq T_B(u_B) \cdot P_I(B) + \int_0^{T_B(u_B)} P_C(B, u_B, t)dt \qquad (33.18)$$

which is equivalent to

$$\int_0^{T_A(u_A)} P_C(A, u_A, t)dt \leq T_B(u_B) \cdot P_I(B) - T_A(u_A) \cdot P_I(A) + \int_0^{T_B(u_B)} P_C(B, u_B, t)dt \qquad (33.19)$$

Under the same assumptions as in Section 33.5.1, powers at *full load* (i.e., during intensive computations) are considered constant and we have $P_C(A, u_A, t) = P_C(A, u_A)$ and $P_C(B, u_B, t) = P_C(B, u_B)$. Moreover, in the context of a scientific application running on a single node, we consider that idle times are negligible, and then $\beta(u_A)$ and $\beta(u_B)$ are both equal to 1. Then, the constraint becomes

$$T_A(u_A) \cdot P_C(A, u_A) \leq T_B(u_B) \cdot P_I(B) - T_A(u_A) \cdot P_I(A) + T_B(u_B) \cdot P_C(B, u_B) \qquad (33.20)$$

which leads to

$$P_C(A, u_A) \leq \frac{T_B(u_B)}{T_A(u_A)} \cdot P_I(B) - P_I(A) + \frac{T_B(u_B)}{T_A(u_A)} \cdot P_C(B, u_B)$$

$$\leq \frac{T_B(u_B)}{T_A(u_A)} \cdot (P_I(B) + P_C(B, u_B)) - P_I(A) \qquad (33.21)$$

inducing a constraint linking together $P_C(A, u_A)$, $P_I(A)$, $P_C(B, u_B)$, and $P_I(B)$.

Let us denote $\alpha = T_B(u_B)/T_A(u_A)$ the speedup of the execution time on node A using computing units u_A with respect to the one on node B using units u_B, then we can rewrite (33.21) as

$$P_C(A, u_A) \leq \alpha \cdot (P_I(B) + P_C(B, u_B)) - P_I(A) \qquad (33.22)$$

Finally, if we denote the respective powers at full load by

$$P_F(A, u_A) = P_C(A, u_A) + P_I(A)$$

and

$$P_F(B, u_B) = P_C(B, u_B) + P_I(B),$$

we obtain

$$P_F(A, u_A) \leq \alpha \cdot P_F(B, u_B) \tag{33.23}$$

which finally gives

$$\alpha \geq \frac{P_F(A, u_A)}{P_F(B, u_B)} \tag{33.24}$$

which expresses a simple constraint over α, only in terms of the powers at full load for the executions on node A and B with respective units u_A and u_B. It can be noticed that the speedup α can be estimated by experimental measures either with small generic benchmarks and the use of a performance model taking the used computing units into account, or by actual executions of the considered application with small instances of the problem.

For example, a good approximation of a performance model for multicore nodes is to measure the speedup α_1 which is the ratio of the execution time with 1 core of node A over the one with 1 core of node B. Then, we deduce α with n_A and n_B cores, respectively, by $\alpha = \alpha_1 \frac{n_B}{n_A}$.

The interest of that general formulation is that it can be used to compare executions on any couple of nodes. For example, it can be used to compare two nodes with different CPUs, or different GPUs, as well as the same node with and without using a GPU.

In that last case, we have $A = B$, $u_A = GPU$ (precisely 1 CPU core and 1 GPU), and $u_B = CPU$ (1 CPU core), and we can rewrite (33.24) as the simplified formulation:

$$\alpha \geq \frac{P_F(GPU)}{P_F(CPU)} \tag{33.25}$$

33.6 Cluster Level Model

In this section, we present the cluster level of the simplified model introduced in the previous section.

Let us consider that we want to choose between two configurations of clusters (A and B), respectively, having N_A and N_B nodes, to run a given application while minimizing the energy consumption. As in the node level section, we specify the computing units respectively used on each node of the two clusters. For the sake of simplicity, and because it is commonly done in practice, we suppose that all the nodes of a given cluster are used in the same way. This means that the same kind of computing units are used. So, in cluster A, units u_A are used (either CPU core(s) or GPU(s)), and in cluster B, units u_B are used. It must be noticed that the two clusters A and B may be the same physical system.

Also, as it has been observed in practice that there is no significant energy overhead in network switches whether there is some traffic or not, we consider in our model that their energy consumption P_{sw} is constant.

Finally, as we perform a comparison between two clusters for the same application, the parameters of that application (size, initial values, etc.) are the same on the two clusters. As a first-order approximation in our modeling, we consider that the application has the same behavior on all the nodes of the used cluster (sequences of idle/communication and full load periods are identical).

However, we have to make a distinction between a synchronous and an asynchronous application as their computational sequences are quite different, and so are their energy signatures. For clarity sake, the asynchronous version is presented first as it is simpler.

33.6.1 Asynchronous Application

In the case of an asynchronous application, the computations are performed uninterruptedly during the execution of the application, in parallel of the communications. This is a rather simple case of parallel execution as it is very similar to the single node model.

In fact, the energy models on clusters A and B take the following forms:

$$E_A(u_A) = T_A(u_A) \cdot \left(\sum_{i=1}^{N_A} P_F^i(A, u_A) + P_{sw} \right)$$
$$E_B(u_B) = T_B(u_B) \cdot \left(\sum_{i=1}^{N_B} P_F^i(B, u_B) + P_{sw} \right) \qquad (33.26)$$

And the execution on cluster A is more energy interesting than the one on cluster B as soon as $E_A(u_A) < E_B(u_B)$, leading to

$$T_A(u_A) \cdot \left(\sum_{i=1}^{N_A} P_F^i(A, u_A) + P_{sw} \right) < T_B(u_B) \cdot \left(\sum_{i=1}^{N_B} P_F^i(B, u_B) + P_{sw} \right) \qquad (33.27)$$

and if we denote, similarly to the node level part, the relative speedup of the execution on cluster A according to the one on cluster B by $\alpha = T_B(u_B)/T_A(u_A)$, then we obtain

$$\alpha > \frac{\sum_{i=1}^{N_A} P_F^i(A, u_A) + P_{sw}}{\sum_{i=1}^{N_B} P_F^i(B, u_B) + P_{sw}} \qquad (33.28)$$

So, by only knowing the number of used nodes, the powers at full load when using either computing units u_A or u_B, and the power of the network switch(es), we can decide which cluster is the most interesting to use. An advantage of that formulation is that the information required to take that decision can be retrieved by small generic benchmarks. The knowledge of the powers of every node in the considered clusters is actually useful even for homogeneous clusters. Indeed, we have observed in practice that there may be significant differences between the powers of similar nodes (see Table 33.1).

In a symmetrical way, (33.28) also specifies a constraint over the minimal relative speedup of cluster A relatively to cluster B in order to get an energy gain. As in the node level context, the speedup α can be estimated without requiring executions of the considered application with full-size instances of problem.

TABLE 33.1 Powers (W) of the Switch and the 16 Nodes of the Cluster

P_{sw}		34.00		
Node id	$P_F(CPU1)$ (1 Core)	$P_F(CPU2)$ (2 Cores)	$P_F(GPU)$	P_I
1	167	167	228	146
2	159	159	228	128
3	159	167	218	133
4	167	174	228	139
5	167	167	228	139
6	159	167	218	133
7	174	182	238	152
8	167	174	238	146
9	152	167	218	133
10	174	182	228	146
11	152	159	228	133
12	167	174	238	139
13	167	167	218	139
14	159	167	228	139
15	174	182	238	159
16	182	190	249	159

Now, in the specific case of deciding whether or not to use the GPUs in a single cluster (implying $A = B$, $N_A = N_B = N$, $P_F^i(A, u_A) = P_F^i(GPU)$ and $P_F^i(B, u_B) = P_F^i(CPU)$), we obtain the following version:

$$\alpha > \frac{\sum_{i=1}^{N} P_F^i(GPU) + P_{sw}}{\sum_{i=1}^{N} P_F^i(CPU) + P_{sw}} \tag{33.29}$$

which can be even more simplified when the power dissipation between the identical nodes is negligible $(P_F^i(X) = P_F^j(X) = P_F(X), \forall i, j \in \{1 \ldots N\})$:

$$\alpha > \frac{P_F(GPU) + \frac{P_{sw}}{N}}{P_F(CPU) + \frac{P_{sw}}{N}} \tag{33.30}$$

33.6.2 Synchronous Application

In the synchronous case, it is commonly assumed that there are, at least partially, some distinct parts of computation and communication phases. From the energy point of view, the main difference between those two parts comes from the fact that during computations the full power of the node is used (P_F), whereas during synchronous blocking communications, the nodes can be considered to be in idle state (P_I). Also, during the potential sequences where communications are overlapped with computations, we consider the energy consumption of the nodes to be at full load (P_F).

The difficult part with synchronous algorithms lies in the necessity to know, at least approximately, the percentage of non-overlapped communications performed during the execution of the algorithm according to the overall execution time. By symmetry, this is equivalent to knowing the percentage of computations $(\beta(u)$ in the complete model in Section 33.5.1). This comes from the distinction in (33.17) of the two different possible phases (computations or communications) during the execution of the application. And, since the application is assumed to have the same behavior on every node of the cluster, as discussed at the beginning of this section, we can rewrite (33.17) for every used node of the cluster, as follows:

$$E_A(u_A) = \int_0^T P_C(A, u_A, t)dt + \int_0^T P_I(A)dt \tag{33.31}$$

And since by definition $P_C()$ is constant (non-null) during computations and null otherwise, the previous equation can be reformulated as

$$E_A(u_A) = T \cdot \beta_A(u_A) \cdot P_C(A, u_A) + T \cdot P_I(A) \tag{33.32}$$

where $\beta_A(u_A)$ is similar to the $\beta(u_i)$ in Section 33.5.1, with the additional precision of the cluster context (configuration A). In practice, a more convenient expression of (33.32) is obtained by replacing $P_C()$ by $P_F()$, the total power at full load:

$$E_A(u_A) = T \cdot (\beta_A(u_A) \cdot P_F(A, u_A) + (1 - \beta_A(u_A)) \cdot P_I(A)) \tag{33.33}$$

Unfortunately, as mentioned at the end of Section 33.5.1, the acquisition of the percentage $\beta_A(u_A)$ requires at least one benchmark of the target application. Moreover, it generally varies in function of the problem size and of the number of nodes in addition to the computing units used.

As a first approximation we consider that for a fixed number of nodes, the computation ratio in synchronous applications slightly changes but does not follow too large variations according to the problem size. In fact, this is rather closely confirmed experimentally, as depicted in Figure 33.6 for one of

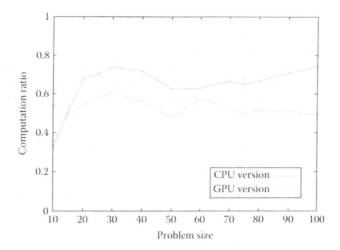

FIGURE 33.6 Computation percentage of the total execution time in function of the problem size for a 17-node cluster. Results are statistics from five executions.

the most difficult cases of application which is an iterative PDE solver (see Section 33.7 for further details). It can be seen that for the CPU version, there are significant variations for the very small problem sizes, then the behavior becomes stable. And for the GPU version, the variations are even more limited. Thus, the assumption of a constant computation ratio is reasonably acceptable for most of the applications. So, we propose to approximate that ratio from a single execution of the considered application. The principle is to choose a problem size as small as possible although not too small to avoid nonrepresentative ratios. In particular cases where larger variability may be expected, a few executions with slightly different problem sizes could be considered to obtain a representative average ratio.

Then, we obtain the following total energy consumption when running the considered application on cluster A with computing units u_A:

$$E_A(u_A) = T_A(u_A) \cdot \left(\beta_A(u_A) \cdot \sum_{i=1}^{N_A} P_F^i(A, u_A) + (1 - \beta_A(u_A)) \cdot \sum_{i=1}^{N_A} P_I^i(A) + P_{sw} \right) \tag{33.34}$$

So, when comparing two clusters A and B, cluster A becomes more interesting than cluster B according to the energy aspect as soon as $E_A(u_A) < E_B(u_B)$, leading to

$$T_A(u_A) \cdot \left(\beta_A(u_A) \cdot \sum_{i=1}^{N_A} P_F^i(A, u_A) + (1 - \beta_A(u_A)) \cdot \sum_{i-1}^{N_A} P_I^i(A) + P_{sw} \right)$$

$$< T_B(u_B) \cdot \left(\beta_B(u_B) \cdot \sum_{i=1}^{N_B} P_F^i(B, u_B) + (1 - \beta_B(u_B)) \cdot \sum_{i=1}^{N_B} P_I^i(B) + P_{sw} \right) \tag{33.35}$$

which finally gives

$$\alpha > \frac{\beta_A(u_A) \cdot \sum_{i=1}^{N_A} P_F^i(A, u_A) + (1 - \beta_A(u_A)) \cdot \sum_{i=1}^{N_A} P_I^i(A) + P_{sw}}{\beta_B(u_B) \cdot \sum_{i=1}^{N_B} P_F^i(B, u_B) + (1 - \beta_B(u_B)) \cdot \sum_{i=1}^{N_B} P_I^i(B) + P_{sw}} \tag{33.36}$$

And we can express the constraint of the energy frontier between clusters A and B only in terms of (1) the number of used nodes, (2) the powers of the nodes (with respective computing units u_A and u_B)

at full load and in idle/communication state, and (3) the power of the network switch(es). Most of this information can be retrieved by small generic benchmarks, except for the $\beta()$ values which require at least one execution of the application, as discussed above.

Now, if we consider a single cluster and we just want to establish whether it is interesting to use its embedded GPUs or not, we can simplify the general notations with $N_A = N_B = N$, $P_F^i(A, u_A) = P_F^i(GPU)$, $P_F^i(B, u_B) = P_F^i(CPU)$, $P_I(A) = P_I(B) = P_I$, $\beta_A(u_A) = \beta(GPU)$, and $\beta_B(u_B) = \beta(CPU)$. According to (33.36), we obtain

$$\alpha > \frac{\beta(GPU) \cdot \sum_{i=1}^{N} P_F^i(GPU) + (1 - \beta(GPU)) \cdot \sum_{i=1}^{N} P_I^i + P_{sw}}{\beta(CPU) \cdot \sum_{i=1}^{N} P_F^i(CPU) + (1 - \beta(CPU)) \cdot \sum_{i=1}^{N} P_I^i + P_{sw}} \tag{33.37}$$

Here again, if the power dissipation between the similar nodes in a same cluster is negligible, we have $P_F^i(X) = P_F^j(X) = P_F(X)$, $P_I^i = P_I^j = P_I$, $\forall i, j \in \{1, \ldots, N\}$, and we obtain the simplified version:

$$\alpha > \frac{\beta(GPU) \cdot P_F(GPU) + (1 - \beta(GPU)) \cdot P_I + \frac{P_{sw}}{N}}{\beta(CPU) \cdot P_F(CPU) + (1 - \beta(CPU)) \cdot P_I + \frac{P_{sw}}{N}} \tag{33.38}$$

33.7 Synchronous and Asynchronous Distributed PDE Solver

Our test application is an iterative PDE solver using the multisplitting-Newton algorithm together with an inner linear solver. We recall that iterative methods perform successive approximations toward the solution of a problem (notion of convergence) whereas direct methods give the exact solution within a fixed number of operations. Although iterative methods are generally slower than direct ones, they generally present the advantage of being less memory consuming. Moreover, they are often the only known way to solve some problems and they are also the only way to express asynchronous algorithms.

Our PDE solver is designed to solve a 3D transport model, which simulates the evolution of the concentrations of chemical species in shallow waters [6]. A system of advection–diffusion–reaction (ADR) has the following form:

$$\frac{\partial c}{\partial t} + A(c, a) = D(c, d) + R(c, t) \tag{33.39}$$

where
   c is the unknown vector of the species concentrations
   $A(c, a)$ is the vector related to the advection
   $D(c, d)$ is the vector of the diffusion
   a is the field of local velocities in the liquid
   d is the matrix of the diffusion coefficients; those last two data are assumed to be known in advance (a may be for example computed by a hydrodynamic model)
   Finally, $R(c, t)$ represents the chemical reactions between the species.

In the following, we consider the problem in three spatial dimensions and with two chemical species. In that case, (33.39) can be written as a system of two PDEs:

$$\begin{pmatrix} \frac{\partial c_1}{\partial t} \\ \frac{\partial c_2}{\partial t} \end{pmatrix} + \begin{pmatrix} \nabla c_1 \times a \\ \nabla c_2 \times a \end{pmatrix} = \begin{pmatrix} \nabla \cdot ((\nabla c_1) \times d) \\ \nabla \cdot ((\nabla c_2) \times d) \end{pmatrix} + \begin{pmatrix} R_1(c, t) \\ R_2(c, t) \end{pmatrix} \tag{33.40}$$

The coupling between the two equations comes from the reaction term R.

33.7.1 Computational Model

First of all, (33.40) is transformed into a discrete time (Euler method) and discrete space (second-order finite differences) ODE system of the form:

$$\frac{dx(t)}{dt} = f(x(t), t) \tag{33.41}$$

where
  x is the mesh of points where the concentrations are computed
  f is the nonlinear function modeling the ADR

By the use of an implicit temporal integration, this ODE system becomes

$$\frac{x(t) - x(t - h)}{h} = f(x(t), t) \tag{33.42}$$

where h is a fixed time step. This equation can then be rewritten as

$$F(x(t), x(t - h), t) = x(t - h) - x(t) + hf(x(t), t) \tag{33.43}$$

and the final problem is to solve $F(x(t), x(t - h), t) = 0$, which can be reformulated in a simpler way by $F(x(t), C(t - h)) = 0$, where $C(t - h)$ represents the constant terms of F at time t.

Using the Newton method, we obtain an iterative scheme to compute $x(t)$:

$$x^{k+1}(t) = x^k(t) - F'^{-1}\left(x^k(t)\right) F\left(x^k(t), C(t - h)\right) \tag{33.44}$$

where
  $x^i(t)$ is the ith iterate of $x(t)$
  $F'\left(x^k(t)\right)$ is the Jacobian matrix of $F\left(x^k(t), C(t - h)\right)$

The equation can be reformulated as

$$F'\left(x^k(t)\right)\left(x^{k+1}(t) - x^k(t)\right) = -F\left(x^k(t), C(t - h)\right) \tag{33.45}$$

Solving that equation requires to solve a linear system at each iteration. However, it can be noticed that in the ADR problem, the Jacobian matrix is sparse and its nonzero terms are scattered just on a few diagonals. The method used to get a parallel version of that algorithm is the *multisplitting-Newton* scheme.

33.7.2 Multisplitting-Newton Algorithm

There are several methods to solve PDE problems, each of them including different degrees of synchronism/asynchronism. The method used in this test application is the multisplitting-Newton [7] which allows for a rather important level of asynchronism. Indeed, it is important to validate our model to get an application that can work either in synchronous or asynchronous mode.

In the computational model described above, the size of the simulation domain can be huge and the domain is then distributed among several nodes of a cluster. Each node solves a part of the resulting linear system and sends the relevant updated data to the nodes that need them. The parallel algorithmic scheme of the method is as follows:

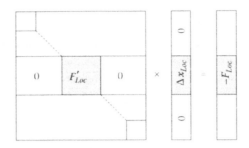

FIGURE 33.7   Local computations associated to the subdomain of one unit.

1. Initialization:
   a. Formulation of the problem under the form of a fixed point problem: $x = T(x), x \in \mathbb{R}^n$ where $T(x) = x - F'(x)^{-1}F(x)$ and F' is the Jacobian
   b. We get $F' \times \Delta x = -F$ with F' a sparse matrix
   c. F' and F are distributed over the computing units
2. Iterative process:
   a. Each unit computes a different part of Δx using the Newton algorithm over its subdomain as can be seen in Figure 33.7
   b. The local elements of x are directly updated with the local part of Δx
   c. The non-local elements of x come from the other units by messages exchanges
   d. F is updated using the entire vector x

33.7.3 Inner Linear Solver

The method described above is a two-stage algorithm in which a linear solver is needed in the inner stage. In fact, most of the time of the algorithm is spent in that linear solver. This is why, in the context of our comparison between CPU and GPU nodes, it is that part of the computations that has been deported on the GPUs. Due to their regularity, those treatments are very well suited to the SIMD architecture of the GPU. Hence, on each computing unit, the linear computations required to solve the partial system are performed on the local GPU while all the algorithmic control, nonlinear computations, and data exchanges between the units are done on the CPU.

The linear solver has been implemented both on CPU and GPU, using the biconjugate gradient algorithm (see [19] for further details). This linear solver was chosen because it performs well on nonsymmetric matrices (on both convergence time and numerical accuracy), it has a low memory footprint, and it is relatively easy to implement on a GPU.

33.7.3.1 GPU Implementation

Several aspects are critical in a GPU: the regularity of the computations and the memory which is of limited amount and the way the data are accessed. In order to reduce the memory consumption of our sparse matrix, we have used a compact representation, depicted in Figure 33.8, similar to the DIA (diagonal) format [18] in BLAS [11], but with several additional advantages. The first one is the regularity of the structure, which allows us to do coalescent memory accesses most of the time. The second one is that it provides an efficient access to the matrix itself as well as to its transpose (using simple array index computations), which is a great advantage as the transpose is required in the biconjugate gradient method.

In order to be as efficient as possible, the shared memory has been used as a cache memory whenever it was possible in order to avoid the slower accesses to the global memory of the GPU. The different kernels used in the solver are divided to make as much as possible data reuse at each call, minimizing by this way

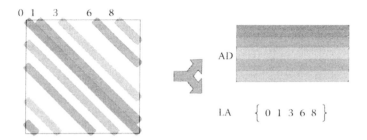

FIGURE 33.8 Compact and regular sparse matrix representation.

the transfers between the global memory and the registers. To get full details on those kernels, the reader should refer to [19].

33.7.4 Asynchronous Aspects

It is quite obvious that over the last few years, the classical algorithmic schemes used to exploit parallel systems have shown their limit. As the most recent systems are more and more complex and often include multiple levels of parallelism with very heterogeneous communication links between those levels, the synchronous nature of the schemes presented previously has become a major drawback. Indeed, synchronizations may noticeably degrade performances in large or hierarchical systems, even for local systems.

For a few years now, asynchronous algorithmic schemes have emerged [2,3,6,9,10,15], and although they cannot be used for all problems, they are efficiently usable for a large part of them. In scientific computing, asynchronism can only be expressed in iterative algorithms, as already mentioned. This condition has strongly motivated the nature of our PDE solver.

The asynchronous feature consists in suppressing any idle time induced by the waiting for the dependency data to be exchanged between the computing units of the parallel system. Hence, each unit performs the successive iterations on its local data with the dependency data versions it owns at the current time. The main advantage of this scheme is to allow for an efficient and implicit overlapping of communications by computations. On the other hand, the major drawbacks of asynchronous iterations are a more complex behavior, which requires a specific convergence study, and a larger number of iterations to reach the convergence. However, the convergence conditions in asynchronous iterations are verified for numerous problems and, in many computing contexts, the time overhead induced by the additional iterations is largely compensated by the gain in the communications [4,8]. In fact, as soon as the frequency of communications relatively to computations is high enough and the communication costs are larger than local accesses, an asynchronous version of an application may provide better performances than its synchronous counterpart.

In the asynchronous version of our PDE solver, the exchanges of parts of the vector x are performed asynchronously. One synchronous global exchange is still required between each time step of the simulation, as illustrated in Figure 33.9.

At the practical level, the main differences with the synchronous version lie in the suppression of some barriers and in the way the communications between the units are managed. Concerning the first aspect, all the barriers between the inner iterations inside each time step of the simulation are suppressed. The only remaining synchronization is the one between each time step as pointed out above.

The communications management is a bit more complex than in the synchronous version as it must enable sending and receiving operations at any time during the algorithm. Although the use of non-blocking communications seems appropriate, it is not sufficient, especially concerning receptions. This is why a multithreaded programming is required. The principle is to use separated threads to perform

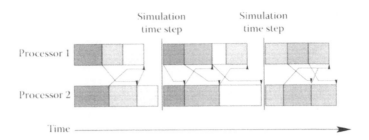

FIGURE 33.9 Asynchronous iterations inside each time step of the computation.

the communications, while the computations are continuously done in the main thread without any interruption, until convergence detection. In our version, we use nonblocking sends in the main thread and an additional thread to manage the receptions. It must be noted that in order to be as reactive as possible, some communications related to the control of the algorithm (the global convergence detection) may be initiated directly by the receiving thread (e.g., to send back the local state of the unit) without requiring any process or response from the main thread. Subsequently to the multithreading, mutexes are necessary to avoid concurrent accesses to data and variables.

Another difficulty brought by the asynchronism comes from the convergence detection. Some specific mechanisms must replace the simple global reduction of local states of the units to ensure the validity of the detection [5]. The most general scheme may be too expensive in some simple contexts such as local clusters. So, when some characteristics of the system are guaranteed (such as bounded communication delay), it is often more pertinent to use a simplified mechanism whose efficiency is better and whose validity is still ensured in that context. Although both general and simplified schemes of convergence detection have been developed for this study, the performances presented in the following sections are those of the simplified scheme, which gave the best results.

33.8 Experimental Validation

In this part, a series of experiments is presented which evaluates the pertinence and accuracy of our model at both node and cluster levels.

33.8.1 Testbed Introduction and Measurement Methodology

All the experiments presented below have been performed on a homogeneous cluster of 16 machines with Intel Nehalem CPUs (four cores + hyperthreading) running at 2.67 GHz, 6 GB RAM, and 1 NVIDIA GeForce GTX 480 card with 768 MB RAM. The OS is Linux Fedora with CUDA 3.0.

In the following experiments, the tested application is the 3D PDE solver described in Section 33.7. Such an application is quite representative of the scientific applications run on a cluster. The results related to that application are an average of five consecutive executions. The small generic benchmarks used to extract the model parameters from the test platform are very simple floating point computations performed either on a specified number of CPU cores or on one CPU core and one GPU. For each benchmark, the conserved power is the maximal one obtained during a period of 30 s. Table 33.1 provides the results obtained with those benchmarks for the test platform.

33.8.2 Node Level

At the node level, we consider the nodes of the cluster separately and we apply our simplified model to compare the PDE solver executions with or without the GPU. In that context, the columns of interest

TABLE 33.2 Model Estimates Compared to Experimental Observations for Every Node of the Cluster

Node id $\longrightarrow$	1	2	3	4	5	6	7	8
Estimated α	1.365	1.434	1.371	1.365	1.365	1.371	1.368	1.425
Ratio	1.090	1.136	1.109	1.166	1.105	1.108	1.106	1.128
Observed α	1.252	1.262	1.236	1.171	1.235	1.238	1.237	1.264
	9	10	11	12	13	14	15	16
Estimated α	1.434	1.310	1.500	1.425	1.305	1.434	1.368	1.368
Ratio	1.110	1.070	**1.194**	1.064	1.031	1.148	1.111	1.085
Observed α	1.292	1.225	1.256	1.339	1.266	1.249	1.231	1.261

in Table 33.1 are only $P_F(CPU1)$ and $P_F(GPU)$. Indeed, our test application uses only one CPU core when run on a single node and either zero or one GPU. Table 33.2 presents, for every node of the cluster, the respective estimations of the α limits deduced from (33.25), the observed ones, and the estimation/observation ratios.

It can be seen that the estimations are quite close to the observed frontiers. However, a global trend of overestimation can be observed in the whole set of estimates (all the ratios are greater than 1). Although that global bias is quite reasonable (mean bias around 0.11), the most important ones are quite significant as they reach a little more than 19% for node 11 (in bold face). In order to give a better idea of the error made in that extreme case, the computing and energy ratios are depicted in function of the problem size for node 11 in Figure 33.10. Such bias may lead to wrong choices near the frontier.

Although a part of that bias may be explained by the fair accuracy of the energy measures, it is not sufficient to explain the whole bias. In fact, the main cause of error comes from the approximations made in the simplified model. Typically, in the case of our test application, the GPU is not used to perform all the computations but only to solve the inner linear systems. Thus, it is only used during a fraction of the total execution time, and considering $P_F(GPU)$ for the entire execution in (33.25) results in an evident overestimation of the required power, and thus of α.

So, when it is possible, evaluating at least the main ratios $\beta(u_i)$ for the target application can substantially enhance the final estimation. As an example, in our test application, we evaluate $\beta(GPU)$ and we obtain the following equation for the estimation of α:

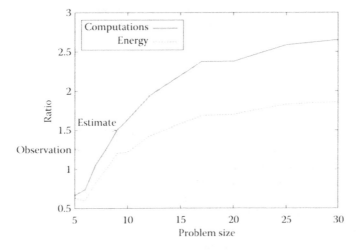

FIGURE 33.10 Computing and energy ratios of the CPU version over the GPU one for the ADR problem on node 11, in function of the problem size (3D cubic domain $x \times x \times x$).

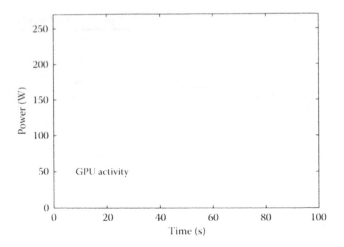

FIGURE 33.11 Power consumption during and after a GPU usage of 20 s on a single node.

$$\alpha \geq \frac{\beta(GPU)P_F(GPU) + (1 - \beta(GPU))P_F(CPU1)}{P_F(CPU1)} \qquad (33.46)$$

To obtain $\beta(GPU)$, we have measured the percentage of GPU usage in the entire execution time. Unfortunately, that measure cannot be used directly for two reasons. The first one is that unlike the full CPU version, the percentage of GPU usage strongly varies with the problem size. In fact, for small problem sizes (between 5^3 and 30^3 here), the GPU is not fully loaded and its usage time does not evolve as fast as the CPU usage with the problem size. We have then to consider an average from a few problem sizes. The second reason is that the energy consumption does not follow discontinuous variations and once a high power has been reached, the return to the lowest power takes some time, as shown in Figure 33.11 with a simple benchmark.

So, to compensate for the slow power decrease after high consumption periods, the measure has to be weighted according to the mean power level after the GPU usage. As the power decrease has a typical step at the middle value between full GPU usage and idle power, a good weighting is the middle value between the measured GPU usage percentage and the total time (i.e., 1), that is:

$$\beta(GPU) = \frac{\beta_{measured}(GPU) + 1}{2}$$

Finally, from (33.46) and the deduced $\beta(GPU)$ for every single node, we obtain a mean ratio between the predicted α and the observed ones of 0.985 and minimal and maximal values respectively of 0.934 and 1.042. Obviously, this is much better than our first estimations as the standard deviation of the ratios are smaller and the ratios are globally centered around 1.

In conclusion, the previous experiments performed on single nodes have shown that the simplified node level model has a slight bias. That bias can be discarded by performing a deeper study and/or benchmark of the target application. However, as discussed in Section 33.5.2, these corrections require application-dependent benchmarks that are not always possible nor desirable to do. Hence, given its relevance, its rather limited estimation bias and its simple practical process, our simplified model can be used in most cases as a good indicator for energy comparisons.

33.8.3 Cluster Level: Asynchronous Mode

In the following experiments, we consider the entire cluster of 16 machines. In the asynchronous case, the columns $P_F(CPU2)$ and $P_F(GPU)$ in Table 33.1 are used. The 2-core version corresponds to the number of threads required in the asynchronous version of the PDE solver.

Here, we had the possibility to get the power of each node in the cluster. However, for homogeneous clusters, if the power information is available only on one node, our model can be adapted by considering that all the $P_F^i()$ are identical. Nonetheless, the user must be aware that this is an approximation which may induce an additional bias in the final predictions of the model.

According to (33.29), we can deduce that the use of GPUs on that cluster will be interesting from the energy point of view as soon as we have

$$\alpha > \frac{\sum_{i=1}^{16} P_F^i(GPU) + P_{sw}}{\sum_{i=1}^{16} P_F^i(CPU2) + P_{sw}}$$

$$> \frac{3669 + 34}{2745 + 34}$$

$$> 1.332 \tag{33.47}$$

So, the GPU version must be at least around 33.2% faster than the CPU one to provide an energy gain on this cluster.

This estimation is confirmed by the experiments. Indeed, it can be seen on Figure 33.12, depicting the computing and energy ratios of the CPU version over the GPU one in function of the problem size for the entire cluster (16 nodes), that the estimated speedup α closely matches the frontier at which the GPU version becomes more interesting than the CPU from the energy point of view.

33.8.4 Cluster Level: Synchronous Mode

In that last context, the columns $P_F(CPU1)$, $P_F(GPU)$ and P_I of Table 33.1 are relevant. The 1-core version corresponds to the synchronous version of the PDE solver, in which there is only one main thread.

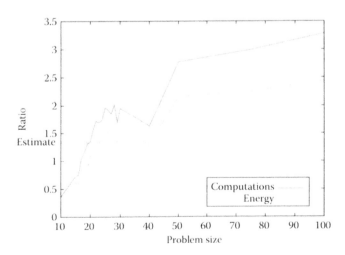

FIGURE 33.12 Computing and energy ratios of the CPU version over the GPU one for the ADR problem on the 16 nodes, in function of the problem size (3D cubic domain $x \times x \times x$).

TABLE 33.3 Computation Ratios
of the Total Execution Time for the
CPU and GPU Versions of the ADR
Problem with a Problem Size of
$30 \times 30 \times 30$

$\beta(CPU)$	$\beta(GPU)$
0.749	0.602

As discussed before, in the synchronous case, generic benchmarks alone are not sufficient to be able to deduce the constraint over α. An evaluation of the percentage of time performed at full load during the execution (i.e., during computations) is required. The following measures of the $\beta()$ have been performed on the target cluster with the target application with a problem size of $30 \times 30 \times 30$.

In fact, complementary measures have shown that the values of $\beta(CPU)$ and $\beta(GPU)$ within a range of problem sizes from 10^3 to 100^3 do not vary very much, and their respective averages are quite close to the ones given in Table 33.3.

According to (33.37), we deduce that it is worth using the GPUs of the cluster as soon as we have

$$\alpha > \frac{\beta(GPU) \sum_{i=1}^{16} P_F^i(GPU) + (1 - \beta(GPU)) \sum_{i=1}^{16} P_I^i + P_{sw}}{\beta(CPU) \sum_{i=1}^{16} P_F^i(CPU) + (1 - \beta(CPU)) \sum_{i=1}^{16} P_I^i + P_{sw}}$$

$$> \frac{0.602 \cdot 3669 + (1 - 0.602) \cdot 2263 + 34}{0.749 \cdot 2646 + (1 - 0.749) \cdot 2263 + 34}$$

$$> 1.217$$

That is, the GPU version should be less energy consuming as soon it is around 21.7% faster than the CPU one.

Similarly to the node level context, there is a slight bias in that estimation with respect to the experimental observation, as can be seen in Figure 33.13. That bias is around 5.9% under the actual observed value of α

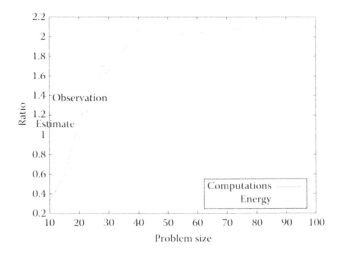

FIGURE 33.13 Computing and energy ratios of a synchronous CPU version over its GPU counterpart for the ADR problem in function of the problem size (3D cubic domain $x \times x \times x$).

(1.293) and may therefore lead to wrong choices of hardware near the energy efficiency frontier between CPU and GPU versions.

Here also, that bias can be explained by the slow power decrease after high consumptions, which are not taken into account in that simplified version. This results in significant underestimations of $\beta(GPU)$ and $\beta(CPU)$, leading to a global underestimation of α. There are two ways for correcting the estimation, which leads to slightly different results but with almost the same final accuracy (final bias under 1%). The former is also the simplest one and the most convenient in practice as it consists in taking into account the slow power decrease directly at the level of $\beta(GPU)$ and $\beta(CPU)$. For the same reasons as in Section 33.8.2, we consider the following corrections:

$$\beta(CPU) = \frac{\beta_{measured}(CPU) + 1}{2} = 0.875$$

$$\beta(GPU) = \frac{\beta_{measured}(GPU) + 1}{2} = 0.801 \tag{33.48}$$

and we obtain

$$\alpha > \frac{0.801 \cdot 3669 + (1 - 0.801) \cdot 2263 + 34}{0.875 \cdot 2646 + (1 - 0.875) \cdot 2263 + 34}$$

$$> 1.301$$

which is only 0.6% over the observation.

The second possible correction is a bit more complex and requires more information about the application. The idea is to get closer to the complete model described in Section 33.5.1. In fact, in the synchronous version of the application, we can distinguish three kinds of activities during the execution: communication (considered similar to idle), CPU-only computations, and GPU computations. This leads to splitting the initial $\beta(GPU)$ into $\beta(GPU)$ on one side and $\beta(CPUo)$ (computations only on CPU) on the other side, leading to the following reformulation of the condition over α:

$$\alpha > \frac{1}{\gamma} \left(\beta(GPU) \sum_{i=1}^{16} P_F^i(GPU) + \beta(CPUo) \sum_{i=1}^{16} P_F^i(CPU1) \right.$$

$$\left. + (1 - \beta(GPU) - \beta(CPUo)) \sum_{i=1}^{16} P_I^i + P_{sw} \right)$$

where γ is the total energy consumption of the CPU version of the application: $\gamma = \beta(CPU) \sum_{i=1}^{16} P_F^i(CPU) + (1 - \beta(CPU)) \sum_{i=1}^{16} P_I^i + P_{sw}$. The final estimation is obtained after having applied the same slow power decrease correction as above to the different $\beta()$. The result is

$$\alpha > \frac{0.565 \cdot 3669 + 0.737 \cdot 2646 + (1 - 0.565 - 0.737) \cdot 2263 + 34}{0.875 \cdot 2646 + (1 - 0.875) \cdot 2263 + 34}$$

$$> 1.281$$

which is 0.9% under the observation.

Finally, as both corrections seem to give very close lower and upper estimations, it may be interesting to use both of them to get a small fuzzy area around the real frontier of energy efficiency between CPU and GPU versions. However, this point has yet to be confirmed with other scientific applications and should be fully investigated in future works.

33.9 Discussion on a Model for Hybrid and Heterogeneous Clusters

As described in Sections 33.5.1 and 33.6, the proposed model directly includes the potential hardware heterogeneity of the used parallel system. We have seen in Section 33.8 that this model allows us to compare different systems or configurations of a single system from the energy point of view. However, the possibility to compare different operating modes of the same application has not been discussed. In particular, it is relevant to compare the efficiencies of the synchronous and asynchronous modes when both of them are available for a given application.

In fact, in previous works [12,13], we performed a whole set of experiments with the ADR application on a heterogeneous cluster composed of two homogeneous clusters (respectively 14 and 17 machines). In those experiments, we compared the computing and energy performances of the synchronous and asynchronous versions of our test application.

As can be seen in Figure 33.14, the frontier between the two versions is not linear and an accurate model is necessary to be able to determine which operating mode is preferable for a given context of use (number of used nodes in each cluster).

The model proposed in this chapter can be quite easily extended to allow for the comparison of operating modes. Indeed, from (33.16), it can be seen that the energy can always be expressed as a product of the execution time and the average electrical power consumed by the application during this time. So, when comparing two contexts of use, whatever those contexts are, it is always possible to deduce a constraint over α that only depends on the respective powers corresponding to those contexts. Thus, it is possible to say that context 2 will be preferable to context 1 as soon as

$$\alpha = \frac{T(\text{context 1})}{T(\text{context 2})} > \frac{\overline{P}(\text{context 2})}{\overline{P}(\text{context 1})} \qquad (33.49)$$

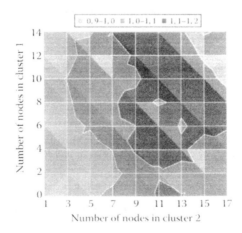

FIGURE 33.14 Speedup of asynchronous vs. synchronous versions of the ADR application with a heterogeneous GPU cluster.

And the comparison of operating modes in the same hardware context can be expressed by

$$\alpha = \frac{T(\text{sync})}{T(\text{async})} > \frac{\overline{P}(\text{async})}{\overline{P}(\text{sync})} \qquad (33.50)$$

Hence, with the formulations of the energy consumptions given in Sections 33.6.1 and 33.6.2, the generic benchmarks on all the nodes of the heterogeneous cluster, and a few application benchmarks, it is possible to provide estimations of α for each heterogeneous configuration. Finally, a comparative map as depicted in Figure 33.14 could be entirely deduced from only a few executions of the application.

Current researches are developed on this topic and a complete experimental study should be proposed in the near future.

33.10 Perspectives: Towards Multikernel Distributed Algorithms and Auto-Adaptive Executions

In this chapter, a complete feedback of our experience in practical and theoretical works over the energy aspects of parallel computing has been proposed as a starting point. Then, our experimental process has been described together with the main experimental issues that can be encountered when measuring energy consumption. Finally, a complete model linking the computing and energy performances has been presented and experimentally validated with a representative scientific application. Also, possible extensions to various contexts have been discussed.

The simplest way to use the model proposed in this chapter is when a user has to choose between several execution contexts and modes. The user only has to get the required information by executing a small set of benchmarks and, according to the results yielded by the model, the user can choose the most suited environment with respect to his needs.

Nevertheless, in many situations, that choice protocol may not be adapted. In fact, the user may not be a specialist in computer science and thus not be able to perform the required benchmarks nor to use the program implementing the model computations. Moreover, it is also probable that the user would not want to do such tasks which are outside his/her main domain of work.

So, it is desirable to provide a system that would implement the entire protocol. However, it is not interesting to insert that choice protocol directly into the application. First of all, this is not possible when the source code of the application is not available. Moreover, even in the opposite case, it may be quite complex and time-consuming to perform such modifications into the code. And it withdraws the advantage of the application independence of the protocol.

Thus, the best solution seems to be an additional system controlling the application execution. The idea is that the user specifies the optimization criteria for the application execution (execution time and/or energy consumption) on that system. In addition, the execution control system (ECS) needs either a multikernel version of the application or a set of different versions, as well as a description of all the available nodes in the parallel system. Provided this, the ECS can automatically run the required generic benchmarks as well as the few application-dependent benchmarks needed to feed the model. It is then able to produce the estimations for the set of possible execution configurations and choose the optimal one with respect to the user's specification. Finally, the adequate version of the application is executed in the corresponding system and configuration.

The complete design and implementation of the ECS is our priority goal in our future works in the domain.

References

1. L. Abbas-Turki, S. Vialle, B. Lapeyre, and P. Mercier. High dimensional pricing of exotic European contracts on a GPU cluster, and comparison to a CPU cluster. In *Parallel and Distributed Computing for Finance (PDCoF09)*, Roma, Italy, May 29, 2009.

2. D. Amitai, A. Averbuch, M. Israeli, and S. Itzikowitz. Implicit-explicit parallel asynchronous solver for PDEs. *SIAM J. Sci. Comput.*, 19:1366–1404, 1998.

3. J. Bahi. Asynchronous iterative algorithms for nonexpansive linear systems. *J. Parallel Distrib. Comput.*, 60(1):92–112, January 2000.

4. J. Bahi, S. Contassot-Vivier, and R. Couturier. Evaluation of the asynchronous iterative algorithms in the context of distant heterogeneous clusters. *Parallel Comput.*, 31(5):439–461, 2005.

5. J. Bahi, S. Contassot-Vivier, and R. Couturier. An efficient and robust decentralized algorithm for detecting the global convergence in asynchronous iterative algorithms. In *Eighth International Meeting on High Performance Computing for Computational Science, VECPAR'08*, pp. 251–264, Toulouse, France, June 2008.

6. J. Bahi, R. Couturier, K. Mazouzi, and M. Salomon. Synchronous and asynchronous solution of a 3D transport model in a grid computing environment. *Appl. Math. Model.*, 30(7):616–628, 2006.

7. J.M. Bahi, S. Contassot-Vivier, and R. Couturier. *Parallel Iterative Algorithms: From Sequential to Grid Computing*. Numerical analysis and scientific computing series. Chapman & Hall/CRC, Boca Raton, FL, 2007.

8. J.M. Bahi, S. Contassot-Vivier, and R. Couturier. Asynchronism for iterative algorithms in a global computing environment. In *The 16th Annual International Symposium on High Performance Computing Systems and Applications (HPCS'2002)*, pp. 90–97, Moncton, New Brunswick, Canada, June 2002.

9. Z. Bai, V. Migallon, J. Penades, and D.B. Szyld. Block and asynchronous two-stage methods for midly nonlinear systems. *Numer. Math.*, 82:1–21, 1999.

10. D.P. Bertsekas and J.N. Tsitsiklis. *Parallel and Distributed Computation*. Prentice Hall, Englewood Cliffs, NJ, 1999.

11. J.J. Dongarra, J. Du Goz, I.S. Duff, and S. Hammarling. A set of level 3 basic linear algebra subprograms. *ACM Trans. Math. Soft.*, 16:1–17, 1990. http://www.netlib.org/blas/

12. S. Contassot-Vivier, T. Jost, and S. Vialle. Impact of asynchronism on GPU accelerated parallel iterative computations. In *PARA 2010 Conference: State of the Art in Scientific and Parallel Computing*, Reykjavík, Iceland, June 2010.

13. S. Contassot-Vivier, S. Vialle, and T. Jost. Optimizing computing and energy performances on GPU clusters: Experimentation on a PDE solver. In J.-M. Pierson and H. Hlavacs, eds. *COST Action IC0804 on Large Scale Distributed Systems, 1st Year*. IRIT, Passau, Germany, 2010. ISBN: 978-2-917490-10-5.

14. A. Fog. Optimizing software in C++: An optimization guide for Windows, Linux and Mac platforms. Technical report, Copenhagen University College of Engineering, Copenhagen, Denmark, September 2009.

15. A. Frommer and D.B. Szyld. On asynchronous iterations. *J. Comput. Appl. Math.*, 123:201–216, 2000.

16. R. Gonzalez and M. Horowitz. Energy dissipation in general purpose microprocessors. *IEEE J. Solid-State Circ.*, 31(9), 1277–1284, September 1996.

17. N. Govindaraju, S. Larsen, J. Gray, and D. Manocha. A memory model for scientific algorithms on graphics processors. In *ACM/IEEE Conference on Supercomputing (SC'06)*, Tampa, FL, November 11–17, 2006.

18. M.A. Heroux. A proposal for a sparse blas toolkit. SPARKER working note #2, Cray Research, Inc., Seattle, WA, 1992.

19. T. Jost, S. Contassot-Vivier, and S. Vialle. An efficient multi-algorithm sparse linear solver for GPUs. In *Parallel Computing: From Multicores and GPU's to Petascale*, Vol. 19 of Advances in Parallel Computing, pp. 546–553. IOS Press, Amsterdam, the Netherlands, 2010.

20. X. Ma, M. Dong, L. Zhong, and Z. Deng. Statistical power consumption analysis and modeling for GPU-based computing. In *Workshop on Power Aware Computing and Systems (HotPower'09)*, Big Sky, MT, October 10, 2009.

21. M. Roufouei, T. Stathopoulos, S. Ryffel, W. Kaiser, and M. Sarrafzadeh. Energy-aware high performance computing with graphic processing units. In *Workshop on Power Aware Computing and Systems (HotPower'08)*, San Diego, CA, December 7, 2008.

22. H. Takizawa, K. Sato, and H. Kobayashi. SPRAT: Runtime processor selection for energy-aware computing. In *Third International Workshop on Automatic Performance Tuning (iWAPT'08)*, in *2008 IEEE International Conference on Cluster Computing (Cluster 2008)*, Tsukuba, Japan, October 1, 2008.

23. W.A. Wulf and S.A. McKee. Hitting the memory wall: Implications of the obvious. *SIGARCH Comput. Archit. News*, 23(1):20–24, 1995.

34

Energy-Efficient Online Provisioning for HPC Workloads

Ivan Rodero
Rutgers University

Manish Parashar
Rutgers University

Andres Quiroz
Xerox Corporation

Francesc Guim
Intel Barcelona

Stephen W. Poole
Oak Ridge National Laboratory

34.1 Introduction

34.1.1 Motivation and Background

Distributed computational infrastructures as well as the applications and services that they support are increasingly becoming an integral part of society and affecting every aspect of life. As a result, ensuring their efficient and robust operation is critical. However, the scale and overall complexity of these systems is growing at an alarming rate (current data centers contain tens to hundreds of thousands of computing and storage devices running complex applications), and issues related to power consumption, air conditioning, and cooling infrastructures are critical concerns impacting reliability as well as operating costs. In fact, power and cooling rates are increasing eightfold every year [1], and are becoming a dominant part of IT budgets. Addressing these costs while balancing multiple dimensions, including performance, quality of service, and reliability is thus an important and immediate task, since it can help service providers increase profitability by reducing operational costs and environmental impact, without significant reduction in the service level delivered.

Virtualized data centers and clouds provide the abstraction of nearly unlimited computing resources through the elastic use of consolidated resource pools, and provide opportunities for higher utilization and energy savings. These platforms are also being increasingly considered for traditional high-performance computing (HPC) applications that have typically targeted grids and conventional HPC platforms.

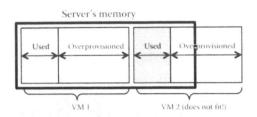

FIGURE 34.1 Overprovisioning may cause poor workload consolidation.

However, maximizing energy efficiency, cost-effectiveness, and utilization for these applications while ensuring performance and other quality of service (QoS) guarantees requires leveraging important and extremely challenging trade-offs. These include, for example, the trade-off between the need to proactively create and provision virtual machines (VMs) on data center resources (requiring timely knowledge of resource requirements) and the need to accommodate the heterogeneous and dynamic resource demands and runtimes of these applications.

Along with the operational costs associated to energy consumption, overprovisioning is another major issue for Cloud providers since it makes workload consolidation more difficult and reduces resource utilization. Figure 34.1 shows a simple example where overprovisioning in terms of memory assigned to VMs prevents allocating two VMs in the same server. Underprovisioning may also result in lower QoS delivered to users, which in turn may cause loss of revenue and users [3].

Existing research has addressed many aspects of the issues described earlier, including, for example, energy efficiency in cloud data centers [6,36], efficient and on-demand resource provisioning in response to dynamic workload changes [17], and platform heterogeneity-aware mapping of workloads [21]. Research efforts have also studied power and performance trade-offs in virtualized [22] and non-virtualized environments [14] considering techniques such as dynamic voltage scaling (DVS) [30]. Thermal implications of power management have also been investigated [11]. However, these existing approaches deal with VM provisioning and the configuration of resources as separate concerns, which can result in inefficient resource configurations, and resource underutilization, which in turn results in energy inefficiencies.

34.1.2 Objectives

The overall objective of this research is to address energy consumption and related costs in an end-to-end perspective using online analysis, workload characterization, and system power modeling. Specifically, in this research, we use online distributed data analysis techniques to effectively process large amounts of monitored data to characterize the system operational state in a timely manner, detect (and model and predict) abnormal operational states, and take appropriate autonomic actions to return the system to a normal state. In the context of this work, abnormal operational states are, for example, those with low energy efficiency and/or high operational costs. To reach a normal operational state, a system may follow different approaches or paths. Therefore, one of the most important challenges of an autonomic system is deciding which is the best of those paths. Figure 34.2 illustrates this problem with the three dimensions that we target in this work: energy efficiency, QoS delivered, and overprovisioning cost.

The decentralized online data analysis approach exploits distributed computing resources within the managed system (i.e., data center). Although centralized data analysis approaches can be more accurate in general, they are found to be unsuitable for large-scale distributed system management because of the costs of centralization in terms of data aggregation, overall responsiveness, and fault tolerance. Since the analyzed monitored system data is naturally distributed, the collective computing power of networked components can be effectively harnessed to provide value-added, system-level services, such as decentralized data analysis.

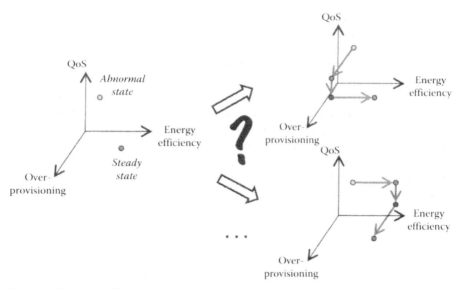

FIGURE 34.2 Different possible paths from abnormal to steady operational state.

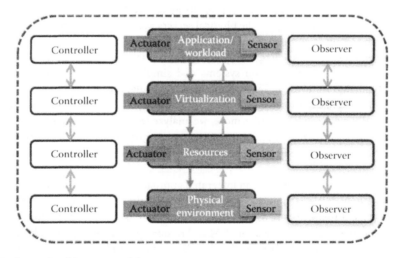

FIGURE 34.3 Layered architecture model.

Our overall architecture is shown in Figure 34.3. Since an integral approach for autonomic management is extremely difficult, we consider a layered model that allows us to reduce complexity by addressing specific problems at different layers and integrating them using cross-layer approaches. The different layers of the model, which belong to different abstract components with different responsibilities, nonetheless have the same common objectives. They work together proactively to enable efficient data center management. Although the complete architecture model includes the physical environment (e.g., temperature, air flow, humidity, etc.), in this work we focus on the three top layers.

34.1.3 Approach and Contributions

In previous work [26], we investigated decentralized online clustering (DOC) and autonomic mechanisms for VM provisioning to improve resource utilization [27]. This work was focused on reducing overprovisioning by efficiently characterizing dynamic, rather than generic, classes of resource require-ments in order to facilitate proactive VM provisioning. Because our clustering analysis was done in a decentralized manner across distributed cloud resources, our approach could deal well with virtualized cloud infrastructures with multiple geographically distributed entry points to which different users submit application jobs with heterogeneous resource requirements and runtimes.

In this work, we extend this concept by exploring workload-aware, just-right dynamic, and proactive provisioning from an energy perspective in the context of a consolidated and virtualized computing platform for HPC applications. Specifically, we use decentralized online clustering in a similar way as before to dynamically characterize the incoming job requests across the platform based on their similarity in terms of system requirements and runtimes. Clustering allows us to identify applications that require similar VM configurations, so that these configurations can be proactively provisioned for subsequent incoming jobs (exploiting the assumption that these jobs' requirements will tend to conform to the configurations of the discovered clusters). Because the configurations given by clustering match the corresponding jobs' resource requirements more closely than static predefined configurations, we can say that our approach performs just-right VM provisioning and resource configuration.

From an energy perspective, clustered jobs with similar requirements are also mapped to the same host system. The homogeneous resource utilization on the hosts means that unused subsystems and components can safely be powered down or downgraded to maximize energy efficiency. This is facilitated by the increasing capability of current servers to provide these specific hardware controls to save energy. In addition to provisioning available resources proactively according to expected resource requests, we also actively and dynamically reconfigure the physical servers according to system power models if none are available with the required configurations.

To evaluate the performance and energy efficiency of the proposed approach, we use real HPC workload traces from widely distributed production systems and simulation tools. As not all of the required information is obtainable from these traces, some data manipulation (see Section 34.4.3) was needed. We present the results from experiments conducted with three different provisioning strategies, different workload types, and system models. Furthermore, we analyze trade-offs of different VM configurations and their impact on performance and energy efficiency. Compared to typical reactive or predefined provisioning, our approach achieves significant improvements in energy efficiency with an acceptable penalty in the QoS.

The overall contribution of this work is an integrated approach that combines online workload-aware provisioning and energy-aware resource configuration for virtualized environments such as clouds. The specific contributions of this work are the following: (1) the extension and application of decentralized online clustering mechanism for energy-aware VM and resource provisioning, to bridge the gap between the two, (2) the analysis of realistic virtualization models for clouds and data centers, and (3) the analysis of performance and energy efficiency trade-offs in cases of HPC and hybrid workloads, aimed at stating (a) the difference with typical reactive or predefined provisioning approaches and (b) the importance of just-right VM provisioning and resource configuration.

The rest of this chapter is organized as follows. Sections 34.2 and 34.3 describe the decentralized online clustering technique and our energy-aware online provisioning approach, respectively. Section 34.4 describes the evaluation methodology that we have followed in our experiments. Section 34.5 presents and discusses the experimental results. Section 34.6 presents the related work, after which we conclude in Section 34.7.

34.2 Distributed Online Clustering (DOC)

Consider each job request represented as a point in a multidimensional space. Each dimension in this space, referred to as an information space, corresponds to one particular resource requirement, or attribute, and the location of a point within the space is determined by the values for each of the attributes

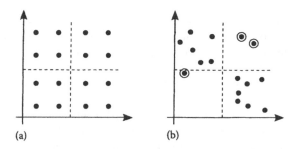

FIGURE 34.4 (a) Uniform distribution of data points in the information space. (b) Point clustering and anomalies among regions; the circled points are potential anomalies.

within the particular request. Our approach for cluster detection is based on evaluating the relative density of points within the information space. In order to evaluate point density, the information space is divided into regions that are analyzed separately. If the total number of points in the information space is known, then a baseline density of a uniform distribution of points can be calculated and used to estimate an expected number of points per region. Clusters are recognized within a region if the region has a relatively larger point count than this expected value. Conversely, if the point count is smaller than expected, then these points may potentially be outliers or isolated points. These concepts are illustrated in Figure 34.4.

The approach described earlier lends itself to a decentralized implementation because each region is assigned to a particular processing node. Also, it can be done online because points are routed to processing nodes as they are produced, and processing nodes analyze the points within their region independently, performing a very lightweight computation (basically comparing point counts), and only communicating with nodes responsible for adjoining regions to deal with boundary conditions and for aggregating cluster data. We assume that the information space (or spaces) is (are) predefined and globally known; however, the assignment of regions to nodes need not be static, and can thus be adapted to the arrival and departure of processing nodes in the network, as long as a mechanism exists to map data points to the node responsible for the region in which the point is located.

The exchange of job requests is supported by a robust content-based messaging substrate [26,33], which also enables the partitioning of the information space into regions by implementing a dynamic mapping of points to processing nodes. The messaging substrate is responsible for getting the information used by the clustering analysis to the distributed processing nodes in a scalable fashion. The substrate essentially consists of a content-based distributed hashtable (DHT) that uses a Peano–Hilbert space-filling curve [32] as a locality-preserving hash function. Content-based DHTs provide an interface that allows network nodes to be addressed by attribute-value pairs, so that, given a point with a value for each dimension (attribute), the DHT can find a route to the node responsible for the region containing that point. Additionally, our messaging substrate is able to resolve multidimensional range queries (in which a range of values is given for each dimension), guaranteeing the discovery of every node whose region intersects with the range of the query. Figure 34.3 shows how points, clusters, and ranges are mapped by the DHT to processing nodes connected by a ring overlay. A detailed analysis of the performance of content-based routing achieved by this substrate can be found in [33].

34.3 Energy-Aware Online Provisioning

In a cloud environment, executing application requests on the underlying resources consists of two key steps: creating VM instances to host each application request, matching the specific characteristics and requirements of the request (VM provisioning); and mapping and scheduling these requests onto distributed physical resources (resource provisioning). In this work, we exploit the concept of clustering-based VM provisioning with two main arguments: we consider energy awareness to the clustering-based VM provisioning, from the perspective of the energy associated to overprovisioning and reprovisioning costs, and we implement a workload-aware resource provisioning strategy based on energy-aware configurations of physical servers, and optimizing the mapping of VMs to those servers using workload characterization.

Our approach is implemented in different steps; each step has different functions, but all have common objectives. The input of our approach is a stream of job requests in the form of a trace with job arrival times and their requirements that we divide into analysis windows. Since we focus on a realistic approach, our policy is characterized by having analysis windows of a fixed duration with variable job arrival times, rather than by having a fixed number of jobs in each window. We consider an analysis window with duration of 1 min, in order to provide a more realistic distribution for job queuing times. Moreover, in order to model the system with multiple entry points for user requests, we have used traces with a high rate of requests (see Section 34.4.3). The different steps of our strategy are shown in Figure 34.5 and summarized as follows:

- *Online clustering.* For each analysis window, it clusters the job requests of the input stream based on their requirements. It returns a set of clusters (groups of job requests with similar requirements) and a set of outliers (job requests that do not match any of the clusters found).
- *VM provisioning.* Using the groups of jobs we provision each job request with a specific VM configuration. To do this, we define a greater number of classes of VMs, each described very finely, rather than a smaller number of coarsely described classes. It also tries to reuse existing VM configurations in order to reduce the reprovisioning costs.
- *Resource provisioning.* We group VMs with similar configurations together and allocate resources for them that are as closely matched as possible to their requirements. To do this, we provision resources using workload modeling based on profiling (see Section 34.4.3), taking into account

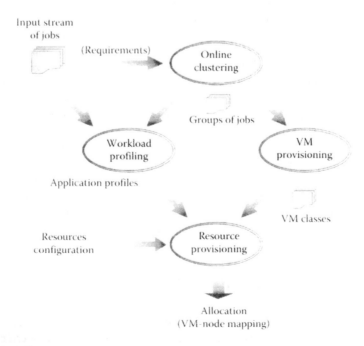

FIGURE 34.5 Overall provisioning schema.

specific hardware configurations available from the underlying servers. Specifically, we apply specific hardware configurations that can save energy (such as low-power modes) for those subsystems that are not required.

Our approach attempts to optimize energy efficiency in the following ways:

- Reducing the energy consumption of the physical resources by powering down subsystems when they are not needed.
- Reducing overprovisioning cost (waste of resources) through efficient, just-right VM provisioning.
- Reducing reprovisioning cost (overhead of VM instantiation) through efficient proactive provisioning and VM grouping.

In the following subsections we describe the previously discussed steps in detail.

34.3.1 Clustering Strategy for Workload Characterization

As described previously, our approach starts with the clustering algorithm, using the requirements from the input stream of job requests, and returning groups of jobs that have similar requirements. To group the incoming requests in different VMs we use DOC (see Section 34.2). The algorithm takes the incoming requests in a given analysis window and creates clusters from groups of requests with similar requirements. When a request does not belong to any of the clusters found, it is said to be an outlier. The clustering algorithm can be done with as many dimensions as wanted, but as the number of dimensions grows, so does the complexity of the space, and results are harder to analyze. The dimensions we take into account in this analysis include the requested execution time (T), requested number of processors (N), requested amount of memory and storage, and network demand. To reduce the search space, we perform the analysis in two steps. The first step runs the clustering algorithm with only two dimensions: the required memory versus a derived value of execution time and number of processors that represents CPU demand (C in Equation 34.1). The value of 100 in Equation 34.1 is a normalization factor that represents the duration in seconds of a reference job.

$$C = \frac{N \times T}{100} \qquad (34.1)$$

Figure 34.6 shows two plots of the clustering results obtained from two different analysis windows. Each rectangle represents a cluster and contains a set of points inside representing job requests that may be grouped together, with stars representing outliers. Both plots have six clusters but there are

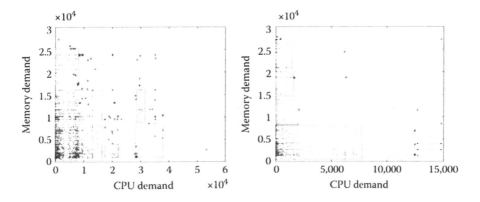

FIGURE 34.6 Clusters of two different analysis windows.

important differences between them. Consequently, the VM classes associated to the different clusters will be different in the two different analysis windows. One of the challenges will be matching clusters in different windows, as is discussed in the following section.

In the second step, the clustering is run on requested storage and network demand over the job requests of each cluster obtained in the first step. With the resulting clusters, we are able to allocate jobs to specific VM types. Using the workload characterization model described in Section 34.4.3, we can also allocate job requests matching similar resource requirements together on the appropriate physical servers. Also, using these clusters and the workload characterization, we are able to proactively configure the subsystems of the physical resources appropriately.

34.3.2 VM Provisioning

In previous work [27], autonomic mechanisms for VM provisioning to improve resource utilization were presented. This approach focused on reducing the overprovisioning that occurs because of the difference between the virtual resources allocated to VM instances and those contained in individual job requests. In particular, DOC was used to efficiently characterize dynamic, rather than generic (such as Amazon's VM types*), classes of resource requirements that can be used for proactive VM provisioning. To address the inaccuracies in client resource requests that lead to overprovisioning, the use of workload modeling techniques and their application to the highly varied workloads of cloud environments were explored. In the VM provisioning mechanism that we proposed, as with most predictive approaches, the flow of arriving jobs was divided into time periods that we call analysis windows. During each window, an instance of the clustering algorithm was run with the jobs that arrived during that window, producing a number of clusters or VM classes. At the same time, each job was assigned to an available VM as it arrived if one was provisioned with sufficient resources to meet its requirements. The provisioning was done based on the most recent analysis results from the previous analysis window. For the first window, the best option was to reactively create VMs for incoming jobs. However, the job descriptions were sent to the processing node network and by the end of the analysis window each node could quickly determine if a cluster existed in its particular region. If so, the node could locally trigger the creation of new VMs for the jobs in the next analysis window with similar resource requirements. According to [23], the time required to create batches of VM in a cloud infrastructure does not differ significantly from the time for creating a single VM instance. Thus, the VMs for each class could be provisioned within the given time window. In order to match jobs to provisioned VMs, the cluster description could be distributed in the node network using the range given by the space occupied by the cluster in the information space. Thus, when a new job arrived, it would be routed to a node that holds descriptors for VMs that had close resource requirements. In this work, we extend this idea with the main objective to reduce overprovisioning cost (due to underutilization) and reprovisioning cost (the delay of configuring and loading a new VM instance). More details regarding the metrics are shown in Section 34.4.4.

VM provisioning is done based on the most recent clustering results from the previous analysis window, and thus it is possible to overlap the clustering computation with the creation of VM batches. Specifically, VM provisioning is performed as follows: given a set of analysis windows $(w_1, \ldots, w_n)$, the job requests belonging to each analysis window $(req_{w_1}, \ldots, req_{w_n})$ sets of clusters $\{(c_{w_1}^1, \ldots, c_{w_1}^m), \ldots, (c_{w_n}^1, \ldots, c_{w_n}^p)\}$ and outliers $(o_{w_1}, \ldots, o_{w_n})$ corresponding to each analysis window (obtained using DOC), assigning a specific VM class to each job request. A specific VM class can be assigned to each job request by identifying the configuration characteristics (e.g., amount of memory) of the cluster to which the job belongs. If the job was found to be an outlier, a new VM class can be

---

* Amazon Elastic Compute Cloud, http://aws.amazon.com/ec2.

reactively provisioned for it. Outliers therefore incur extra overhead, but by definition are expected to be the exception rather than the norm. We differentiate two different steps:

1. For the first analysis window, the algorithm reactively creates VMs classes for all incoming job requests. In case of an outlier, the algorithm provisions it with a new VM class configured with the requirements of the job request. In case of a job request that matches a cluster, the algorithm provisions it with a VM class configured with the requirements of that cluster. It means that the provisioned VM class is configured with the highest requirements of the cluster in order to be able to accommodate all requests of that cluster in that VM. Although the above procedure does result in some overprovisioning, it reduces the overhead for reprovisioning with new VM classes. Once a job request has been provisioned with a VM class, that VM class becomes available in the system.

2. For subsequent analysis windows, if the job request matches a cluster, the algorithm provisions it with an instance of an existing VM class configured for that cluster. Otherwise, the algorithm tries to correlate the job request with other existing VM classes to find the closest match. This reduces the reprovisioning cost because the job request can usually be provisioned with an existing VM instance, therefore saving the VM instantiation delay.

To match a given job request with the previously defined VM classes, we use the corners of the clusters' bounding boxes (i.e., the area or space occupied by the clusters) in the two-dimensional space described in the previous section. If the requirements of a job are completely covered by the top right corner of an existing cluster, then it can be provisioned with the corresponding VM class, because the resource configuration is enough to meet the job's requirements. However, as discussed previously, some level of overprovisioning may occur. If the job request matches one or more clusters of the previous analysis window, the algorithm selects the VM class of lowest resource configuration. If the job request does not match any cluster from the previous analysis window but matches a cluster in the current analysis window the algorithm provisions it with a VM class configured with the requirements of that cluster. If the job request is an outlier in the current analysis window, it provisions the job request with a VM class meeting the job request requirements.

Figure 34.7 illustrates a simple scenario with three cluster from two different analysis windows (win_{i-1}, which is the previous analysis window and win_i, which is the current one). Requests of $cluster_1$ of win_i can be mapped to the VM class that satisfied the requirements of cluster of win_{i-1}. However, since some job requests within $cluster_2$ of win_i are not within the area of the cluster of win_{i-1} (shadowed area in Figure 34.7), not all job requests of $cluster_2$ of win_i can be satisfied by the VM configuration of the cluster of win_{i-1}.

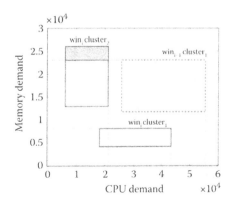

FIGURE 34.7 Example of matching clusters of two different analysis windows.

34.3.3 Resource Provisioning and Configuration

Once a job request has been provisioned with a specific VM class, it is provisioned with a specific VM instance. Since we proactively create VM batches, we try to reuse existing VM instances from the previous analysis window rather than creating new ones. Furthermore, we allocate physical servers to the VMs that are as closely matched as possible to their requirements. In contrast to other typical approaches that allocate job requests with nonconflicting, i.e., dissimilar, resource requirements together on the same physical server, our policy is to allocate job requests with similar resource requirements together on the same physical server. This allows us to downgrade the subsystems of the server that are not required to run the requested jobs in order to save energy. To do this, we consider specific configurations of the physical servers' subsystems to reduce their energy demand. Specifically it follows an energy model that leverages previous research on energy-efficient hardware configurations (e.g., low-power modes) in four different dimensions:

1. *CPU speed using dynamic voltage scaling (DVS).* We are able to reduce the energy consumed by those applications that are, for example, memory-bound [13].
2. *Memory usage.* For those applications that do not require high memory bandwidth we consider the possibility of slightly reducing the memory frequency or possibly shutting down some banks of memory in order to save power [5].
3. *High-performance storage.* It may be possible to power down unneeded disks (e.g., using flash memory devices that require less power) or by spinning-down disks [7].
4. *High-performance network interfaces.* It may be possible to power down some network subsystems (e.g., Myrinet interfaces) or using idle/sleep modes.

We have implemented two different resource provisioning strategies: a static approach where physical servers maintain their initial subsystem configuration, and a dynamic one that allows the physical servers to be reconfigured dynamically.

In the static strategy the input parameters are the VM class associated to the job request and the resource requirements of a job request, the available physical servers, and the existing VMs in each server. The resource requirements of the job request are the CPU, memory, storage and network demand, respectively. To select the most appropriate VM instance to run the requested job, it first discards the physical servers that do not match the resource requirements of the job request. If a physical server that matches the job request requirements is not available, the job request cannot be provisioned with a VM instance; thus, it does not return any VM instance at that time. In this case, if we follow a first-come-first-serve (FCFS) scheduling policy with the static approach, a request may remain queued (thus blocking all following queued jobs) until a server with the required configuration becomes available. If there is available a physical server meeting the job request requirements, the static strategy selects a server that can host VM instances with sufficient configured resources to run the job request. To do this, it discards the servers that do not match the requirements of the job request's VM class in terms of required CPU, memory, disk, and network. If a server that matches the requirements of the job request's VM class is not available, we create a new VM instance reactively on the physical server with lowest power requirements (i.e., with the most subsystems disabled or in low-power mode) and hosting the fewest job requests, and we provision the job request with the new VM instance. If there are various servers hosting available VMs that match the job request's VM class, we first employ the VMs allocated in servers hosting the fewest job requests, lowest power requirements, and with the lowest resource configuration (e.g., memory configured).

In the dynamic strategy, when required physical resources are unavailable, we reconfigure an available physical server to provide the appropriate characteristics and then provision it. Specifically, we can reconfigure servers if they are idle, but if there are no idle servers available, we can reconfigure only those servers that are configured to use fewer subsystems than those that are requested (if a server is configured to deliver high memory bandwidth, we cannot reconfigure it to reduce its memory frequency, since that would negatively impact jobs already running on it. However, if a server is configured with reduced memory frequency, we can reconfigure it to deliver full memory bandwidth without negatively impacting running jobs). Moreover, since we still do not consider VM migration, we try to fill servers with requests of similar types. Not only does this efficiently load servers, it allows more servers to remain fully idle, which allows them to be configured to host new job requests.

34.4 Evaluation Methodology

To evaluate the performance and energy efficiency of the proposed approach, we use real HPC workload traces from widely distributed production systems. Since not all of the required information is obtainable from these traces, some data manipulation was needed, such as the trace preprocessing described in Section 34.4.3. In order to separate concerns and because the batch simulation of clustering over all analysis windows is a time-consuming task, we perform the experiments in two steps. First, we compute the clusters for each analysis window (driven by the job arrival times) using the distributed clustering procedure over the input trace. Afterward, the original trace is provided as input to the simulator, along with the configuration files and an additional trace file that contains requirements and the output of the clustering for each job request (analysis window, cluster number, corners of the cluster, etc.). We obtain the results from the simulator output and process the output traces post-mortem.

34.4.1 Simulation Framework

For our experiments, we have used the kento-perf simulator (formerly called Alvio [10]), which is a C++ event driven simulator that was designed to study scheduling policies in different scenarios, from HPC clusters to multi-grid systems. In order to simulate virtualized cloud data centers we have extended the framework with a virtualization (VM) model. Due space limitations, in this work we only describe the main characteristics of this model. Our VM model considers that different VMs can run simultaneously on a node if it has enough resources to create the VM instances. Specifically, a node must have an available CPU for each VM instance and enough memory and disk for all VMs running on it. This means that the model does not allow sharing a CPU between different VMs at the same time. Although this limitation in the model may penalize our results, some simplification was needed to reduce the model's complexity. Moreover, as we have not modeled VMs with multiple CPUs yet, we configure N VMs with a single CPU each in order to provision a VM for an application request that requires N CPUs. However, we have considered requests with high demand for memory in order to experience memory conflicts between different VMs within the physical nodes. Therefore, the model simulates the possible conflicts and restrictions that may occur between VMs with multiple CPUs. We have modeled the overhead of creating VM instances with a fixed time interval of 1 min, which is the duration of the analysis windows. It allows us to overlap the clustering computation with the creation of VM instances. We have also modeled four predefined classes of VMs for the reference reactive approach (using 0.5, 1, 2, and 4 GB of memory). We have simulated a homogeneous cluster system based on servers with four CPUs and 8 GB of RAM each.

34.4.2 System Power Model

Our server power model is based on empirical data, the technical specifications from hardware vendors, and existing research. The actual power measurements were performed using a "Watts Up? .NET" power meter. The meter was attached between the wall power and an Intel Xeon-based server with a quad-core processor, 8 GB of RAM, two disks, and both Gigabit Ethernet and Myrinet network interfaces. Using our measurements and existing research [18,24] (e.g., to obtain the power required by a subsystem scaling from the total server power), we configured the simulations with the power required for the different subsystems and the switch latencies shown in Table 34.1. The model has some simplifications, such as

TABLE 34.1   Power Requirements (in Watts) and Delay
Associated to Different Server Subsystems

Subsystem	Running	Low	Idle	Latency
CPU	155 W	105 W	85 W	0.01 s
Memory	70 W	30 W	—	0.06 s
Disk	50 W	—	10 W	2 s
NIC	15 W	—	5 W	0.15 s
Others	110 W	—	—	—
Total	400 W	—	—	—

using a coarse grain level for switch latencies (we use longer latencies) and using a fixed switch latency between different power modes of a subsystem. Specifically, for the CPU we consider three different states: running mode, i.e., C0 C-state and highest P-state ("Running" in Table 34.1), low-power mode, i.e., C0 C-state and the deepest P-state ("Low" in Table 34.1), and idle state, i.e., (C-state different to C0) ("Idle" in Table 34.1). For the memory subsystem, we consider two states (regular and low-power mode). We estimate the memory energy consumption from the power models based on RDRAM memory throttling discussed in [12,16]. Although the techniques described in [12,16] are not available in our physical testbed, we consider low-power modes for memory to be supported by modern systems. For the storage subsystem, we used actual power measurements and the disk characteristics such as in [25], but assuming newer technology. For the network subsystem we consider two different states: regular and idle. The idle state corresponds to the hardware sleep state in which the network interface is in listening mode.

Since we assume that modern systems use power management techniques within the operating system, we consider low-power modes in our simulations when the resources are idle. As well as taking into account the power required by the previous subsystems, we also include in our model the power required by other components such as motherboard, fans, etc. Therefore, some fixed amount of power is always required, independently of the specific physical server configuration used. However, we do not consider the power required for cooling and to manage external elements. Although this model is not completely accurate with respect to applications' execution behaviors, it gives us a base framework to evaluate the possibilities of our approach.

34.4.3 Workloads and Characterization Model

In order to model the heterogeneous nature of virtualized cloud infrastructures with multiple geographically distributed entry points, we required traces with different application requirements and high arrival rates. In the present work, we have used traces from the Grid Observatory,* which collects, publishes, and analyzes data on the behavior of the EGEE Grid.* This trace meets our needs because it currently produces one of the most complex public grid traces and meets our need. The frequency of request arrivals is much higher in contrast to other large grids such as Grid5000.

Since the traces are in different formats and include data that is not used, they are preprocessed before being input to the simulation framework. First, we convert the input traces to Standard Workload Format (SWF).* We also combine the multiple files of which they are composed into a single files. Then, we clean the trace in SWF format in order to eliminate failed jobs, cancelled jobs, and anomalies. Finally, we generate the additional requirements that are not included in the traces. For instance, traces from Grid Observatory do not include required memory, and the mapping between gLite (i.e., the EGEE Grid middleware) and the Local Resource Management Systems (LRMS) that contains such information is not available. Thus, we included memory requirements following the distribution of a trace from Los Alamos National Lab

---

* Grid Observatory, http://www.grid-observatory.org/
* Enabling Grid for E-sciencE, http://www.eu-egee.org/
* Parallel Workload Archive, http://www.cs.huji.ac.il/labs/parallel/workload/

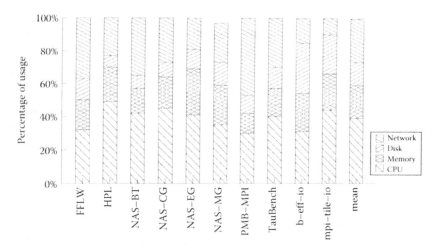

FIGURE 34.8 Percentage of CPU, memory, disk, and network usage per benchmark.

(LANL) CM-5 log, and we included the disk and network demand based on a workload characterization (see following section) using a randomized distribution. Although the memory requirements are obtained from a dated trace, we scaled them to the characteristics of our system model. We estimate that similar results can be obtained using a resource requirement distribution from other traces or recent models.

In order to perform an efficient mapping of the incoming job requests to physical servers, we need their requirements in terms of subsystems utilization. Specifically, we consider CPU utilization, requested memory and storage, and communication network usage. As the traces found from different systems do not provide all the information needed for our analysis, we needed to complete them using a model based on benchmarking a set of applications. We profiled standard HPC benchmarks with respect to behaviors and subsystem usage on individual computing nodes rather than using a synthetic model. To do this we used different mechanisms such as commands from the OS (e.g., iostat or netstat) and PowerTOP from Intel. A comprehensive set of HPC benchmark workloads has been chosen. Each stresses a variety of subsystems—compute power, memory, disk (storage), and communication (network). The relative proportion of stress on each subsystem of each benchmark is shown in Figure 34.8. After calculating the average percentage of CPU, memory, storage, and network usage for each benchmark, we randomly assign one of these 10 benchmark profiles to each request in the input trace, following a uniform distribution.

34.4.4 Metrics

We evaluate the impact of our approach on the following metrics: makespan (workload execution time, which is the difference between the earliest time of submission of any of the workload tasks, and the latest time of completion of any of its tasks), energy efficiency (based on both static and dynamic energy consumption, and energy delay product), resource utilization (based on CPU utilization), average request waiting time (as a QoS metric), and overprovisioning cost (to show the accuracy with which the different data analysis approaches can statically approximate the resource requirements of a job set). We define overprovisioning cost for resource R (e.g., memory) in each analysis window i as the average difference between requested (Q) and provisioned (P) resources for the set of jobs j in that window (see Equation 34.2).

$$O_i^R = \frac{1}{N} \sum_{j=1}^{N} \left(P_{ij}^R - Q_{ij}^R \right) \tag{34.2}$$

34.5 Results

We have conducted our simulations using different variants of the provisioning strategies, workloads, and system models described in the previous sections. Specifically, we have evaluated the following provisioning strategies:

- **REACTIVE**: Implements the reactive or predefined provisioning implemented by typical cloud providers, such as Amazon EC2. Specifically, it provisions with four classes of VMs as described in Section 34.4.1. It follows a FCFS scheduling policy to serve requests (in the same order of arrival), and implements a First-Fit resource selection policy to provision physical servers. This means that it maps the provisioned VMs to the first available physical servers that match the requests' requirements.
- **STATIC**: Implements our proposed provisioning approach based on clustering techniques with a static approach to configure the resources. It means that the servers configurations remain constant; therefore a physical server can host only applications that match its specific characteristics. Specifically, we consider eight classes of configurations (one for each possible combination of their subsystems configuration) and we model the same amount of servers with each one.
- **DYNAMIC**: Implements our proposed provisioning approach similarly to the STATIC strategy but implementing dynamic resource reconfigurations when they are necessary. This means that when there are not available resources configured with the requested configuration, it reconfigures servers reactively in order to service new requests.

We have generated three different variants of the workload described in Section 34.4.3 in order to cover the peculiarities of three different scenarios:

- **HPC**: Follows the original distribution of the EGEE workload described in the previous section, which is mostly composed of HPC applications with long execution times and high demand for CPUs.
- **SERVICE**: Follows an application pattern of shorter execution times and fewer resource requirements (i.e., service applications characteristics), but respecting the original trace's arrival rate and distribution.
- **MIXED**: Includes both HPC and SERVICE profiles in order to address the heterogeneous nature of clouds. We will consider this workload variant as the reference one to obtain conclusions because it covers a wider range of application request profiles.

We have conducted experiments with the different workloads on a simulated setup of more than 2000 servers. The experiments attempt to state that our proposed policies can save energy consumption while maintaining a similar execution and waiting times (QoS given to users). The experiments also attempt to state that the proposed strategies do not fall in underutilization of the resources while they reduce the overprovisioning cost.

Figures 34.9 through 34.11 show the relative makespan, energy consumption, Energy Delay Product (EDP), utilization, and average waiting time results, respectively. For readability, the results are normalized to the results obtained with the REACTIVE approach for each workload variant. Although in some HPC systems waiting times may be longer, we consider average waiting time rather than average response time because in our results the execution time is the leading portion of the response time, which is probably the most appropriate metric to measure the QoS given to users.

Figures 34.9a and 38.6 show that with the REACTIVE strategy, the delays are shorter than those obtained with our approach, and with the STATIC strategy, they are much longer than with DYNAMIC. This is explained by the fact that using the STATIC provisioning strategy with available resources, matching the request requirements is harder because the number of resources with the same subsystem configuration is fixed. Since the scheduling strategy follows a FCFS policy to serve the incoming requests, the average waiting times are much longer with the STATIC approach and therefore the makespan

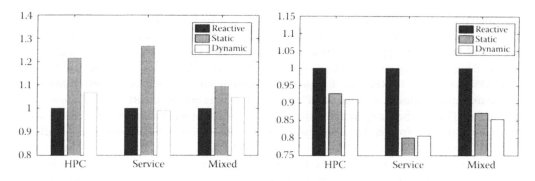

FIGURE 34.9 (a) Relative makespan. (b) Relative energy consumption.

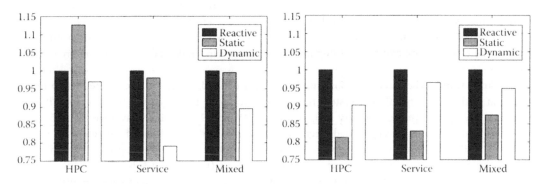

FIGURE 34.10 (a) Relative energy delay product. (b) Relative utilization.

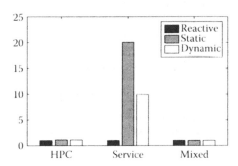

FIGURE 34.11 Relative average waiting time.

also becomes longer (more than 5% on average). However, the same strategies present different results depending on the workload type. With SERVICE workload the trends of the results are different than those observed with the other workloads. In particular, makespan is shorter with DYNAMIC than with REACTIVE using MEDIUM and LARGE system models while the average waiting time is much shorter with the REACTIVE approach. This is explained by the fact that although runtimes are much shorter in relation to other workloads, the overhead of creating new VM instances is much larger. Moreover, with the STATIC approach the results are worse because resource limitations cause job blocking. However, the delays shown in Figure 34.11 are also very long with STATIC and DYNAMIC strategies because we have not considered the overheads of creating new VM instances as part of the waiting time and these overheads are critical in this scenario. We will consider SERVICE workload as a specific use case and MIXED workload as the reference one for our comparisons.

Although the makespan is shorter with the REACTIVE approach, the energy efficiency obtained with our proposed strategy is better than that obtained with the REACTIVE approach. Specifically, both STATIC and DYNAMIC approaches obtain between 8% and 25% energy savings with respect to the REACTIVE approach (depending on the scenario). With the DYNAMIC approach, the EDP is between 2% and 25% lower than the REACTIVE approach (depending on the scenario). However, although the energy consumption is lower with the STATIC approach with respect to the REACTIVE approach, the EDP is higher. The specific configuration of the resource subsystems is the main source of energy savings. Our proposed strategy provides the largest energy savings with respect to the REACTIVE approach when the SERVICE workload is used. In this scenario, the overhead of creating VMs (which also consumes energy) has a strong impact on energy consumption. Furthermore, the energy efficiency is slightly worse with the DYNAMIC approach with larger system models due to higher overprovisioning costs. However, in smaller systems the energy efficiency is better with the DYNAMIC approach because the makespan is longer with STATIC and therefore, with STATIC, the servers consume less energy, but for a longer time. Although a larger system should consume more energy, the workload may be executed in a shorter time, and thus the predominant factor of energy efficiency is not the system model. In contrast, the type of workload has a strong impact on the energy efficiency as well as the provisioning strategy.

In contrast to the other metrics whose results are very dependent on the size of the system model, utilization is similar with different system models and workload types. This is because quantitatively the utilization of the resources is very high (between 59% and 93% with the average over 85%) due to the fact that systems are very loaded (high job arrival rate and high demand for resources such as memory). In Figure 34.10b we can appreciate that by using the REACTIVE approach the utilization is higher because configurations without resource restrictions (in terms of subsystems configuration) facilitates mapping VMs to physical servers. In contrast, with the STATIC approach the utilization is lower because, during the periods of time that specific subsystem configurations are not available, some resources may remain idle. We estimate that the utilization could be lower if we would consider VM configurations with different numbers of CPUs. However, in the presented results, memory conflicts in the allocation of VMs limit resource utilization. Although we have presented results that focus on the resource utilization ratio, memory usage is more efficient with our provisioning approach in contrast to REACTIVE provisioning. Figure 34.12 shows the average overprovisioning cost of the analysis window for each provisioning strategy (Bezier curve).

The memory overprovisioning cost difference between REACTIVE and our approach is quite large (4.5–6.3 times lower with our approach in contract to the REACTIVE one). Specifically, the over-provisioning cost is 390 MB on average for the REACTIVE approach, and 61 and 86 MB on average for

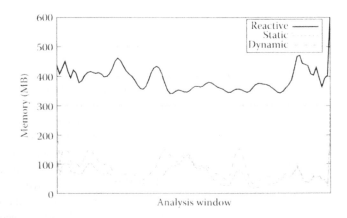

FIGURE 34.12   Average difference between requested memory for jobs and the provisioned one.

our STATIC and DYNAMIC approach, respectively. Furthermore, the DYNAMIC approach presents a higher level of overprovisioning, and the points in Figure 34.12 are less concentrated with respect to STATIC. This is due to the fact that some overprovisioning situations may occur when dynamically reconfiguring resources that are already hosting requests with less resource requirements. To improve this particular inefficiency, we suggest migrating VMs when a resource is required to be reconfigured resulting in nodes hosting requests that best fit their configurations.

We have also conducted another experiment to analyze trade-offs between different VM configurations and the impact on the previously analyzed metrics. Figure 34.13 presents the results obtained using the REACTIVE approach with MIXED workload. The results are given using different numbers of VM classes, but all of them are in the same range of memory configuration (up to 4 GB, but as the number of classes increases, the grain of VM configurations is finer). They are normalized to the results obtained with fewer numbers of VM classes. The results of each metric shown in Figure 34.13 follow the same pattern—better results are obtained when the number of VM classes increases. The metrics affected most are makespan and average waiting time (from 2% to 8% approximately). This improvement stems from increased possibilities for fitting finer-grained VMs to servers with limited resources. For example, if a server has only 8 GB of memory, and the only available VM classes deliver 2 and 4 GB, then the server could only serve four 2 GB VMs at maximum, or two 4 GB VMs. But if the granularity of VM memory size is made finer, say, to 512 MB, 1 GB, 1.5 GB up to 4 GB, then the combinations of VMs that a server can run increase. Using a larger number of VM classes also results in better utilization and energy efficiency but the improvements are not very significant (around 2% better in the best case).

Figure 34.14 shows the results obtained from the evaluation of the same setup of Figure 34.12 but considering different sizes of VMs rather than different numbers of VM classes. Specifically, it considers three different configurations: SMALL_VMs (using up to 4 GB of memory), MEDIUM_VMs (using up to 6 GB of memory), and LARGE_VMs (using up to 8 GB of memory). It allows us to evaluate the impact of overprovisioning on performance. Since the nodes are modeled with 8 GB of memory and the memory demand is very high, LARGE_VM configurations cause difficulties for efficiently allocating VMs. Specifically, larger VMs cannot be allocated to the same amount of resources achieving the same performance due to higher resource conflicts (i.e., memory conflicts). Therefore, in addition to increasing both makespan and average waiting time, energy efficiency suffers. However, utilization remains higher with smaller VMs because the system is more loaded and, as a result, the resources are more fully used. This is consistent with the results shown in Figure 34.9 and discussed previously. Thus, the trend is to perform worse when the size of the VM configurations increases.

We can conclude that using a larger set of VM classes with finer grained configurations tends to provide better results, which supports our argument that using just-right VM provisioning can improve QoS, resource utilization, and energy efficiency.

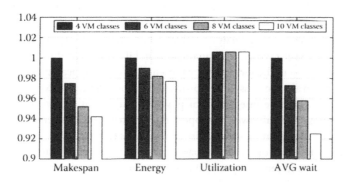

FIGURE 34.13 Relative results using different number of VM classes.

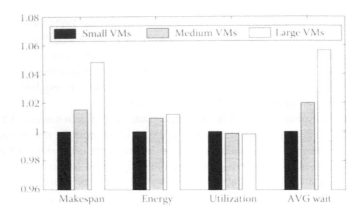

FIGURE 34.14   Relative results using different VM configuration sizes.

34.6 Related Work

Different projects have identified the need for energy efficiency in cloud data centers. Bertl et al. [6] state the need for energy efficiency for information technologies and reviews the usage of methods and technology currently used for energy efficiency. They show some of the current best practice and relevant literature in this area, and identify some of the remaining key research challenges in cloud computing environments. Abdelsalam et al. [2] propose an energy-efficient technique to improve the management of cloud computing environments. The management problem is formulated as an optimization problem that aims at minimizing the total energy consumption of the cloud, taking SLAs into account. Srikantaiah et al. [36] study the interrelationships between energy consumption, resource utilization, and performance of consolidated workloads.

Recent research has focused extensively on efficient and on-demand resource provisioning in response to dynamic workload changes. These techniques monitor workloads experienced by a set of servers on virtual machines and adjust the instantaneous resources availed by the individual servers or VMs. Menasce and Bennani [17] propose an autonomic controller and showed how it can be used to dynamically allocate CPUs in virtualized environments with varying workload levels by optimizing a global utility function using beam design. Nathuji et al. [21] consider the heterogeneity of the underlying platforms to efficiently map the workloads to the best fitting platforms. In particular, they consider different combinations of processor architecture and memory subsystem.

Several research efforts propose methods to jointly manage power and performance. One of the most used techniques to save energy in the last decades is DVS. Researchers have developed different DVS scheduling algorithms and mechanisms to save energy to provision resources under deadline restrictions. Chen et al. [8] address resource provisioning proposing power management strategies with SLA constraints based on steady-state queuing analysis and feedback control theory. They use server turn off/on and DVS for enhancing power savings. Ranganathan et al. [30] highlight the current issue of underutilization and overprovisioning of the servers. They present a solution of peak power budget management across a server ensemble to avoid excessive overprovisioning considering DVS and memory/disk scaling. Nathuji et al. [22] investigate the integration of power management and virtualization technologies. In particular they propose VirtualPower to support the isolated and independent operation of virtual machine and control the coordination among virtual machines to reduce the power consumption. Rusu et al. [31] propose a cluster-wide on/off policy based on dynamic reconfiguration and DVS. They focus on power, execution time, and server capacity characterization to provide energy management. Kephart et al. [9,14] address the coordination of multiple autonomic managers for power/performance trade-offs by using a utility function approach in a nonvirtualized environment.

A large body of work in data center energy management addresses the problem of the request distribution at the VM management level in such a way that the performance goals are met and the energy consumption is minimized. Song et al. [34] propose an adaptive and dynamic scheme for adjusting resources (specifically, CPU and memory) between virtual machines on a single server to share the physical resources efficiently. Kumar et al. [15] present vManage, a practical coordination approach that loosely couples platform and virtualization management toward improving energy savings and QoS, and reducing VM migrations. Soror et al. [35] address the problem of optimizing the performance of database management systems by controlling the configurations of the virtual machines in which they run. Laszewski et al. [38] present a scheduling algorithm for VMs in a cluster to reduce power consumption using DVS.

Several approaches have also been proposed for energy efficiency in data centers, including other additional factors such as cooling and thermal considerations. Moore et al. [20] propose a method to infer a model of thermal behavior to automatically reconfigure the thermal load management systems, thereby improving cooling efficiency and power consumption. They also propose in [19] thermal management solutions focusing on scheduling workloads considering temperature-aware workload placement. Bash and Formen [4] propose a policy to place the workload in areas of a data center that are easier to cool resulting in cooling power savings. Tang et al. [37] formulate and solve a mathematical problem that maximizes the cooling efficiency of a data center. Bianchini et al. [11] propose emulation tools for investigating the thermal implications of power management. In [29], they present C-Oracle, a software prediction infrastructure that makes online predictions for data center thermal management based on load redistribution and DVS. Raghavendra et al. [28] propose a framework which coordinates and unifies five individual power management solutions (consisting of HW/SW mechanisms).

The majority of the existing approaches address resource provisioning to save energy consumption. Different mechanisms are considered to save energy both at the resource level (DVS in the servers, dynamic reconfiguration of the servers, etc.) and at virtualization level (VM scheduling, migration, etc.). Although resources should be provisioned and managed efficiently, it can have important limitations if the VM provisioning is not performed appropriately (i.e., due to overprovisioning). Other work addresses VM provisioning, but they focus on VM scheduling and dynamic configurations for adjusting resource sharing and consequently improving overprovisioning and utilization. In contrast to the existing solutions, our approach combines proactive VM provisioning with dynamic classes of resources and workload characterization to provision and configure resources to bridge their gaps and optimize energy efficiency while ensuring performance and QoS guarantees.

34.7 Conclusion

In this work, we addressed the autonomic system management of virtualized cloud data centers for HPC workloads with the focus on energy consumption and cost optimization from an end-to-end management point of view with the help of workload characterization, online data analysis and models Specifically, we investigated as a use case an energy-aware online provisioning approach for consolidated and virtualized computing platforms. We explored workload-aware proactive provisioning from an energy perspective, focusing on the use of clustering techniques to bridge the gap between VM provisioning and resource provisioning that negatively impacts energy efficiency. We evaluated our approach based on simulations, using workload traces from widely distributed production systems. Compared to typical reactive or predefined provisioning, our approach achieves significant improvements in energy efficiency (around 15% on average) with an acceptable penalty in QoS (less than 5% in workload execution time), which is a good trade-off between user expectation and cost savings. It supports our argument that just-right dynamic and proactive provisioning using decentralized clustering techniques can improve energy efficiency with minimal degradation of QoS. We also observed that both using a larger number of VM classes and using a larger set of VM classes with finer grained configurations can provide better results.

Acknowledgments

The research presented in this work is supported in part by the National Science Foundation (NSF) via grant numbers IIP 0758566, CCF-0833039, DMS-0835436, CNS 0426354, IIS 0430826, and CNS 0723594, by the Department of Energy via grant number DE-FG02-06ER54857, by The Extreme Scale Systems Center at ORNL and the Department of Defense, and by an IBM Faculty Award, and was conducted as part of the NSF Center for Autonomic Computing at Rutgers University. This material was based on work supported by the NSF, while working at the Foundation. Any opinion, finding, and conclusions or recommendations expressed in this material are those of the author and do not necessarily reflect the views of the NSF. The authors would like to thank the Grid Observatory, which is part of the EGEE-III EU project INFSO-RI-222667.

References

1. Report to congress on server and data center energy efficiency. Technical report, U.S. Environmental Protection Agency, August 2007.
2. H. Abdelsalam, K. Maly, R. Mukkamala, M. Zubair, and D. Kaminsky. Towards energy efficient change management in a cloud computing environment. In *International Conference on Autonomous Infrastructure, Management and Security*, Enschede, the Netherlands, pp. 161–166, 2009.
3. M. Armbrust, A. Fox, R. Griffith, A. D. Joseph, R. Katz, A. Konwinski, G. Lee, D. Patterson, A. Rabkin, I. Stoica, and M. Zaharia. A view of cloud computing. *Communications of the ACM*, 53:50–58, April 2010.
4. C. Bash and G. Forman. Cool job allocation: Measuring the power savings of placing jobs at cooling-efficient locations in the data center. In *USENIX Annual Technical Conference*, Santa Clare, CA, pp. 363–368, 2007.
5. H. Ben Fradj, C. Belleudy, and M. Auguin. Multi-bank main memory architecture with dynamic voltage frequency scaling for system energy optimization. In *EUROMICRO Conference on Digital System Design*, Goatia, pp. 89–96, 2006.
6. A. Bertl, E. Gelenbe, M. Di Girolamo, G. Giuliani, H. De Meer, M. Q. Dang, and K. Pentikousis. Energy-efficient cloud computing. *The Computer Journal*, 53(7):1045–1051, 2010.
7. T. Bisson, S. A. Brandt, and D. D. Long. A hybrid disk-aware spin-down algorithm with i/o subsystem support. In *IEEE International Performance, Computing, and Communications Conference*, New Orleans, LA, pp. 236–245, 2007.
8. Y. Chen, A. Das, W. Qin, A. Sivasubramaniam, Q. Wang, and N. Gautam. Managing server energy and operational costs in hosting centers. In *ACM SIGMETRICS International Conference on Measurement and Modeling of Computer Systems*, Banff, Albuta, Canada, pp. 303–314, 2005.
9. R. Das, J. O. Kephart, C. Lefurgy, G. Tesauro, D. W. Levine, and H. Chan. Autonomic multi-agent management of power and performance in data centers. In *International Joint Conference on Autonomous Agents and Multiagent Systems*, Estaril, Portugal, pp. 107–114, 2008.
10. F. Guim, J. Labarta, and J. Corbalan. Modeling the impact of resource sharing in backfilling policies using the alvio simulator. In *IEEE International Symposium on Modeling, Analysis and Simulation of Computer and Telecommunication Systems*, Gete Island, Greece, pp. 145–150, 2007.
11. T. Heath, A. P. Centeno, P. George, L. Ramos, Y. Jaluria, and R. Bianchini. Mercury and freon: Temperature emulation and management for server systems. In *International Conference on Architectural Support for Programming Languages and Operating Systems*, Sam Jose, CA, pp. 106–116, 2006.
12. I. Hur and C. Lin. A comprehensive approach to dram power management. In *International Conference on High-Performance Computer Architecture*, Salt Lake City, UT, pp. 305–316, 2008.
13. C. Isci, G. Contreras, and M. Martonosi. Live, runtime phase monitoring and prediction on real systems with application to dynamic power management. In *IEEE/ACM International Symposium on Microarchitecture*, Orlando, FL, pp. 359–370, 2006.

14. J. O. Kephart, H. Chan, R. Das, D. W. Levine, G. Tesauro, F. Rawson, and C. Lefurgy. Coordinating multiple autonomic managers to achieve specified power-performance tradeoffs. In *International Conference on Autonomic Computing*, Jacksonville, FL, p. 24, 2007.

15. S. Kumar, V. Talwar, V. Kumar, P. Ranganathan, and K. Schwan. vmanage: Loosely coupled platform and virtualization management in data centers. In *International Conference on Autonomic Computing*, Barcelona, Spain, pp. 127–136, 2009.

16. X. Li, R. Gupta, S. V. Adve, and Y. Zhou. Cross-component energy management: Joint adaptation of processor and memory. *ACM Transactions on Architecture and Code Optimization*, 4(3):28–37, 2007.

17. D. A. Menasce and M. N. Bennani. Autonomic virtualized environments. In *International Conference on Autonomic and Autonomous Systems*, Silicon Valley, CA, p. 28, 2006.

18. L. Minas and B. Ellison. *Energy Efficiency for Information Technology: How to Reduce Power Consumption in Servers and Data Centers*. Intel Press, Santa Clara, CA, 2009.

19. J. Moore, J. Chase, P. Ranganathan, and R. Sharma. Making scheduling "cool": Temperature-aware workload placement in data centers. In *Annual Conference on USENIX Annual Technical Conference*, Anaheim, CA, pp. 10–15, 2005.

20. J. D. Moore, J. S. Chase, and P. Ranganathan. Weatherman: Automated, online and predictive thermal mapping and management for data centers. In *International Conference on Autonomic Computing*, Dublin, Ireland, pp. 155–164, 2006.

21. R. Nathuji, C. Isci, and E. Gorbatov. Exploiting platform heterogeneity for power efficient data centers. In *International Conference on Autonomic Computing*, Jacksonville, FL, p. 5, 2007.

22. R. Nathuji and K. Schwan. Virtualpower: Coordinated power management in virtualized enterprise systems. In *ACM SIGOPS Symposium on Operating Systems Principles*, Stevenson, WA, pp. 265–278, 2007.

23. S. Osternmann, A. Iosup, N. Yigibasi, R. Prodan, T. Fahringer, and D. Epema. An early performance analysis of cloud computing services for scientific computing. Technical report, Delft University of Technology, Delft, the Netherlands, December 2008.

24. J. Pfluenger and S. Hanson. Data center efficiency in the scalable enterprise. *Dell Power Solutions*, February 2007.

25. E. Pinheiro and R. Bianchini. Energy conservation techniques for disk array-based servers. In *International Conference on Supercomputing*, Malo, France, pp. 68–78, 2004.

26. A. Quiroz, N. Gnanasambandam, M. Parashar, and N. Sharma. Robust clustering analysis for the management of self-monitoring distributed systems. *Cluster Computing*, 12(1):73–85, 2009.

27. A. Quiroz, H. Kim, M. Parashar, N. Gnanasambandam, and N. Sharma. Towards autonomic workload provisioning for enterprise grids and clouds. In *IEEE/ACM International Conference on Grid Computing*, Banff, Alberta, Canada, pp. 50–57, 2009.

28. R. Raghavendra, P. Ranganathan, V. Talwar, Z. Wang, and X. Zhu. No "power" struggles: Coordinated multi-level power management for the data center. *SIGOPS Operating Systems Review*, 42(2):48–59, 2008.

29. L. Ramos and R. Bianchini. C-oracle: Predictive thermal management for data centers. In *International Symposium on High-Performance Computer Architecture*, Salt Lake City, UT, pp. 111–122, 2008.

30. P. Ranganathan, P. Leech, D. Irwin, and J. Chase. Ensemble-level power management for dense blade servers. *SIGARCH Computer Architecture News*, 34(2):66–77, 2006.

31. C. Rusu, A. Ferreira, C. Scordino, and A. Watson. Energy-efficient real-time heterogeneous server clusters. In *IEEE Real-Time and Embedded Technology and Applications Symposium*, Sam Jose, CA, pp. 418–428, 2006.

32. H. Sagan. *Space-Filling Curves*. Springer-Verlag, New York, 1994.

33. C. Schmidt and M. Parashar. Flexible information discovery in decentralized distributed systems. In *Proceedings of the 12th IEEE International Symposium on High Performance Distributed Computing*, Seattle, WA, p. 226-235, 2003.

34. Y. Song, Y. Sun, H. Wang, and X. Song. An adaptive resource flowing scheme amongst VMS in a VM-based utility computing. In *IEEE International Conference on Computer and Information Technology*, Fubeshime, Japan, pp. 1053–1058, 2007.

35. A. A. Soror, U. F. Minhas, A. Aboulnaga, K. Salem, P. Kokosielis, and S. Kamath. Automatic virtual machine configuration for database workloads. In *ACM SIGMOD International Conference on Management of Data*, British Columbia, Canada, pp. 953–966, 2008.

36. S. Srikantaiah, A. Kansal, and F. Zhao. Energy aware consolidation for cloud computing. In *USENIX Workshop on Power Aware Computing and Systems*, Sam Diego, CA, 2008.

37. Q. Tang, S. Kumar, S. Gupta, and G. Varsamopoulos. Energy-efficient thermal-aware task scheduling for homogeneous high-performance computing data centers: A cyber-physical approach. *IEEE Transactions on Parallel Distribution Systems*, 19(11):1458–1472, 2008.

38. G. von Laszewski, L. Wang, A. J. Younge, and X. He. Power-aware scheduling of virtual machines in DVFS-enabled clusters. In *IEEE International Conference on Cluster Computing*, Herablion, Greece, pp. 1–10, 2009.

35

Exploiting Heterogeneous Computing Systems for Energy Efficiency

Wei Chen
University of Sydney

Young Choon Lee
University of Sydney

Albert Y. Zomaya
University of Sydney

35.1 Introduction

The volume of data and computation has grown dramatically in the past few decades not only in traditional computing tasks (in science, engineering, and business), but also in personal information processing needs. This ever-increasing demand on computing capacity has pushed computer hardware industry and other computing-related service providers to deliver high-performance computing facilities. It is only recent that the energy efficiency in those high-performance computing resources has brought a lot of attention; in fact, this is only a slight change from the traditional viewpoint of resource utilization. Clearly, hardware approaches including low-power microprocessors and solid-state drives (SSDs) can substantially improve energy efficiency. However, efficient resource utilization still remains as the key factor and challenge to resolve (or at least alleviate) the more general issue of datacenter inefficiency.

It is now more or less common sense that typical datacenters (e.g., clouds) have their resource utilization at less than 30% (Barroso and Holzle 2007). What is more, idling resources consume a significant portion of power. For example, an idling server computer draws more than 50% of the peak power (Srikantaiah et al. 2008). This idle power dissipation issue may become more serious when considering other factors, such as cooling and maintenance. The 1960s' concept of virtualization has gained its momentum with the so-called energy-efficient/green computing. The prevalence of multicore processors and virtualization techniques has great influence in this regard. Server virtualization and consolidation using (live) virtual machine (VM) colocation and migration has become a very appealing energy efficiency solution. While this solution helps reduce idle power dissipation in particular by decreasing the number of active physical machines (nodes), it also brings about some other issues mostly in performance. Note that server consolidation can be more broadly seen as a resource allocation problem.

In datacenter environments, it is rather common to encounter various unforeseen circumstances including resource failures and sudden workload surges. Besides, workloads (or applications) and resources are heterogeneous in nature. Therefore, resource allocation (using virtualization) should take into account those factors of dynamism and heterogeneity in both applications and resources in order to "optimize" energy efficiency.

In this chapter, we particularly focus on the exploitation of heterogeneity of applications and resources from the viewpoint of resource allocation. Specifically, we review current practices with respect to energy models in the datacenter and its individual resources, and discuss issues in this regard with a practical example.

The rest of the chapter is organized as follows: Section 35.2 overviews the current energy efficiency status and related issues in datacenters; Section 35.3 reviews and discusses energy models derived from different resource components of the datacenter; Section 35.4 presents our preliminary study on energy-aware workflow scheduling using a recent energy-aware computer system model with hybrid computing resources; and Section 35.5 summarizes and concludes the chapter.

35.2 Power Consumption in Datacenters

To exploit opportunities for energy efficiency, we start by studying where and how the power is consumed in datacenters (clouds). Nowadays, the scale of these computer systems has grown to be very large to accommodate ever-increasing computing and storage demands in a flexible and cost-effective manner. Here, energy consumption has become a major limiting factor in the effective use and expansion of these systems. What is more, in such systems not all power is used for IT facilities. In fact, quite a little of energy is lost in power distribution, mechanical load, and cooling.

35.2.1 Power Usage Effectiveness and Datacenter Infrastructure Effectiveness

The Green Grid has defined two terms to measure the energy efficiency of datacenters: power usage effectiveness (PUE) and datacenter infrastructure effectiveness (DCiE).

$$\text{PUE} = (\text{Total Facility Power})/(\text{IT Equipment Power})$$

$$\text{DCiE} = (\text{IT Equipment Power})/(\text{Total Facility Power}) \times 100\%$$

PUE is the total facility power delivered to the property line of the datacenter over that used by critical loads, such as servers and communication equipments. It indicates how many watts have to be consumed in order to get 1 W to the actual computing equipments. DCiE is the reciprocal of PUE and shows the percentage of power that is really used for computation. Both terms represent the same information that reflects the energy overhead in the datacenter building infrastructure.

PUE is poor in most average datacenters. A report from the U.S. EPA (Environmental Protection Agency 2007) estimated that in 2006, the PUE of most enterprise datacenters was as low as 2.0 or even 3.0. In other words, every watt of power used for computation requires an extra 1 or 2 W for power distribution, cooling, and other auxiliary equipment; DCiE is less than 50% in these cases. In the report, the U.S. EPA also expected that in 2011, PUE could drop to 1.9 according to the efficiency improvement trend at the time, or 1.7 with better operational practices. With the best practice or state-of-the-art approaches, the EPA predicted that the PUE ratio could reach 1.5 or 1.4. In the meantime, some exceptionally good PUE ratios have been "advertised" from a few industry-leading facilities. The principal architect of datacenter infrastructure in Microsoft, Kushagra Vaid, mentioned in his invited talk at the 2010 Hotpower workshop (http://www.usenix.org/events/hotpower10) that the current PUE of Microsoft datacenters is between 1.5 and 1.2, and the fourth-generation modular datacenter will be in the market soon with 1.05–1.15 of PUE. Google also reported their measurement results (http://www.google.com/corporate/green/datacenters/measuring.html) on 10 large-scale Google datacenters from 2008 to 2010. The results show that PUE was mostly between 1.1 and 1.3.

35.2.2 Server PUE

PUE is widely adopted in measuring datacenter energy efficiency as a whole, but it does not include the energy (in)efficiency of individual IT devices. That is, the power delivered to each server or communication component is not completely used for actual computation. A substantial portion of power is dissipated for the power supply to servers, voltage regulator modules (VRMs), and cooling fans. An analogous term "server PUE" (SPUE) can be used to express this energy overhead:

$$SPUE = (Server\ Input\ Power)/(Computation\ Useful\ Power).$$

Here, the computation useful power is that consumed by electronic components for computation, such as CPU, memory, disk, network, motherboard, and other circuit.

SPUE indicates the energy efficiency in a single server, and like datacenter PUE, SPUE is poor in most average servers. A test by EPRI PEAC Corporation (made in 2004 to improve the efficiency of the server power supplies used in datacenters) shows that the average efficiency of a typical server power supply is 75% (less than 70% at low load that is the typical load of a datacenter at most time) (Ton and Fortenbury 2005). The same inefficiency is also common in the secondary power supplies such as VRMs in many motherboards. Climate savers computing initiative (CSCI) estimates that SPUE of servers popularly used in datacenters is between 1.6 and 1.8. More than 30% of input power may be lost in electrical conversion. However, a state-of-the-art SPUE is expected to be less than 1.2.

The product of multiplication by datacenter PUE and server PUE (PUE × SPUE) gives the true energy overhead in the datacenter facilities and equipments. This is usually referred as the true (or total) PUE (TPUE). In an ideal situation, when both PUE and SPUE are 1.2, there would be 70% of power ultimately used in computation components: CPU, dynamic random-access memory (DRAM), hard disk drive (HDD), motherboard, network, and so on. In the next subsection, we will discuss how the power is distributed in different computation components.

35.2.3 Power Usage Breakdown in a Typical Server

The power consumed in different components of a server can vary significantly depending on the system configuration and the workload feature. Barroso and Holzle (2009) reported the distribution of peak power measured on a typical server (deployed in one of Google's datacenters built in 2007). Approximately, CPUs consume 33% of the total power, followed by DRAM of 30%. Hard disk and networking devices consume relatively less power, 10% and 5%, respectively. The remaining 22% of power is consumed by other components, such as motherboard, fans, and power supply.

Tsirogiannis et al. (2010) also present their measurement on the power consumption of system components. An HP xw8600 workstation running a 64 bit Fedora 4 Linux with kernel 2.6.29 was used for the measurement. They designed a set of microbenchmarks that use typical database operations to exercise hardware components from idle to full utilization. Similarly, their measurement shows the two Intel Xeon CPUs consume the most power in the server: more than 50% under fully loaded and 30% during idle time. The 4 × 4 GB RAM and system board also consume a substantial part of total power. While CPU is again the most power-consuming component in a datacenter server, it is no longer the merely dominant focus in power efficiency improvement. There is significant portion of power consumed by other components, especially memories and system board. Therefore, it is essential to have a comprehensive view to more effectively deal with the energy efficiency of datacenter servers.

35.3 Energy Models

To facilitate the power management in large-scale high-performance computing environments, it is important to know how the energy efficiency is affected by other factors in system usage. We pay attention to two significant factors: resource utilization and server performance. Mathematical models are built in this section to study their relationship with the power consumption.

35.3.1 Resource Utilization and Power Consumption

We first study the energy model over resource utilization. The computation workload of a server is primarily concerned with CPU, memory, hard disk, and network component. As a result, these major power consumers (CPU, memory, and hard disk) and their energy models are of our particular focus in the following. We also give an overall model of power consumption over the load levels of a server considering cooling fans and other circuits.

35.3.1.1 CPU Power over Utilization

Typically, a CPU's power consumption increases linearly with its utilization. However, the idle power (starting point of this linearity) accounts for a substantial portion of the peak power.

Figure 35.1 gives a typical model of CPU power over utilization. The power consumption P_n at any specific CPU utilization ($n\%$) can be calculated as

$$P_n = (P_{peak} - P_{idle})\frac{n}{100} + P_{idle}$$

where
P_{peak} is the maximum power consumption at full utilization
P_{idle} is the power consumption when CPU is at the idle state

In many CPUs, the idle power may be up to 50% of the peak power, resulting in serious energy inefficiency particularly when the computation workload is low.

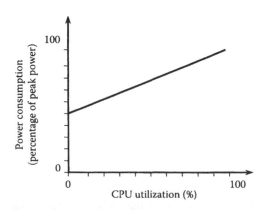

FIGURE 35.1   CPU utilization and power consumption.

To save idle power, CPU can be commanded to enter a low-power mode (C-state). Modern CPUs have several power modes, from halt to deep sleep and even deep power down. This advancement in CPU power management saves energy because the clock signal and power to idle units are cut inside CPUs. The more units are stopped, the more energy is saved, but the more time is required to wake-up a CPU to be fully operational. C1 (halt) and C2 (stop grant) states can save up to 70% of CPU's maximum power consumption, and the wake-up time is very short (in nanoseconds). C4 (deeper sleep) state saves 98% of peak power, but it requires a few hundreds of microseconds to wake-up a CPU. With the new technique C6 (deep power down) state, a CPU's internal voltage can be reduced to zero. The wake-up time from C6 state is much longer because it does not preserve the CPU context. This transition time might become a non-negligible overhead in some applications sensitive to response time.

35.3.1.2 Memory Power Consumption

Memory is another major power consumer in a datacenter server. As the data volume of applications (such as in Google and Facebook) continuously increases, more memory modules are installed in a server (more broadly, in the datacenter) to improve the performance of these memory intensive applications, and this much affects the power consumption of datacenters.

Memory used in current servers is packaged in dual in-line memory modules (DIMMs). Power consumed by DIMMs is typically measured in active and idle standby states. Minas and Ellison (2009) collected the data about DIMMs power consumption from the publicly available datasheets by some typical vendors. The data show a 4 GB capacity DIMM commonly uses 8–13 W power when active. The idle power ranges from 30% to 60% of its active power.

35.3.1.3 Power Consumption in Storage System

The storage system of a datacenter server is mainly composed of HDDs. A typical server may be installed with four disks that consume several dozen watts of power. Some large enterprise applications need external storage systems requiring significant amount of power. Again, idle power is the main source of energy inefficiency in HDDs; they constantly draw power to keep disks spinning even when there is no data activity. In many products, this idle power may exceed 50% of the peak power.

An alternative device that may be used in the storage system instead of HDDs is SSDs. Apart from its appealing performance in access time and latency, one outstanding benefit of SSDs over traditional electromechanical HDDs is the high energy efficiency. SSDs consume less power in operating and cooling, and they use almost no power when idle. Although SSDs are still much more expensive than HDDs, the gap becomes narrower especially when the cost in energy is considered. Now, SSDs have begun to be deployed in some large datacenters, such as MySpace (Stokes 2009).

35.3.1.4 Energy-Proportional Computing

Energy-proportional computing (Barroso and Holzle 2007) is an imagination that the computation workload should consume power in proportion to the utilization level. Under such an assumption, Barroso and Holzle argue that such a linear relationship (without a constant part) would keep energy efficiency on a high level across the most activity range.

Figure 35.2 shows an example of energy-proportional computing. We assume a server consumes 10% of peak power at idle time and the power usage increases linearly with the resource utilization. The energy efficiency is calculated by *utilization/power*. The figure shows energy efficiency being remained more than 50% even at a low utilization.

However, there is always a substantial constant part in most power-utilization curves of real servers (Dawson-Haggerty et al. 2009); they consume much power even at idle time. Figure 35.3 shows the case if this idle power reaches 50% of the peak power. It is clear that energy efficiency decreases quickly especially when resource utilization is low.

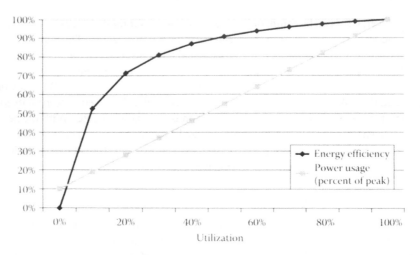

FIGURE 35.2   Energy efficiency by proportional computing.

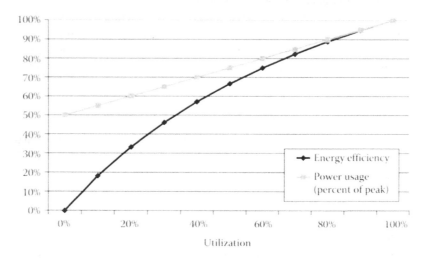

FIGURE 35.3   Low-energy efficiency in poor energy proportionality.

The main reason for this poor energy proportionality is that computation components (CPU, memory, and hard disk) and motherboard always consume quite an amount of power even at idle time. By applying a better power-saving C-state (C6, deep power down), a recent report (Vaid 2010) states that the idle power consumption can be reduced to about 30% in servers at Microsoft's datacenter.

35.3.2 Computation Performance and Power Consumption

Another important factor that affects a server's power consumption is its performance. Generally, higher performance might cost more power in computation. In this subsection, we study the energy models over a server's computation performance, especially the performance of CPUs.

35.3.2.1 Power Performance Function and Dynamic Voltage and Frequency Scaling

Complementary metal-oxide semiconductor (CMOS) is the technology for constructing integrated cir-cuits, such as CPU, memory, and other digital logic circuits. The power consumption of a CMOS-based CPU is defined to be the summation of capacitive, short-circuit, and leakage power. The capacitive power

TABLE 35.1 Dynamic Voltage Scaling in Intel Pentium M Processor 1.6 GHz

Frequency	Supply Voltage (V)
1.6 GHz	1.484
1.4 GHz	1.420
1.2 GHz	1.276
1.0 GHz	1.164
800 MHz	1.036
600 MHz	0.956

(dynamic power dissipation) is the most significant factor of power consumption. The capacitive power (P_c) is defined as

$$P_c = ACV^2 f$$

and

$$f = \frac{k(V - V_t)^2}{V}$$

where

A is the number of switches per clock cycle
C is the total capacitance load
V is the supply voltage
f is the frequency that is roughly in proportion with V (k is the constant of circuit and V_t is the threshold voltage)

Therefore, it is clear that the higher the supply voltage, the higher performance (frequency), but also the more power consumption. Approximately, the CPU frequency increases linearly with supply voltage, but the power consumption increases exponentially with supply voltage.

Recent CPUs are capable of obtaining a better trade-off between performance and power consumption using dynamic voltage and frequency scaling (DVFS). This technique enables CPUs to dynamically adjust voltage supply levels (VSLs) aiming to reduce power consumption. These power-performance states are referred as P-states of a processor. Table 35.1 shows the P-states supported by Intel Pentium M Processor 1.6 GHz.

By lowering VSLs or CPU frequency, a server can effectively reduce the power consumed by CPUs. Furthermore, P-state transition is much faster than to wake-up a CPU from deep sleep. Therefore, this technique is quite feasible for latency-sensitive applications.

35.3.2.2 Energy Consumption Model

The power-performance function gives a simple relationship between CPU power and the supply voltage or frequency. However, CPUs are not the sole component consuming energy. A substantial part of power is also used in memory, disk, and system board, which does not change with the supply voltage or frequency of the CPU. Therefore, we model the total load power of a server as two parts: the power of CPU that can be adjusted by supply voltage and a constant power consumed by other components:

$$P_{load} = P_{cpu} + P_{const} = ACV^2 f + P_{const}$$

This is the amount of power consumed when the server is under load. For the power at idle time, we assume the idle power of CPU can be ignored because of effective power-saving C-state, and the power decreasing in other components (memory and disk) is also ignorable compared to the total idle power on the system board. Thus, the idle power of the server can be estimated as

$$P_{idle} = P_{const}.$$

Based on the aforementioned functions of load power and idle power, we calculate the energy efficiency of servers under different CPU supply voltages or frequencies. We consider two resource usage situations in our model. In the first situation, it is not feasible to shut down servers to save power in idle time. This situation is common with Web service applications because the rebooting overhead can be of much risk in performance (e.g., response time). And, it may also be a case in tightly coupled clusters where memories and disks are highly shared by processor nodes. In this situation, idle power must be considered into energy efficiency.

For a given computation work W, we assume $T_{computation} = W/f$. Then, for certain time duration T under consideration, the idle time $T_{idle} = T - T_{computation}$. The energy consumed during time T is

$$E = P_{load}T_{computation} + P_{idle}T_{idle}$$
$$= (P_{cpu} + P_{const})T_{computation} + P_{const}(T - T_{computation}).$$
$$= P_{cpu}T_{computation} + P_{const}T$$
$$= ACV^2W + P_{const}T$$

It is clear that the lower the CPU supply voltage, the lower energy consumption.

In the second situation, we assume the servers can be put into a deep sleep or even shut down when idling, especially when the cost of wake-up is relatively negligible compared with the total response time. This is the case with some large-scale applications, such as scientific workflow or BoT applications. As the idle power consumption is saved by shutting down, energy consumption is

$$E = P_{load}T_{computation} = (P_{cpu} + P_{const})W/f = ACV^2W + P_{const}W/f.$$

The energy consumption in this function includes two parts: CPU power that increases with the square of supply voltage and that consumed by other components during the computation time that increases inversely with CPU frequency. When CPU voltage is scaled down, less energy is consumed by CPU, but more is wasted by other components.

To illustrate how the energy efficiency is affected by CPU supply voltage in this situation (when idle power can be saved), we calculate the energy consumption by CPU and other components to execute a given work W under different supply voltage levels. We adopt the P-states shown in Table 35.1 as an example set. We assume the constant power consumed by other components at idle time may up to 50% or 30% of the peak power. Figure 35.4 shows the energy consumption (normalized) of CPU and the whole server (plus that of other components) when the constant power is 50% and 30% of the peak power, respectively.

Although the energy efficiency of CPU can still be increased by lowering the supply voltage as expected, the story is quite different if from the whole server's perspective. As the constant power takes 50% of the peak power, the energy efficiency of the whole server is high when CPU supply voltage is high. This is because the energy saved on other components by shortening the execution time counterbalances that more consumed on CPU. In the case that the constant power is 30% of the peak power, the sweet point is in the middle of voltage scale.

35.3.2.3 Wimpy Server

Traditionally, high performance is the first design concern in large-scale distributed systems. Servers in these systems are equipped with high speed CPUs, large multilevel caches, and high capacity disks. But high-performance servers are also much more power consuming. The energy efficiency is low in these systems, especially when the resource utilization is low; a substantial amount of power is consumed by memory and motherboard at idle time.

Another issue that may account for the energy inefficiency of these servers is the imbalance steps of technical improvement among different components (CPU, DRAM, LAN, and HDD). Patterson (2004)

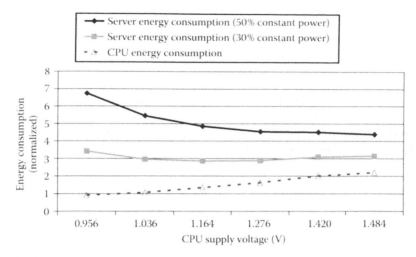

FIGURE 35.4 Energy consumption of CPU and the whole server.

points out that the bandwidth of CPU (MIPS [million instructions per second]) improved 1.5 times per year, which is faster than that of DRAM, LAN, and disk (MB/s): 1.3–1.4 times per year. This performance gap between CPU and other components often causes low CPU utilization (especially in I/O-bound data-intensive applications) and compromises performance improvement of the whole system. On the other hand, it also worsens energy proportionality of the server because high-performance components consume more power even at idle time.

An alternative server design (with low-power microprocessors or wimpy servers) was proposed in recent years. Hamilton (2009) compared three types of wimpy servers against the traditional high-performance one, when they just reach 60% CPU load (this is a typical load in many datacenters). The experimental results show that wimpy servers have much higher energy efficiency (performance per joule and performance per dollar) than that of high-performance servers. Andersen et al. (2009) proposed a new cluster architecture that couples low-power CPUs to small amounts of local flash storage. This design balances the computation and I/O capabilities, and therefore can operate more efficiently for specific I/O-bound workloads. An excellent product in this domain is SM10000 by SeaMicro that uses the Intel Atom CPU. The benchmark test shows that it delivers much higher performance in the metrics of work done per joule or per dollar.

35.4 Exploitation of Energy Efficiency from System Heterogeneity

As resource utilization plays a key role in energy efficiency in large-scale computer systems like datacenters, there are a number of advantages to tackle this energy efficiency issue in the operations management level using energy-aware scheduling and resource allocation strategies. In this section, we discuss scheduling and resource allocation approaches that exploit system heterogeneity for the energy saving purpose.

35.4.1 Heterogeneous Workloads

Internet services and large-scale computing tasks are two typical application models in distributed computing systems (e.g., clouds, grids, and clusters). Here, we first study characteristics of these applications in terms particularly of their heterogeneity in workloads and review some noteworthy software-based energy saving approaches in this regard.

Various Web-based services, such as Internet search, electronic markets, online chatting, online gaming, have become part of daily life. Such services are usually delivered by service providers who own and/or operate these services in their datacenters (e.g., clouds).

Workloads on these Internet services are primarily related to customer requests. Typically, these loads greatly fluctuate. For example, a trace of request rates (Chase et al. 2001) on the Website www.ibm.com shows it is a constant peak-and-valley model during workdays and weekends. The peak load could be three to five times more than the lowest workload during off-peak time. Similar patterns are found in many other applications such as logged users in Windows Live Messenger (Chen et al. 2008) and requests on the supercomputer in Kyoto University (Hikita et al. 2008).

To guarantee the quality of these services (such as latency and response time), the service provider must provision enough resources to meet requests on the peak time; hence, overprovisioning. These redundantly provisioned resources waste much energy particularly during off-peak time.

Large-scale computing tasks, such as science and engineering applications, are another common application model requiring a massive volume of resources (processing, storage, and networking capacities). In essence, these applications are dynamic and heterogeneous in nature and deployed in datacenters. The large-scale computing task model can be further categorized into Bag-of-Tasks (BoT) applications and workflow applications. The main distinction between them is intertask communication (or task dependence).

A BoT application consists of individual independent tasks that can run without any intertask communications or dependencies. These tasks are usually queued in a waiting list and scheduled on demand. A workflow application, on the contrary, consists of tasks with data dependencies (or precedence constraints). These tasks must run in a partial order, which is often represented by a directed acyclic graph (DAG; see Figure 35.5), although loops and conditional branching may be applied in some complicated applications.

Generally, the computational workload of each task in an application (BoT or workflow) may be quite different. This imbalance in workload is more outstanding in workflow applications because of intertask dependency. For example, in Figure 35.5, there should be more slack time for task 2 in scheduling because task 5 must wait until task 4 is finished. The heterogeneous load of an application provides opportunity for power-aware scheduling, for example, non-urgent tasks can be assigned to low-performance resources (or low-performance states of resources) that are usually more energy efficient.

35.4.2 Dynamic Server Provisioning Using Virtualization Techniques

Idle power is a major reason for energy inefficiency in clusters or datacenters that host Internet services. An effective way to minimize idle power is to consolidate workloads on a smaller number of servers and shut down others. The approach improves the resource utilization and saves the energy on servers that run under a low workload.

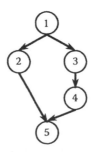

FIGURE 35.5   A simple example of DAG.

35.4.2.1 Virtualization and Server Consolidation

The traditional architecture design of Internet services is to divide functions into separate components, and each is hosted in its own security domain: a dedicated server (Web server, application server, database server, etc.). To guarantee the latency or response time of services at peak time, each server is sized to meet the maximum expected workload; this makes the resources underutilized during most of the time and wastes much energy in idle power.

VM techniques (Figueiredo et al. 2003) provide a way to merge multiple applications from different servers to one shared server; that is, several logical servers can be deployed on the same physical infrastructure. Virtualization separates each logical server within the boundaries of its own security domain. They can have their own unique configuration of the running environment (operating system, namespace, etc.) and do not interfere with each other.

With virtualization techniques, services (or VMs) can be consolidated to fewer physical servers during the off-peak time, and unnecessary servers may be shut down to save energy. When the workload increases, servers can be woken up again to guarantee the quality of services. Dynamic migration techniques (Chase et al. 2003) make server consolidation more flexible to rapidly changing workloads.

Server clustering is another common mechanism adopted in many Internet services for the purpose of scalability and availability. In cluster-based services, the data and software are deployed on a cluster of servers, and a redirecting front-end is used to switch requests to an appropriate server in the pool. Traditionally, the switch policy is to balance request traffic across the servers for the best performance and availability.

But when the workload is low and most servers are underutilized, energy efficiency becomes the main concern in request switch. The system can adopt a power-aware policy in load dispatching that concentrates requests on fewer running servers, and other excess servers can be turned to a low-power state or even shut down to save energy. A preconfigured number of active servers always runs as long as utilization stays below a certain threshold to satisfy the quality of service. When the volume of requests exceeds this threshold, one or more inactive servers are woken up to distribute new requests.

35.4.2.2 Utilization, Performance, and Energy Consumption

The aforementioned server consolidation problem can be seen as a multidimensional bin packing problem. Physical servers are bins and applications or requests are items need to be packed in the bins. Different component resources of a physical server, such as CPU, memory, disk, are different dimensions of the bin. An optimal solution for the problem is the one that uses the minimal number of servers.

But server consolidation is not simply a multidimensional bin packing problem. First, the consolidation of several VMs onto a single physical server influences resource utilization in a nontrivial way. Services may put changing loads on different components (CPU, memory, disk, etc.), resulting in the resource utilization fluctuating in an unexpected pattern.

Second, the degradation in performance must be considered in server consolidation. As several applications or requests are consolidated into the same physical server, the conflicts in resource access will definitely influence the performance. The degradation must be controlled in an acceptable level, so that the quality of service can still be guaranteed.

Third, the energy efficiency of consolidated services is affected by both resource utilization and server performance, but they change in a nontrivial manner. When resource utilization is low, idle power is the main reason for inefficiency. But if the utilization is too high, the degradation in performance will make the execution time much longer; this also compromises the energy efficiency. Usually, the energy consumption per job shows a U-shape curve along the utilization of resources. There is a sweet point of energy consumption at a certain level of resource utilization.

Therefore, server consolidation is not simply squeezing applications or requests on the minimal number of servers. The change of resource utilization and the degradation in service performance must be carefully considered, so that it can best improve the energy efficiency but without too much compromising in the quality of services.

35.4.2.3 Energy-Aware Server Consolidation Approaches

VM techniques play a key role in workload management (or server consolidation). These techniques—with the prevalence of multicore processors in particular—have greatly eased and boosted parallel processing. Specifically, those multicore processors can effectively provide resource isolation to VMs. Energy reduction is a primary motivation in server consolidation.

Srikantaiah et al. (2008) proposed a power-aware server consolidation heuristic algorithm, which makes decisions based on the information obtained from experimental data, that is, profiling data of performance degradation and energy efficiency under varying utilization states on different resource dimensions (CPU, memory, disk, etc.). In the first step, the algorithm determines the optimal point of resource utilization that minimizes energy consumption per job and also guarantees the quality of service. That optimal point can be identified from the profiling data. When a service needs to be instantiated or a new request arrives, the required resource utilization is calculated in each dimension; and for a particular server in consideration, the resultant utilization is identified with that calculation. The key of the algorithm lies in using an effective heuristic for bin packing, resulting in the desired workload distribution across servers. Srikantaiah et al. used a scalar metric in their heuristic, that is, the n-dimensional Euclidean distance between the current resource utilization to the optimal point. This metric is adopted as the core decision-making determinant of the algorithm. Specifically, a new service instance or request is assigned to a server that maximizes the sum of the Euclidean distances at all servers. This heuristic is based on the intuition that the server resource can be better utilized if it leaves the maximum empty space to the fullest (the optimal utilization point) after the current resource allocation. In the case that all the servers are almost full, and therefore, the new service instance or request cannot be allocated, a new server is woken up, and all allocations are rearranged.

In Kansal et al. (2010), a utility analytic model for Internet-oriented server consolidation is proposed. The model considers servers being requested for services like e-books database or e-commerce Web services. The main performance goal is the maximization of resource utilization to reduce energy consumption with the same quality of service guarantee as in the use of dedicated servers. The model introduces the impact factor metric to reflect the performance degradation of consolidated tasks.

Server consolidation mechanisms developed in Torres et al. (2008), and Nathuji and Schwan (2007) deal with energy reduction using different techniques, especially Torres et al. (2008). Unlike typical server consolidation strategies, the approach used in Torres et al. (2008) adopts two interesting techniques, memory compression and request discrimination. The former enables the conversion of CPU power into extra memory capacity to allow more (memory intensive) servers to be consolidated, whereas the latter blocks useless/unfavorable requests (coming from Web crawlers) to eliminate unnecessary resource usage. The VirtualPower approach proposed in Nathuji and Schwan (2007) incorporates server consolidation into its power management combining "soft" and "hard" scaling methods. These two methods are based on power management facilities (e.g., resource usage control method and DVFS) equipped with VMs and physical processors, respectively.

A recent study on VM consolidation (Jung et al. 2010) has revealed (or confirmed) that live VM migration may even lead to significant performance degradation; thus, VM migration should be modeled explicitly taking into account the current workloads (i.e., application characteristics) and their changes.

35.4.3 Dynamic Power Management

Server consolidation can effectively reduce energy waste on idle power during off-peak time. But the drawback of this approach is also very clear. First, it induces a high latency in waking up machines when the workload exceeds the threshold. This may be intolerable in some e-commerce services because slow response may make them lost customers. Second, the significant overhead in service migration degrades the system's ability to react to transient load disturbances and bursts. Third, services and data may be highly distributed in some applications for the purpose of high parallelism and reliability, but consolidation would compromise the system performance in these aspects. An alternative power-aware approach in these datacenters is to lower the voltage of CPUs when the workload is not high.

35.4.3.1 Dynamic Voltage with System Load

The power-performance function shows that low-power CPUs consume much less energy than high-performance ones. This feature is especially useful when servers must be kept power on, such as the situation discussed earlier. DVFS has been widely supported in modern CPUs. The low cost in voltage or P-state changing makes it quite feasible for latency-sensitive applications.

The basic idea of the approach is to adjust the voltage of CPUs according to the fluctuation in system load. When workload is high, a high voltage is set to provide high-performance services; whereas during the off-peak time, lower voltage is supplied to save energy. The system can set checkpoints to monitor workload changing. The sampling rate should be chosen carefully according to the application features. High-rate sampling introduces excessive overhead in voltage adjusting. But if the rate is too low, it is hard to be sufficiently responsive to load variations.

35.4.3.2 QoS-Aware Power Supply

An important issue in dynamic power supply is that the degradation of performance must be considered when lowering the voltage of CPUs. A natural way to determine when the voltage should be adjusted is to check the current quality of services. For example, the system can lower the voltage of CPUs when all the requests are responded to in time. Conversely, the voltage must be increased if continual deadline missing is reported. But a problem of this method is that the system must tolerate certain deadline misses before it responds with an increase in the CPU voltage.

Another solution for the issue is to predict the schedulability from the current utilization of resources. Abdelzaher and Lu (2001) studied the schedulability conditions upon resource utilization for aperiodic tasks (requests). Under the optimal arrival-time-independent scheduling policy, they proved that the deadlines of all aperiodically arriving tasks can be guaranteed if utilization remains below a bound of 58.6% ($1/(1 + \sqrt{1/2})$) at all times. This utilization bound is a sufficient schedulability condition: exceeding it does not necessarily mean deadline missing. Sharma et al. (2003) use this bound as a control set point in their adaptive algorithm for dynamic voltage scaling. It allows servers to save energy but still effectively maintain the delay constraints on the services.

35.4.4 Hybrid Resources

In an attempt to design more energy-efficient computer systems, wimpy servers (using low-power microprocessors) are proposed in studies (Andersen et al. 2009; Hamilton 2009). This design is very suitable for cold storage workload and highly partitionable workload, but clearly not for every application. The measurements by Chun et al. (2009) show high-performance servers are more energy efficient in some computation-intensive applications. It is also not suitable to distribute poorly scaling workloads on a cluster of wimpy servers, because the overhead in startup and the complex intercommunication may exclude benefits that can be obtained by using low-power nodes (Lang et al. 2010). Another issue is that wimpy servers may not be able to meet rigorous QoS requirements in response time or workflow makespan, especially when resource competition is quite high. As a result, a hybrid design that combines low-power servers with high-performance ones might be a better choice to handle diverse applications under different resource utilization states. In this subsection, we study several scheduling approaches that explore heterogeneity in cluster resources for energy efficiency.

35.4.4.1 Hybrid Datacenter

Datacenters could be built to include both high-performance servers and low-power ones. A crucial problem is how to schedule tasks to servers with different performance characteristics in terms of both processing capacity and energy consumption.

Chun et al. (2009) investigated two possible scheduling approaches in hybrid datacenters. In the first approach, tasks may be migrated between high-performance servers and low-power servers. Initially, the system only uses low-power servers while the high-performance servers are in deep sleep. When the workload (or the number of concurrent tasks) exceeds a threshold, the high-performance servers will be woken up and all the tasks are migrated to them for execution. At this time, the low-power servers are transferred into deep sleep. Tasks will be migrated back to low-power servers when the workload falls below the threshold. In the other approach, no task migration is considered—all the tasks are finished in the same server where they started. In this case, the low-power servers always run but new tasks may be scheduled on the high-performance ones if the workload exceeds the threshold. The performance and feasibility of such approaches are studied in a simple scenario where one traditional high-performance server upon Xeon CPU and one wimpy server by Atom are combined together. The experimental results show that this simple hybrid solution achieves good energy proportionality and low latency in response time.

35.4.4.2 Power-Aware Workflow Scheduling on Hybrid Clusters

Workflow applications can be scheduled on a hybrid cluster for the purpose of energy saving. The rationale behind this scheduling approach is that heterogeneity in tasks of a workflow job (i.e., different temporal constraints) can be exploited using heterogeneous resources to effectively capture the trade-off between application completion time and energy consumption. Because of the imbalance workload among different parallel paths (as shown in Figure 35.5), some tasks have certain slack time in scheduling and they may run on a slower computing node without influencing the makespan of the whole workflow. A hybrid cluster, mixed with high-performance nodes and low-power ones, provides such an opportunity for power-aware workflow scheduling. The high-performance nodes guarantee the workflow jobs can be finished before their deadlines, whereas the low-power ones can be used to reduce energy consumption of tasks that have slack time in particular.

35.4.4.2.1 Scheduling Approach

We propose an approach (Algorithm 1) that schedules workflow jobs on a hybrid cluster with the consideration of energy saving. We assume each workflow job comes with a deadline. The scheduler should guarantee the deadline of a workflow job possibly with the minimum energy consumption.

When a workflow job arrives, the scheduler first calculates the execution deadline for each task based on the temporal constraints. Our previous study (Chen et al. 2010) shows details in calculating task deadlines. Then, all the tasks of a workflow job are ranked and put into a priority list according to their deadlines. For each task in the list, the most energy-efficient computing node is allocated in that the node is capable of meeting the deadline of that task with the minimum energy consumption among all nodes. We implement a rescheduling mechanism (Chen et al. 2010) in the approach to optimize resource allocations hoping for more workflow jobs to be finished before their deadlines.

35.4.4.2.2 Experiments and Results

Simulation experiments were carried out to study the performance of our power-aware scheduling approach in terms of deadline satisfiability and energy consumption. For the sake of simplicity, we assume workflow jobs in our experiments are all I/O-bound data processing tasks. There are three different configurations of computing nodes used in our experiments, high-performance nodes, low-power nodes, and middle-class nodes. The normalized performance, power, and price of computing nodes are assumed to be that shown in Table 35.2 (Vaid 2010).

Algorithm 35.1 Power-Aware Workflow Scheduling

```
Input a workflow job Output scheduling of the job
Calculate deadline for each task
Rank tasks and put them into a priority list
for each task in the list do
  Calculate energy consumption of the task on each node
  Rank nodes into a list by energy consumption (increasing order)
  for each node in the list do
    if the task can be finished before its deadline on the node then
      Schedule the task on the node
  end if
  end for
  if scheduling fails on all nodes then
      for each node in the list do
        Reschedule tasks to make the new task finish before its deadline
      end for
      if rescheduling fails on all nodes then
        Schedule the task on the earliest finish time
        if this finish time > job's deadline then
          Reject the job
    end if
    end if
  end if
end for
```

TABLE 35.2 Performance, Power, and Price of
Different Computing Nodes

Computing Node	Performance	Power	Price
High-performance node	1.18	1.56	1.6
Middle-class node	1.07	1.24	1.1
Low-power node	1.0	1.0	1.0

We simulated three hypothetical clusters. The first cluster has 1000 high-performance computing nodes. The second cluster is built totally with low-power computing nodes. As the price of a high-performance node is 1.6 times more than that of the low-power node, the number of nodes in the second cluster is set to 1600. The third cluster is a hybrid cluster consisting of 800 low-power nodes, 364 middle-class nodes, and 250 high-performance nodes.

We applied our power-aware scheduling approach to a stream of randomly produced workflow jobs (with various sizes and parallelism degrees). We compared the scheduling performance over these different clusters. Specifically, the comparison metrics include acceptance rate (the percentage of successful workflow jobs in terms of their deadlines) and energy consumption per unit of workload.

In our experiments, workflow jobs are submitted using a Poisson process, and the average interarrival time represents resource competition among different workflow jobs. Figure 35.6a shows acceptance rate changes when workflow jobs are scheduled onto different clusters. Clearly, the clusters with more computing nodes (i.e., the low-power cluster and the hybrid cluster) perform better in acceptance rate; however, the poor deadline satisfiability with the low-power cluster is its major limiting factor. The hybrid cluster makes up for this defect; it has a good acceptance rate across a wide range of resource competition. Furthermore, the energy efficiency of hybrid cluster is quite near that of the low-power cluster (as shown in Figure 35.6b).

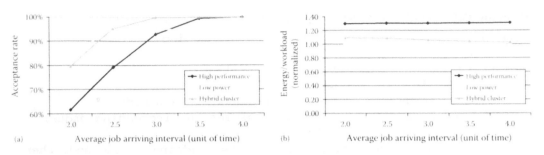

Average job arriving interval (unit of time)   (b)   Average job arriving interval (unit of time)

FIGURE 35.6 Performance comparison among different clusters. (a) Acceptance rate and (b) energy efficiency.

35.5 Summary and Conclusion

In this chapter, we reviewed and discussed energy efficiency issues in datacenters with a focus on exploitation of heterogeneity. Energy wastage (from low resource utilization caused primarily by idle servers) in datacenters has been identified and reconfirmed as a major source of datacenter inefficiency. Such inefficiency is a serious concern not only to datacenter operation and expansion, but also to the environment. Energy models of different datacenter resources were also discussed. Due to the fact that applications and resources in datacenters are highly dynamic and heterogeneous in nature, it is more appropriate to make resource allocation decisions explicitly taking into account those characteristics. We have exemplified this with a resource allocation strategy. Based on preliminary results from our example solution, the exploitation of heterogeneity in datacenters can help make significant energy savings.

References

Abdelzaher, T. F. and Lu, C. (2001). Schedulability analysis and utilization bounds for highly scalable real-time services. In *Proceedings of the Seventh IEEE Symposium on Real-Time Technology and Applications*. 30 May 2001, Taipei, Taiwan, pp. 15–25.

Andersen, D. G., Franklin, J., Kaminsky, M., Phanishayee, A., Tan, L., and Vasudevan, V. (2009). FAWN: A fast array of wimpy nodes. In *Proceedings of the 22nd ACM Symposium on Operating Systems Principles (SOSP)*. 11 Oct 2009, Big Sky, Montana, USA, pp. 1–14.

Barroso, L. A. and Holzle, U. (2007). The case for energy-proportional computing. *IEEE Computer*, 40(12): 33–37.

Barroso, L. A. and Holzle, U. (2009). The datacenter as a computer: An introduction to the design of warehouse-scale machines. In *Synthesis Lectures on Computer Architecture*, ed. M. D. Hill, the Morgan & Claypool Publishers San Francisco, USA, pp. 10.

Chase, J. S., Anderson, D. C., Thakar, P. N., Vahdat, A. M., and Doyle, R. P. (2001). Managing energy and server resources in hosting centers. In *Proceedings of the Eighth ACM Symposium on Operating Systems Principles (SOSP)*. 21 Oct 2001, Chateau Lake Louise, Banff, Canada, pp. 103–116.

Chase, J. S., Irwin, D. E., Grit, L. E., Moore, J. D., and Sprenkle, S. E. (2003). Dynamic virtual clusters in a grid site manager. In *Proceedings of IEEE International Symposium on High Performance Distributed Computing*. 22 June 2003, Seattle, Washington, USA, pp. 90–100.

Chen, W., Fekete, A., and Lee, Y. C. (2010). Exploiting deadline flexibility in grid workflow rescheduling. In *Proceedings of the 11th ACM/IEEE International Conference on Grid Computing* 25 Oct 2010, Brussels, Belgium.

Chen, G., He, W., Liu, J., Nath, S., Rigas, L., Xiao, L., and Zhao, F. (2008). Energy-aware server provisioning and load dispatching for connection-intensive internet services. In *Proceedings of the*

Fifth USENIX Symposium on Networked Systems Design and Implementation (NSDI'08). 16 April 2008, San Francisco, California, USA, pp. 337–350.

Chun, B., Iannaccone, G., Katz, R., Lee, G., and Niccolini, L. (2009). An energy case for hybrid datacenters. In *Workshop on Power Aware Computing and Systems (HotPower '09)* 10 Oct 2009, Big Sky, Montana, USA.

Dawson-Haggerty, S., Krioukov, A., and Culler, D. E. (2009). Power optimization—A reality check. Technical Report No. UCB/EECS-2009-140, Electrical Engineering and Computer Sciences, University of California at Berkeley, Berkeley, CA.

Environmental Protection Agency, ENERGY STAR Program. (2007). Report to Congress on server and data energy efficiency. http://www.energystar.gov/ia/partners/prod_development/downloads/EPA_ Datacenter_Report_Congress_Final1.pdf (Accessed December, 2007).

Figueiredo, R. J., Dinda, P. A., and Fortes, J. A. B. (2003). A case for grid computing on virtual machines. In *Proceedings of 23rd IEEE International Conference on Distributed Computing Systems*. 19 May 2003, Providence, Rhode Island, USA, pp. 550–559.

Hamilton, J. (2009). Cooperative expendable micro-slice servers (CEMS): Low cost, low power servers for internet-scale services. In *Proceedings of Fourth Biennial Conference on Innovative Data Systems Research (CIDR)* 4 Jan 2009, Asilomar, California, USA.

Hikita, J., Hirano, A., and Nakashima, H. (2008). Saving 200 kW and $200 K/year by power-aware job/machine scheduling. In *Proceedings of IEEE International Symposium on Parallel and Distributed Processing (IPDPS)* 14 April 2008, Miami, Florida, USA.

Jung, G., Hiltunen, M. A., Joshi, K. R., Schlichting, R. D., and Pu, C. (2010). Mistral: Dynamically managing power, performance, and adaptation cost in cloud infrastructures. In *Proceedings of the 2010 IEEE 30th International Conference on Distributed Computing Systems (ICDCS 2010)*. 21 June 2010, Genova, Italy, pp. 62–73.

Kansal, A., Zhao, F., Liu, J., Kothari, N., and Bhattacharya, A. (2010). Virtual machine power metering and provisioning. In *Proceedings of ACM Symposium on Cloud Computing (SOCC 2010)*. New York: ACM pp. 39–50.

Lang, W., Patel, J. M., and Shankar, S. (2010). Wimpy node clusters: What about non-wimpy workloads? In *Proceedings of the Sixth International Workshop on Data Management on New Hardware (DaMoN 2010)*. 6 June 2010, Indianapolis, Indiana, USA, pp. 47–55.

Minas, L. and Ellison, B. (2009). The problem of power consumption in servers. http://www.infoq.com/ articles/power-consumption-servers, posted by Intel on 1 April, 2009.

Nathuji, R. and Schwan, K. (2007). VirtualPower: Coordinated power management in virtualized enter-prise systems. In *Proceedings of Twenty-First ACM SIGOPS Symposium on Operating Systems Principles (SOSP'07)*. New York: ACM. pp. 265–278.

Patterson, D. A. (2004). Latency lags bandwidth. *Communications of the ACM*, 47(10):71–75.

Sharma, V., Thomas, A., Abdelzaher, T., Skadron, K., and Lu, Z. (2003). Power-aware QoS management in web servers. In *Proceedings of the 24th IEEE International Symposium on Real-Time Systems*. 3 Dec 2003, Cancun, Mexico, USA, pp. 63–72.

Srikantaiah, S., Kansal, A., and Zhao, F. (2008). Energy aware consolidation for cloud computing. In *Workshop on Power Aware Computing and Systems (HotPower '08)* 7 Dec 2008, San Diego, California, USA.

Stokes, J. (2009). Latest migrations show SSD is ready for some datacenters. http://arstechnica.com/ business/news/2009/10/latest-migrations-show-ssd-is-ready-for-some-datacenters.ars, published on Ars Technica Openforum, 19 October, 2009.

Ton, M. and Fortenbury, B. (2005). *High Performance Buildings: Datacenters Server Power Supplies*. Lawrence Berkeley National Laboratories and EPRI.

Torres, J., Carrera, D., Hogan, K., Gavalda, R., Beltran, V., and Poggi, N. (2008). Reducing wasted resources to help achieve green data centers. In *Proceedings of Fourth Workshop on High-Performance, Power-Aware Computing (HPPAC'08)*, 14 April 2008, Miami, Florida, USA.

Tsirogiannis, D., Harizopoulos, S., and Shah, M.A. (2010). Analyzing the energy efficiency of a database server. In *Proceedings of ACM International Conference on Management of Data (SIGMOD'10)*, 6 June 2010, Indianapolis, Indiana, USA, pp. 231–242.

Vaid, K. (2010). Datacenter power efficiency: Separating fact from fiction. *Invited Talk on Workshop on Power Aware Computing and Systems* 3 Oct 2010, Vancouver, British Columbia, Canada.

36

Code Development of High-Performance Applications for Power-Efficient Architectures

Khaled Z. Ibrahim
Lawrence Berkeley National Laboratory

36.1 Introduction

Achieving extreme-scale computing requires ultra power-efficiency of the computing elements. Power efficiency is usually achieved by cutting transistor budget from hardware structures that exploit data locality such as caches. They are replaced with software-managed local-store to maintain performance. It can also require removing hardware structures that exploit instruction-level parallelism, such as out-of-order execution units, relying on the support for vector execution units. Power efficiency generally leads to complicating software development. Heterogeneous systems provide a trade-off that combines complex processor cores with power-efficient accelerators to handle multiple code types.

Power-efficiency has become one of the main design constraints for high-performance machines because the annual cost of powering and cooling a computer may easily exceed its acquisition cost. Being friendly to the environment is another factor that recently comes into play. The main challenge that power-efficient architectures face is how to execute the large volume of legacy codes existing today and how to automate software development to improve the productivity of generating new codes for these architectures. This chapter will outline how power-efficient architectures are currently built and what difficulties exist in migrating current programming models to these architectures.

New programming models, introduced to efficiently exploit these architectures, are prohibitively complex. Additionally, the stability of traditional architectures, in contrast with power-efficient architectures, typically discourages the high-performance computing society from migrating their codes to these architectures.

We plan to outlines the major attributes of the programming models for power-efficient architectures. Specifically, we will discuss their newly introduced memory models and the need for explicit instruction-level parallelism in the software layer.

We will focus on presenting methodologies in developing codes for power-efficient architectures, including manual migration, automated migration (compiler-based), and autotuning frameworks. The trade-offs between these approaches will be presented and their prospects for future power-efficient computing will be discussed.

36.2 Evolution of Architecture and Programming Paradigms

Software paradigms and microprocessor architectures followed an evolution path where the gap between the two yielded improved performance and productivity. As shown in Figure 36.1, improving the

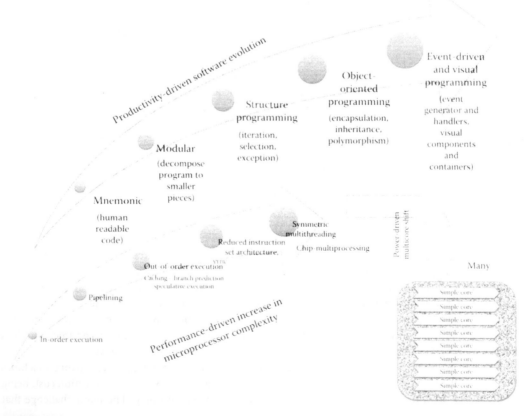

FIGURE 36.1   Trends in the evolution of software paradigms and microprocessor complexity.

productivity of the software layer contributed to increasing its complexity from unstructured assembly code to visual programming. The outcome is a more complex software code, with many functionalities, in less development time. Computer architecture used to provide hardware structures to manage software complexities for the sake of improving performance. For instance, structured code usually carries loops; a branch predictor can reduce the impact of the introduced branches on a pipelined architecture. Data prefetchers try to capture spatial locality, the tendency to access contiguous data, which appears in most loop constructs. Caches try to exploit temporal locality, the tendency of the program to reference recently accessed data.

High-performance computing mostly followed the evolution of software paradigms up to structured programming, for the sake of having a manageable gap with the underlying hardware. The object-oriented paradigm is less popular in high-performance computing because of its performance implications and the wealth of the legacy structured codes that are used by the scientific community. On the computer architecture side, the path to improving performance faces a dramatic change in direction because of the power wall. Improving the single microprocessor performance becomes increasingly difficult; having on-chip multiprocessing is unavoidable. The architectural idea of having many less-powerful computing processors has become very popular in the past few years. The most obvious path is to replicate cores within the socket to improve power efficiency, which was adopted by most microprocessor manufacturers. Having more specialized cores, with different computational characteristics, has started to surface as a viable architectural design for efficient execution.

36.2.1 High-Performance Computing Evolution and Power Efficiency

Several milestones in computing performance were successfully achieved. The latest, the petaflops milestone, was mostly driven by technological advances in CMOS technology and to a lesser extent by power efficiency; exa-scale computing will require rethinking power efficiency in every aspect of the system design, including microprocessor cores, caches, floating-point units, interconnection to the memory and between nodes, disk power, etc. [18,32]. The evolution of system designs could probably bring peak performance to the exa-scale within a decade, but sustained performance of the machine will rely on how well the software will exploit the new hardware designs.

In the context of high-performance computing, one of the main performance metrics is floating-point operations per second (flops). Floating-point units (FPUs) typically use less than 0.5% [18] of the total machine power. The power consumed by other components of the system needs to be reduced significantly because even if we have 4x scaling down of the FPU power consumption, an exa-scale machine will need 4 MW [18] for FPUs alone. It will be difficult to keep the FPU power percentage as it is today for an exa-scale system, but cutting transistors from other hardware structures should be carefully done to avoid hurting the sustained performance. For large-scale systems, microprocessors typically consume a significant portion of the whole system power. For instance, in the Roadrunner supercomputer, microprocessors consume more than 65% of the total system power—up to 75% with Linpack; a small percentage of that goes to the computing FPUs while caches and the main CPU—based on AMD Opteron processor—consume the rest. The cache hardware structure usually consumes a large portion of the power consumed by the microprocessor (can exceed 70%).

36.3 Power-Efficient Architectures versus Conventional Architectures

The improvement of single-core performance slowed down significantly because of the power wall, and computer architects started targeting alternative architectures to improve the performance per watt. Conventional architectures continued its evolution path by introducing multicore architectures, which have replicated cores design on-chip. Heterogeneous architectures, based on accelerators, such as graphic processing units (GPUs) and IBM PowerXCell 8i, started to prove their strong potential and viability for computing because of their ability to deliver not only high performance but also power efficiency. On the software side, most of the potential performance is achieved by manual coding, customized autotuning, and to a lesser extent by automatic code generation. Productive and transparent migration of legacy code to these architectures is still a challenging problem and in some cases endangers the likelihood of survival for some of these architectures.

As of November 2010, the top machines in the Green500 list [34] are based on the accelerators (IBM PowerXCell and GPUs) and the top machine in terms of performance in the top 500 supercomputers [35] is Tanhai, which uses the NVIDIA Fermi GPU. The following sections we will shed more light on multicore architectures and heterogeneous (accelerator-based) architectures.

36.3.1 Conventional Architectures

Most of the advances in microprocessor and system architectures were targeting the improvement of single-thread performance. Most of these performance-enhancing hardware constructs require a large amount of power to get them functioning.

Microarchitectural designs introduce out-of-order execution of instructions if no true dependency exists between them, thus improving the throughput. Pipelined execution is another technique to improve throughput by dividing the execution of an instruction into multiple stages, each requiring a shorter cycle time. Pipelining improves the system throughput by increasing chip frequency.

Microprocessor can also speculatively execute instructions such that even dependent instructions, or those dependent on the outcome of a branch instruction, can run without waiting for the dependency to be resolved. In such a case, the system state should be stored to recover in case of mis-speculation. Although the potential for improving performance is high, the required budget of power is usually higher than the performance benefit.

Another notable technique to improve performance is to cache data that are referenced by the application. Many well-written applications exhibit temporal locality, which is the tendency to reference data that were recently referenced. Applications may also have spatial locality, which is the tendency to reference contiguous data in memory. Temporal locality is usually well captured by caches by storing recently referenced data. Spatial locality can be captured by prefetchers (hardware-based or software-based) or increasing the granularity of the data brought from the memory (cache line size). Caches are usually faster to access than conventional memory, thus the higher the locality of the data references, the less dependent the application becomes on the memory subsystem performance. Traditionally, caches are designed in layers of hierarchy, one to three layers; they may require up to 70% of the total chip power. As the performance of single core becomes more difficult to improve, most chip manufacturers started introducing multicore architectures, where replicated complex cores share the same socket. The performance improvement becomes more dependent on the software layer to express parallelism by multiple threads. Figure 36.2 shows a typical multicore architecture that exists on most computing systems nowadays. These multicore chips are usually the building blocks of high-performance machines. In this figure, each core has its own level-1 cache, while the level-2 cache is shared by all cores. L1 caches are usually coherent. The L2 cache can be totally shared or can be split between core groups to guarantee performance isolation.

36.3.2 Power-Efficient Architectures

The path to power efficiency can go through cutting transistors from performance-enhancing hardware structures that do not contribute to performance in proportion to their power requirement. Most of these structures do not add functionalities (or instructions) to the processor and they work transparently to improve the performance. Candidates are caches, out-of-order execution engine, speculative execution

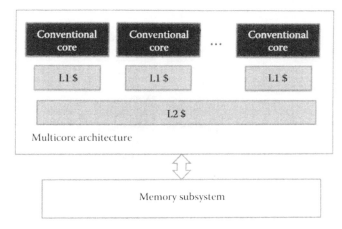

FIGURE 36.2   Multicore architecture: a complex core design is replicated within a socket.

engine, etc. Building a microprocessor with pure function units will be very harsh to a software developer in terms of performance. A power-efficient architecture could provide the support for vector processing. Carefully written code can directly exploit the power of these architectures. The cache hierarchy can be replaced with an on-chip memory that the software can control for managing data brought from the main memory. The advantage of such systems is that a small on-chip memory capacity can provide the performance of a larger cache, thus saving power. Software-managed local memory is equivalent to a fully associative cache, while a typical cache is usually set associative.

The difficulty of writing codes to power-efficient architectures is due to the burden put on the software developer to manage all aspects of code optimizations that affect performance. For instance, the developers need to understand the intrinsic operations needed to manage memory transfers. They need to express parallelism in an explicit manner in the software, as the hardware will not extract parallelism using out-of-order execution and speculation techniques. From the system developers point of view a code is better written in a high level language that guarantees portability of the code across architectures. Additionally, because of the large volume of codes that most disciplines have, it is not easy to switch from one programming paradigm to another.

Ideally, the additional complexity imposed by power-efficient architectures should be digested by an automatic code generation environment. Compilers tend to help in automating the process, but the challenge that any new architecture poses to a compiler is the inability of the compiler to well tune the performance while lacking the runtime information about the code at compile time. For instance, memory locations for two memory address names may be pointing to overlapping memory region (aliasing), which prevents the compiler from applying many optimizations, and usually causes the compiler to generate less-efficient code for the sake of correctness. Additionally, the compiler cannot change the data layout, which may be needed to generate vectorizable code. Optimal data layout can also be a challenge even to experienced programmers.

36.3.2.1 Accelerator-Based Architectures

Building a computer of pure functional units, although it may look the most power efficient, is challenged by the difficulty to achieve a reasonable performance. Heterogeneity is a middle-way solution, in which multiple specialized architectures coexist in the system. For code segments that cannot be optimized for power-efficient architectures, a conventional core is used. For codes that can be rewritten for specialized cores, a power-efficient accelerator is provided. Figure 36.3 shows a heterogeneous architecture where specialized cores coexist with a conventional main core on a single socket. The specialized cores can be used in the power-efficient acceleration of the codes running on the main processor. A notable example of these

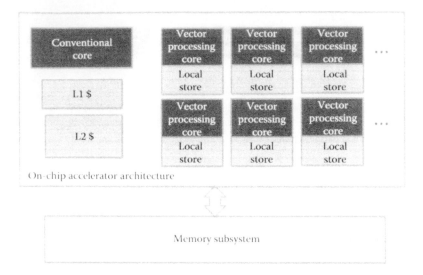

FIGURE 36.3   Design of heterogeneous computational resource with on-chip specialized accelerators.

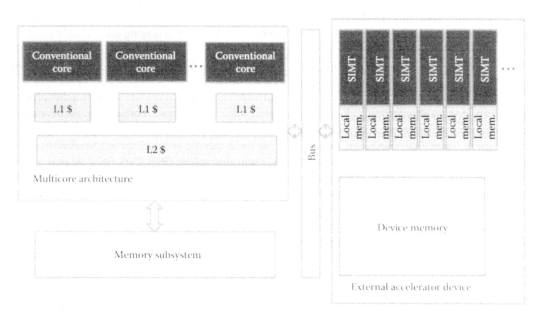

FIGURE 36.4   Heterogeneous computational resource in system design with off-chip specialized accelerator device.

architectures is the IBM PowerXCell that powers 4 of the top 10 green computers in the Green500 [34] list as of November 2010.

Another attractive heterogeneous design is to connect a power-efficient architecture to the main processor through an external bus. As shown in Figure 36.4, the specialized multiprocessor cores have a separate memory space, which is connected as an external device to the main processor. Graphic processing units are adopting such a scheme and are gaining momentum with computational scientists in many disciplines. The top supercomputing machine in the world, as of November 2010 [35], is based on this architecture. GPU-based systems constitute half of the top 10 green machines, as shown in Table 36.1.

TABLE 36.1 Top 10 Computers in the Green500 List, November 2010. Most of These Supercomputers are Based on Accelerators

Rank	Processor—Site	kW	Gflops/W
1	Blue Gene/Q Prototype—IBM T.J. Watson	38.8	1.68
2	G7 Xeon 6C X5670, NVIDIA GPU—Tokyo Institute of Technology	1243.8	0.96
3	Core i3 2.93 Ghz Dual Core, NVIDIA C2050—NCSA	36.0	0.93
4	SPARC64 VIIIfx 2.0 GHz—RIKEN	58.0	0.83
5	PowerXCell 8i, 3.2 GHz—Forschungszentrum Juelich	57.5	0.77
5	PowerXCell 8i, 3.2 GHz—Universitaet Regensburg	57.5	0.77
5	PowerXCell 8i, 3.2 GHz—Universitaet Wuppertal	57.4	0.77
8	Opteron 2.1 GHz, ATI Radeon GPU—Universitaet Frankfurt	385.0	0.74
9	Xeon 6C X5660 2.8GHz, nVidia Fermi—GeorgiaTech	94.4	0.67
10	nVidia GPU—National Institute for Environmental Studies	117.1	0.63

These architectures allow handling applications with mixed workload characteristics. For instance, codes that are control-flow intensive, such as operating systems, compilers, etc., can run on the general-purpose core, while codes that stress floating-point function units can run on the accelerator. The programming of these architectures is one of the main challenges to the wide adoption of these architectures.

Another architecture technology widely used in accelerators is field programmable gate array (FPGA), which is an array of logic gates that can be programmed to perform a user-specific function. The clock rate for FPGAs is usually low, 100–550 MHz, which leads to low power requirements. FPGAs can be used as floating-point accelerators, digital signal processors, communication assists, etc. Two of the big players on the FPGA market are Altera [3] and Xilinx [41]. Most users use third party products that comprise development environment in addition to the FPGA chips. The power-efficiency argument of the FPGA is based on providing flexibility in creating a reconfigurable-computing environment that matches application needs.

FPGAs are usually programmed using a special hardware description language (HDL) such as verilog, or VHDL. Programming can be a tedious job, but its overhead can be amortized by frequent subsequent uses.

The advantages of FPGAs are the their power-efficiency compared with general-purpose processors and their ease of customization. The adoption of FPGA in high-performance supercomputing for floating-point acceleration is usually limited because of the relative low performance of these accelerators compared with state-of-the-art accelerators (GPUs and PowerXCell). Additionally, efficient customization of FPGAs cannot be done when a wide variety of workloads are used, as those used on a typical supercomputer. Multiple workloads are usually used on these systems, each using multiple kernel routines. Application-specific designs are the best candidates for exploiting the efficiency of FPGAs, as will be discussed in more detail in Section 36.5.3.

36.4 Programming Approaches for Power-Efficient Architectures

Code development productivity, portability, and efficiency are very critical to the success and the survival of an architecture. Performance, although important, cannot guarantee the pervasiveness of any novel architecture. In this section, we will discuss the different programming approaches to accelerator-based architectures. We will show the processes involved in these approaches as well as the difficulties and limitations imposed on them.

Recently, an intensive amount of research targeted code generation for accelerators to leverage their potential performance and efficiency. Code generation is either done manually [4,5,7,9,16,22–24,28,33,

38,40], through an automated process [2,8,25,27], or through searching large optimization space [31,39]. Manual code rewriting has shown best the potential of accelerators for scientific computing. The power efficiency argument is currently dominated by system solutions using IBM PowerXCell for floating-point acceleration.

Automatic code migration to accelerators has also been prototyped successfully, at least at the functional level by multiple research projects and commercial compilers, for instance, by IBM Watson [27], Portland PGI Compiler [29], CAPS Enterprise [21] compiler, and CellSS [10]. Some studies [27] showed the difficulties that an automated compilation process can face, especially for applications with irregular or complex memory access patterns, such as NAS NBP [26] FT and CG, making the memory handling of data transfers a bottleneck. Some compilers do not provide a fully automated process. For instance, the CellSS compiler [10] relies on manual data layout to allow managing the accelerator memory while automating code scheduling on the IBM Cell accelerator. Automatic code migration of legacy code to GPUs [8,25] is somewhat more successful because of the large local memory that GPUs have.

Autotuning is another approach to achieve the best performance for a wide set of architectures. Autotuning is traditionally needed when it is difficult to reason about the performance associated with optimizations because of the following factors: first, proprietary hardware architectures (out-of-order execution, speculative execution, unknown prefetching policy, proprietary branch predictor, proprietary cache associativity, etc.); second, proprietary compilers and instruction set architectures (ISAs), especially with GPUs, where manufacturers hide architectures or compilation techniques to guarantee portability; finally, the complexity of the interaction between optimizations in a way that makes the performance not easily comprehensible.

Autotuning usually requires an expertise beyond what is needed by for code optimization, because it requires exhaustive knowledge of the optimization space, good heuristics to search the huge optimization search space, and the ability to correctly produce multiple variants of the code. These complexities are addressed by few successful autotuning libraries such as ATLAS [37], Spiral [30], FFTW [17], and OSKI [36], but they are not targeting newly emerging power-efficient architectures.

The autotuning technique was recently applied for power-efficient architectures such as IBM Cell [40], but modestly because of the deterministic behavior of the accelerator, and complexity of the memory model thus requiring to *manually* write all the code variants that are mostly tied to a specific problem specification and size. We believe that problems handled by autotuners should be avoided in power-efficient systems where we need a transparent and deterministic execution environment (compiler and architecture) to achieve efficiency; a good example for such an environment is the one associated with IBM PowerXCell [12].

The two main challenges in migrating legacy code to power-efficient heterogeneous multicore are the disparity between the legacy code memory model and the new memory models introduced by these architectures, and the difficulty in expressing parallelism in the software layer to allow vector execution.

We will discuss manual code migration to accelerator-based architectures, which has proved the performance and power efficiency benefits of using these architectures, but at the cost of reducing productivity. We will also discuss automatic migration with the help of compilers.

36.4.1 Manual Coding for Accelerator-Based Architectures

Programming for accelerator-based architectures relies on offloading computational-intensive part of the code to the accelerator. Multiple operations are involved in this process: First, the working set should be guaranteed to fit in the accelerator memory. The memory transfer from and to the accelerator memory should be explicitly managed. Second, the code needs to be vectorized to run efficiently on the accelerator, especially because accelerators do not provide out-of-order execution to extract parallelism from the code. The code fragment in Figure 36.5 shows a matrix–matrix multiplication code fragment. In the following discussion we will discuss the steps needed to execute (or to offload) this code to an accelerator.

```
for i ← 0 to N do
  for j ← 0 to M do
    C[i, j] ← 0
    for k ← 0 to L do
      C[i, j] ← C[i, j] + A[i, k] * B[k, j]
    end for
  end for
end for
```

FIGURE 36.5   Multiplying matrices on a general-purpose CPU.

Accelerator constrained memory handling: The first step in migrating a code to an accelerator is to refactor the code such that it works on data tiles that can fit in the accelerator memory. The dataset tiles can then be processed one at a time. For instance, the input matrices in the matrix–matrix multiplication can be divided into $T \times T$ tiles, that can fit in the accelerator memory. Figure 36.6 shows the tiled version of the code in Figure 36.5. The original routine is split into two routines: a driver routine, Figure 36.6, and an innermost kernel routine, Figure 36.7.

This transformation can be used on general-purpose cache-based CPUs for performance reasons, such as increasing the hit rate in the cache. In accelerator-based architectures, this transformation has a correctness dimension, as the code will not be able to run if it does not fit in the limited accelerator memory. Unlike general-purpose processors, having virtual memory that is larger than the physical memory is not supported.

Data movement across memory hierarchy: The second step is to manage the memory transfer to and from the accelerator special memory. These memory transfers can be managed by the accelerator itself, using for instance, direct memory access (DMA) get and put operations. This technique is adopted by the IBM PowerXCell processor. Alternatively, the transfers can be managed by the host CPU, as in the case of GPUs. In this case, memory copying operations are managed by the host CPU. Figure 36.8 shows the memory transfer code snippet added to the driver routine for the offloaded kernel routine.

```
for i ← 0 to N/T do
  for j ← 0 to N/T do
    C[i, j] ← 0
    for k ← 0 to N/T do
      MatMultiplyBlock(C[i, j], A[i, k], B[k, j])
    end for
  end for
end for
```

FIGURE 36.6   Tiled matrix multiplication. A, B, and C are formatted as four-dimensional arrays. The first two dimensions are the block indices. The second two provide indices within a block. Tile size, T, is chosen such that the source and the destination dataset can fit in the accelerator local memory.

```
proc MatMultiplyBlock(OUT  C_t, IN  A_t, IN  B_t)
for o ← 0 to T do
  for p ← 0 to T do
    for q ← 0 to T do
      C_t[o, p] ← C_t[o, p] + A_t[o, q] * B_t[q, p]
    end for
  end for
end for
end proc
```

FIGURE 36.7   *MatMultiplyBlock* kernel: matrix multiplication of two sub-matrices.

```
for i ← 0 to N/T do
  for j ← 0 to N/T do
    for k ← 0 to N/T do
      copy memory from host to device for A[i,k] B[k,j]
      Device routine of MatMultiplyBlock(C[i,j], A[i,k], B[k,j])
      copy memory from device to host C[i,j]
    end for
  end for
end for
```

FIGURE 36.8 Managing memory transfers for matrix–matrix multiplication on accelerator-based architectures. The driver routine is usually run on the main CPU, as with the GPU architecture. It can also run on the accelerator processor using DMA, as with the IBM PowerXCell architecture.

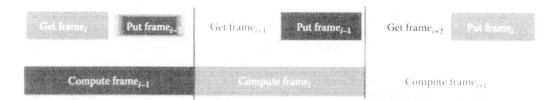

FIGURE 36.9 Double buffering technique to overlap computation with communication in accelerator-based architectures.

The figure shows a simple memory transfer mechanism, which toggles between computations and memory transfers; it does not overlap them. An efficient technique, called double buffering, splits the accelerator memory into two regions. The memory transfers are issued for the buffer that is not being processed. By the time of finishing the computation on a buffer, the memory transfer of the second one is hopefully complete. This technique has the potential of doubling the performance, if computation and memory transfer times perfectly overlap. Figure 36.9 shows double buffering for three computation steps. Double buffering pipelines the computation and the memory access while taking care of dependency.

Code vectorization: The final step is to vectorize the code such that a simple vector processor can execute the code efficiently. The vectorization can be done either explicitly or implicitly. Explicit vectorization is done through loop unrolling, then using intrinsics to operate on parallel data chunks. The condition for vectorizing the code is to have a similar operation on multiple contiguous data. Data layout may be changed to guarantee this spatial locality. Figure 36.10 shows a pseudo code of the vectorized accelerator code in Figure 36.7. This technique is usually suitable for the IBM PowerXCell.

Implicit parallelization is a model associated with single instruction multiple thread model (SIMT), which is used on GPUs. In this model, different threads are assigned different data to operate upon, but the same instruction is executed by all threads. Good spatial locality is needed for optimal performance, but instead of having spatial locality for the data accessed by

```
for o ← 0 to T do
  for p ← 0 to T do
    for q ← 0 to T/v_width step v_width do
      C_t[o,p] ← C_t[o,p] + A_t[o, q : q + v_width − 1] * B_t[q + v_width − 1, p]
    end for
  end for
end for
```

FIGURE 36.10 Explicit vectorized matrix–matrix multiplication with the vectorization width v_{width}.

tidx, tidy are the thread coordinates in a thread block
$o \leftarrow tidx$
$p \leftarrow tidy$
for $q \leftarrow 0$ to T **do**
$\quad C_t[o,p] \leftarrow C_t[o,p] + A_t[o,q] * B_t[q,p]$
end for

FIGURE 36.11 Implicit vectorized matrix–matrix multiplication with single instruction multiple threads model. The thread ids—*tidx, tidy*—are used to identify the accessed data.

each thread, the locality should be maintained across threads. In essence, the data accessed by different threads should be contiguous in memory to achieve coalesced memory access by the thread group. The code outlined in Figure 36.11 is the SIMT vectorized version of the kernel code in Figure 36.7. This execution model is adopted by CUDA and OpenCL programming paradigms.

36.4.1.1 Limitations of Manual Coding

The outlined steps to port codes for accelerator-based architectures show two main challenging aspects. First, the code needs significant transformations, even for simple codes. Given the large volume legacy codes and their diverse characteristics, it is usually difficult to migrate all types of codes to these architectures.

Second, the programming model is usually not unified across architectures. For instance, code vectorization for GPUs relies on SIMT model (implicit vectorization), while for IBM PowerXCell, code vectorization is explicit using intrinsic instructions. Data layout with GPU needs to allow coalescing across threads. In contrast with IBM PowerXCell, the memory should be contiguous for the access of each thread. In general, code portability across these architectures is difficult.

The additional uncertainty of the survival of these architectures on high performance computing platform in the future besides the prevalence of the legacy programming model makes the transition to these architectures very hesitant and slow.

Additional efforts should be exerted if a code does not have a regular access pattern, for instance, with sparse matrices representation and graph data structures. A code with control flow instructions should be modified to remove most of them. Data layout should be thought carefully for more complex codes, not only to optimize for the performance on the accelerator, but also to avoid hurting codes that operate on the same data but on the general-purpose CPU. Memory transfers should be minimized given the high cost associated with them. This may require algorithmic modification to leverage the computational power of the accelerator and to reduce the dependence of performance on the slow memory transfers.

Nowadays, many efforts are geared toward automating the process of code migration of legacy code to these accelerator-based architectures. In the following sections, we will present the methodology and limitations of these techniques.

36.4.2 Automatic Code Generation for Accelerators Using Compilers

Automatic code generation for accelerator attempts to hide the details of the migration process that were presented in the previous section. Several companies such as CAPS enterprise [21] and Portland Group PGI [29] provide automation software for GPUs. For the IBM PowerXCell processor, several research prototypes [10,27] such as CellSS, and commercial products, such as Gadae [19], are addressing the automatic code generation problem.

Common to these solutions are the following features:

- The source code needs to be annotated with pragmas specifying candidate codes to be executed on the accelerator.

- The compiler generates the code responsible for transferring data to and from the accelerator memory.
- The compiler generates the code of the kernel routines that execute on the accelerator and tries to vectorize the code when possible.
- Given that the kernel can run on the host CPU or the accelerator, some compilers generate different variants of the code for multiple target architectures. Depending on runtime information, such as dataset size, system configuration, and availability of the accelerator, the most appropriate variant of the code is executed.

For the code shown in Figure 36.5, the programmer needs to annotate the code for the compiler to be able to offload the computation to the accelerator. The annotated code, as shown in Figure 36.12, may need to be refactored to ensure that the data transferred to the accelerator will fit in its local memory. In the figure, the algorithm is shown with tiling to access the limited memory on the accelerator.

As shown in Figure 36.13, the compiler usually has a front-end that uses the input annotation to generate a separate source for a driver routine and kernel routines. Different back-end compilers are used for the CPU and the accelerator. The driver routine is compiled for the main CPU, while the kernel routines are compiled both for the CPU and the accelerator. The objective of having multiple variants of the kernel routines is to allow portability across different platforms. The driver code can detect at startup whether the accelerator is available or not.

The process of automating the code migration is not universal. The code needs to satisfy certain conditions to be compiled for the accelerator. In the following section, we will discuss that in detail.

36.4.2.1 Code Requirements

The following requirements should be satisfied in the code annotated for compilation to an accelerator:

Loop iteration independence parametrized: The compiler relies on the programmer input asserting that the candidate code does not carry dependency. Loop carried dependency typically prevents parallelization.

Footprint fitness within the accelerator local memory: The annotated code should fit in the accelerator memory. Otherwise, the programmer should refactor the code manually before handing

```
for i ← 0 to N/T do
  for j ← 0 to N/T do
    Cₜ ← C|i,j|
    reset Cₜ to zero
    for k ← 0 to N/T do
      Aₜ ← A|i,k|
      Bₜ ← B|k,j|

      pragma kernel to offload to accelerator
      begin offload
        for o ← 0 to T do
          for p ← 0 to T do
            for q ← 0 to T do
              Cₜ|o,p| ← Cₜ|o,p| + Aₜ|o,q| * Bₜ|q,p|
            end for
          end for
        end for
      end offload

    end for
  end for
end for
```

FIGURE 36.12   Legacy code annotated for automatic compilation to accelerator-based architectures.

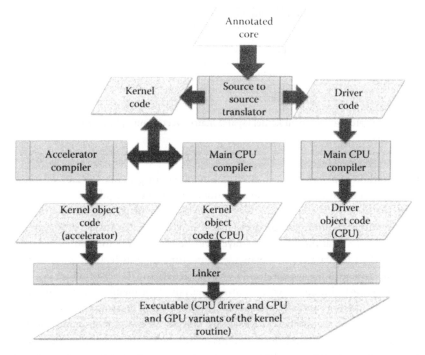

FIGURE 36.13 Compilation process for automatic migration of annotated legacy code to accelerator-based architectures.

it to the compiler. A common technique is tiling the computation into blocks and executing the computation in stages. While the tiling requirement is an optimization for cache-based system, for accelerators it is a necessity for correctness. Otherwise, data will not fit in the special accelerator memory and runtime errors or address wrapping may occur.

The compiler cannot generally do tiling on the data, and the job is usually left to the programmer. The tiling block size can be parametrized if performance tuning is required.

Avoiding pointer aliasing: Aliasing, which is having multiple names for the same memory location, is normally allowed in many high-level programming languages such as the C language. The possibility of aliasing limits the compiler ability to optimize the code. The programmer needs to assert and communicate to the compiler that such pointer aliasing does not exist. In the C language, the keyword strict can be used to convey that.

36.4.2.2 Restrictions

Automatic code generation for accelerators cannot handle some common programming constructs. In the following section, we will discuss these constructs and why they are typically challenging for any automatic code migration for accelerators.

Use of indirection arrays: The first case that is typically difficult to handle is when the code uses indirection array to access the dataset. This code construct is very common in dealing with sparse matrices. For instance, Figure 36.14 shows a code snippet of sparse matrix-vector multiplication that uses compressed-row format. The difficulties associated with this code are multiple folds: First the compiler cannot guarantee that the iterations of the outer loop are independent. Additionally, the compiler cannot strip-mine (tile) the dataset access in a

```
for i ← 0 to it_max do
    for j ← 0 to row_last − row_first do
        q[j] ← 0
        for k = rowstr[j] to rowstr[j + 1] do
            q[j]+=a[k] * B[CollideX[k]]
        end for
    end for
end for
```

FIGURE 36.14 Conjugate gradient main kernel (sparse matrix-vector multiplication). Sparse matrix is stored in a compressed-row format.

way that guarantees a perfect hit in the local memory. In general, the lack of runtime information about the access pattern makes it difficult for the compiler to provide an efficient solution.

A conservative solution can be done using an additional software management layer of the limited device local memory. This layer will be responsible for issuing memory transfer commands when the data are not available. Unfortunately this technique hinders the power-efficiency advantage of the accelerator by severely penalizing its performance.

Manual coding techniques involve runtime analysis of the memory access pattern to generate a precomputed memory transfer metadata. The overhead of creating these metadata is usually amortized by the large number of iterations that use these precomputed metadata. If the indirection array changes every iteration, this technique will be more difficult to adopt.

Branch removal or transformation: Branches are usually very expensive on most power-efficient architectures because of the lack of branch prediction and speculative execution. On GPUs, divergent branches can lead to serializing the execution of an SIMT block. In many cases, when branches exist in the innermost of a parallelizable loop, simple equivalent code transformations can help in removing them. For instance the code in Figure 36.15 may not execute efficiently on an accelerator because of the conditional code inside the loop. Generally, the programmer needs to refactor these types of codes to achieve efficient execution. For instance, the code in Figure 36.16 trades conditional execution with additional flops that are much cheaper in execution time on accelerators.

Dependency, coherency, and atomics: In many parallel codes, a loop may carry conflicting access to a common data structure. To guarantee serialized access, the programmer may use critical sections or atomics to protect shared data. These constructs are problematic on accelerators. For optimal performance, the program should run in a lock-free manner. GPUs, for instance, have very expensive atomics operations. In the IBM PowerXCell, no hardware coherency is maintained between local memory storage in the on-chip accelerators (synergistic processing elements).

```
for i ← 0 to N do
    if A[i] > 0 then
        B[i] ← B[i] + A[i]
    end if
end for
```

FIGURE 36.15 Conditional execution within a loop.

```
for i ← 0 to N do
    s ← sign(A[i])
    B[i] ← B[i] + (1 − s) × A[i]
end for
```

FIGURE 36.16 Control-flow free execution for the code in Figure 36.15.

```
for i ← 0 to N do
  for j ← 0 to f(i) do
    compute something on A[i, j]
  end for
end for
```

FIGURE 36.17 Loop with complex bounds.

Nested calls: Except for inlinable calls, nested calls are usually problematic for accelerator compilers because of their side effects, uncertainty about their memory footprint, etc. On most accelerators, there is a limited stack space, and performance is severely penalized if the stack size requirement is high.

Code size: Another side effect of limited fast memory space on accelerators and the difficulty to manage large memory space transparently is the need to have a limited code size. If a large code is to execute on the accelerator, the code should be divided into multiple kernels, as with GPUs. Alternatively, code overlays should be managed by the software, as with the IBM PowerXCell.

Nested structure and data layout: Nested structures can complicate the compiler attempt to align data for optimal performance. Many legacy codes have nested data structures leading to the need for manual intervention before passing the code to the compiler.

Variable loop bounds: Loop bounds should be runtime constants. The bound of a loop should not be a function of an outer loop index. This restriction aims at allowing the compiler to statically partition the data to fit in a limited memory. Figure 36.17 shows a code snippet with the innermost loop bound depending on the outermost loop limit. The lack of clear loop bounds usually limits the compiler ability to analyze the memory footprint.

The aforementioned limitations shed the light on the areas that need improvement on future generations of compiler technologies. Currently, numerous research groups and companies are actively engaging on addressing these issues.

36.5 Other Approaches for Achieving Power Efficiency

While accelerator-based architectures demonstrate efficiency in terms of peak computational power to the consumed power, which is what is usually advertised by system vendors, the most representative metric for the application developers is the sustained performance to the consumed power. Sustained performance depends on how well an application works on a particular architecture. For instance, while many applications can perform well on accelerator-based architectures, others, especially with irregular access patterns, can perform very poorly on these architectures. Generally, we cannot rely on the flop/watt metric for power efficiency if the peak values are solely considered. Additionally, the LinPack rating of the machine, used for Top500 and Green500, can be representative of a special class of algorithms that are compute intensive. LinPack Top500 could be different from sparse methods Top500 (if such a rating exists).

In addition to the accelerator-based architectures introduced earlier, code optimization on conventional architectures can help in achieving power-efficient sustained performance for many applications.

Customized designs is an alternative approach that usually involves building an efficient solution for a particular application with a reasonable power budget. This is typically useful for grand challenge problems, where solutions cannot be achieved with today's technologies.

In the following sections, we will shed some light on these approaches.

36.5.1 Performance Tuning and Energy Saving

Improving the performance of an application on a target architecture can potentially decrease the energy consumed to reach a solution. Most computing machines have a static energy consumption that is proportional to the application execution time. Reducing application runtime through optimizations reduces the amount of energy consumed to run the application.

Achieving optimal performance may require developing a large space of code variants and checking their interaction with the target architecture. One example of design space exploration is to try different block sizes (tile sizes) to search for an optimal interaction with the cache system, see Figure 36.7. Even though the cache size may be known, the effective size of the cache is affected by the cache associativity, the sharing with other OS activities, and the effectiveness of the hardware prefetcher typically existing on most modern microprocessors.

Some techniques such as loop unrolling may also have a large number of possibilities that require exploration for optimal performance. Their effectiveness depends on the underlying architecture and the back-end compiler. Deeper unrolling exposes more parallelism but also exerts more pressure on the microprocessor register file.

Few successful examples exist for autotuned libraries including ATLAS [37], Spiral [30], FFTW [17], and OSKI [36]. Exploiting this technique in other disciplines is very important in designing power-efficient systems. A typical challenge to this approach is the extensive expertise needed both for the algorithmic and the architectural sides.

36.5.2 Tunable Architectures to Application Behavior

Commonly, microprocessor designs consider a wide spectrum of use scenarios. They are constructed to perform well on average for most typical uses. This usually leads to overdesigning many architectural hardware structures. Typically a microprocessor provides a single design point. The designer usually considers how much bandwidth to maintain with the memory and how much computational power is provided by the function units. The cache system is also designed with fixed capacity, associativity, and line size.

The algorithm running on an architecture may not match the design point of that architecture. For instance, microprocessor ratio of flop/s to memory bandwidth could be 0.5 byte/flop, but the algorithm may require 1.0 byte/flop. In this case, the processor may stall most of the time waiting for the data supplied from memory. Obviously, if processor clock speed is reduced the performance of the application may not be affected but the power consumption will be significantly reduced. For this, dynamic voltage/frequency scaling technique [14,15,20] can prove beneficial in saving power.

Some algorithms do not utilize the cache fully, and their effective dataset may require a small cache space. Using a large cache can be a waste of power, given that caches can consume up to 70% of the total processor power. Turning off some of these cache lines can significantly save power without affecting the performance for these algorithms.

In general, tuning the architecture to the application needs can improve power-efficiency of the system. A typical challenge to this approach is that an application usually changes its requirements while navigating through the different phases of program execution.

36.5.3 Codesign Techniques

Another approach for power efficiency is to have a custom design that fits the need of a particular application. Several historical prototypes were built for special applications such as Lattice Quantum Chromo Dynamics (QCD), including QCDSP [13], QCDOC [11]. Recently, the QPACE [6] machine adopted an accelerator-only design with a custom interconnect, targeting several problems in physics, especially Lattice QCD.

Another notable project, green-flash [1], considers building a special machine for climate modeling. Given the enormous computational power needed for this application and the relatively small number of kernels that can be used, a special machine is sought to achieve an optimal performance and efficiency. To realize a machine that simulates the climate accurately within a reasonable power budget, the compute engine needs to be as efficient as the processors used in our consumer electronics, such as cell phones, which are powered by batteries. The extreme-scale performance of this application can be achieved via increasing the degree of parallelism. Such a design can meet the demand of this application while achieving unprecedented efficiency.

The drawback of this approach is the high overhead for building such machines, while being customized, or tuned, for a single application or a small set of applications.

In general, some designs are aggressive in designing all aspects of the systems, while some others build on commodity chips or components to cut the cost and the efforts involved.

36.6 Summary

Power-efficient designs can be achieved using custom designs, or heterogeneous architectures with specialized accelerators. The biggest challenge that power-efficient architectures are facing is how to program them in a productive way. Factors that should be considered are the current programming models and the large volume of legacy codes that can potentially be used on these architectures. While manual coding showed efficiency of using these architectures, the wide adoption of these architectures relies heavily on how to automate porting legacy code to these architectures. Most developers are still sticking to their conventional programming models because of the ubiquity of the legacy programming model tools that evolved with the evolution of computing systems, the difficulty to maintain multiple variants of the code, and the uncertainty about the future of power-efficient processors. Tool-based migration to power-efficient architectures have made significant improvements, but still have some limitations. Fortunately, both academia and industry are trying to build better tools to make transparent utilization of power-efficient architectures a reality.

Glossary

GPU graphic processing unit.

IBM PowerXCell IBM Cell processor is a heterogeneous chip design with one main CPU and eight on-chip synergistic processing elements.

Accelerator A specialized processor connected to a main CPU that efficiently executes a certain class of codes.

Heterogeneous architectures An architecture comprising multiple architectural designs.

Multicore architectures A homogeneous architecture where computing cores are replicated within a socket.

Local store A limited on-chip memory that is managed by the software.

Green computing Computing systems that target power and energy efficiency as one of their main design goals.

References

1. A new breed of supercomputers for improving global climate predictions. http://www.lbl.gov/cs/html/greenflash.html
2. R. Allen and K. Kennedy. Automatic translation of FORTRAN programs to vector form. *ACM Transactions on Programming Languages and Systems*, 9(4):491–542, 1987.
3. Altera Corporation programmable logic solutions. http://www.altera.com

4. D. A. Bader and V. Agarwal. FFTC: Fastest Fourier transform for the IBM cell broadband engine. *Proceedings of the 14th IEEE International Conference on High Performance Computing (HiPC)*, Goa, India, pp. 172–184, December 2007.

5. D. A. Bader, V. Agarwal, K. Madduri, and S. Kang. High performance combinatorial algorithm design on the cell broadband engine processor. *Parallel Computing*, 33(10–11):720–740, 2007.

6. H. Baier, H. Boettiger, M. Drochner, N. Eicker, U. Fischer, Z. Fodor, G. Goldrian, S. Heybrock, D. Hierl, T. Huth, B. Krill, J. Lauritsen, T. Lippert, T. Maurer, J. McFadden, N. Meyer, A. Nobile, I. Ouda, M. Pivanti, D. Pleiter, A. Schfer, H. Schick, F. Schifano, H. Simma, S. Solbrig, T. Streuer, K. H. Sulanke, R. Tripiccione, T. Wettig, and F. Winter. Status of the QPACE project. *PoS LATTICE* 2008(arXiv:0810.1559):039, Williamsburg, VA, p. 7, October 2008.

7. F. Banterle and R. Giacobazzi. A fast implementation of octagon abstract domain on graphics hardware. *Proceedings of the 14th International Static Analysis Symposium (SAS '07)*, Kongens Lyngby, Denmark, pp. 315–332, August 2007.

8. M. M. Baskaran, U. Bondhugula, S. Krishnamoorthy, J. Ramanujam, A. Rountev, and P. Sadayappan. A compiler framework for optimization of affine loop nests for GPGPUs. *The 22nd Annual International Conference on Supercomputing (ICS '08)*, Island of Kos, Greece, pp. 225–234, 2008.

9. R. G. Belleman, J. Bedorf, and S. P. Zwart. High performance direct gravitational *N*-body simulations on graphics processing units II: An implementation in CUDA. *Journal of New Astronomy*, 13(2):103–112, 2008.

10. P. Bellens, J. M. Perez, F. Cabarcas, A. Ramirez, R. M. Badia, and J. Labarta. CellSS: Scheduling techniques to better exploit memory hierarchy. *Scientific Programming*, 17(1–2):77–95, 2009.

11. P. A. Boyle, D. Chen, N. H. Christ, M. Clark, S. D. Cohen, C. Cristian, Z. Dong, A. Gara, B. Joů, C. Jung, C. Kim, L. Levkova, X. Liao, G. Liu, R. D. Mawhinney, S. Ohta, K. Petrov, T. Wettig, and A. Yamaguchi. Hardware and software status of QCDOC. *Nuclear Physics B–Proceedings Supplements*, 129–130:838–843, 2004.

12. Cell SDK 3.1. http://www.ibm.com/developerworks/power/cell/index.html, October 2007.

13. D. Chen, P. Chen, N. H. Christ, R. G. Edwards, G. Fleming, A. Gara, S. Hansen, C. Jung, A. Kahler, S. Kasow, A. D. Kennedy, G. Kilcup, Y. Luo, C. Malureanu, R. D. Mawhinney, J. Parsons, C. Sui, P. Vranas, and Y. Zhestkov. QCDSP machines: Design, performance and cost. *Proceedings of the 1998 ACM/IEEE Conference on Supercomputing*, Orlando, FL, pp. 1–6, 1998.

14. K. Choi, R. Soma, and M. Pedram. Dynamic voltage and frequency scaling based on workload decomposition. *The 2004 International Symposium on Low Power Electronics and Design (ISLPED '04)*, Newport Beach, CA, pp. 174–179, 2004.

15. G. Dhiman and T. S. Rosing. Dynamic voltage frequency scaling for multi-tasking systems using online learning. *The 2007 International Symposium on Low Power Electronics and Design (ISLPED '07)*, Portland, OR, pp. 207–212, 2007.

16. G. I. Egri, Z. Fodor, C. Hoelbling, S. D. Katz, D. Nogradi, and K. K. Szabo. Lattice QCD as a video game. *Computer Physics Communications*, 177(8):631–639, 2007.

17. M. Frigo and S. G. Johnson. FFTW: An adaptive software architecture for the FFT. *The 1998 IEEE International Conference on Acoustics, Speech and Signal Processing*, Seattle, WA, vol. 3, 1381–1384, May 1998.

18. A. Gara. Energy efficiency challenges for exascale computing. *The ACM/IEEE Conference on Supercomputing: Workshop on Power Efficiency and the Path to Exascale Computing*, Austin, TX, November 2008.

19. Gedae and the cell broadband engine (CBE). http://www.gedae.com/ibm.php

20. S. Herbert and D. Marculescu. Analysis of dynamic voltage/frequency scaling in chip-multiprocessors. *The 2007 International Symposium on Low Power Electronics and Design (ISLPED '07)*, Portland, OR, pp. 38–43, 2007.

21. HMPP Workbench, a directive-based compiler for hybrid computing. http://www.caps-entreprise.com/hmpp.html

22. K. Z. Ibrahim and F. Bodin. Implementing Wilson-Dirac operator on the cell broadband engine. *The 22nd ACM/SIGARCH International Conference on Supercomputing*, Island of Kos, Greece, pp. 4–14, June 2008.

23. K. Z. Ibrahim and F. Bodin. Efficient SIMDization and data management of the Lattice QCD computation on the cell broadband engine. *Journal of Scientific Computing*, 17(1–2):153–172, 2009.

24. K. Z. Ibrahim, F. Bodin, and O. Pene. Fine-grained parallelization of lattice QCD kernel routine on GPUs. *Journal of Parallel and Distributed Computing*, 68(10):1350–1359, 2008.

25. S. Lee, S.-J. Min, and R. Eigenmann. OpenMP to GPGPU: A compiler framework for automatic translation and optimization. *The 14th ACM SIGPLAN Symposium on Principles and Practice of Parallel Programming (PPoPP '09)*, Raleigh, NC, pp. 101–110, 2009.

26. NAS Parallel Benchmarks (NPB). http://www.nas.nasa.gov/resources/software/npb.html

27. K. O'Brien, K. O'Brien, Z. Sura, T. Chen, and T. Zhang. Supporting OpenMP on cell. *International Journal of Parallel Programming*, 36(3):289–311, June 2008.

28. J. D. Owens, D. Luebke, N. Govindaraju, M. Harris, J. Krüger, A. E. Lefohn, and T. J. Purcell. A survey of general-purpose computation on graphics hardware. *Eurographics 2005, State of the Art Reports*, Dublin, Ireland, pp. 21–51, August 2005.

29. PGI Accelerator Compilers. http://www.pgroup.com/resources/accel.htm

30. M. Puschel, J. M. F. Moura, J. R. Johnson, D. Padua, M. M. Veloso, B. W. Singer, J. Xiong, F. Franchetti, A. Gacic, Y. Voronenko, K. Chen, R.W. Johnson, and N. Rizzolo. SPIRAL: Code generation for DSP transforms. *Proceedings of the IEEE*, 93(2):232–275, February 2005.

31. S. Ryoo, C. I. Rodrigues, S. S. Stone, S. S. Baghsorkhi, S.-Z. Ueng, J. A. Stratton, and W. M. W. Hwu. Program optimization space pruning for a multithreaded GPU. *The Sixth Annual IEEE/ACM International Symposium on Code Generation and Optimization (CGO '08)*, Boston, MA, pp. 195–204, 2008.

32. S. Scott. Building effective, power-efficient systems over the next decade. *The ACM/IEEE Conference on Supercomputing: Workshop on Power Efficiency and the Path to Exascale Computing*, Austin, TX, November 2008.

33. J. Spray, J. Hill, and A. Trew. Performance of a lattice quantum chromodynamics kernel on the cell processor. *Computer Physics Communications*, 179(9):642–646, 2008.

34. The Green500 list. http://www.green500.org/

35. The top 500 supercomputer sites. http://www.top500.org/

36. R. Vuduc, J. W. Demmel, and K. A. Yelick. OSKI: A library of automatically tuned sparse matrix kernels. *Journal of Physics: Conference Series*, 16(1):521–530, 2005.

37. R. C. Whaley, A. Petitet, and J. J. Dongarra. Automated empirical optimizations of software and the ATLAS project. *Parallel Computing*, 27(1–2):3–35, 2001.

38. S. Williams, J. Carter, L. Oliker, J. Shalf, and K. Yelick. Lattice Boltzmann simulation optimization on leading multicore platforms. *IEEE International Symposium on Parallel and Distributed Processing (IPDPS '08)*, Miami, FL, pp. 1–14, April 2008.

39. S. Williams, L. Oliker, R. Vuduc, J. Shalf, K. Yelick, and J. Demmel. Optimization of sparse matrix-vector multiplication on emerging multicore platforms. *Parallel Computing*, 35(3):178–194, 2009.

40. S. Williams, J. Shalf, L. Oliker, S. Kamil, P. Husbands, and K. Yelick. Scientific computing kernels on the cell processor. *International Journal of Parallel Programming*, 35(3):263–298, 2007.

41. Xilinx programmable chips. http://www.xilinx.com/

37

Experience with Autonomic Energy Management Policies for JavaEE Clusters

Daniel Hagimont
University of Toulouse

Laurent Broto
University of Toulouse

Aeiman Gadafi
University of Toulouse

Noel Depalma
Joseph Fourier University

37.1 Introduction

Nowadays, in response to the continuous growth of computing needs, many organizations rely on clusters to obtain the required computing power. Two strategies are generally followed: either building and managing a cluster within the organization itself, or externalizing its computing infrastructure which is therefore managed by a third-party hosting provider.

In both cases, the energy used by servers and data centers is significant. According to the U.S. Environmental Protection Agency (EPA) [1], energy consumption of these infrastructures is estimated to have doubled between 2000 and 2006 and the development of hosting centers will amplify this tendency. Minimization of the energy consumption has become a crucial issue for organizations that manage such infrastructures.

Autonomic management systems [2] have been proposed as a solution for the management of distributed infrastructures to automate tasks and reduce administration burden. Such systems can be used to deploy and configure applications in a distributed environment. They can also monitor the environment and react to events such as failures or overloads and reconfigure applications accordingly and autonomously.

In this chapter, we report on an experiment which consists in using an autonomic management system to dynamically adapt the managed environment according to its submitted load in order to save energy. The application we considered is a clustered JavaEE application and we relied on an autonomic management platform—called TUNe [3]—that we formally implemented. We experimented with two scenarios.

In a first scenario, we consider a dedicated cluster scenario (the cluster is dedicated to run the JavaEE application). We use TUNe to intervene at the application level, i.e., to adapt the application architecture according to the submitted load. In the second scenario, we consider a hosting center scenario (the cluster can be used to host any kind of application). We exploit virtual machines (VMs) and more precisely VM migration, in order to ship VMs on more or less physical machines according to the monitored load. In both cases, unused machines can be dynamically turned off to save energy. We show that our autonomic management system can be used to implement both scenarios and to implement energy-aware management policies.

The rest of the chapter is structured as follows: Section 37.2 presents the context of our work and our motivations. Section 37.3 presents the energy-aware management policies we designed. Section 37.4 describes the TUNe autonomic management system and Section 37.5 the implementation of energy management policies with TUNe. Our experimental results are reported in Section 37.6. After an overview of related works in Section 37.7, we conclude with Section 37.8.

37.2 Context

In this section, we introduce the application and the two scenarios we consider in our experiments.

37.2.1 E-Commerce JavaEE Applications

In our experiments, we considered an e-commerce application structured with the JavaEE model. The Java Enterprise Edition (JavaEE) platform defines a model for developing Web applications [4] in a multitiered architecture. Such applications are typically composed of a Web server (e.g., Apache [5]), an application server (e.g., Tomcat [6]), and a database server (e.g., MySQL [7]). Upon an HTTP client request, either the request targets a static Web document, in this case the Web server directly returns that document to the client, or it refers to a dynamically generated document, in that case the Web server forwards the request to the application server. When the application server receives a request, it runs one or more software components (e.g., Servlets, EJBs) that query a database through a JDBC driver (Java DataBase Connection driver). Finally, the resulting information is used to generate a Web document on-the-fly that is returned to the Web client.

In this context, the increasing number of Internet users has led to the need for highly scalable and highly available services. To deal with high loads and provide higher scalability of Internet services, a commonly used approach is the replication of servers in clusters. Such an approach usually defines a particular software component in front of each set of replicated servers, which dynamically balances the load among the replicas (e.g., MySQLProxy). Here, different load-balancing algorithms may be used, e.g., Random, Round-Robin, etc.

In such an architecture, an energy-performance trade-off is difficult to find. Use of more machines will achieve a good performance, e.g., minimize the response time, but at the same time will consume more energy. The administrator of such application must find the best degree of replication for each tier, which should be sufficient to tolerate load peaks, but not too high to prevent energy wasting. This task is very complex, especially with the variation of the load in such an application.

37.2.2 JavaEE Clustering in Practice

As mentioned in the introduction, we consider two scenarios to use a cluster.

37.2.2.1 Dedicated Cluster

In this first case, the enterprise is responsible for the management of its own physical infrastructure and also of the applications which are installed and launched on it. The administrator of this cluster has a direct access to hardware resources as well as applications. Energy management in this case will directly influence the organization's budget.

37.2.2.2 Hosting Center

In this second scenario, the organization decides to delegate the management of the physical infrastructure. It relies on a hosting center which provides access to the required computing resources. The provider is expected to meet the QoS requirements of its customers.

Regarding administration, we can here distinguish two different roles: the client administrator who works for the organization and who has the responsibility to install the applications on the infrastructure provided by the provider, and the provider administrator who works for the hosting center and who has the responsibility to manage the physical infrastructure without necessarily knowing the applications running on the cluster.

The motivations are also different from these two points of view: From the client's point of view it is interesting to save money by paying for the resources actually used for its application, and from the provider's point of view it is interesting to save energy by using the minimum resources (physical machines) that satisfy the client QoS needs.

37.2.3 Dynamic Adaptation

In the context of a clustered JavaEE application, minimizing energy consumption while meeting the performance needs cannot be achieved with a static allocation (once at launch time without any runtime allocation). With static allocation, it is difficult for the application administrator to decide on the degree of replication of each tier. If the degree of replication of one tier is too low, this tier will saturate and may become a bottleneck for the whole application, leading to a global slowdown. If the degree of replication is too high, the allocated machines are underused, which leads to energy wasting.

Therefore, it is interesting to benefit from a dynamic adaptation of the software environment according to the load to avoid energy wasting and meet performance needs. This adaptation is a means to dynamically allocate or free machines, i.e., to dynamically turn cluster nodes on (to be able to efficiently handle the load imposed on the system and guarantee the performance of the application) and off (to save power under lighter load).

Turning machines off, and especially on, is quite costly. Indeed, we measured that such an operation takes about 45 s in our experimental environment (Section 37.6.1). Instead, we rely on suspension to RAM, which allows to suspend and resume the activity of a machine at a low cost (about 4 s on the average for resuming a machine) while saving as much energy as if it were turned off. Suspend-to-RAM stores information on the system configuration, the open applications, and the active files in main memory (RAM) while most of the system's other hardware is turned off. When a machine is suspended, only the RAM and the network device are powered on.

We will see in the next sections that such dynamic adaptations can be implemented with the different scenarios we consider: a dedicated cluster or a hosting center.

37.3 Energy Management Policies

The main idea we are exploring in this chapter is to rely on an autonomic management system in order to save energy in a cluster by dynamically allocating or de-allocating resources to a JavaEE application according to the load submitted to that application. We identified two ways to implement such energy management for a clustered JavaEE application.

In the first solution we encapsulate the JavaEE tiers and we reconfigure the JavaEE architecture autonomously to increase/decrease the number of replicas used by the application according to the received load. This solution corresponds to the case of a dedicated cluster (Section 37.2.2.1) because the administrator can manage the application level.

In the second solution, we exploit VM technologies and more precisely VM migration, in order to locate JavaEE tiers on more or less physical machines according to the load. Therefore, we don't reconfigure the JavaEE application, but relocate the VMs on which the JavaEE tiers are running. This solution is more convenient in the case of a hosting center (Section 37.2.2.2), as applications running on the cluster are usually unknown to the administrator.

37.3.1 Solution 1: Dynamic Replication in a Dedicated Cluster

This approach consists in adding or removing tier replicas according to the monitored load of each tier. The application administrator chooses for each tier the number of initial replicas that should be launched at deployment time. Each replica is deployed and launched on a separate node. The administrator also specifies the reconfiguration policy that should be applied. A reconfiguration policy generally takes the form of a threshold that a monitored load should not exceed (or a monitored quality of service that should be maintained). Monitoring is performed by sensors that notify events whenever a constraint is violated.

According to defined thresholds, the probe will eventually generate an event to add or remove a replica. The reaction to an *overload* event is as follows:

- Identify the overloaded tier.
- Allocate a new machine from a pool of free nodes and turn it on via *wake-on-lan* notification.
- Deploy and launch a new tier replica on that machine (e.g., a MySQL server).
- Reconfigure the load balancer associated with this tier (e.g., a restart of MySQLProxy) to integrate the new replica.

Conversely, the reaction to an *underload* event is as follows:

- Identify the underloaded tier.
- Select a tier replica to remove from the application.
- Reconfigure the load balancer to remove the replica.
- Stop and undeploy the tier replica.
- De-allocate the machine that was hosting the replica, free the machine via *suspend-to-ram* notification.

This solution is quite generic as it can be used with different applications as long as they rely on a master-slave scheme with replicated servers and load balancers.

37.3.2 Solution 2: Virtual Machine Consolidation in a Hosting Center

In order to provide a solution that does not require any knowledge of the managed applications, the second solution exploits the notion of virtualization and migration capabilities.

In this case, we consider that the cluster administrator does not know which applications will be deployed on the cluster. He provides to clients an access to VMs. The client is responsible for deploying its application on these VMs.

We consider also that the client (the administrator of the JavaEE architecture) statically allocates the machines (VMs) on which the JavaEE tiers will be deployed. He must choose for each tier the maximum number of instances in the worst case (at the maximum load). Then each server instance is created and deployed on a VM. Depending on the provider policy, the client may be billed only for the actually used physical resources.

The autonomic management system is used to deploy and manage the VMs. It deploys them on the minimal number of nodes able to accept the load. The cluster administrator must specify a reconfiguration policy that should be applied when the load increases/decreases. The sensors in this case are used to monitor the status of the physical machines. These sensors notify the autonomic management system in case of overload/underload.

The reaction to an *overload* event is as follows:

- Identify the most loaded physical machine.
- Select a VM to migrate on that machine.
- Identify a (underloaded) physical machine that can host the selected VM.
- If all the machines are busy, allocate a new machine from a pool of free machines and turn it on via *wake-on-lan* notification.
- Migrate the selected VM toward this machine.

Conversely, the reaction to an *underload* event is as follows:

- Identify the least loaded physical machine.
- Select a VM to migrate on that machine.
- Identify a physical machine that can host the selected VM.
- If one was found, migrate the selected VM toward this machine.
- If the migrated VM was the last one on the physical machine, turn it off via *wake-on-lan* notification.

In the next section, we present the TUNe autonomic administration system and the implementation of the aforementioned policies on TUNe.

37.4 TUNe Autonomic Administration System

An autonomic management system is a system that provides support for the management of applications. It generally covers the life cycle of the applications it manages: deployment, configuration, launching, and dynamic management as well. Dynamic management implies that the system should allow monitoring of the execution and reaction to events such as failures or load peaks, in order to adapt the managed application accordingly and autonomously.

Many works in this area have relied on a component model to provide such an autonomic system support. The basic idea is to encapsulate (wrap) the managed elements (legacy software) in software components and to administrate the environment as a component architecture. Then, the administrators can benefit from the essential features of the component model, encapsulation, deployment facilities, and reconfiguration interfaces, in order to implement their autonomic management processes.

Autonomic management, an area in which we have been working for several years, can be obviously very helpful for implementing dynamic adaptations in a cluster for energy management. It can be used to monitor applications in a cluster and whenever it is required, allocate a new machine, deploy the required software component on that node, and reconfigure the application in order to integrate this new component. It can also be used to monitor machines and migrate VMs according to the monitored load. Notice here that autonomic management has also been much used to implement repair behaviors to tolerate software or hardware failures.

The work presented in this chapter relies on an autonomic management platform called TUNe [3] developed in the same research group. The following sections present the TUNe system.

37.4.1 Design Principles of TUNe

Several researches conducted in the autonomic management area follow an approach that we call component-based management (introduced in the previous section). The TUNe system implements this approach (relying on the fractal component model [8]), but it evolved to provide higher-level formalisms for all administration tasks (wrapping, deployment, reconfiguration).

Regarding the initial configuration of a deployed architecture, our approach is to introduce a UML profile for graphically describing configuration and deployment architecture (such a specification is called the *configuration schema*). The introduced formalism is more intuitive and abstract (than traditional architecture languages) as it describes the general organization of the configuration (types of components and interconnection pattern) in intension, instead of describing in extension all component instances that have to be deployed. This is particularly interesting for applications where tens or hundreds of servers have to be deployed. Thus, any component in the configuration schema is instantiated into many instances, and bindings between component instances are computed from the described interconnection pattern. Therefore, the configuration schema is used to generate a component-based architecture that we call the *System Representation* (SR) of the administrated application. In the described configuration schema, each component has a set of attributes that describes the configuration of the encapsulated software and a set of methods to configure and reconfigure the component. Similarly to software, the description of the environment in which software will be deployed is based on the same UML profile. A cluster of physical machines is represented by a Node component. This later encapsulates the characteristics of the set of machines it represents. Any Node component is instantiated in the SR into several Node component instances. A special software component attribute indicates to TUNe the target cluster in which it has to be deployed. The deployment process instantiates an implicit binding in the SR between each software component instance and its deployment Node component instance. For example, Figure 37.1a depicts the J2EE configuration schema used in our dedicated cluster solution.

Regarding wrapping, the instantiated components in the SR are wrappers of the administrated legacy software. TUNe introduces a Wrapping Description Language (WDL) to specify the behavior of wrapper components. WDL allows defining methods (for software and Node components) that can be invoked to start, stop, configure, or reconfigure the wrapped component. Each method declaration indicates a method implementation in a Java class (this implementation is specific to the encapsulated software but most of them can be reused for different software) and the attributes from the configuration schema that should be passed to the method upon invocation. WDL allows navigation in the configuration schema, which means that in a wrapper description, the specification of a parameter to be passed to a method can be a path (following bindings between components) to a component attribute, this path starting from the wrapper component.

Regarding reconfiguration, our approach is to introduce a very simple specific language called Reconfiguration Description Language (RDL), to define workflows of operations that have to be executed for reconfiguring the managed environment. An event is generated either by a specific monitoring component (e.g., a probe in the configuration schema) or by a wrapped legacy software which already includes its own monitoring functions. A specific API allows a Java method associated with a wrapper to send such an event to TUNe, which will trigger the execution of a reconfiguration. RDL only provides basic language constructs: variable assignment, component method invocation, and *if...then...[else]* statement. It also allows addition or removal of component instances, which is the only way to modify the architecture of the SR. One of the main advantages of RDL is that reconfigurations can only produce an (concrete) architecture which conforms with the configuration schema, thus enforcing reconfiguration correctness.

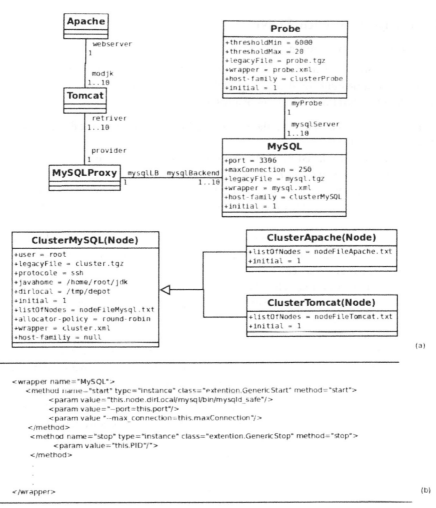

FIGURE 37.1   Dedicated cluster solution with TUNe.

37.4.2 Management with TUNe

37.4.2.1 Configuration Schema

Two kinds of attributes and methods can be associated with components: static and instance. Static attributes have the same value for all instances of the component while instance attributes are specific to an instance. We identified three categories of attributes in TUNe:

1. Attributes introduced by users in their configuration schemas. Since the description is in intension, they are always static.
2. Static attributes introduced by TUNe. They are required by TUNe to implement administration and have to be assigned by users. These attributes are
   a. *initial*: defines the number of instances to be created and deployed by TUNe.
   b. *legacyFile*: the archive file which includes the software to deploy (binary code and configuration files).
   c. *wrapper*: the file which contains the description of the wrapper (a WDL description).
   d. *host-family*: specifies the cluster (a Node component) on which the software component instances will be deployed.
3. Instance attributes introduced by TUNe are
   a. *srName*: a unique name associated with the component instance, which is generated by TUNe at creation time.
   b. *nodeName*: the real name (IP address or DNS name) of a Node component instance. A Node component behaves as a node allocator. The wrapper of a Node component has to implement a method which initializes this *nodeName* attribute for all its instances and methods to allocate or free nodes at runtime (more details are given in the following).

Links defined in the configuration schema have the same semantic as associations in a UML description [9]: Roles are used to identify link extremities. An implicit binding, whose roles are named *node* and *software*, is automatically created by TUNe in the SR between a software component instance and the Node component instance on which it is deployed. TUNe allows to use these links to navigate inside the SR when writing a wrapper or a reconfiguration program. For example, to obtain the MySQL port value from the Apache component wrapper (in Figure 37.1a), the navigation path is *this.modjk.provider.mysqlBackend.port*. Notice that multiple links return a collection of components or attributes. In this case, TUNe provides a *select* operator to get one (random) element from a collection of elements, and a *size* operator to get the number of elements in the collection. Notice also that inheritance can be used in a configuration schema to factorize general attributes in a top component, as depicted in Figure 37.1a.

37.4.2.2 Execution

At deployment time, TUNe creates the Node component instances according to the configuration schema and invokes the *initAlloc()* static method of each Node component wrapper. This method is responsible for initializing the *nodeName* attribute for each Node component instance and initializing the allocator's data structures. Then, each time a software component instance is created, TUNe invokes the *allocate()* static method of the Node component on which the software has to be deployed. This method allocates a node and returns its *srName*. The software component instance is linked with the Node component instance in the SR. By implementing Node component wrappers, a user can define specific node allocation strategies.

This aforementioned deployment process only affected the SR. A special reconfiguration program (called the *starter*) is launched in order to effectively deploy, configure, and start the legacy software which composes the application. This program usually invokes wrapper methods which deploy (i.e., copy binaries on remote hosts) and configure (i.e., update configuration files) for the legacy software with attributes defined in the configuration schema. It finally invokes methods which start the processes associated with the software elements.

A reconfiguration program can invoke any method on the component instances. When a reconfiguration is triggered by an event generated by a probe, the *this* variable references this probing component instance, which is the starting point for navigating in the SR and accessing component instances and their attributes and methods. TUNe also introduces particular methods for modifying the SR:

- *aComponent.add(aNode)*: this static method creates a new instance in the SR. The created instance is bound with the *aNode* Node component instance and it is integrated in the SR according to the interconnection pattern defined in the configuration schema.
- *anInstance.remove()*: this instance method removes the target instance from the SR.

37.4.2.3 Wrappers

The WDL allows describing in a simple XML file the methods which implement the behavior of a wrapper. Figure 37.1b shows the WDL description of the wrapper for the MySQL software. The name of the method will be used in the reconfiguration program. For each declared method, the *class* and *method* attributes reference the Java implementation of the method. For example, the *start()* method is implemented by the *start_with_pid_linux* method of the *extention.GenericStart* Java class. This method takes three parameters. The *this* key word references the current wrapped component instance.

We present in the next section the implementation of our two considered scenarios.

37.5 Policy Implementations with TUNe

In this section, we present the diagrams used by TUNe to implement our solutions.

37.5.1 Solution 1: Dynamic Replication in a Dedicated Cluster

In our experiments, we focus on resource allocation for the MySQL tier. Figure 37.1a shows the configuration schema for this JavaEE application (the attributes are detailed only for the MySQL tier). The figure describes the organization of Node components, defining a different cluster of machines for each tier (*ClusterApache, ClusterTomcat, ClusterMySQL*). As many attributes of these Node components are similar, we use inheritance to factorize many common attributes. The top part of the figure describes the JavaEE software architecture, which initially includes a single MySQL server (but reconfigurations may add new servers). A Probe component (a unique instance) is used to monitor the MySQL tier: it defines threshold attributes which indicate the minimum and the maximum accepted load on MySQL tier (underloads and overloads generate an event).

The deployment of the overall architecture is done into two steps. In the first step, Node components are created in the SR and since the *host-family* attribute is *null*, TUNe does not perform any physical deployment (this is the bootstrap). TUNe invokes the *initAlloc()* method to initialize Node component instances. In this solution, the *initAlloc()* method (implemented in the wrapper) uses a file of physical machine DNS names, the path to this file being given by the *listOfNodes* attribute. At the end of this step, Node component instances are created and initialized and the allocators associated with the Node components are also initialized.

In the second step, TUNe deploys, configures, and starts JavaEE servers according to the configuration schema and the *starter* program (Figure 37.1c). This later ensures that (1) binary code are deployed, (2) software configuration files are well generated, and (3) servers are started following an adequate order (MySQL, MySQLProxy, Tomcat, Apache, and Probe). Notice that some configuration operations are performed simultaneously (*parallel* and *branch* statements).

Only one MySQL server is initially deployed. Based on the MySQL tier monitored load, the Probe component can generate two types of event: *fixMySQLOverload* if the MySQL tier is overloaded and *fixMySQLUnderload* if the MySQL tier is underloaded (Figure 37.1c). We detail in the following the reconfiguration program associated with the *fixMySQLOverload* event:

- *this.stop()* identifies the Probe component instance. This action stops the probing component to prevent the generation of multiple events.
- *newNode=ClusterMySQL.allocate()* asks a new machine to the allocator of the *ClusterMySQL* component. The *srName* of the allocated machine is stored in the *newNode* variable.
- *newMySQL=MySQL.add($newNode)* adds a new instance of the MySQL component in the SR. This instance is bound with the *newNode* Node instance. The *srName* of this server is stored in the *newMySQL* variable.
- *$newMySQL.deploy()* ... invokes the *deploy, configure* and *start* methods on the new instance.
- *this.mysqlServer.select.mysqlLB.restart()* selects one MySQL component instance (monitored by the Probe) and restart the MySQLProxy component instance it is linked with.
- *this.stop()* restarts the Probe instance.

Similarly, the reconfiguration program associated with the *fixMySQLUnderload* event reduces the number of MySQL server instances when the MySQL tier is underloaded. The program selects one MySQL server and removes it from the architecture.

37.5.2 Solution 2: Virtual Machine Consolidation in a Hosting Center

An important aspect of TUNe is that nodes are themselves managed as software components. Therefore, VMs can be wrapped and managed by TUNe. Therefore, the implementation of this solution is decomposed into three stages: (1) physical Node components are configured, (2) VMs are deployed on physical machines, and (3) JavaEE servers are deployed on VMs. In this solution, only MySQL servers (our study tier) are deployed on VMs, the others use physical machines.

Stage 1: The *Cluster* Node component represents the set of physical nodes as in Solution 1 with a simple modification: its allocation strategy returns a single machine at deployment time and returns a new machine at reconfiguration time. Concretely, after the deployment of VMs, the reconfiguration strategy changes from *singleNode* to *round-robin*. This strategy allows to initially deploy VMs on a single machine.

Stage 2: We introduce a new Node component called *ClusterVM*, to manage VMs (see Figure 37.2a). Its *initAlloc()* method configures VM instance DNS names (*nodeName*) with a list of names stored in a file (*listOfNodes* attribute) as in Solution 1. *ClusterVM* implements a simple pool allocation strategy: round-robin. This strategy allows to deploy application server instances on different VMs. Regarding *ClusterVM*'s wrapper (Figure 37.2b), it declares methods that implement hypervisor calls to manage VM. The *startVM* method launches a VM. Its parameters are, respectively, the VM image path, the configured name of the VM, and the name of the machine on which the VM is launched. Similarly, the *shutdown* method turns off a VM. Finally, the *migrate* method migrates a VM from one machine to another.

Stage 3: The deployment of the JavaEE architecture is the responsibility of the client of the hosting center, and the administrator of the hosting center is not aware of the applications which are deployed on the allocated VMs. In the implementation of Solution 2, we considered that TUNe was also used by the client to deploy his JavaEE architecture on top of the VMs (which are managed by TUNe at the hosting center level).

Regarding monitoring, we implemented a Probe component which provides advanced monitoring functions. This Probe component monitors the load of each machine and is aware of which VMs run on which node (thanks to the SR). It can therefore maintain a full cartography of the load of each node/VM. This Probe component provides the following monitoring functions:

- *getMostLoadedNode()*: returns the most loaded node.
- *getMostLoadedNode(aVM)*: returns the most loaded node which can host the *aVM* virtual machine.

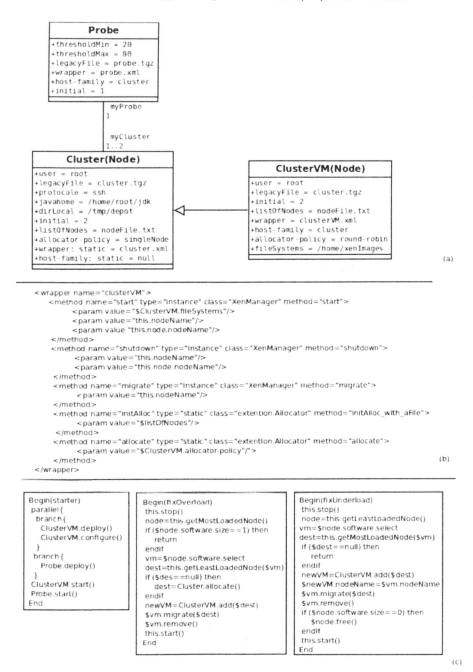

FIGURE 37.2 Hosting center solution with TUNe.

- *getLeastLoadedNode()*: returns the least loaded node.
- *getLeastLoadedNode(aVM)*: returns the least loaded node which can host the *aVM* virtual machine.

Reconfiguration programs (Figure 37.2c) in this solution essentially consist in the migration of VMs. When the probe detects that a machine is overloaded, the *fixOverload* program identifies this overloaded machine (*node* variable). It then selects a VM running on this machine (*vm* variable) and looks for the least loaded machine which can receive the migration of that VM (*dest* variable). Such a migration will better balance the load between the hosts (especially if physical machines do not host the same number of VMs). If such a migration is not possible, a new machine is allocated to receive the migration.

In order to maintain the SR, a VM component instance is added and associated with the destination host. The *nodeName* attribute of this instance is updated and the VM is effectively migrated. Finally, the old VM instance (associated with the left node) is removed.

In this section, we described the implementation of the two considered policies with TUNe. The next section reports on our experiments with these scenarios.

37.6 Evaluation

In this section, we present the evaluation of our energy autonomic management policies implemented with TUNe.

37.6.1 Evaluation Settings

37.6.1.1 Testbed Application

The evaluation has been realized with RUBiS [3], a JavaEE application benchmark based on servlets, which implements an auction site modeled over eBay. It defines 26 Web interactions, such as registering new users, browsing, buying or selling items. RUBiS also provides a benchmarking tool that emulates Web client behaviors and generates a tunable workload. This benchmarking tool gathers statistics about the generated workload and the Web application behavior.

37.6.1.2 Software Environment

The nodes run version 2.6.30 of the Linux kernel. The JavaEE application has been deployed using open source middleware solutions: Jakarta Tomcat 6.0.20 for the Web and servlet servers, MySQL 5.1.36 for the database servers, MySQLProxy 0.7.2 for the database load-balancer, and Apache 2.2.14 for the application server load-balancer. We used RUBiS 1.4.3 as the running JavaEE application. These experiments have been realized with Sun's JVM JDK 1.6.0.05. We used the MySQL Connector/J 5.1 JDBC driver to connect the database load-balancer to the database servers.

37.6.1.3 Hardware Environment

The experimental evaluation was carried out using the Grid'5000 experimental testbed.* The experiments required up to nine nodes: one node for the TUNe management platform, one node for the Web server tier (*Apache*), one node for the servlet tier (*Tomcat*), one node for the database load-balancer (*MySQLProxy*), up to three nodes for database servers (*MySQL*), and two nodes for RUBiS client emulators (which emulate up to 3000 clients). The number of nodes actually used during these experiments varies, according to the dynamic changes of the workload which result in dynamic reconfigurations. All the nodes are connected through a 100 Mbps Ethernet LAN to form a cluster.

37.6.1.4 Test Configuration

In this evaluation, we provide measurements for the database replicated tier only. The RUBiS benchmark is configured to send read-only queries. The parameters that control dynamic reconfigurations are shown in Table 37.1.

For the first policy (dedicated cluster), the probe monitors the average latency of requests sent to the MySQL tier. It triggers a *fixMySQLOverload* event whenever the observed latency exceeds 6000 ms. We

---

* Initiative from the French Ministry of Research through the ACI GRID incentive action, INRIA, CNRS, and RENATER and other contributing partners (see https://www.grid5000.fr).

TABLE 37.1 Test Configuration

Parameter Name	Dedicated Cluster Solution (Latency) (ms)	Hosting Center Solution (CPU) (%)
thresholdMax	6000	80
thresholdMin	20	20

consider here that this is the maximal acceptable waiting time for a client of the Web server. Notice here that nodes allocated to the database tier may see their CPU saturated before reaching this 6000 ms threshold (the degradation of the latency is linear). Conversely, a *fixMySQLUnderload* event is triggered when the observed latency gets lower than 20 ms.

For the second policy (hosting center), the probe monitors the average CPU usage of the machines which host VMs, which should be kept between 20% and 80%. It triggers *fixOverload* and *fixUnderload* events accordingly. Notice here that with this policy, the latency will be kept at its minimal level since physical nodes allocated to this tier will never be saturated.

37.6.1.5 Evaluation Scenario

We aim at showing that dynamic allocation and deallocation of nodes in response to workload variations allows energy saving.

In our benchmark, the load increases progressively up to 3000 emulated clients: 50 new clients every 30 s. We configured RUBiS client to run for 31 min, so the total time of the experiments is about 1 h.

We compare the QoS (e.g., latency) variable and the energy consumption (e.g., occupancy rate of machines) in two situations:

- *Static configuration*: We measure the performance (regarding QoS and energy consumption) with one, two, and three servers. Fewer servers will save energy but with a degradation of QoS. More servers will optimize QoS, but with energy wasting.
- *Dynamic configuration*: TUNe is used to dynamically adapt the number of servers as described earlier. Therefore, quality of service is guaranteed without wasting energy. We then compare the two solutions described in Section 37.3.

37.6.2 Static Configuration

The first experiment we conducted was to measure the energy consumption of one node according to its load. To do this we used a framework that collects energy usage information in Grid'5000.* The results are given in Table 37.2.

From these measurements, we can see that keeping a machine on when it is not used has a high cost. This means that powering off machines is the most effective way to save energy.

In a second experiment, we measured the CPU usage and the average latency of the MySQL tier when it is statically configured with one, two, and three MySQL servers. These measurements are given in Figures 37.3 and 37.4.

TABLE 37.2 Energy Consumption for One Node

CPU Usage (%)	Energy Consumption (W)
0	204
50	211
100	224

---

* For more information see http://www.ens-lyon.fr/LIP/RESO/Projects/GREEN-NET/

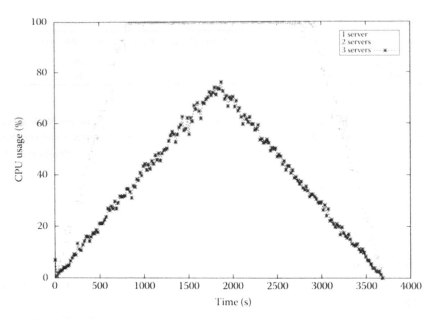

FIGURE 37.3   CPU usage with a static configuration.

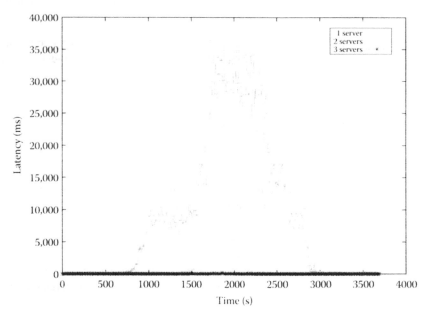

FIGURE 37.4   Latency with a static configuration.

We observe that when we use one database server, its CPU becomes saturated between times 750 and 2950 (Figure 37.3). This saturation has an impact on the quality of service as shown by Figure 37.4. Indeed, when the server is saturated, the latency increases at the same time. We observe the same behavior when we use two servers (between times 1500 and 2200). The configuration with three database servers can maintain the quality of service for all the experiment's time (the curve is totally flat).

In conclusion, for this experiment, maintaining the quality of service requires three database servers, but these three servers are not required during the whole experiment, which leads to energy wasting.

37.6.3 Solution 1: Dynamic Replication in a Dedicated Cluster

In this first solution (dedicated cluster) with the same workload, TUNe dynamically reconfigures the application according to the monitored latency. It adds or removes MySQL servers between the defined thresholds (Table 37.1) and turns cluster nodes on or off, so that we only use nodes (and consume energy) when needed.

Figure 37.5 shows the latency, CPU load, and number of used nodes during the experiment with the first solution. When the observed latency reaches the maximal threshold (6000 ms), a new machine is allocated and a new MySQL server is added to the architecture. The observed latency consequently falls down to a lower value (still higher than the minimal threshold). Such a reconfiguration occurs twice at times 1030 and 1930. We also observe that the average CPU on that tier saturates before reaching the maximal threshold. When the monitored latency gets below the minimal threshold, a database replicas is removed (if possible), which increases the average CPU load on the remaining replicas.

37.6.4 Solution 2: Virtual Machine Consolidation in a Hosting Center

In this section, we present an evaluation on the case of hosing center solution.*

As described before, TUNe is used in this case to administrate the VMs independently from the software running on these VMs. In this experiment, we use three VMs. These VMs are initially deployed on one physical machine. When the cpu load reaches the threshold defined by the administrator, the probe triggers a *fixOverload* event. TUNe reacts by migrating VMs according to the policy defined by the administrator. It can be observed in Figure 37.6 where four migrations occur: two in response to *fixOverload* events at times 570 and 890 (when the average CPU load reaches 80%) and two in response to *fixUnderload* events at times 1420 and 1590 (when the average CPU load gets lower than 20%).

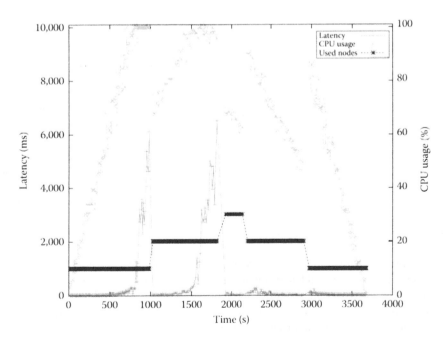

FIGURE 37.5   Dynamic reconfiguration in the dedicated cluster solution.

---

* Note that this experiment was carried out in a different cluster of Grid'5000, which explains why the duration of the experiment is different than the one presented in the previous section.

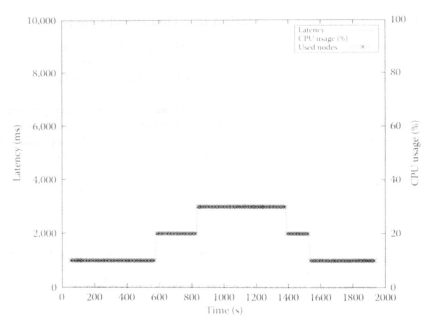

FIGURE 37.6   Dynamic reconfiguration in the hosting center solution.

We can see that in this scenario, the latency is kept at its lower level since VMs are migrated so that physical machines are never saturated. Only very little latency variations are observed when VM migrations are performed.

37.7 Related Work

Many works have addressed the issue of power management. Most of them have focused on energy management for electronic devices powered by electric battery, and few have addressed this issue in grid or cluster infrastructures.

Many research works that aim at managing energy for a single processor system can be used to optimize energy consumption independently on each node of a cluster. We can cite for example the work of Elnozahy et al. [10], which proposes and evaluates a strategy to reduce the energy consumption in a cluster by changing the frequency of each CPU independently. In the same vein, we can cite other researches which target the optimization of other hardware components like memory [11], disk [12], etc.

Most of the research that focuses on cluster-wide energy management deals with resource allocation. We can mention some examples of such works:

- *Load balancing*: In this category, we can cite the work of Pinheiro et al. [13] who developed an algorithm which makes load balancing decisions by considering both the total load imposed on the cluster and the power and performance implications of turning nodes off.
- *Virtualization*: In this category, we can cite the work of Hermenier et al. [14] who developed a system which relies on virtual machine migration for transparently relocating any application in the cluster. The placement policy takes into account the CPU and memory usages, in order to concentrate the workload on fewer nodes of the cluster, thus allowing unused nodes to be shut down. We are currently cooperating with them to integrate our autonomic management system with their work.

- *Simulation*: We can here cite the work of Khargharia et al. [15]. They present a theoretical framework and methodology for autonomic power and performance management in data centers. They rely on simulation to apply their approach to two case studies, a multi-chip memory system and a high-performance server cluster.

Our work is orthogonal to these contributions. While most of the works made in this domain are specific to energy management, our autonomic computing approach is generic, as it can be used to define any management policy for any distributed software architecture. The field of energy management was not previously addressed by TUNe, but the experiments reported in this chapter show that TUNe can by used to define an energy management autonomic policy.

The closest work to ours is that of Das et al. [16] who proposed a multi-agent approach for managing power and performance in server clusters by turning off servers under low-load conditions. Instead of relying on components and architectures, their autonomic system follows a multi-agent paradigm.

37.8 Conclusion

In this chapter, we reported on our experiments in using an autonomic management system to provide energy-aware resource management in a cluster. We improve energy management of clusters by dynamically resizing the active server set to take into account varying workload conditions. We illustrate our approach with two scenarios: the case of a dedicated cluster, and the case of a hosting center.

The experiments that we conducted show that the autonomic computing approach meets the need of energy-aware computing, which can be summarized as follows: minimize power consumption, without affecting the performance of the system. As shown in our experiments, we were able to reduce the power consumption in a cluster while maintaining the performance of the system.

This chapter reported on preliminary works. In the near future, we aim at evaluating more deeply our prototype through more elaborated power management policies, which would include other parameters. We also plan to work in the HPC research area and especially with MPI applications which are based on message passing. We plan to modify our approach to take into account MPI process CPU usage and to collocate processes which communicate intensively.

References

1. U.S. Environmental Protection Agency (EPA). Report to Congress on server and data center energy efficiency, August 9, 2007.
2. J.O. Kephart and D.M. Chess. The vision of autonomic computing. In *IEEE Computer Magazine*, 36(1):41–50, 2003.
3. L. Broto, D. Hagimont, P. Stolf, N. Depalma, and S. Temate. Autonomic management policy specification in TUNE. In *SAC'08: Proceedings of the 2008 ACM Symposium on Applied Computing*, pp. 1658–1663, Fortaleza, Ceara, Brazil, 2008.
4. Oracle. Java EE. 2011. http://www.oracle.com/technetwork/java/javaee/overview/index.html
5. The Apache Software Foundation. Apache HTTP server documentation. HTTP Server Project. 2011. http://httpd.apache.org/
6. The Apache Software Foundation. Tomcat documentation. The Apache Jakarta Project. 2011. http://tomcat.apache.org/tomcat-6.0-doc/
7. MySQL. MySQL Enterprise Edition. 2011. http://www.mysql.com/
8. E. Bruneton, T. Coupaye, M. Leclercq, V. Quema, and J.-B. Stefani. The fractal component model and its support in Java. In *Software—Practice and Experience, special issue on "Experiences with Auto-adaptive and Reconfigurable Systems"*, 36(11–12):1257–1284, September 2006.

9. Object Management Group, Inc. *Unified Modeling Language (UML) 2.1.2 Superstructure*, Needham, MA, November 2007. Final Adopted Specification.

10. E.N. (Mootaz) Elnozahy, M. Kistler, and R. Rajamony. Energy-efficient server clusters. In *Proceedings of the 2nd Workshop on Power-Aware Computing Systems*, pp. 179–196, Cambridge, MA, 2002.

11. B. Khargharia, S. Hariri, and M.S. Yousif. Autonomic power and performance management for computing systems. In *Autonomic Computing, 2006. ICAC'06. IEEE International Conference on*, pp. 145–154, Dublin, Ireland, 2006.

12. E. Pinheiro and R. Bianchini. Energy conservation techniques for disk array-based servers. In *Proceedings of the 18th Annual International Conference on Supercomputing, ICS'04*, pp. 68–78, Saint Malo, France, 2004. ACM: New York.

13. E. Pinheiro, R. Bianchini, E.V. Carrera, and T. Heath. Load balancing and unbalancing for power and performance in cluster-based systems. In *Workshop on Compilers and Operating Systems for Low Power*, pp. 4-1–4-8, Barcelona, Spain, 2001.

14. F. Hermenier, N. Loriant, and J.-M. Menaud. Power management in grid computing with Xen. In *Frontiers of High Performance Computing and Networking ISPA 2006 Workshops*, pp. 407–416, Sorrento, Italy, 2006.

15. B. Khargharia, S. Hariri, and M.S. Yousif. Autonomic power and performance management for computing systems. *Cluster Computing*, 11(2):167–181, 2008.

16. R. Das, J.O. Kephart, C. Lefurgy, G. Tesauro, D.W. Levine, and H. Chan. Autonomic multi-agent management of power and performance in data centers. In *AAMAS '08: Proceedings of the 7th International Joint Conference on Autonomous Agents and Multiagent Systems*, pp. 107–114, Richland, SC, 2008. International Foundation for Autonomous Agents and Multiagent Systems.

VIII

Data Centers and Large-Scale Systems

38

Power-Aware Parallel Job Scheduling

Maja Etinski
Barcelona Supercomputing Center

Julita Corbalan
Barcelona Supercomputing Center

Jesus Labarta
Barcelona Supercomputing Center

38.1 Introduction

This chapter deals with power management in high-performance computing (HPC) systems. Power consumption of battery operated devices has been an issue for a long time. Over the last decade power consumption of HPC systems has emerged as a significant problem that even limits future development. Striving for performance has resulted in an enormously high peak power draw. The struggle for performance is reflected in the Top500 list [21] of the 500 most powerful supercomputers. The number one ranked system of the June 2010 list, Jaguar [20], comprises an incredible number of almost 225,000 cores and it brings the theoretical peak capability to 2.3 petaflop/s. Jaguar requires almost 2.8 times the electric power of its predecessor Roadrunner, the number two on the same list. This difference translates into millions of dollars per year in operating costs. Power estimates for exascale computer power dissipation range from many tens to low hundreds of megawatts [14]. Hence power consumption is one of the most important design constraints for HPC centers nowadays. Besides a tremendous increase in cost of ownership, power-awareness in HPC centers is motivated by other reasons such as system reliability and environmental footprint. Recently, the Top500 list has been accompanied by the Green500 list ranking the most energy efficient supercomputers in FLOPS/w.

In an HPC center power reduction techniques may be motivated by operating costs, system reliability and environmental concerns. One might want to decrease power dissipation, accepting a certain performance penalty in return. But there are other reasons for power management in HPC environments; for instance, a power constraint might be imposed by existing power/cooling facilities. In such a situation, the main goal is to maximize performance for a given power budget.

Processor power consumption is a significant portion of total system power. Though the portion is system and load dependent, roughly speaking it makes a bit more than a half of the total system power when the system is under load [12]. The DVFS (dynamic voltage frequency scaling) technique is commonly used to manage CPU power. A DVFS-enabled processor supports a set of frequency-voltage pairs, i.e., gears. Running a processor at lower frequency/voltage results in lower power/energy consumption. Lower frequency normally increases application execution time; thus, frequency scaling should be applied carefully, especially in HPC systems as their main goal is still performance.

Though supercomputers are ranked by achieved number of FLOPS, in daily operation of an HPC center user satisfaction is not determined only by this performance. User satisfaction in an HPC center does not depend only on job execution time but on its wait time as well. The job scheduler has a complete view of the HPC system: it is aware of running jobs and current load, jobs in the wait queue and their wait times, and available resources. Therefore, a job scheduler can estimate job performance loss due to frequency scaling. A job scheduler implements a job scheduling policy and in conjunction with resource selection policy it manages system resources at job level. Since power has appeared as an important resource, it is natural to enable job schedulers to manage power. Work presented in this chapter deals with power-aware parallel job scheduling, which has been proposed recently [3–5].

38.2 Overview of HPC Power Reduction Approaches

Due to its importance, power consumption in HPC systems has been investigated extensively over the last decade. The work presented in this chapter targets CPU power consumption as it takes up a high portion of total system power. Furthermore, it is based on the DVFS technique, widely used to manage CPU power. It is important to note that the work is at the level of job scheduling, implying coarse grain DVFS application. Parallel job scheduling is described in the next section and it should be distinguished from other levels of scheduling, such as task scheduling. Task scheduling for multicore processors has been investigated as well for energy efficiency in related work [1].

Other HPC power reduction works can be divided into two main groups depending on the granularity they deal with. The first group targets power/energy consumption of HPC applications. Power profiling of parallel applications under different conditions is done in some works [11,22]. Although they just report power and execution time on specific platforms for different gears or numbers of nodes, they give a valuable insight in relations between CPU frequency and power and execution time. There are power reduction systems based on previous application profiling [9,16,32]. Several runtime systems that apply DVFS in order to reduce energy consumed per application are implemented [17,23,28]. These systems are designed to exploit certain application characteristics like load imbalance of MPI applications or communication-intensive intervals. Therefore, they can be applied only to certain jobs.

The second group of works deals with system/CPU power management of whole workload instead of only one application. Lawson et al. aim to decrease supercomputing center power dissipation by powering down some nodes [27]. The EASY backfilling is used as the job scheduling policy. In this manner job performance is affected seriously in cases of high load. Two policies are proposed to determine the number of active nodes of the system. The first policy proposed is a two-level policy that fluctuates the number of active processors between the maximum number of processors and a system-specific minimum number of processors. In the presence of fluctuating workload conditions, the two level policy does not behave well. Online simulation policy executes multiple online simulations assuming different numbers of active processors. The system chooses the lowest number of active processors whose computed average slowdown satisfies the predefined service level agreement (SLA). An empirical study on powering down some of the system nodes is done [15]. A resource selection policy used to assign processors to a job is designed in order to pack jobs as densely as possible and accordingly to allow powering down unused nodes. Kim et al. propose a power aware scheduling algorithm for bag-of-tasks applications with deadline constraints on DVFS enabled clusters [26]. It gives a frequency scaling algorithm for a specific type of job scheduling with deadline constraints that is not common in HPC centers. There is one more work on energy consumption optimization for computational grids based on cooperative game theoretical technique [25]. There are works on energy efficiency of server clusters. Fan et al. explore the aggregate power usage characteristics of a large collection of servers [6]. The authors also investigate the possibility of energy savings using DVFS that is triggered based on CPU utilization. Elnozahy et al. propose policies for server clusters that adjust the number of nodes online as well as their operating frequencies according to the load intensity [30]. Pinheiro et al. also decrease power consumption by turning down cluster nodes under low load [31]. Since shutting a node of their system takes approximately 45 s and bringing it back up takes approximately 100 s, it is not recommended to simply shut down all unused nodes.

First we are going to give a short introduction to parallel job scheduling in the next section. It is followed by a high level model of CPU frequency scaling impact on power consumption and application performance. Section 38.5 explains and evaluates two energy saving policies. Finally, the policy that maximize performance for a given power budget is described in Section 38.6. Our conclusions related to power management in HPC systems are exposed in the end.

38.3 Parallel Job Scheduling

A parallel job scheduler determines how to share resources of a parallel machine among jobs submitted to the system. Scheduling at a parallel job level has two dimensions (2D), number of processors requested and execution time. A user of a supercomputing center is obliged to submit the number of requested processors and normally requested time when submitting a job. Parallel job scheduling policy determines the order in which arrived jobs are executed from the waiting queue. On the other hand, a resource selection policy selects which available resources will be used for the job chosen for execution. The earlier used parallel job scheduling policy, first come first served (FCFS), has been replaced by policies based on backfilling that improve system utilization. The idea of backfilling is presented in the next section.

38.3.1 EASY Backfilling Policy

Backfilling-strategies are a set of policies designed to eliminate the fragmentation typical for the FCFS policy. With the FCFS policy a job cannot be executed before previously arrived ones, although there might be holes in the schedule where it could run without delaying the others. Backfilling policies allow a job to run before previously arrived ones, only if its execution does not delay jobs waiting with a reservation. There are many backfilling policies classified by characteristics such as the number of reservations and the prioritization algorithm used in the backfilling queue. The number of reservations determines how many jobs in the head of the wait queue will be allocated such that later arrived jobs cannot delay them. When there are fewer jobs in the wait queue, then reservations jobs are executed in FCFS order. If all reservations are used, the algorithm tries to *backfill* jobs from a second queue (the backfilling queue) where jobs are potentially sorted in an order different from submission time.

The EASY-backfilling is one of the simplest, but is still a very effective backfilling policy. The EASY backfilling queue is sorted in FCFS order and the number of reservations is set to 1. With EASY backfilling a job can be scheduled for execution in two ways that are presented by two functions *MakeJobReservation(J)* and *BackfillJob(J)* in an EASY backfilling implementation. The reservation for the first job in the wait queue is made with *MakeJobReservation(J)*. If at its arrival time there are enough processors, *MakeJob Reservation(J)* will start a job immediately. Otherwise, it will make a reservation for the job based on submitted user estimates of already running job runtimes. With backfilling policies users are expected to provide runtime estimates in order to allow the scheduler to exploit the unused fragments. It is in the user's interest to give an accurate estimate of the runtime as an underestimation leads to killing the job, while an overestimation may result in a long wait time. The EASY-backfilling is executed each time a job is submitted or when a job finishes, making additional resources available for jobs in the wait queue.

If there is already a reservation made with **MakeJobReservation**, **BackfillJob(J)** tries to find an allocation for a job *J* from the backfilling queue such that the reservation is not delayed. It means that the job requires no more than the currently free nodes and will terminate by the reservation time, or it requires no more than the minimum of the currently free nodes and the nodes that will be free at the reservation time. Jobs scheduled in this way are called *backfilled* jobs.

38.3.2 Metrics

Response time, slowdown and bounded slowdown are frequently used metrics in the evaluation process of parallel job scheduling policies [7]. Response time is defined as total wall-clock time from the moment of job submission until its completion time. This time consists of: the waiting time that the job *J* spends waiting for execution (*WaitTime(J)*) and the running time (*RunTime(J)*) during which it is executing on processing nodes. The waiting time itself is used as a job performance metric as well. As job runtimes can vary a lot therefore there is a large variance in response times. There are metrics that take job runtimes into account. Slowdown of a job **J** presents ratio of the job response time and its runtime:

$$Slowdown(J) = \frac{WaitTime(J) + RunTime(J)}{RunTime(J)}. \tag{38.1}$$

Though the slowdown, metric takes into account job runtime when measuring job delay, still a job with short runtime can have very high slowdown in spite of an acceptable wait time. A new metric, bounded slowdown(BSLD) has been proposed to avoid this effect of very short jobs on slowdown statistics. It is equal to the following:

$$BoundedSlowdown(J) = max\left(\frac{WaitTime(J) + RunTime(J)}{max(Th, RunTime(J))}, 1 \right). \tag{38.2}$$

A job with runtime shorter than the threshold *Th* is assumed to be very short and its BSLD has value 1—perfect slowdown. In today's supercomputing workloads, a job shorter than 10 min can be assumed to be very short [24]. Accordingly, in following policy evaluations the threshold *Th* is set to 10 min. As frequency scaling affects job runtime we have defined BSLD of a job executed at reduced frequency:

$$BoundedSlowdown(J, f) = max\left(\frac{WaitTime(J) + RunTime(J) * P_f(J, f)}{max(Th, RunTime(J))}, 1 \right) \tag{38.3}$$

where $P_f(J, f)$ is the penalty factor that determines how much the job runtime increases when the CPU frequency is reduced to *f* (described in Section 38.4.2). BSLD metric gives bounded ratio between time spend in system and job runtime. Defined in this way, *BoundedSlowdown(J, f)* reflects performance loss due to frequency scaling.

38.4 DVFS Modeling at Job Level

In this section we explain how DVFS affects processor power consumption and job runtime. We are concerned with CPU power dissipation while the rest of system power is assumed not to vary much with load. Furthermore, CPU power presents major portion of total system power consumption: for instance, when the system observed in related work is under load, CPU power is 56% of system power [12].

38.4.1 Power Model

CPU power consists of dynamic and static power. Dynamic power depends on the CPU switching activity while static power presents various leakage powers of the MOS transistors. The dynamic component equals to

$$P_{dynamic} = ACfV^2 \tag{38.4}$$

where A is the activity factor, C is the total capacity, f is the CPU frequency, and V is the supply voltage. Here, it is assumed that all applications have same average activity factor, i.e., load-balanced applications. Hence, dynamic power is proportional to the product of the frequency and the square of the voltage.

Regarding power of idle processors there are two possible scenarios. Idle processors can be at the lowest available frequency, as is the case with the OnDemand Linux governor. In this work, the activity factor of idle processors is assumed to be 2.5 lower than the activity of a running processor. This value is based on measurements from related work [8,22]. According to our model and used DVFS gear set, this idle power is equal to 21% of the nominal running power. The other idle power scenario assumes that idle processors are in a low power mode in which power consumption is negligible. These low power modes are supported by modern processors. Moreover, a design of a whole system that consumes negligible power while idling has been proposed [29].

According to [2], static power is proportional to the voltage

$$P_{static} = \alpha V \tag{38.5}$$

where the parameter α is determined as a function of the static portion in the total CPU power of a processor running at the top frequency. Static power also depends on CPU temperature, but for need of high level power modeling this dependence can be neglected. In our experiments static power makes 25% of the total active CPU power at the highest frequency. All the parameters are platform dependent and therefore adjustable in configuration files used as input files of the simulator (Section 38.5.3).

We have used the DVFS gear set given in Table 38.1. The last row of the table presents normalized average power dissipated per processor running a job for each frequency/voltage pair.

In this power management approach, the CPU frequency assigned to a job is the same over the whole execution. Therefore, overheads due to transitions between different frequencies are negligible because the transitions occur at very coarse grain. The same applies for entering or exiting low power modes as all these transitions are measured in milliseconds or less.

38.4.2 Execution Time Model

Running a job at lower frequency reduces CPU power consumption but it increases the job runtime. Usually, due to non-CPU activity (memory accesses and communication latency) the increase in time is not proportional to the change in frequency. The β model, introduced by Hsu and Kremer [18] and investigated by Freeh et al. [10], gives the application slowdown compared to the CPU slowdown. We refer to this slowdown as the penalty function $P_{J,f}$:

$$P_f(J, f) = T(f)/T(f_{max}) = \beta(f_{max}/f - 1) + 1 \tag{38.6}$$

Different jobs experience different execution time penalty depending on their CPU-boundedness. The β parameter has values between 0 and 1. Theoretically, if an application would be completely CPU bound then its β would be equal to 1. β equals to 0 means that execution time is completely insensitive to a

TABLE 38.1 DVFS Gear Set

f(GHz)	0.8	1.1	1.4	1.7	2.0	2.3
V(V)	1.0	1.1	1.2	1.3	1.4	1.5
Norm(P)	0.28	0.38	0.49	0.63	0.80	1.0

change of frequency. β of an application can be computed for a reduced frequency f according to formula (38.6):

$$\beta(f) = \frac{T(f)/T(f_{max}) - 1}{f_{max}/f - 1}. \tag{38.7}$$

The β value of an application can vary only slightly for different values of frequency f. The largest difference that was observed between any two β values for same application but different frequency was 5% ([10]). Therefore the β value is an application characteristic and it does not depend on the amount the frequency was reduced by. β values used in this work are extrapolated from measurement results reported in related work [10]. While sequential applications from the NAS, SPEC INT, and SPEC FP suites have averages of 0.40, 0.59, and 0.71, respectively, parallel benchmarks from the NAS PB suite have a variety of β values from 0.052 of FT class A to 0.466 of SP class C running on eight nodes. Keeping in mind that the nodes were connected by very slow 100 Mb/s network applications have shown less sensitivity to frequency scaling than they would have with a faster network. Hence, we have assumed β values to be slightly higher. Generally, an application running on more nodes spends more time in communication resulting in lower β compared to an execution on fewer processors. We have generated β for each job according to the following normal distributions:

- If the number of processors is less or equal to 4: $N(0.5, 0.01)$
- If the number of processors is higher than 4 and less or equal to 32: $N(0.4, 0.01)$
- If the number of processors is higher than 32: $N(0.3, 0.0064)$

38.5 Energy Saving Policies

Policies presented in this section use frequency scaling to reduce CPU power consumption while controlling overall system performance penalty. Executing a job at lower CPU frequency necessarily increases the job runtime. As explained in Section 38.4.2, impact of frequency scaling on execution time is normally less than proportional to reduction in frequency. Such a penalty in performance might be acceptable for users but this artificial increase in load caused by longer runtimes can decrease performance further by increasing wait times of the jobs in the wait queue. Thus, it is important to apply frequency scaling carefully when the load is not very high as otherwise it leads to high performance degradation.

Two ways to control performance penalty are described in this section. One way to control performance penalty is to predict BSLD at different frequencies of the job being scheduled. This approach is used in Section 38.5.1. The other way uses system utilization as a proxy when to use frequency scaling. It is described in Section 38.5.2.

38.5.1 BSLD-Driven Parallel Job Scheduling

38.5.1.1 Frequency Assignment Algorithm

This policy upgrades the EASY backfilling described in Section 38.3.1. It is extended by a CPU frequency assignment algorithm. In order to perform frequency scaling in a controlled manner, a job's *predicted* BSLD is checked (BSLD metric was defined in Section 38.3.2). We define *predicted* BSLD in the following way:

$$PredBSLD(J) = max\left(\frac{WaitTime(J) + RequestedTime(J) * P_f(J, f)}{max(Th, RequestedTime(J))}, 1\right) \tag{38.8}$$

where *WaitTime* is the job wait time according to the current schedule. *RequestedTime* presents a run time estimate submitted by the user and $P_f(J, f)$ is the penalty factor that depends on CPU frequency and job CPU boundedness. How frequency scaling affects job runtime was described in Section 38.4.2. Here, the job requested time is used instead of the runtime as at the moment of scheduling the runtime is not known. The frequency assignment algorithm selects the lower frequency f for the job J execution only if its predicted BSLD at the frequency f is less than previously set *BSLDthreshold*. Otherwise, the nominal, highest available frequency will be assigned to the job. The scheduler iterates starting from the lowest available CPU frequency trying to schedule a job such that the BSLD condition is satisfied. If it cannot be scheduled at the lowest frequency, the scheduler tries with higher ones.

There is one more condition that has to be satisfied in order to run a job at reduced frequency. A job will be run at reduced frequency only if there are no more than *WQthreshold* jobs in the wait queue (jobs waiting for execution) at the moment of frequency assignment. Otherwise, the highest frequency F_{top} will be selected. This parameter provides an additional control of frequency scaling impact on other jobs still waiting in the wait queue.

38.5.1.2 Integration with the EASY Backfilling

With EASY backfilling, a job can be scheduled in two manners as explained in Section 38.3.1. Here, we describe how the frequency scaling algorithm is incorporated into the scheduling policy. If a job is at the head of the wait queue, it is allocated with *MakeJobReservation(J)*. Depending on current resource availability, the job will be sent to execution immediately or a reservation will be made for it. The other way to schedule a job is with *BackfillJob(J)* function. It is called when there is already a job with a reservation. *BackfillJob(J)* tries to find an allocation for the job J such that the reservation is not violated. *MakeJobReservation(J)* and *BackfillJob(J)* algorithms are shown in Figures 38.1 and 38.2 respectively.

```
MakeJobReservation(J)
if (WQsize ≤ WQthreshold) then
    for f = F_lowest to F_nominal do
        Alloc = findAllocation(J,f);
        if (satisfiesBSLD(Alloc, J, f)) then
            schedule(J, Alloc);
            break;
        end if
    end for
else
    Alloc = findAllocation(J,F_nominal)
    schedule(J, Alloc);
end if
```

FIGURE 38.1 BSLD-driven policy: Making job reservation.

```
BackfillJob(J)
if (WQsize ≤ WQthreshold) then
    for f = F_lowest to F_nominal do
        Alloc = TryToFindBackfilledAllocation(J,f);
        if (correct(Alloc) and satisfiesBSLD(Alloc, J, f)) then
            schedule(J, Alloc);
            break;
        end if
    end for
else
    Alloc = TryToFindBackfilledAllocation(J,F_nominal)
    if (correct(Alloc) and satisfiesBSLD(Alloc, J,F_nominal)) then
        schedule(J, Alloc);
    end if
end if
```

FIGURE 38.2 BSLD-driven policy: Backfilling job.

38.5.2 Utilization-Driven Parallel Job Scheduling

This approach uses lower frequencies when the system is not highly loaded, avoiding an increase in job wait times. As the algorithm is utilization driven, we call this policy UPAS—utilization-driven power-aware scheduling. The frequency assignment algorithm is interval-based. The interval duration is denoted as T hereafter. A job started during the interval I_j will be run at CPU frequency that depends on the previous interval I_{j-1} utilization. Utilization of the jth time interval, U_j, is equal to

$$U_j = \frac{\sum_{k=1}^{N_{jobs}} Proc_k * RunTime_k}{N_{proc} * T} \tag{38.9}$$

where

$Proc_k$ is the number of processors of the kth job that has been executed during the interval I_j
$RunTime_k$ is its execution duration in the interval I_j
N_{proc} is the total number of processors of the system

The CPU frequency assigned to a job depends on two utilization thresholds U_{upper} and U_{lower}. Utilization of previous interval is compared against the thresholds, and depending on the results, CPU frequency is assigned to the job. As in the previous policy, the additional threshold $WQ_{threshold}$ is used to enable better control over performance loss. It prevents the scheduler from frequency scaling when there are many jobs in the wait queue. If there are more than $WQ_{threshold}$ jobs in the wait queue no DVFS will be applied. A job J_k started during the jth interval runs at the frequency determined according to (38.10). WQ_{size} presents the current number of jobs in the wait queue.

$$freq(J_k) = \begin{cases} f_{top} & \text{for } U_{j-1} \geq U_{upper} \text{ or } WQ_{size} > WQ_{threshold}, \\ f_{upper} & \text{for } U_{lower} \leq U_{j-1} < U_{upper} \text{ and } WQ_{size} \leq WQ_{threshold}, \\ f_{lower} & \text{for } U_{j-1} < U_{lower} \text{ and } WQ_{size} \leq WQ_{threshold}. \end{cases} \tag{38.10}$$

f_{upper} and f_{lower} are two frequencies selected from the supported DVFS gear set on the used platform. The CPU frequency assumed in scheduling before the start time was determined according to (38.10) at submission time. The new requested/run time at the assigned frequency presents the requested/run time scaled by a factor determined according to the execution time model (Section 38.4.2). As the load can change, at the job start time CPU frequency is computed again. If the new frequency differs from the previously assigned one, the job scheduler is informed about the new requested time as rescheduling may be needed.

38.5.3 Evaluation

A parallel job scheduling simulator, Alvio [13], has been upgraded to support DVFS enabled clusters and the power management policy. Alvio is an event-driven C++ simulator that supports various backfilling policies. A job scheduling policy interacts with a resource selection policy which determines how job processes are mapped to the processors. In the simulations First Fit is used as the resource selection policy. The next section describes workloads used in the evaluation. It is followed by Section 38.5.3.2 that gives values of policy parameters used in simulations.

38.5.3.1 Workloads

Cleaned traces of five logs from Parallel Workload Archive [19] are used in the simulations. A cleaned trace does not contain flurries of activity by individual users that may not be representative of normal usage. Table 38.2 summarizes workload characteristics. The number in workload name presents the number of processors that the system comprises. In the evaluation process, we have simulated 5000 job

TABLE 38.2 Workloads

Workload—#CPUs	Jobs (K)	AvgBSLD	Utilization	AvgLR
CTC—430	20–25	4.66	70.09	1.61
SDSC—128	40–45	24.91	85.33	8.17
SDSCBlue—1152	20–25	5.15	69.17	2.31
LLNLThunder—4008	20–25	1	79.59	0.80
LLNLAtlas—9216	10–15	1.08	75.25	0.94

part of each workload. The parts are selected so that they do not have many jobs removed. Simulated workload parts and the average BSLD value and system utilization when the traditional EASY backfilling (no DVFS) is used are given in the table. The column *AvgLR* presents average load requested—total number of requested processors normalized with respect to the system size.

Each job is represented by a single line containing various data fields. We have used the following data fields: job number, submit time, run time, requested time, and requested number of processors.

The CTC log contains records for IBM SP2 located at the Cornell Theory Center. The log presents a workload with many large jobs but with a relatively low degree of parallelism. SDSC and SDSC-Blue logs are from the San Diego Supercomputing Center. The SDSC workload has fewer sequential jobs than the CTC workload, while run time distribution is very similar. In the SDSC-Blue workload there are no sequential jobs, so each job is assigned to at least eight processors. The LLNL-Thunder and LLNL-Atlas workloads contain several months worth of accounting records in 2007 from systems installed at Lawrence Livermore National Lab. Thunder was devoted to running large numbers of smaller to medium jobs, while the Atlas cluster was used for running large parallel jobs. More information about workloads can be found in the Parallel Workload Archive [19].

38.5.3.2 Policy Parameters

The frequency assignment algorithm of the BSLD-driven parallel job scheduling policy has two parameters, *BSLDthreshold* and *WQthreshold*. We have tested three values for *BSLDthreshold*: 1.5, 2 and 3. Values used for *WQthreshold* are 0, 4, 16, NO. Zero means no DVFS will be applied if there is a job waiting on execution. Then, we have tested two less restrictive thresholds for the wait queue size, 4 and 16 jobs. The last used threshold sets no limit on the wait queue size. It means that CPU frequency is assigned only based on the predicted BSLD. In the evaluation of this policy, all jobs have the same β value and it is equal to 0.5.

Parameter values for both policies are given in Table 38.3. Same values of the *WQthreshold* are used for evaluation of the UPAS policy. The rest of the utilization-driven algorithm parameters are the interval duration T, two utilization thresholds U_{upper} and U_{lower}, and reduced frequencies f_{upper} and f_{lower}. Analyzing workload utilizations we have decided to set U_{upper} to 80% and U_{lower} to 50%. We have set the higher reduced frequency f_{upper} to 2.0 GHz, the first lower frequency from the used DVFS gear set. The lower reduced frequency f_{lower} is set to 1.4 GHz since, among lower frequencies from the DVFS set, it has

TABLE 38.3 Policy Parameters

Parameter	Policy	Value(s)
BSLDthreshold	BSLD-driven	1.5, 2, 3
WQthreshold	BSLD-driven, utilization-driven	0, 4, 16, NO
U_{lower}	Utilization-driven	50%
U_{upper}	Utilization-driven	80%
f_{lower}	Utilization-driven	1.4 GHz
f_{upper}	Utilization-driven	2.0 GHz
T	Utilization-driven	10 min

the best ratio between energy reduction and penalty in execution time. Two values for the parameter T, 10 min and 1 h, are tested in the simulations. As the difference in results is 1% or less, the algorithm is not very interval duration sensitive. The 10-min interval shows slightly better results in both energy and performance. Hence here we present results for the 10-min interval.

38.5.3.3 Results

The reduction in CPU energy consumption achieved with the BSLD-driven job scheduling policy can be seen in Figure 38.3. The first graph presents computational CPU energy (idle processors do not dissipate power) while the second graph gives energy results assuming that idle processors consume energy according to the model described in Section 38.4.1. Notice that there is almost no difference between the two graphs as each presents values normalized to their corresponding original values.

All workloads except SDSC shows an energy decrease of about 10% or more depending on the thresholds used. For the least restrictive combination of parameters (*BSLDthreshold* = 3 and *WQthreshold* = NO LIMIT) savings in computational energy are up to 22%. The SDSC workload has the worst original performance; its average BSLD without frequency scaling is 24.91. Hence the proposed policy with used *BSLDthreshold* values cannot lead to an energy decrease.

For a fixed *BSLDthreshold* value increasing the wait queue limit gives frequency schedule that results in lower energy. But for a fixed wait queue limit, an increase in *BSLDthreshold* does not necessary result in higher energy savings as one might expect. For instance, the LLNLThunder workload saves 8.95% of computational energy for *BSLDthreshold* = 1.5 and *WQthreshold* = 4, while for the same *WQthreshold* and *BSLDthreshold* = 2, it saves 3.79% of computational energy. In the first case 1219 jobs are run at lower frequency, while that number in the second case is 854. A higher *BSLDthreshold* can lead to fewer reduced jobs in total due to an increase in wait time of jobs that arrive after frequency scaling is applied. The number of jobs run at lower frequency for each workload and parameter combination is shown in Figure 38.4.

Besides the number of jobs run at reduced frequency, their size and frequency used are important as well. For instance, when using the policy with *BSLDthreshold* = 2 and *WQthreshold* = NO LIMIT for the

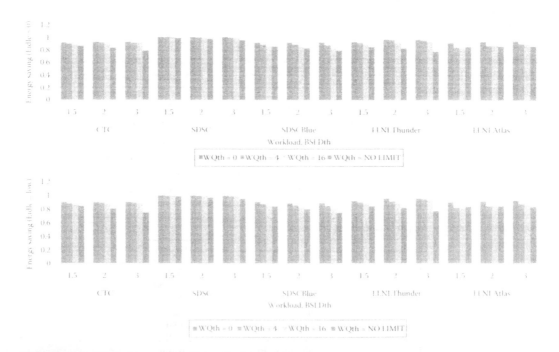

FIGURE 38.3   BSLD-driven policy: Normalized energy.

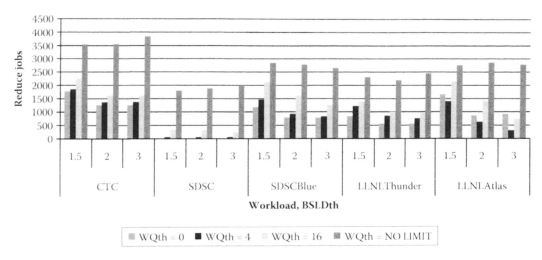

FIGURE 38.4 BSLD-driven policy: Number of jobs run at reduced frequency.

SDSCBlue workload, 2778 jobs run at lower frequency. With *BSLDThreshold* = 3 and *WQthreshold* = NO LIMIT, the policy executes 2654 jobs at lower frequency. In this case greater energy savings are achieved.

Average BSLD values for all parameter combinations are given in Figure 38.5. As shown in Table 38.2, the average BSLD values obtained with the traditional EASY backfilling vary significantly for different workloads. The SDSC workload has the highest original average BSLD 24.91. The best one is the LLNLThunder's average BSLD (= 1). Majority of LLNLThunder jobs are shorter than the threshold from the formula (38.2), hence their BSLD is 1. Therefore, different workloads may have high variation in average BSLD range.

The most aggressive parameter combination *BSLDthreshold* = 3 and *WQthreshold* = NO LIMIT penalizes the most the average BSLD but it gives the highest energy savings. However, penalty in performance is not always directly proportional to energy savings. For example, *BSLDthreshold* = 1.5 and *WQthreshold* = 0 threshold combination for LLNLAtlas gives better energy and performance than *BSLDthreshold* = 2 and *WQthreshold* = 0. Frequency scaling affects performance of the job to which it

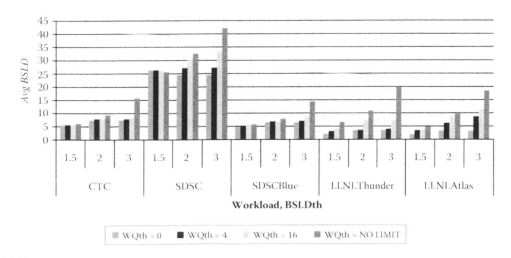

FIGURE 38.5 BSLD-driven policy: Average BSLD.

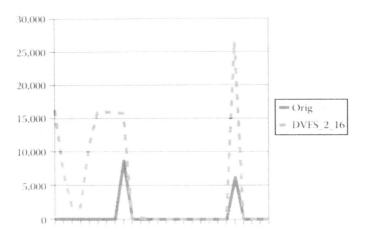

FIGURE 38.6   BSLD-driven policy: Zoom of SDSCBlue wait time behavior.

is applied. Furthermore, it can additionally affect jobs that will be executed after the reduced job as their wait time may increase.

Figure 38.6 shows the job wait time in seconds as a function of the job submit time. The lower line presents wait time without frequency scaling and the upper one shows wait time for $BSLDthreshold = 2$ and $WQthreshold = 16$ parameter combination. It can be observed that in general wait time with frequency scaling is much higher than without it.

Energy consumption with the utilization-driven policy is given in Figure 38.7. The upper graph presents reduction in computational CPU energy (energy needed to execute jobs) and the lower graph gives results assuming that idle processors consume some power, as explained before. Both graphs give energy consumption values normalized with respect to their corresponding original values without DVFS, assuming that idle processors do not consume or consume energy, respectively.

The highest energy savings are achieved for LLNLAtlas, CTC and SDSCBlue workloads, 12% in the most aggressive case ($WQ_{threshold} =$ NO LIMIT). Per workload system utilizations are given in Table 38.2. Since SDSC and LLNLThunder have higher system utilization, there is less opportunity to scale down CPU frequencies. Therefore, their energy savings are modest, 4% and 8% in the best case. Relative energy savings of two energy scenarios are almost equal. Regarding the impact of the $WQ_{threshold}$ parameter, the highest energy savings are achieved when no limit is imposed. Differences are 4%–8% between using the limit of zero jobs and no limit. However for bigger systems, LLNLThunder and LLNLAtlas, there is almost no difference for 4, 16 and no limit thresholds implying that their wait queues are usually shorter. Hence, $WQ_{threshold}$ needs to be set according to the workload.

Average BSLD values are given in Figure 38.8. Here we present values normalized to the average BSLD workload obtained with the traditional EASY backfilling (no frequency scaling). Relative penalty in performance is not always proportional to achieved savings. Although less restrictive frequency scaling per workload (higher the $WQ_{threshold}$ parameter) results in an increase in average BSLD value, similar energy savings for different workloads can result in different relative performance penalty. Here the increase in the average BSLD comes from two reasons as before. According to how we define BSLD for jobs executed at reduced frequency by the formula (38.3), scaling a job frequency down increases its BSLD as it increases the job runtime. Moreover, as in this way system load is artificially increased, certain jobs might wait longer for execution. Longer wait time means higher BSLD, and that is the second reason for the performance decrease.

LLNLAtlas and CTC achieve similar relative energy savings with the UPAS policy but LLNLAtlas suffers from higher relative performance penalty. LLNLAtlas jobs wait on average much shorter on execution than CTC jobs. Hence considering how BSLD metric is defined (38.2), the artificial

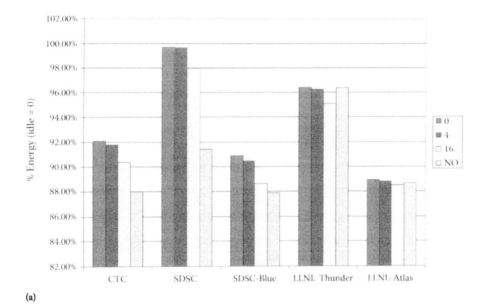

(a)

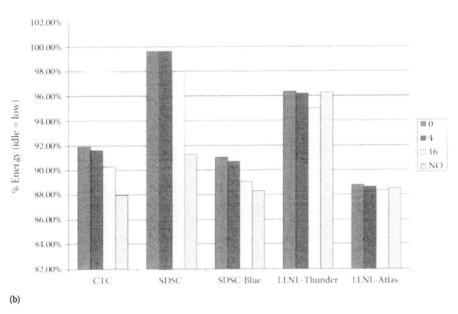

(b)

FIGURE 38.7 Utilization-driven: Normalized energies. (a) $P_{idle} = 0$. (b) $P_{idle} > 0$.

increase in computational load due to frequency scaling affects more relative performance in the case of LLNLAtlas.

The SDSCBlue workload shows an improvement in both performance and energy. As frequency scaling changes job run times, with backfilling it can give a different schedule. There are more backfilled jobs when UPAS is applied to the SDSCBlue workload. Their BSLD improvement gives an improvement in the overall performance. The number of jobs that have been run at reduced frequency for different values of $WQ_{threshold}$ is shown in Figure 38.9. The first graphic of Figure 38.10 presents SDSCBlue's utilization behavior during time. The second one shows how average frequency of running CPUs follows changes in utilization while the last graphic gives the number of jobs in the wait queue.

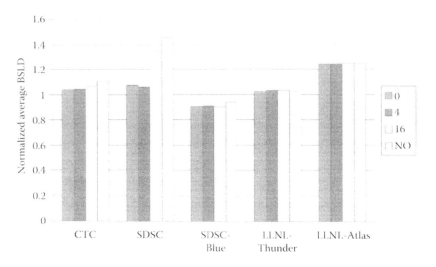

FIGURE 38.8   Utilization-driven: Performance results.

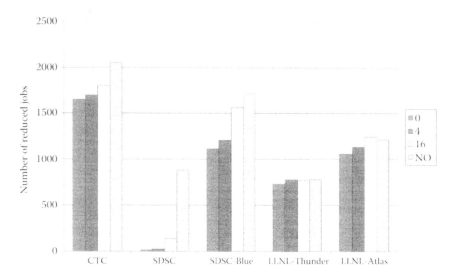

FIGURE 38.9   Utilization-driven: Number of jobs executed at lower frequency.

Utilization-driven policy has been designed as a conservative approach to reduce energy consumed by HPC centers. Hence penalty in job performance is not very high. On the other hand, energy savings are modest. Depending on parameter combination, the BSLD policy gives higher energy savings compared to the utilization-driven policy. Unfortunately, the higher savings are followed by an additional performance loss.

The loss of performance due to frequency scaling with both policies depends on the workload. As one may expect, fewer loaded workloads are more suitable for this energy-performance trade-off, while high loaded ones cannot save energy without seriously affecting performance. For instance, since the SDSC workload is very loaded, a low decrease in CPU energy consumption results in a high performance penalty. Other workloads are much easier to manage, as they have lower utilization and fewer jobs in the wait queue.

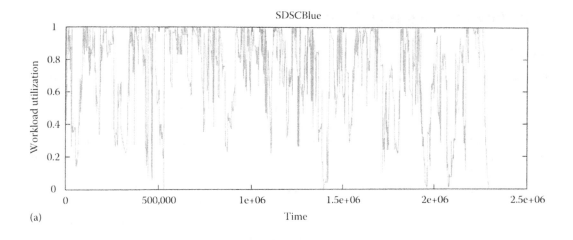

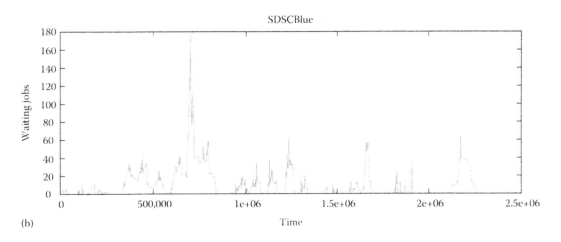

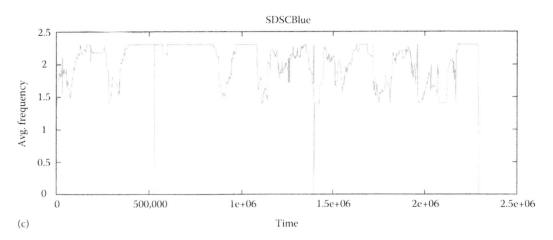

FIGURE 38.10 Utilization-driven: SDSCBlue behavior. (a) Utilization. (b) Average frequency. (c) Waiting jobs.

38.6 Power Budgeting Policy

38.6.1 Policy

In this section we give a power budgeting modification of the EASY backfilling policy. It improves job performance under a given power budget using frequency scaling. Since only one CPU frequency is assigned to a job for a whole execution, we have decided to define our policy as *power conservative*. In a similar way that work conservative scheduling policies manage CPUs [33], we keep a certain amount of power anticipating new arrivals. This concept implies that we start to apply DVFS before a job cannot be started because of the power constraint. On the other hand, when there is no danger of overshooting the power limit, DVFS should not be applied in order to maintain execution times achieved at the nominal frequency. Having always in mind that this policy should be integrated in an HPC center, the main aim is to control the performance. Hence, the CPU frequency is determined depending on the job's *predicted* BSLD 1.8. Table 38.4 gives a list of the variables used in the DVFS management policy. The *BSLDth* threshold controls DVFS application. By changing the value of this threshold, we can control DVFS aggressiveness. Higher *BSLDth* values allow more aggressive DVFS application that includes use of the lowest available CPU frequencies. Jobs consume less at lower frequencies, allowing for more jobs to run simultaneously. Setting *BSLDth* to a very low value prevents the scheduler from running jobs at reduced frequencies. In order to run at reduced frequency f, a job has to satisfy the *BSLD condition* at frequency f meaning that its predicted BSLD at the frequency f is lower than the current value of *BSLDth*. The value of *BSLDth* is changed dynamically, depending on the actual power draw as presented in the Equation 38.11. *BSLDth* is set based on current power consumption $P_{current}$, which includes power consumed by already running jobs, and power that would be consumed by the job that is being scheduled at the given frequency f. P_{lower}, and P_{upper} are thresholds that manage *closeness* to the power limit. When CPU power consumption overpasses P_{lower}, it means that processors consume a considerable amount of power. When P_{upper} is overshot, it is high probability that soon it would not be possible to start a job due to the power constraint. The power thresholds determine the *BSLDth* threshold (38.11).

Hence, when instantaneous power is not high, no frequency scaling will be

$$BSLDth = \begin{cases} 0 & \text{for } U_{j-1} \geq U_{upper} \text{ or } WQ_{size} > WQ_{threshold}, \\ BSLD_{lower} & \text{for } P_{lower} \leq P_{current} < P_{upper}, \\ BSLD_{upper} & \text{for } P_{current} \geq P_{upper}. \end{cases} \tag{38.11}$$

applied as predicted BSLD according to definition (38.8) is always higher than 1. When the power consumption starts to increase, *BSLDth* increases as well leading to frequency scaling. If power draw

TABLE 38.4   Variables Used within the Policy and Their Meaning

Variable	Description
BSLDth	Current BSLD target
$P_{current}$	Current CPU power draw (W)
P_{lower}	User-specified bound above which frequency scaling is enabled
P_{upper}	User-specified CPU power bound for aggressive frequency scaling
$BSLD_{lower}$	BSLD target when $P_{lower} \leq P_{current} < P_{upper}$
$BSLD_{upper}$	BSLD target when $P_{current} \geq P_{upper}$

almost reaches the limit, *BSLDth* is increased even more to force aggressive frequency reduction using the lowest available frequencies. As explained before, with the EASY backfilling policy, a job is scheduled with one of the two functions: ***MakeJobReservation(J)*** and ***BackfillJob(J)***. These functions modified for the PB-guided scheduling are shown in Figures 38.11 and 38.12 respectively. With the PB-guided scheduling it is not sufficient anymore to find enough free processors to make a job allocation. An allocation has to satisfy

```
1: MakeJobReservation(J)
2: if alreadyScheduled(J) then
3:     annulateFrequencySettings(J);
4: end if
5: scheduled ← false;
6: shiftInTime ← 0;
7: nextFinishJob ← next(OrderedRunningQueue);
8: while (!scheduled) do
9:     f ← Flowest
10:    while f < Fnominal do
11:        Alloc = findAllocation(J,currentTime + shiftInTime,f);
12:        if (satisfiesBSLD(Alloc, J, f) and satisfiesPowerLimit(Alloc, J, f) ) then
13:            schedule(J, Alloc);
14:            scheduled ← true;
15:            break;
16:        end if
17:    end while
18:    if (f == Fnominal) then
19:        Alloc = findAllocation(J,currentTime + shiftInTime, Fnominal)
20:        if (satisfiesPowerLimit(Alloc, J,Fnominal))
           then
21:            schedule(J, Alloc);
22:            break;
23:        end if
24:    end if
25:    shiftInTime ← FinishTime(nextFinishJob) − currentTime;
26:    nextFinishJob ← next(OrderedRunningQueue);
27: end while
```

FIGURE 38.11   PowerBudgeting policy: Making job reservation.

```
1: BackfillJob(J)
2: if alreadyScheduled(J) then
3:     annulateFrequencySettings(J);
4: end if
5: f ← Flowest
6: while f < Fnominal do
7:     Alloc = TryToFindBackfilledAllocation(J,f);
8:     if (correct(Alloc) and satisfiesBSLD(Alloc, J, f) and
        satisfiesPowerLimit(Alloc,J,f)) then
9:         schedule(J, Alloc);
10:        break;
11:    end if
12: end while
13: if (f == Fnominal) then
14:    Alloc = TryToFindBackfilledAllocation(J,Fnominal)
15:    if (correct(Alloc) and satisfiesPowerLimit(Alloc, J,Fnominal)) then
16:        schedule(J, Alloc);
17:    end if
18: end if
```

FIGURE 38.12   PowerBudgeting policy: Backfilling job.

the power constraint and the BSLD condition if the job will run at reduced frequency (Figure 38.11—line 12) or only the power constraint if scheduled for execution at the nominal frequency (Figure 38.11—line 20). The scheduler iterates starting from the lowest available CPU frequency trying to schedule a job such that the BSLD condition is satisfied at that frequency. If it is not possible to schedule it at the lowest frequency, the scheduler tries with higher ones. Forcing lower frequencies is especially important when there are jobs waiting for execution because of the power constraint (although there are available processors). On the other hand when the load is low, jobs will be prevented from low frequencies by the lower *BSLDth* value. If none of the allocations found in an iteration of the allocation search satisfy all the conditions, then in the next iteration the allocation search looks for an allocation starting from the moment of the next expected job termination (estimated according to requested times). *BackfillJob(J)* tries to find an allocation that does not delay the head of the queue and satisfies the power constraint. It also checks the BSLD condition when assigning a reduced frequency.

38.6.2 Evaluation

In this section we evaluate performance and CPU energy consumption when the PB-guided policy is applied to four of the workloads described in Section 38.5.3.1. As a baseline we have assumed a policy that enforces the same power budget without DVFS. With the baseline policy, jobs are scheduled with the EASY backfilling with a power constraint that prevents a job from being started if it would violate the power constraint. Taking into account average system utilizations without a power constraint (Table 38.2), we have decided to set power budgets to 80% of the maximum CPU power of the corresponding system. Maximum CPU power is consumed when all system processors are busy running at the nominal frequency. Percentage of time that a workload spends above the power budget with the EASY backfilling without power constraint is given in Table 38.5. Imposing an 80% power budget with the baseline policy decreases job performance tremendously. The power constraint severely penalizes average job wait time. For instance, the average wait time of the CTC workload without power budgeting is 7,107 s, while with the baseline power budgeting, it becomes 26,630 s. The LLNLThunder average wait time originally was 0 s, and in the baseline case it has been increased to 7037 s. Hence BSLD job performance degradation is very high. The policy parameters P_{lower} and P_{upper} are set to 60% and 90% of the workload power budget, respectively. After some initial tests we have decided to use the average BSLD of the workload without power constraints (avg(BSLD)) for the parameter $BSLD_{lower}$. In this way the $BSLD_{lower}$ value is set to a workload dependent value. The parameter $BSLD_{upper}$ is two times higher; it is set to $2 * avg(BSLD)$. Original average BSLD values without power budgeting are given in Table 38.2. The policy parameters are same for all reported results. Job β value is needed at the moment of scheduling in order to predict its BSLD. In practice, β value of a job is unknown when the job is being scheduled. Therefore, we investigate what is the impact of assuming the worst β case at the scheduling time. Section 38.6.2.1 presents the oracle case when β values are available at the scheduling time. It is followed by the worst case β evaluation.

38.6.2.1 Oracle Case

The average BSLD per workload for the baseline and our power control policy is given in Figure 38.13a (lower values are better). Although application of DVFS increases job runtime, having more jobs executing at the same time reduces wait time. Average wait times of baseline and PB-guided policies are given in

TABLE 38.5 Time Spent with Power Consumption over 80% of the Maximal Power Draw

CTC	SDSC	SDSC-Blue	LLNL-Thunder
72%	95%	74%	89%

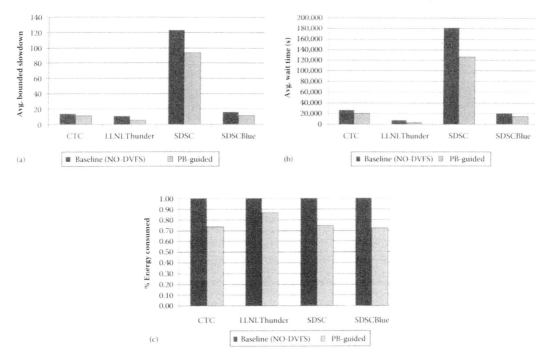

FIGURE 38.13 Original system size. (a) Performance—Avg.BSLD. (b) Performance—Avg.WT. (c) Normalized energy.

Figure 38.13b. Our policy under the power constraint improves significantly overall job performance for all workloads. In the case of the LLNLThunder workload, performance is almost twice better. Figure 38.13c shows CPU energies consumed per workload with the two policies. The values are normalized with respect to the case when all jobs are run at the nominal frequency. The energy consumed with the PB-guided policy is significantly reduced as a result of DVFS use in the PB-guided policy. Baseline energy is equal to 1, since it assumes that all jobs are executed at the nominal frequency. Figure 38.14 gives system utilization and normalized instantaneous power of the baseline (the upper graphics) and the PB-guided (the lower graphics) policies for the LLNLThunder workload over its execution. Instantaneous power is normalized with respect to the power budget. The PB-guided policy has slightly lower instantaneous power, and the workload execution takes shorter time. Furthermore, system utilization is higher (in Figure 38.14b—it is always lower than 80%). With the PB-guided policy, utilization reaches 100% running jobs at reduced frequencies.

38.6.2.2 Impact of Unknown Beta

Interestingly, we have observed that assuming more conservative values of β, better performance is achieved. Requested time used at scheduling is changed using the conservative $\beta = 1$, while new runtime is determined with the real β value. In this way more inaccuracy is introduced to job requested times. It has been remarked that inaccurate estimates can yield better performance than accurate ones [34]. By multiplying user estimates by the factor $\beta * (f_{max}/f - 1) + 1$ ($\beta = 1$) jobs with long runtimes can have large runtime overestimation at schedule time, leaving at runtime larger "holes" for backfilling shorter jobs. As a result average slowdown and wait time may be lower. In Table 38.6 results are given for the case when β values are not known in advance (no β) and for the oracle case when it is assumed that they are known in advance (β). Given values are normalized with respect to corresponding baseline case values. All workloads expect SDSC-Blue achieve better performance (**Normalized Avg.BSLD**) when β values

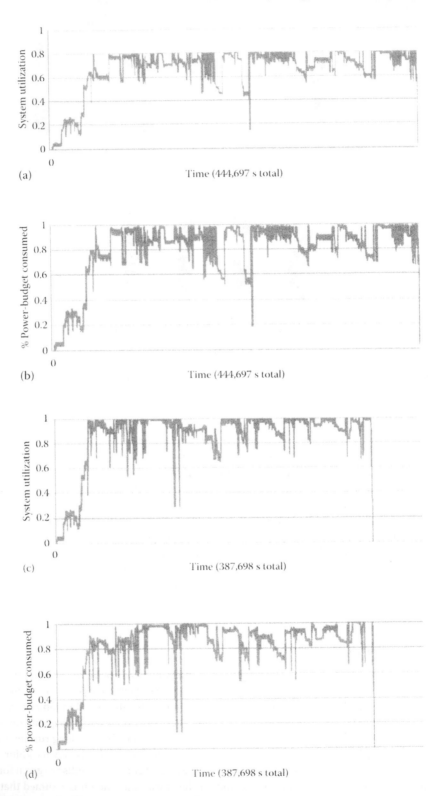

FIGURE 38.14 Comparison of baseline and PB—guided policies. (a) Baseline—system utilization. (b) Baseline—instantaneous power. (c) PB-Guided—system utilization. (d) PB-Guided—Instantaneous Power.

TABLE 38.6 Comparison of Unknown and Known β

Workload	Normalized Avg.BSLD		Normalized Avg.WT		Avg.Freq		Backfilled Jobs	
	No β	β	No β	β	No β	β	No β	β
CTC	0.80	0.79	**0.63**	0.75	1.5	1.4	3924	3923
LLNLThunder	0.33	0.51	**0.28**	0.47	2.07	1.9	3830	3651
SDSC	0.62	0.76	**0.53**	0.70	1.6	1.5	4103	3944
SDSC-Blue	0.86	0.75	**0.73**	0.76	1.5	1.4	3611	3550

are not known in advance. This can be explained by an increase in the number of backfilled jobs that have been observed (**Backfilled Jobs**).

When the most conservative case of β = 1 is assumed, assigned frequencies are higher on average (see column **Avg.Freq**). As higher frequencies have lower penalty on runtimes, it presents the second reason for better performance with unknown β. Energy savings are slightly higher when β values are known at the moment of scheduling. This is explained again by lower frequency selection in the oracle case, as jobs running at lower frequencies consume less CPU energy.

38.7 Conclusions

In this chapter, we have explained how to reduce CPU energy in an HPC environment. We have targeted CPU energy consumption as processors consume approximately half of the total system power consumption. Furthermore, it is easy to manage CPU power using the widespread DVFS technique. The same technique is used in this chapter to improve job performance for a given power budget. When there is no power limit imposed, frequency scaling results in performance degradation. There are two reasons for performance degradation: first—lower frequency increases job runtime; and second—higher runtimes increase load and may result in longer wait times. It is important to control both ways of performance degradation in order to achieve balanced performance-energy trade-off. Two policies proposed in this chapter provide manageable control over performance loss.

On the other hand, frequency scaling can improve job performance under a power limit. Assuming that a system contains more processors than the number that can run at nominal frequency under a given power budget, frequency scaling improves performance, allowing more jobs to run simultaneously. In this way, in spite of runtime increase, overall job performance measured in BSLD metric improves. As power is becoming a precious resource, we can expect to see systems with more processors than available power. In these systems, the power budgeting policy proposed here would help to improve user satisfaction.

References

1. I. Ahmad, S. U. Khan, and S. Ranka. Using game theory for scheduling tasks on multi-core processors for simultaneous optimization of performance and energy. In *IEEE International Symposium on Parallel and Distributed Processing (2008)*, Miami, FL, pp. 1–6, March 2008. IEEE.
2. J. A. Butts and G. S. Sohi. A static power model for architects. In *Proceedings of the 33rd Annual IEEE/ACM International Symposium on Microarchitecture, 2000. MICRO-33*, Monterey, CA, pp. 191–201, 2000.
3. M. Etinski, J. Corbalan, J. Labarta, and M. Valero. BSLD threshold driven power management policy for HPC centers. In *Proceedings of the IEEE International, Parallel and Distributed Processing Symposium, Workshops and PhD Forum*, Atlanta, GA, pp. 1–8, April 2010.

4. M. Etinski, J. Corbalan, J. Labarta, and M. Valero. Optimizing job performance under a given power constraint in HPC centers. In *Green Computing Conference, 2010 International*, Chicago, IL, pp. 257–267, August 2010.

5. M. Etinski, J. Corbalan, J. Labarta, and M. Valero. Utilization driven power-aware parallel job scheduling. *Computer Science Research and Development*, 25/2010:207–216, Dordrecht, the Netherlands: Springer, August 2010.

6. X. Fan, W.-D. Weber, and L. A. Barroso. Power provisioning for a warehouse-sized computer. In *ISCA '07: Proceedings of the 34th Annual International Symposium on Computer Architecture*, pp. 13–23, San Diego, CA, 2007. New York: ACM.

7. D. G. Feitelson. Metrics for parallel job scheduling and their convergence. In *Revised Papers from the Seventh International Workshop on Job Scheduling Strategies for Parallel Processing, JSSPP '01*, pp. 188–206, Cambridge, MA, 2001. London, U.K.: Springer-Verlag.

8. X. Feng, R. Ge, and K. W. Cameron. Power and energy profiling of scientific applications on distributed systems. In *Proceedings of the 19th IEEE International Parallel and Distributed Processing Symposium, 2005*. Denver, CO, p. 34, April 2005.

9. V. W. Freeh and D. K. Lowenthal. Using multiple energy gears in MPI programs on a power-scalable cluster. In *PPoPP '05: Proceedings of the Tenth ACM SIGPLAN Symposium on Principles and Practice of Parallel Programming*, pp. 164–173, Chicago, IL, 2005. New York: ACM.

10. V. W. Freeh, D. K. Lowenthal, F. Pan, N. Kappiah, R. Springer, B. L. Rountree, and M. E. Femal. Analyzing the energy-time trade-off in high-performance computing applications. *IEEE Transactions on Parallel and Distributed Systems*, 18(6):835–848, 2007.

11. V. W. Freeh, F. Pan, N. Kappiah, D. K. Lowenthal, and R. Springer. Exploring the energy-time tradeoff in MPI programs on a power-scalable cluster. In *Proceedings of the 19th IEEE International Parallel and Distributed Processing Symposium*, pp. 4a–4a, April 2005.

12. R. Ge, X. Feng, S. Song, H.-C. Chang, D. Li, and K. W. Cameron. Powerpack: Energy profiling and analysis of high-performance systems and applications. *IEEE Transactions on Parallel Distributed Systems*, 21:658–671, May 2010.

13. F. Guim and J. Corbalan. A job self-scheduling policy for HPC infrastructures. In *JSSPPS '08: Proceedings of the Workshop on Job Scheduling Strategies for Parallel Processing*, Cancun, Mexico, pp. 51–75. Springer-Verlag, 2008.

14. P. Henning and A. B. White Jr. Trailblazing with Roadrunner. *Computing in Science and Engineering*, 11:91–95, 2009.

15. J. Hikita, A. Hirano, and H. Nakashima. Saving 200 KW and 200 K dollars per year by power-aware job and machine scheduling. In *Proceedings of the IEEE International Symposium on Parallel and Distributed Processing, IPDPS 2008*, pp. 1–8, April 2008.

16. Y. Hotta, M. Sato, H. Kimura, S. Matsuoka, T. Boku, and D. Takahashi. Profile-based optimization of power performance by using dynamic voltage scaling on a PC cluster. In *Proceedings of the IEEE International Symposium on Parallel and Distributed Processing*, Rhodes Island, Greece, 0:340, 2006.

17. C. H. Hsu and W. C. Feng. A power-aware run-time system for high-performance computing. In *Proceedings of the ACM/IEEE Supercomputing, SC'05*, Seattle, WA, p. 1, 2005.

18. C.-H. Hsu and U. Kremer. The design, implementation, and evaluation of a compiler algorithm for CPU energy reduction. In *PLDI '03: Proceedings of the ACM SIGPLAN 2003 Conference on Programming Language Design and Implementation*, pp. 38–48, San Diego, CA, 2003. New York: ACM.

19. http://www.cs.huji.ac.il/labs/parallel/workload/. Parallel workload archive.

20. http://www.nccs.gov/computingresources/jaguar/. The Jaguar supercomputer.

21. http://www.top500.org/. The Top500 list.

22. S. Kamil, J. Shalf, and E. Strohmaier. Power efficiency in high performance computing. In *IPDPS 2008: IEEE International Symposium on Parallel and Distributed Processing, 2008*. Miami, FL, pp. 1–8, April 2008.

23. N. Kappiah, V. W. Freeh, and D. K. Lowenthal. Just in time dynamic voltage scaling: Exploiting inter-node slack to save energy in MPI programs. In *Proceedings of the 2005 ACM/IEEE Conference on Supercomputing*, Seattle, WA, 0:33, 2005.

24. R. Kettimuthu, V. Subramani, S. Srinivasan, T. Gopalsamy, D. K. Panda, and P. Sadayappan. Selective preemption strategies for parallel job scheduling. *International Journal of High Performance Computing and Networking*, 3(2/3):122–152, 2005.

25. S. U. Khan and I. Ahmad. A cooperative game theoretical technique for joint optimization of energy consumption and response time in computational grids. *IEEE Transactions on Parallel and Distributed Systems*, 20:346–360, March 2009.

26. K. H. Kim, R. Buyya, and J. Kim. Power aware scheduling of bag-of-tasks applications with deadline constraints on DVS-enabled clusters. *CCGRID 2007: Seventh IEEE International Symposium on Cluster Computing and the Grid*, Rio de Janeiro, Brazil, pp. 541–548, May 2007.

27. B. Lawson and E. Smirni. Power-aware resource allocation in high-end systems via online simulation. In *ICS '05: Proceedings of the 19th Annual International Conference on Supercomputing*, Cambridge, MA, pp. 229–238, 2005. New York: ACM.

28. M. Y. Lim, V. W. Freeh, and D. K. Lowenthal. Adaptive, transparent frequency and voltage scaling of communication phases in MPI programs. In *Proceedings of the 2006 ACM/IEEE Conference on Supercomputing*, Tampa, FL, 0:14, 2006.

29. D. Meisner, B. T. Gold, and T. F. Wenisch. Powernap: Eliminating server idle power. *SIGPLAN Notices*, 44(3):205–216, 2009.

30. E. N. (Mootaz) Elnozahy, M. Kistler, and R. Rajamony. Energy-efficient server clusters. In *Proceedings of the Second Workshop on Power-Aware Computing Systems*, Cambridge, MA, pp. 179–196, 2002.

31. F. Pinheiro, R. Bianchini, E. V. Carrera, and T. Heath. Load balancing and unbalancing for power and performance in cluster-based systems. In *Workshop on Compilers and Operating Systems for Low Power*, Barcelona, Spain, 2001.

32. B. Rountree, D. K. Lowenthal, S. Funk, V. W. Freeh, B. R. de Supinski, and M. Schulz. Bounding energy consumption in large-scale MPI programs. In *SC '07: Proceedings of the 2007 ACM/IEEE Conference on Supercomputing*, Reno, NV, pp. 1–9, 2007. New York: ACM.

33. R. Smirni, E. Rosti, E. Smirni, G. Serazzi, and L.W. Dowdy. Analysis of non-work-conserving processor partitioning policies. In *IPPS '95 Workshop on Job Scheduling Strategies for Parallel Processing*, Santa Barbara, CA, pp. 165–181. Springer-Verlag, 1995.

34. D. Tsafrir and D. G. Feitelson. The dynamics of backfilling: solving the mystery of why increased inaccuracy may help. In *IEEE International Symposium on Workload Characterization (IISWC)*, San Jose, CA, pp. 131–141, October 2006.

39

Toward Energy-Efficient Web Server Clusters

Yu Cai
Michigan Technological University

Xinying Zheng
Michigan Technological University

39.1 Introduction

Green computing is to support personal and business computing needs in a green and sustainable manner, such as minimizing the strain and impact on resources and the environment. Computing systems, particularly enterprise data centers and high-performance cluster systems, consume a significant amount of power. Not only does this increase operational costs, it also places an increasing burden on businesses that generate and distribute energy. For example, the power consumption of enterprise data centers in the U.S. doubled between 2000 and 2005 [1]. In 2005, U.S. data centers consumed 45 billion kW-h; roughly 1.2% of the total amount of United States electricity consumption, resulting in utility bills of $2.7 billion [4]. In 2006, the U.S. Congress passed bills to raise the IT industry's role in energy and environmental policy to the national level [2]. Furthermore, it is estimated that servers consume 0.5% of the world's total electricity, which if current demand continues, is projected to quadruple by 2010 [42]. The Environmental Protection Agency recently reported that the energy used by data centers by the year 2011 is estimated to cost $7.4B [1]. Some analysts predicted that IT infrastructure will soon cost more on power consuming than the hardware itself [31]. The alarming growth of reducing data center power consumption has led to a surge in research activity.

The server system accounts for 56% of total power consumption of an Internet data center; its power draw can vary drastically with data center utilization. Therefore, much previous work has been carried out to reduce power consumption in a server cluster system. Much of the existing work on power management in server clusters relies heavily on heuristics or feedback control [26,27,29,42]. An important principle in green computing is to ensure energy consumption proportionality, which states that the energy consumption should be proportional to the system workload [5]. This idea can effectively improve the energy efficiency in real-life usage. Energy proportionality is relatively hard to achieve on a standalone server because of hardware constraints. However, it is possible to achieve energy proportionality on a server cluster, since we can control the number of active and inactive nodes in a server cluster.

Energy management involves a trade-off between power and performance. Many researchers formulate the energy issue as an optimization problem. Two main mechanisms are commonly applied for energy savings: dynamic voltage and frequency scaling (DVFS) dynamically change the frequency and voltage of servers to produce energy savings [33,40]; Vary-on vary-off (VOVF) uses server turn on/off mechanisms for power management [14,18,36]. Some thought is given to integrating both DVFS and VOVF mechanisms together [9]. Applying DVFS and VOVF simultaneously requires careful consideration due to transition overhead, which not only leads to performance degradation, but also reduces the life cycle of hardware components. However, the transition overhead was not well studied in the literature.

This chapter introduces three theoretical frameworks for managing server power consumption. The rest of this chapter is organized as follows: Section 39.2 reviews related work; Section 39.3 proposes an energy proportionality model for energy savings in a server cluster; Section 39.4 formulates an optimization power model and a double-control-periods (DCP) model to achieve energy efficiency; In Section 39.5, a constrained Markov decision process is formulated, which provides an off-line optimization solution for energy savings. The last section concludes this chapter.

39.2 Related Work

Green computing–related terms in the literature include green IT, sustainable computing, energy efficiency, energy savings, power aware, power savings, and energy proportional. In this section, we review related techniques applied to both single server and server clusters, respectively.

39.2.1 Single Server

Green computing techniques for a single server focus on microprocessors, memories, and disks. Current microprocessors allow power management by DVFS. DVFS works because reducing the voltage and frequency provides substantial savings in power at the cost of slower program execution. Some research ties the scheduler directly to DVFS [13,34,35]. Most work deals exclusively with meeting real-time scheduling deadlines while conserving energy. Traditionally, many power management solutions rely heavily on heuristics. Recently, feedback control theoretical approaches for energy efficiency have been proposed by a number of researchers. On a single server, recent works [38,44] proposed power control schemes based on feedback control theory. For standalone servers, Lefurgy and Wang [27] propose a power management scheme that is based on feedback control. This improves on traditional techniques by providing greater accuracy and stability. Thermal management is another issue in power-aware computing, since temperature is a by-product of power dissipation [17]. Recent research demonstrated that dynamic thermal management (DTM) can respond to thermal conditions by adjusting the power consumption profile leased on a chip based on temperature sensor feedback [6,38]. Research work on memory is often combined with processors and disks. In [15], the authors used open-loop control to shift power between processor and memory to maintain a server power budget. A large portion of the power budget of servers goes into the I/O subsystem, the disk array in particular. Many disk systems offer multiple power modes and can be switched to a low power mode when not in use to achieve energy savings. Such techniques had been proposed in [20,43]. Sudhanva et al. [19] presented a new approach called DRPM to modulate disk speed dynamically, and a practical implementation was provided for this mechanism.

39.2.2 Server Clusters

In recent years, power management has become one of the most important concerns with server clusters. Some methods proposed on a single server can be extended to server clusters. In [33,40], the authors presented similar ways of applying DVFS and cluster reconfiguration using threshold values, based on the utilization of the system load to keep the processor frequencies as low as possible, with fewer active nodes. In [42], the authors extended the feedback control scheme to clusters. Power has been used as a tool for application–level performance requirements. Sharma et al. [37] proposed feedback control schemes to control application–level quality of service requirements. Dovrdis et al. [11] presented a feedback controller to manage the response time in server clusters. Some researchers applied DTM on an entire data center rather than individual servers or chips. Sharma et al. [36] laid out policies for workload placement to promote uniform temperature distribution using active thermal zones. VOVF is a dynamic structure configuration mechanism to ensure energy-aware computing in server clusters, which turns nodes on and off to adjust the number of active servers by the workload. Other work had been carried out based on VOVF [14,18,36]. Another group developed power saving techniques for connection-oriented servers [8]. The authors tested server provisioning and load dispatching on the MSN instant messaging framework, and evaluated various load skewing techniques to trade off between energy savings and quality of service. The energy-related budget accounts for a large portion of total storage system cost of ownership. Some studies tried multispeed disks for servers [7,19]. Other techniques were introduced to regulate data movement. For example, the most used data can be transferred to specific disks or memory; thus other disks can be set to a low power mode [39]. Recently, the Markov model was further studied in server clusters for energy savings. In [16], a three-speed disk Markov power model is formulated for the disk systems, and prediction schemes were proposed for achieving disk energy savings. A CMDP is constructed for power and performance control in Web server clusters in [30]. This work is similar to ours, but DVFS mechanisms were not applied to reduce power consumption, leaving some room to further reduce energy consumption by employing DVFS techniques in the CMDP model.

Virtualization is another key strategy for reducing power consumption in enterprise networks. With virtualization, multiple virtual servers can be hosted on a smaller number of more powerful physical servers, using less electricity. In [23], researchers demonstrated a method to efficiently manage the aggregate platform resources according to the relative importance (class-of-service) of the guest virtual machines (VM) for both the black-box and the VM-specific approach. The use of virtualization for consolidation is presented in [32], which proposes a dynamic configuration approach for power optimization in virtualized server clusters and outlines an algorithm to dynamically manage the virtualized server cluster. Following the same idea [28], aims to reduce virtualized data center power consumption by supporting VM migration and VM placement optimization while reducing human intervention, but no evaluation is provided. Other work [41] also proposes a virtualization-aware adaptive consolidation approach; the energy costs for the execution of a given set of applications were measured. They use correlation techniques in order to predict usage, while we use machine learning to predict application power and performance. Their techniques do not currently include powering off any under-utilized parts of the system.

39.3 Achieving Energy Proportionality in Web Server Clusters

In this section, we propose an energy proportionality model in a server cluster and study its performance on service differentiation. We further investigate the transition overhead based on this model. The simulation results show that the energy proportionality model can provide controllable and predictable quantitative control over power consumption with theoretically guaranteed service performance.

39.3.1 Energy Proportional Model

Much of the existing work on power management in server clusters relies heavily on heuristics or feedback control [26,27,29,42]. An important principle in green computing is to ensure energy consumption proportionality, which states that the energy consumption should be proportional to the system workload [5]: $P = a\lambda + b$ (*ideally* $b = 0$). For example, when there is no or little workload, the system should consume little or no energy; when workload increases, energy consumption should increase proportionally until the system reaches the full workload. This idea can improve energy efficiency in real-life usage although energy proportionality is relatively hard to achieve on a standalone server because of hardware constraints. However, it is possible to achieve energy proportionality on a server cluster since we can control the number of active and inactive nodes. On typical Web servers and Web clusters, a system workload can be described by the request arrival rate λ. Let M be the total number of servers in the cluster, and Λ be the maximum arrival rate for the cluster. $\sum m$ is the total number of active servers. The total energy consumption of a server cluster is

$$P = \sum m * P_{ac} + (M - \sum m) * P_{in} \qquad (39.1)$$

where
  P_{ac} is the power consumption of fully active nodes
  P_{in} is the power consumption of inactive nodes

Based on the energy proportional model, we have

$$\frac{P}{\lambda} = \frac{P_{max}}{\Lambda} * r \qquad (39.2)$$

where $P_{max} = M * P_{ac}$. The parameter r adjusts the energy consumption curve. The rationale of using this parameter is as follows: ideally $r = 1$. However, we can adjust it to satisfy different performance constraints. With the help of (39.1), we can rewrite Equation (39.2) as

$$\sum m = \left(\frac{P_{ac}}{\Lambda/M} * r - M * P_{in} \right) / (P_{ac} - P_{in}) \qquad (39.3)$$

Here Λ/M is the maximum number of jobs that a single cluster node can handle. Ideally $P_{in} = 0$, which indicates that a server consumes no energy when it is running in inactive mode. For simplicity, we suppose $P_{in} = 0$ in this chapter; this assumption will not affect the performance of our model. The total number of active servers $\sum m$ is determined by the system workload λ:

$$\sum m = \frac{\lambda}{\Lambda/M} * r \qquad (39.4)$$

Based on (39.4), the number of servers might not be an integer. We will set the number to no less than $\sum m$, the minimum number of servers running in fully active mode.

39.3.2 Performance Metrics

One important and commonly used quality of service (QoS) metric used in conjunction with Internet servers is slowdown, which is defined as the division of waiting time by service time. Another commonly used performance metric is request time, which is the sum of waiting time and service time. We choose slowdown and request time as performance metrics in our model because they are related to both waiting time and service time. Our theoretical framework is built along the line of the previous service differentiation models presented in [10,46–48]. In our network model, a heavy-tailed distribution of packet size is used to describe Web traffic. Here we assume that the service time is proportional to the packet size. The packet interarrival time follows exponential distribution with a mean of $1/\lambda$, where λ is the arrival rate of incoming packets. A set of tasks with size following a heavy-tailed bounded Pareto distribution are characterized by three parameters: α, the shape parameter; k, the shortest possible job; p, the upper bound of jobs. The probability density function can be defined as

$$f(x) = \frac{1}{1 - (k/p)^\alpha} \alpha k^\alpha x^{-\alpha-1} \tag{39.5}$$

where, $\alpha, k > 0, k \leq x \leq p$. If we define a function:

$$K(\alpha, k, p) = \frac{\alpha k^\alpha}{1 - (k/p)^\alpha} \tag{39.6}$$

then we have

$$E[X] = \int_k^p f(x)dx = \begin{cases} \frac{K(\alpha,k,p)}{K(\alpha-1,k,p)} & \text{if } \alpha \neq 1; \\ (lnp - lnk)K(\alpha,k,p) & \text{if } \alpha = 1. \end{cases} \tag{39.7}$$

According to Pollaczek–Khinchin formula, the average waiting time for the incoming packets is

$$E[W] = \frac{\lambda E[X^2]}{2(1 - \lambda E[X])} \tag{39.8}$$

After applying a round-robin dispatching policy, the expected slowdown in an M/G/m queue on an Internet server cluster is

$$E[S] = E[W]E[X^{-1}] = \frac{\lambda E[X^2]E[X^{-1}]}{2(m - \lambda E[X])} \tag{39.9}$$

The expected request time with the incoming job rate λ is

$$E[R] = E[W] + E[X] = \frac{\lambda E[X^2]}{2(m - \lambda E[X])} + E[X] \tag{39.10}$$

39.3.3 Servers Allocation on Service Differentiation

In a cluster system, the incoming requests are often divided into N classes. Each class may have different QoS requirements. We assume m_i is the number of active server nodes in class i, and λ_i is the arrival rate in class i. We adopt a relative service differentiation model where the QoS slowdown factor between different classes is based on their predefined differentiation parameters.

$$\frac{E[S_i]}{E[S_j]} = \frac{\delta_i}{\delta_j} \tag{39.11}$$

where $1 \leq i, j \leq N$.

We assume class 1 is the highest class and set $0 < \delta_1 < \delta_2 < \cdots < \delta_N$. Higher classes receive better service (less slowdown) [37]. Based on the above energy proportionality and service differentiation model, according to formulas (39.4) and (39.18), we can derive the server allocation scheme in a cluster system as

$$m_i = \lambda_i E[X] + \frac{\tilde{\lambda}_i \sum_{i=1}^N \lambda_i \left(\frac{M}{\Lambda}r - E[X]\right)}{\sum_{i=1}^N \tilde{\lambda}_i} \tag{39.12}$$

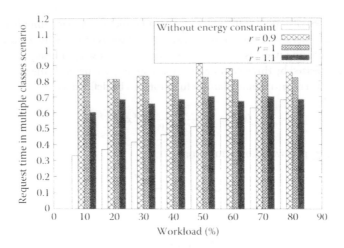

FIGURE 39.1 Comparison of request time in higher priority class between nonenergy proportional model and energy proportional models. *r* is set differently according to different requirements of performance in a multiple classes scenario.

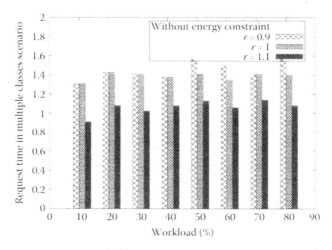

FIGURE 39.2 Comparison of request time in lower priority class between nonenergy proportional model and energy proportional models. *r* is set differently according to different requirements of performance in a multiple classes scenario.

Here m_i is the number of active servers in class i, and $\tilde{\lambda}_i = \lambda_i/\delta_i$ is the normalized arrival rate. The first term of formula (39.12) ensures that the subcluster in class i will not be overloaded. The second term is related to arrival rates, differentiation parameters, and r.

39.3.4 Evaluation

The GNU Scientific Library is used for stochastic simulation. We compare the performance metrics in Figures 39.1 through 39.3. The number of classes is normally two or three [45,49]. We set the target slowdown ratio $\delta_2{:}\delta_1 = 2{:}1$. We observe that the model can achieve desirable proportionality of slowdown differentiation with request time constraints. Figure 39.4 also compares the energy consumptions for proportional and nonproportional models in a multiple classes scenario. We can achieve better energy efficiency under low workload, which leads to high energy saving in a server cluster. We also study the

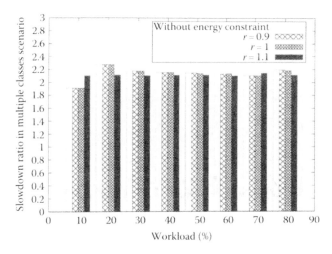

FIGURE 39.3 Comparison of slowdown ratio between nonenergy proportional model and energy proportional models.

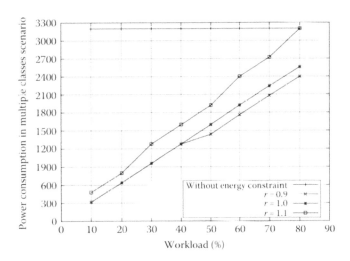

FIGURE 39.4 Comparison of power consumption between nonenergy proportional model and energy proportional model.

influence on performance caused by transition overhead, under a different time frame. A relatively few spare servers were added to compensate for the transition overhead. Figure 39.5 evaluates the power consumption based on our model under real workload conditions using trace data. The system arrival rate is the same as shown in Figure 39.6. The power consumption is dynamically changed as the workload changes. With little more power consumption, we can achieve better performance, and eliminate the effect of transition overhead.

39.4 Optimal Server Provisioning and Frequency Adjustment in Web Server Clusters

This section explores the benefits of DVFS and server number controlling VOVF for power management in a server cluster. The online energy management strategy is formulated as a minimization problem: The

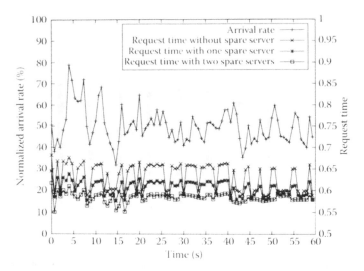

FIGURE 39.5   Request time when adding two spare servers based on energy proportional model in a single class scenario.

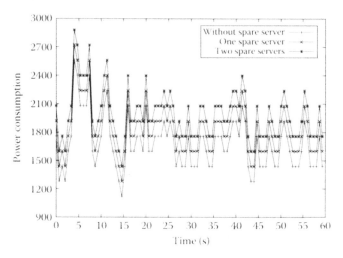

FIGURE 39.6   Power consumption when adding two spare servers based on energy proportional model in a single class scenario.

optimal solution is obtained by selecting the proper number of active servers running at f_i while request time is kept within a predetermined threshold. We formulate the problem in the following two scenarios: single class and multiple classes.

39.4.1 Optimal Power Management Modeling

In this section, we use request time for our performance metric since it is the greatest concern from the user's perspective. Our theoretical framework is built on a server cluster system with uniformed servers. Let M be the total number of servers. Each server has N levels of adjustable frequency $f_i(1 \leq i \leq N)$, where $f_1 < f_2 < f_3 < \dots < f_N$. Since all the incoming requests are CPU bounded, higher operating frequencies lead to greater server processing capacity, which can be represented as $c_i = af_i$. c_i. We adopt the same queue model as we described in Section 39.3.2. Given an M/G/1 queue on a server, X is service time, X' is the service time under a given capacity c, we have

$$E[X'] = \frac{1}{c}E[X] \tag{39.13}$$

$$E[X'^2] = \frac{1}{c^2}E[X^2] \tag{39.14}$$

The average waiting time for the incoming packets under capacity c in a single server can be represented as

$$E[W'] = \frac{\lambda E[X^2]}{2c(c - \lambda E[X])} \tag{39.15}$$

When applying a round-robin dispatching policy, the packet arrival rate of a node is λ/m. The processing capacity is always proportional to the operating frequency. The expected request time for any server in a server cluster can be calculated as

$$E[R] = \frac{\lambda E[X^2]}{2af_i(af_i m - \lambda E[X])} + \frac{1}{af_i}E[X] \tag{39.16}$$

where $1 \leq i \leq N, 1 \leq m \leq M$.

In a single class scenario, we assume all the incoming requests are classified into just one class. In other words, the same QoS should be met. Here the threshold β, sets to bound the average request time. Let M be the total number of identical servers in the system, and m is the number of active server nodes handling the incoming requests. We solve the following:

$$\text{Min}: \quad \sum P = \sum P_{single} + \sum P_{insingle}$$

$$\text{S.t. } E[R] = \frac{1}{m}\sum_{j=1}^{m}\left(\frac{\lambda E[X^2]}{2af_i(af_i m - \lambda E[X])}\right.$$

$$\left. + \frac{1}{af_i}E[X]\right) \leq \beta \tag{39.17}$$

where

$1 \leq m \leq M, 1 \leq i \leq N.$

$P_{insingle}$ is the power consumption of an inactive server

Here we do not consider the transition power, because it is considerably less important in a typical Internet server workload, since load fluctuation occurs on a larger timescale [21]. The above optimization is nonlinear and discrete in terms of the decision variables for both objective and constraint. One feasible way is to consider a finite number of frequencies and servers to determine the optimal solution. However, the complexity is $O(N^M)$ when considering M servers and N levels of adjustable frequency. It is can be reduced to $O(MN)$ after applying a coordinated voltage scaling approach, in which all active servers are assigned equal frequencies. Previous research adopted the same strategy and showed that a coordinated voltage scaling approach can provide substantially higher savings [9,12].

In a cluster system, the incoming requests are often divided into W classes. Each class may require a different QoS according to its priority. Here, we adopt a relative service differentiation model where the QoS factors (request time between different classes) are based on their predefined differentiation parameters.

$$\frac{E[R_i]}{E[R_j]} = \frac{\delta_i}{\delta_j} \tag{39.18}$$

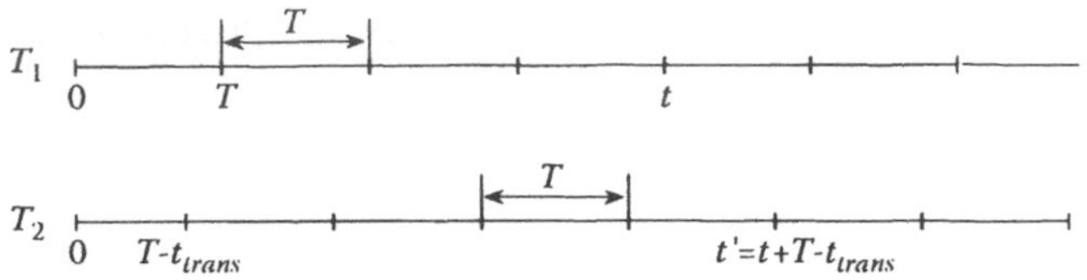

FIGURE 39.7　Double control periods.

In formula (39.18), where $1 \leq i,j \leq W$, we assume class 1 is the highest class and set $0 < \delta_1 < \delta_2 < \cdots < \delta_W$. Again, higher classes receive better service, i.e., a shorter request time. We solve the optimization problem in the multiple classes scenario as follows:

$$\text{Min}: \sum P = \sum_{j=1}^{W} m_j P_{single} + \left(M - \sum_{j=1}^{W} m_j \right) P_{insingle}$$

$$\text{S.t. } E[R_j] = \frac{\lambda_j E[X^2]}{2af_{ij}(af_{ij}m - \lambda_j E[X])} + \frac{1}{af_{ij}} E[X] \leq \beta \delta_j \tag{39.19}$$

The problem can be solved by decomposing it into W single class optimization problems.

39.4.2 Overhead Analysis

The proposed model is a continual optimization process; here the overhead caused by frequency adjustment is ignored since it is very small [22]. We assume frequency adjustment is accomplished instantly. However, the transition time between server inactive mode to active mode cannot be ignored, as this will greatly influence performance. This is especially true when the workload increases during the next control period, which may lead to an increase of active servers. This is not always the case; workload increase could also result in a decreasing the number of active servers. The fact is, we dynamically optimize the server number and frequency together; fewer servers may not be an optimal solution even though the workload decreases in the next control period. Thus, it is necessary to estimate the cost of transition overhead. In general, transition times depend on the processor and other hardware constraints. We propose a DCP model to compensate for the transition overhead, allowing for better performance.

The basic idea of the DCP model is shown in Figure 39.7. "Double" stands for two control periods denoted as T_1 and T_2, respectively. The control interval for both periods is identical: $T_1 = T_2 = T$. The number of active servers and frequency adjustments occur at the beginning of the T_1 control period. Control period T_2 turns on additional servers for the next T_1 control period. The two control periods are designed with a control time difference: $t_{diff} = T - t_{trans}$, where t_{trans} is the transition time when a server node shifts from inactive to active mode. A schematic of the DCP model is shown in Figure 39.8.

The workload predictor anticipates the incoming request arrival rate for both control periods: $\lambda_{T1}(t)$ and $\lambda_{T2}(t' - T)$. Here $t' = t + T - t_{trans}$. At the beginning of every T_2 control period, the optimal solution calculator uses the formula $\lambda_{T2}(t' - T)$ to compute the best solution. To avoid a redundant optimization process, we record the optimization solution $S_{T2}(t' - T)$. At each beginning of T_1 control period, DCP will first check the requests' arrival rate variance of $\lambda_{T1}(t)$ and $\lambda_{T2}(t' - T)$. If $\lambda_{T1}(t) - \lambda_{T2}(t' - T) \leq \gamma$, DCP adopts the solution $S_{T1}(t) = S_{T2}(t' - T)$ instead of recalculating the server provisioning and frequency adjustments according to $\lambda_{T1}(t)$. This strategy enhances computational efficiency. Additional servers $\sum m_{\lambda T2}(t') - \sum m_{\lambda T1}(t)$ will be turned on at the beginning of each T_2 control period if more servers are required for the next T_1 control period. Each server will be set to the lowest frequency f_1 for energy efficiency. Additional servers have sufficient time t_{trans} to shift from inactive to active mode.

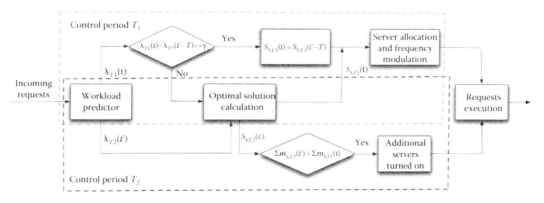

FIGURE 39.8 Designing of the DCP model.

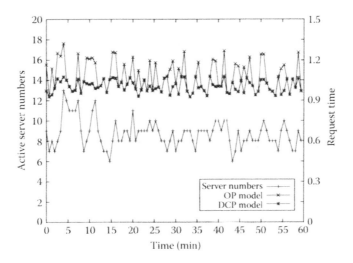

FIGURE 39.9 Comparison of request time of real workload data trace between OP model and DCP model in a single class scenario.

The DCP model takes advantage of workload characteristics. As we mentioned before, workload fluctuations occur on a larger timescale, which means $\lambda_{T1}(t)$ and $\lambda_{T2}(t' - T)$ are close enough for optimization prediction.

39.4.3 Evaluation

In the simulation, the optimization model is evaluated using both stochastic workload and real workload trace data. Only the simulation results with real workload data are shown. Figure 39.9 illustrates the performance based on an optimization (OP) model and a DCP model in a single class scenario. The request arrival rate and job size are normalized. As shown in Figure 39.9, performance degradation is clearly seen in the OP model; this is caused by increasing the number of active servers. The DCP model improves performance significantly. Figure 39.10 compares the power consumption of our model under real workload trace data. The OP model and DCP model are further evaluated in a multiple class scenario as shown in Figures 39.11 and 39.12 under real workload trace data.

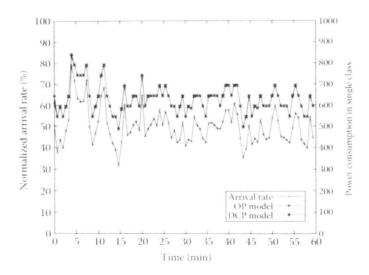

FIGURE 39.10   Comparison of power consumption between OP model and DCP model in a single class scenario.

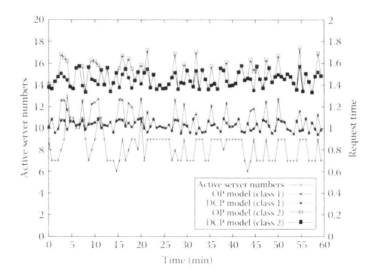

FIGURE 39.11   Comparison of request time between OP model and DCP model in a multiple classes scenario.

39.5 A CMDP Model for Power Management in a Web Server Cluster

Power and performance are of the utmost importance when designing a power-aware computing system. One of the most effective methods in dealing with the trade-off between power and performance is to formulate an optimization problem. A power controller can allocate resources in the computing system according to an optimal solution. This is often accomplished by a periodical online optimization process. If a combined DVFS and VOVF strategy is used to obtain the optimal power conservation solution in a server cluster, the problem is NP-completed to solve exactly. The online computation complexity can be reduced to $O(MF)$ given a server cluster with M homogeneous servers and F levels of adjustable frequency levels by applying a coordinated voltage scaling approach [9]. It is still time consuming due to the computation

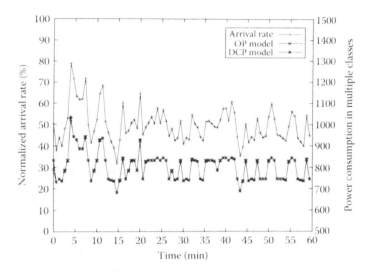

FIGURE 39.12 Comparison of power consumption between OP model and DCP model in a multiple classes scenario.

complexity. In this work, a constrained Markov decision process (CMDP) is constructed to achieve energy savings. The significant advantage of our CMDP model is our optimization solution is computed off-line, which greatly reduces the online computation time. The online computation complexity is reduced to $O(1)$ after applying a deterministic CMDP policy.

39.5.1 CMDP Modeling

39.5.1.1 State Space *X* and Action Space *A*

In our CMDP model, let M be the total number of servers. Each server has F levels adjustable frequency $f_i (1 \leq i \leq F)$, where $f_1 < f_2 < f_3 < ... < f_F$. The composite state of the CMDP is defined as a tuple $x = \{s, m, f_i\} \in X$, where s represents the number of jobs waiting in a queue. $s \in \{0, 1, 2, \cdots, S\}$. S is the maximum queue length. If more than S jobs need to be stored in the queue, the queue is blocked and some jobs will be lost. The decision maker should try to avoid queue blocking as much as possible. We assume there is at least one server in active mode to handle the incoming requests. $m \in \{1, \cdots, M\}, f_i \in \{f_1, \cdots, f_F\}$. m and f_i represent the number of servers in active mode running at the frequency level of f_i. f_1 and f_F are the minimum and maximum frequency levels, respectively, when a server is in active mode. A coordinated voltage scaling approach is applied in our model, in which all active servers are assigned an equal frequency level. A set of composite actions are defined as $a \in A$:

$$a = \{(M_a, F_a);$$

$$M_a \in \{-(M-1), -(M-2), \cdots, 0, 1, \cdots, M-1\},$$

$$F_a \in \{-(F-1), -(F-2), \cdots, 0, 1, \cdots, F-1\}\} \tag{39.20}$$

where
M_a is the number of servers to be switched between active and sleep modes
F_a is the frequency adjustment for active servers

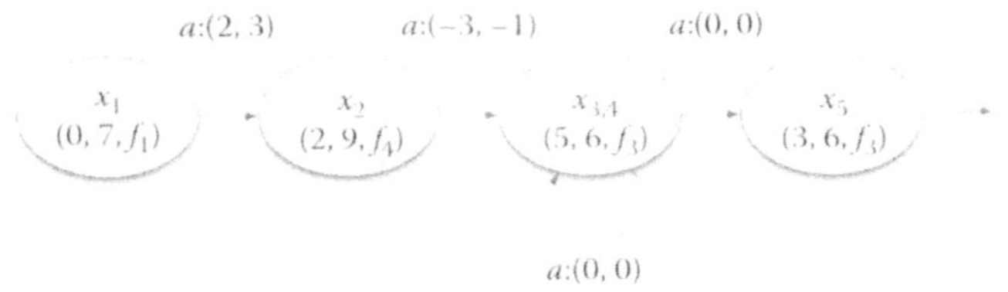

$a{:}(2, 3)$ $a{:}(-3, -1)$ $a{:}(0, 0)$

x_1 x_2 $x_{3,4}$ x_5
$(0, 7, f_1)$ $(2, 9, f_4)$ $(5, 6, f_3)$ $(3, 6, f_3)$

$a{:}(0, 0)$

FIGURE 39.13 Sequential state transition.

In this case, the action for $M_a > 0$ indicates that M_a servers are switched from sleep mode to active mode, and $M_a = 0$ indicates that there is no change in the number of servers in active and sleep modes. $M_a < 0$ indicates that $|M_a|$ servers are switched from active mode to sleep mode [30]. F_a gives the similar action for frequency adjustment of the active servers. For example, $F_a = -(F - 3)$ corresponds that all the servers in active modes adjust their frequency from f_i to f_{i-F+3}, for $i > F + 3$. We denote by $A(x) \subset A$ actions that are available at state x. Set $K = \{(x, a) : x \in X, a \in A(x)\}$ to be the set of state-action pairs. A sequential state and action transition scheme is shown in Figure 39.13.

39.5.1.2 Transition Probability P_{xay}

P_{xay} is the probability in CMDP of moving from state $x = \{s, m, f_i\}$ to $y = \{s', m', f_i'\}$ if action $a = \{M_a, F_a\}$ is taken. Given a composite action $a = \{M_a, F_a\}$ and current state $x = \{s, m, f_i\}$, the active server number m' and frequency level f_i' in next state are deterministic. Since they can be easily derived, $m' = m + M_a$, $f_i' = f_{i+F_a}$. However, the queue length is nondeterministic. Consider a single queue with a buffer of finite size S. In each control period, the probability of i jobs arriving and waiting in the queue is defined as $Pi(i)$, d jobs leave and finish processing with a probability of $Pd(d)$. The probability $P_{s,s'}$ when the queue length changes from s to s' can be obtained:

$$P_{s,s'}(a) = \begin{cases} \displaystyle\sum_{d=0}^{max(m,m')} Pd(d) * Pi(d + s' - s); \ s' \geq s \\ \displaystyle\sum_{i=0}^{I_{max}} Pi(i) * Pd(i + s - s'); \quad s' < s \end{cases} \tag{39.21}$$

where I_{max} is the maximum number of jobs that can arrive in a control period. We define $E[Service]$ as the average service time. Job size follows a bounded Pareto distribution with the average of $E[job]$. Here we assume that the service time is proportional to the packet size and inversely proportional to the server processing capacity $c = f_i/f_F$. Higher processing capacity means faster processing speed. So the average processing capacity is $c_{ea} = f_{\lfloor \frac{1+F}{2} \rfloor}/f_F$. The average service time for the incoming requests can be obtained from

$$E[service] = \frac{E[job]}{c_{ea}} \tag{39.22}$$

The probability that a server finishes processing a job in one control period T is defined as follows:

$$\beta = \begin{cases} 1; & if \ \dfrac{T}{E[service]} \geq 1; \\ \dfrac{T}{E[service]} & otherwise \end{cases} \tag{39.23}$$

The packet interarrival time follows exponential distribution with a mean of $1/\lambda$, where λ is the average arrival rate of incoming packets. The probability of i jobs arriving in a control period can be obtained from

$$Pi(i) = \frac{e^{-\lambda T}(\lambda T)^i}{i!} \tag{39.24}$$

Given at least s jobs in the queue and also m to m' servers in active mode for $m, m' \in \{1, 2, \cdots, M\}$, the probability that d jobs are finished processing in a time slot when action $a = \{M_a, F_a\}$ is taken can be obtained from

$$Pd(d) = \begin{cases} \frac{Tr}{T} * \binom{m}{d} (\beta)^d (1-\beta)^{m-d} + \frac{T-Tr}{T} * \binom{m'}{d} (\beta)^d (1-\beta)^{m'-d} ; M_a \geq 0 \\ \binom{m'}{d} (\beta)^d (1-\beta)^{m'-d} ; \ M_a < 0 \end{cases} \tag{39.25}$$

Where $d \in \{0, 1, 2, \cdots, min(S, m')\}$.

The transition probability P_{xuy} can be finally summarized as

$$P_{xay} = \begin{cases} P_{s,s'}(a); \ if \ m' = m + M_a \ and \ f_i' = f_{i+F_a} \\ 0; \quad otherwise \end{cases} \tag{39.26}$$

39.5.1.3 Objective and Constraints

We adopt a cubic power model in our theoretical framework. The power consumption for a single node at frequency f is $P_{act} = c_0 + c_1 f^3$. We define the power consumption as the objective measure in our CMDP model. The immediate power consumption is consisted of three parts: active server power consumption, inactive server power consumption, and transition power caused by server number allocation. The power consumption of the whole system in a control period T can be expressed as follows:

$$C(x_T, a_T) = \begin{cases} P_{act} * (m * T + M_a * (T - Tr)) + (M - m - M_a) * P_{in} * T + \\ M_a * P_{trans} * Tr. \quad M_a \geq 0; \\ \\ (m - |M_a|) * T * P_{act} + ((M - m) * T + |M_a|(T - Tr)) * P_{in} + \\ |M_a| * P_{trans} * Tr, \quad M_a < 0. \end{cases} \tag{39.27}$$

Little's law relates two important measurements: average waiting time and average number of jobs waiting in a queue in a service system [24]. It states that the long-term average number of jobs L in a stable system is equal to the long-term average arrival rate λ multiplied by the long-term average waiting time of a job in the system W: $L = \lambda W$. Therefore, we can derive the immediate waiting time from the long-term waiting time according to Little's law. After applying a round-robin dispatching method, the immediate request time considering the server busy probability can be expressed as follows [25]:

$$R(x_T, a_T) = \begin{cases} \dfrac{s - \rho}{\sum_{i=1}^{Imax} i * Pi(i)} + \dfrac{E[service]}{c} \quad s - \rho > 0; \\ \\ 0. \quad Otherwise \end{cases} \tag{39.28}$$

The performance measurement involves all the state information. For any state $x = \{s, m, f_i\}$, job blocking could occur when there are more than S jobs that need to be stored in the queue, which can be expressed as $i + s - d \geq S$. The job blocking probability Pb can be obtained from

$$Pb(x_T, a_T) = \begin{cases} 0, & if \quad S - s + 1 \geq I_{max} \\ \sum_{d=0}^{max(m,m')} Pd(d) * Pi(S + d - s + 1); & (for \quad S + d - s + 1 \leq I_{max}) \end{cases} \tag{39.29}$$

The optimization problem can be formulated as minimizing the power consumption with request time and job blocking constraints, both the objective and constraints are related to state and actions.

$$\text{Min}: \quad C_{ea}^n(x_T, a_T)$$

$$\text{S.t.} \quad R_{ea}^n(x_T, a_T) \leq R_{max} \tag{39.30}$$
$$Pb_{ea}^n(x_T, a_T) \leq B_{max}$$

where
   R_{max} is the maximum average request time
   B_{max} is the maximum average blocking probability

39.5.2 Control Policy and Linear Programming

The most critical part of CMDP is to specify the policy by which the controller chooses action at different states. In our CMDP model, we first obtain a nondeterministic policy, then the nondeterministic policy is transferred to a deterministic policy by applying a maximal action probability strategy. We will explain how to obtain the nondeterministic policy first.

Let $f(a)$ is the probability distribution which gives the probability of taking action a at state x under nondeterministic policy $a = \pi(x)$. The optimization problem in (39.30) is then transferred to obtaining the optimal nondeterministic policy $a = \pi^*(x)$:

$$\text{Min}: \quad C_{ea}^n(\pi)$$

$$\text{S.t.} \quad R_{ea}^n(\pi) \leq R_{max} \tag{39.31}$$
$$Pb_{ea}^n(\pi) \leq B_{max}$$

Linear programming (LP) is a technique for the optimization of a linear objective function, subject to linear equality and linear inequality constraints. The optimal nondeterministic policy of CMDP can be obtained by transferring it to an LP problem [3]. Let $\rho(x, a)$ denotes the steady state probability over the set of state-action pairs corresponding to the optimal nondeterministic policy π^*. It has the property the objective and constraint in (39.31) corresponding to the optimal policy, which can be expressed by the immediate power and performance cost with respect to the probability $\rho(x, a)$. The CMDP can be transferred to a LP problem as follows:

$$\text{Min}: \quad \sum_{x \in X} \sum_{a \in A(x)} C(x, a) \rho(x, a)$$

$$\text{S.t.} \quad \sum_{x \in X} \sum_{a \in A(x)} R(x, a) \rho(x, a) \leq R_{max}$$

$$\sum_{x \in X} \sum_{a \in A(x)} Pb(x, a) \rho(x, a) \leq B_{max} \tag{39.32}$$

$$\sum_{a \in A(y)} \rho(y, a) = \sum_{x \in X} \sum_{a \in A(x)} P_{xay} \rho(x, a)$$

$$\sum_{x \in X} \sum_{a \in A(x)} \rho(x, a) = 1$$

Let $\rho^*(x, a)$ is the optimal nondeterministic solution of the LP in (39.32). The corresponding optimal nondeterministic policy of CMDP can be expressed as [3]:

$$f(a = \pi^*(x)) = \frac{\rho^*(x, a)}{\sum_{a \in A(x)} \rho^*(x, a)} \tag{39.33}$$

for $\sum_{a \in A(x)} \rho^*(x, a) > 0$.

Otherwise, we specify a performance guaranteed first policy:

$$f(a = \pi^*(x)) = \begin{cases} 1; & \text{for } M_a = max(0, s - m), \text{ and } Fa = F - i \\ 0. & \text{otherwise} \end{cases} \tag{39.34}$$

The deterministic policy $f_d(a)$ can be obtained from the nondeterministic policy. We define our deterministic policy $f_d(a = \pi_d * (x)) = 1$ if $f(a = \pi^*(x))$ is the maximal probability for all $a \in A(x)$, otherwise $f_d(a = \pi_d * (x)) = 0$. In each control period, our online power controller can easily make an action decision according to current state and the deterministic policy. This is a one-to-one mapping with the online computation complexity of $O(1)$.

39.5.3 Evaluations

In order to make the results more apparent, we define the processing capacity of the server cluster in a control period as

$$Capacity = \begin{cases} (m * T + (T - Tr) * Ma) * f_{i+Fa} & Ma \geq 0 \\ (m + Ma) * f_{i+Fa}; & \text{otherwise} \end{cases} \tag{39.35}$$

Higher processing capacity means faster processing ability but more energy consumption. As we mentioned before, for each state $x = \{s, m, f_i\}$, m and f_i are deterministic given an action a. We only present the nondeterministic queue length in each control period instead of state x to verify the correctness of our deterministic policy. Figure 39.14a illustrates the queue length and corresponding processing capacity in 200 control periods. As we can see from the figure, the processing capacity increases as the number of queue length increases, which means the power controller chose an action to increase the processing capacity for achieving quality of service as the number of jobs in the queue increases. On the contrary, the power controller reduced the active server number and frequency level corresponding to low processing capacity if the number of jobs in the queue is small for more energy savings. In each control period, the power controller is trying to minimize the energy consumption and guarantee the performance together. Figure 39.14 further studies the power consumption in 200 control periods. The maximum request time is set to $R_{max} = 8$ and average job size is $E[job] = 5$. As shown in Figure 39.14b, the CMDP model can achieve significant energy savings under low workload (small queue length), which leads to large amounts of energy savings in a server cluster.

To further evaluate our CMDP model, we vary the number of servers in the server cluster. We compare the performance and power consumption between our CMDP model and VOVF-based CMDP in [30]. Figure 39.15 shows the average request time as the number of servers changes from 2 to 9. As shown in the figure, both CMDP and VOVF-based CMDP models can achieve desirable performance. Figure 39.16 compares the power consumption. On achieving the same performance, our CMDP model with combined VOVF and DVFS strategy can provide more energy savings than the CMDP model with only VOVF strategy. The performance is not affected by changing the server number in the server cluster.

The power consumption and performance are given in Figures 39.17 and 39.18 as the average job size is changed. Given the same average job size, the power consumption increases as the number of servers in the server cluster increases, which can be explained easily: Inactive servers still consume a certain amount

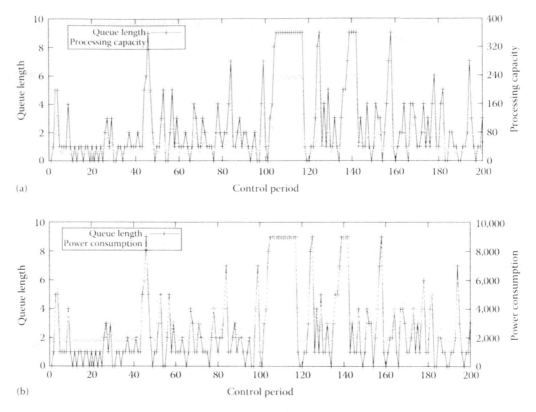

(a)

(b)

FIGURE 39.14 Performance evaluation. (a) Instant processing capacity and queue length in 200 control periods. (b) Instant power consumption and queue length in 200 control periods.

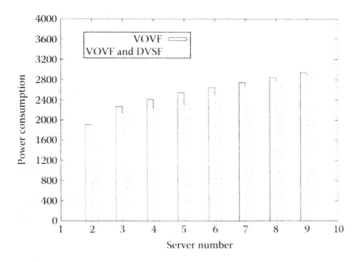

FIGURE 39.15 Comparison of request time between VOVF and combined VOVF, DVFS CMDP models. The number of servers in the server cluster is varied between 2 and 9.

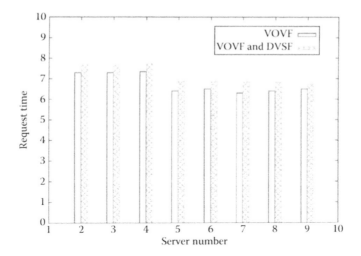

FIGURE 39.16 Comparison of power consumption between VOVF and combined VOVF, DVFS CMDP models. The number of servers in the server cluster is varied between 2 and 9.

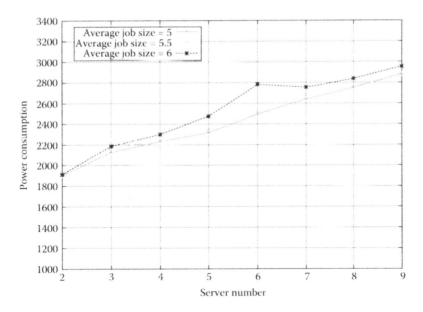

FIGURE 39.17 Comparison of power consumption when the average job size is set to be 5, 5.5, and 6, respectively.

of energy. So the server number in a cluster should be carefully designed in order to achieve more energy savings. In Figure 39.18, the performance constraint can be met when the average job size varies between 5 and 6; however, a larger job size needs more energy consumption to meet the desirable performance. Because larger job size will potentially increase the waiting time and queue length, in order to meet the same performance constraint, more servers and higher frequency levels are required to process the incoming jobs.

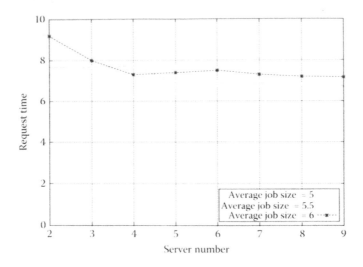

FIGURE 39.18 Comparison of request time when the average job size is set to be 5, 5.5, and 6, respectively.

39.6 Conclusion

Energy management has now become a key issue in server clusters and data centers. This chapter aims to provide effective strategies to reduce power consumption and the impact of performance degradation. However, there are still some new challenges for energy management with the recent developments in data centers and server systems. First, the implementation of low-power active states only exists in few devices, which makes it extremely hard to dynamically control the power consumption solution for those devices. The widely recognized energy proportionality concept provides a new solution in this domain; we can expect more innovation for energy proportional hardware design; second, most large-scale server systems are heterogeneous because of system upgrade, replacement, and capacity increases. No literature has so far proposed a systematic framework with effective power and performance models. Therefore more research is necessary in this area. Finally, the rise of cloud computing has made virtualization a hot research area. Research on energy-efficient virtualized servers is still in its preliminary stage, and we can foresee more research work.

References

1. U.S. Environmental Protection Agency. Report to Congress on server and data center energy efficiency. Washington, DC, August 2007.
2. U.S. Congress. House Bill 5646. To study and promote the use of energy efficient computer servers in the United States. http://www.govtrack.us/congress/bill.xpd?bill=h109-5646 retrieved: February 14, 2008.
3. E. Altman. *Constrained Markov Decision Processes*, Chapman & Hall/CRC, Boca Raton, FL, 1999.
4. J. S. Aronson. Making IT a positive force in environmental change. *IT Professional*, 10(1):43–45, January 2008.
5. L. Barroso and U. Holzle. The case for energy-proportional computing. *Computer*, 40(12):33–37, December 2007.
6. D. Brooks and M. Martonosi. Dynamic thermal management for high-performance microprocessors. *High-Performance Computer Architecture*, Nuevo Leone, Mexico, pp. 171–182, January 2001.

7. E. Carrera, E. Pinheiro, and R. Bianchini. Conserving disk energy in network servers. *Proceedings of the 17th Annual International Conference on Supercomputing*, San Francisco, CA, pp. 86–97, January 2003.

8. Y. Chen, A. Das, W. Qin, A. Sivasubramaniam, and Q. Wang. Managing server energy and operational costs in hosting centers. *Proceedings of the 2005 ACM SIGMETRICS International*, Alberta, Canada, pp. 303–314, January 2005.

9. G. Chen, W. He, J. Liu, S. Nath, L. Rigas, and L. Xiao. Energy-aware server provisioning and load dispatching for connection-intensive Internet services. *Proceedings of the 5th USENIX Symposium on Networked Systems Design and Implementation*, San Francisco, CA, pp. 337–350, January 2008.

10. C. Dovrolis and P. Ramanathan. A case for relative differentiated services and the proportional differentiation model. *Network*, 13:26–34, January 1999.

11. C. Dovrolis, D. Stiliadis, and P. Ramanathan. Proportional differentiated services: Delay differentiation and packet scheduling. *Proceedings of the Conference on Applications*, Cambridge, MA, pp. 12–26, January 1999.

12. E. Elnozahy, M. Kistler, and R. Rajamony. Energy-efficient server clusters. *Power-Aware Computer Systems*, LCNS 2325, Springer-Verlag Berlin, Heidelberg, pp. 179–197, January 2003.

13. M. Elnozahy, M. Kistler, and R. Rajamony. Energy conservation policies for web servers. *Proceedings of the 4th Conference on USENIX Symposium on Internet Technologies and Systems*, Seattle, WA, p. 8, January 2003.

14. X. Fan, W. Weber, and L. Barroso. Power provisioning for a warehouse-sized computer. *Proceedings of the 34th Annual International Conference on Architecture*, San Diego, CA, pp. 13–23, January 2007. B-2-2-2.

15. W. Felter, K. Rajamani, T. Keller, and C. Rusu. A performance-conserving approach for reducing peak power consumption in server systems. *Proceedings of the 19th Annual International Conference on Supercomputing*, Cambridge, MA, pp. 293–302, January 2005.

16. R. Garg, S. W. Son, M. Kandemir, P. Raghavan, and R. Prabhakar. Markov model based disk power management for data intensive workloads. *Cluster Computing and the Grid, IEEE International Symposium on*, 0:76–83, 2009.

17. R. Graybill and R. Melhem. *Power Aware Computing*. Kluwer Academic/Plenum Publishers: New York, January 2002.

18. R. Guerra, J. Leite, and G. Fohler. Attaining soft real-time constraint and energy-efficiency in web servers. *Proceedings of the 2008 ACM Symposium on Applied Computing*, Ceará Fortaleza, Brazil, pp. 2085–2089, January 2008.

19. S. Gurumurthi, A. Sivasubramaniam, and M. Kandemir. DRPM: Dynamic speed control for power management in server class disks. *Computer Architecture*, San Diego, CA, pp. 169–179, January 2003.

20. D. Helmbold, D. Long, T. Sconyers, and B. Sherrod. Adaptive disk spin-down for mobile computers. *Mobile Networks and Applications*, 5:285–297, January 2000.

21. T. Horvath and K. Skadron. Multi-mode energy management for multi-tier server clusters. *Proceedings of the 17th International Conference Parallel Architectures and Compilation Techniques*, Toronto, Ontario, Canada, pp. 270–279, January 2008.

22. T. Imada, M. Sato, Y. Hotta, and H. Kimura. Power management of distributed web servers by controlling server power state and traffic prediction for QoS. *IEEE International Symposium on Parallel and Distributed Processing*, Miami, FL, pp. 1–8, January 2008.

23. M. Kesavan, A. Ranadive, A. Gavrilovska, and K. Schwan. Active coordination (ACT)—Toward effectively managing virtualized multicore clouds. *2008 IEEE International Conference on Cluster Computing*, Tsukuba, Japan, pp. 23–32, January 2008.

24. C. W. Lee and M. S. Andersland. Average queueing delays in ATM buffers under cell dropping. *Global Telecommunications Conference, 1998. GLOBECOM 98. The Bridge to Global Integration. IEEE*, Sydney, New South Wales, Australia, vol. 4, pp. 2440–2445, 1998.

25. H. W. Lee, S. H. Cheon, and W. J. Seo. Queue length and waiting time of the M/G/1 queue under the D-policy and multiple vacations. *Queueing Systems: Theory Applications*, 54(4):261–280, 2006.
26. C. Lefurgy, K. Rajamani, F. Rawson, and W. Felter. Energy management for commercial servers. *Computer*, 36:39–48, January 2003.
27. C. Lefurgy, X. Wang, and M. Ware. Server-level power control. *Autonomic Computing, 2007. ICAC '07. Fourth International Conference on*, Jacksonville, FL, pp. 4–4, May 2007.
28. L. Liu, H. Wang, X. Liu, X. Jin, W. B. He, Q. B. Wang, and Y. Chen. Greencloud: A new architecture for green data center. *Proceedings of the 6th International Conference Industry Session on Autonomic Computing and Communications Industry Session, ICAC-INDST'09*, New York, pp. 29–38, 2009. ACM.
29. Y. Lu and G. D. Micheli. Operating-system directed power reduction. *Proceedings of International Symposium on Low Power Electronics and Design*, Rapallo, Italy, pp. 37–42, January 2000.
30. D. Niyato, S. Chaisiri, and L. B. Sung. Optimal power management for server farm to support green computing. *CCGRID'09: Proceedings of the 2009 9th IEEE/ACM International Symposium on Cluster Computing and the Grid*, Shanghai, China, pp. 84–91, 2009. IEEE Computer Society: Washington, DC.
31. ORACLE. http://www.sun.com/aboutsun/environment/.
32. V. Petrucci, O. Loques, and D. Moss. Dynamic configuration support for power-aware virtualized server clusters, 2009.
33. E. Pinheiro, R. Bianchini, E. Carrera, and T. Heath. Dynamic cluster reconfiguration for power and performance. *Compilers and Operating Systems for Low Power*, Kluwer Academic Publishers: Norwell, MA, January 2001.
34. J. Pouwelse, K. Langendoen, and H. Sips. Energy priority scheduling for variable voltage processors. *Proceedings of the 2001 International Symposium on Low Power*, Huntington Beach, CA, pp. 28–33, January 2001.
35. G. Quan and X. Hu. Energy efficient fixed-priority scheduling for real-time systems on variable voltage processors. *Design Automation Conference*, Las Vegas, NV, pp. 828–833, January 2001.
36. R. Sharma, C. Bash, C. Patel, and R. Friedrich. Balance of power: Dynamic thermal management for Internet data centers. *IEEE Internet Computing*, 9(1):42–49, January 2005.
37. V. Sharma, A. Thomas, T. Abdelzaher, K. Skadron, and Z. Lu. Power-aware QoS management in web servers. *24th IEEE Real-Time Systems Symposium*, Cancun, Mexico, pp. 63–73, January 2003.
38. K. Skadron, T. Abdelzaher, and M. R. Stan. Control-theoretic techniques and thermal-RC modeling for accurate and localized dynamic thermal management. *International Symposium on High Performance Computer Architecture*, San Antonio, TX, pp. 17–28, February 2002.
39. M. Song. Energy-aware data prefetching for multi-speed disks in video servers. *Proceedings of the 15th International Conference on Supercomputing*, Augsburg, Germany, pp. 755–758, January 2007.
40. M. Vasic, O. Garcia, J. Oliver, P. Alou, and J. Cobos. A DVS system based on the trade-off between energy savings and execution time. *Control and Modeling for Power Electronics, 2008. COMPEL 2008. 11th Workshop on*, Zurich, Switzerland, pp. 1–6, July 2008.
41. A. Verma, P. Ahuja, and A. Neogi. Power-aware dynamic placement of HPC applications. *Proceedings of the 22nd Annual International Conference on Supercomputing, ICS'08*, Island of Kos, Greece, pp. 175–184, 2008. ACM: New York.
42. X. Wang and M. Chen. Cluster-level feedback power control for performance optimization. *Proceedings of Symposium on High-Performance Computer Architecture*, Salt Lake City, UT, pp. 101–110, January 2008.
43. A. Weissel, B. Beutel, and F. Bellosa. Cooperative I/O-a novel I/O semantics for energy-aware applications. usenix.org
44. Q. Wu, P. Juang, M. Martonosi, L. Peh, and D. Clark. Formal control techniques for power-performance management. *IEEE Micro*, 25(5):52–62, January 2005.

45. L. Zhang. A two-bit differentiated services architecture for the Internet. Request for Comments: 2638 (Informational), January 1999.

46. X. Zhou, Y. Cai, C. Chow, and M. Augusteijn. Two-tier resource allocation for slowdown differentiation on server clusters. *Parallel Processing*, 2005:31–38, January 2005.

47. X. Zhou and C. Xu. Harmonic proportional bandwidth allocation and scheduling for service differentiation on streaming servers. *Parallel and Distributed Systems, IEEE Transactions on*, 15(9):835–848, September 2004.

48. X. Zhou, Y. Cai, G. Godavari, and C. Chow. An adaptive process allocation strategy for proportional responsiveness differentiation on web servers. *Web Services, 2004. Proceedings. IEEE International Conference on*, San Diego, CA, pp. 142–149, June 2004.

49. H. Zhu, H. Tang, and T. Yang. Demand-driven service differentiation in cluster-based network servers. *INFOCOM 2001. Twentieth Annual Joint Conference of the IEEE Computer and Communications Societies. Proceedings. IEEE*, Anchorage, AK, vol. 2, pp. 679–688, March 2001.

40

Providing a Green Framework for Cloud Data Centers

Andrew J. Younge
Indiana University

Gregor von
Laszewski
Indiana University

Lizhe Wang
Indiana University

Geoffrey C. Fox
Indiana University

40.1 Introduction

For years, visionaries in computer science have predicted the advent of utility-based computing. This concept dates back to John McCarthy's vision stated at the MIT centennial celebrations in 1961.

> "If computers of the kind I have advocated become the computers of the future, then computing may someday be organized as a public utility just as the telephone system is a public utility... The computer utility could become the basis of a new and important industry."

Only recently have the hardware and software become available to support the concept of utility computing on a large scale.

The concepts inspired by the notion of utility computing have combined with the requirements and standards of Web 2.0 [11] to create cloud computing [14,20,36]. Cloud computing is defined as, "a large-scale distributed computing paradigm that is driven by economies of scale, in which a pool of abstracted, virtualized, dynamically-scalable, managed computing power, storage, platforms, and services are delivered on demand to external customers over the Internet." This concept of cloud computing is important to distributed systems because it represents a true paradigm shift [47] within the entire IT infrastructure. Instead of adopting the in-house services, client–server model, and mainframes, Clouds push resources out into abstracted services hosted en masse by larger organizations. This concept of distributing resources is similar to many of the visions of the Internet itself, which is where the "clouds" nomenclature originated, as many people depicted the Internet as a big fluffy cloud one connects to.

While cloud computing is changing IT infrastructure, it also has had a drastic impact on distributed systems itself. Gone are the IBM mainframes of the 1980s which dominated the enterprise landscape. While some mainframes still exist, they are used only for batch-related processing tasks and are relatively unused for scientific applications as they are inefficient at floating point operations. As such, they were replaced with Beowulf clusters [59] of the 1990s and 2000s. The novelty of supercomputing is that instead of just one large machine, many machines are connected together and used to achieve a common goal, thereby maximizing the overall speed of computation. Clusters represent a more commodity-based supercomputer, where off-the-shelf CPUs are used instead of the highly customized and expensive processors in supercomputers. Supercomputers and clusters are best suited for large-scale applications such as particle physics, weather forecasting, climate research, molecular modeling, bioinformatics, and physical simulations, to name a few. These applications are often termed "grand challenge" applications and represent the majority of scientific calculations done on supercomputing resources today. However, recently these scientific clusters and supercomputers are being subverted by the cloud computing paradigm itself.

Many scientists are realizing the power that clouds provide on demand, and are looking to harness the raw capacity for their own needs without addressing the daunting task of running their own supercomputer. While many production-level grid computing systems have looked to provide similar services for the past 10 years through services such as the Open Science Grid [55] and TeraGrid [21], the success of such grid systems has been mixed. Currently we are on the cusp of a merger between the distributed grid middleware and cutting-edge cloud technologies within the realm of scientific computing. Recent projects such as the NSF FutureGrid Project [2] and the DOE Magellan Project [3] aim to leverage the advances of grid computing with the added advantages of clouds. Yet these projects are at a research-only stage and in their infancy; the success of cloud computing within the scientific community is still yet to be determined.

40.1.1 Motivation

As new distributed computing technologies like clouds become increasingly popular, the dependence on power also increases. Currently it is estimated that data centers consume 0.5% of the world's total electricity usage [32]. If current demand continues, it is projected to quadruple by 2020. In 2005, the total energy consumption for servers and their cooling units was projected at 1.2%, the total U.S. energy consumption and doubling every 5 years [4,45]. The majority of the energy used in today's society is generated from fossil fuels which produce harmful CO_2 emissions. Therefore, it is imperative to enhance the efficiency and potential sustainability of large data centers.

One of the fundamental aspects of virtualization technologies employed in cloud environments is resource consolidation and management. Using hypervisors within a cluster environment allows for a number of stand-alone physical machines to be consolidated to a virtualized environment, thereby requiring less physical resources than ever before. While this improves the situation, it often is inadequate. Large cloud deployments require thousands of physical machines and megawatts of power. Therefore, there is a need to create an efficient cloud computing system that utilizes the strengths of the cloud while minimizing its energy and environmental footprint.

In order to correctly and completely unify a "green" aspect to the next generation of distributed systems, a set of guidelines is needed. These guidelines must represent a path of sustainable development that can be integrated into data center construction and management as a whole. While the framework provided in this paper represents many promising ways to reduce power consumption, true sustainable development also depends on finding a renewable and reliable energy source for the data center itself. When combined, many of today's limits in the size of data centers will begin to deteriorate.

40.1.1.1 Economic Costs

Many supercomputers and large-scale data centers are operated at a power envelope on the scale of megawatts or even tens of megawatts. Throughout the past, many of these resources were operated at an institutional level, where such power concerns were not dealt with directly by those that operate or administer such a data center. However recently these energy requirements have grown so large that institutions are starting to feel the economic burden. As such, the time of the energy "blank check" is over, and energy efficiency for the sake of economic stability is now one of the top concerns across all institutions and resource providers alike.

This recent trend is not only being realized in small institutions trying to minimize costs during an economic recession, but also in large, national-scale laboratories operating multimegawatt facilities. Many supercomputing and chip manufacturers are now focused on performance per dollar, not just power.

40.1.1.2 Environmental Concerns

While the economic costs of operating the world's data centers can be extremely staggering as a whole, there is also another important consideration; the environment. As shown in Figure 40.1, nearly 70% of the energy consumed in the United States comes from nonrenewable fossil fuels [10]. As current data center energy consumption is estimated at 2.4%, the overall CO_2 emissions due to the data centers represent sobering reality to the environmental impact the industry has created.

To properly address the sustainability of data centers, it would make sense to focus primarily on the sources of energy. However, achieving the goal of producing 70% of the United States, current power generation from renewable energy sources is a task that will take at least a generation, if not many generations. In the meantime, it is imperative to focus on the data centers themselves to improve efficiency, not only for economic reasons, but also for the environment.

Therefore, improving the energy efficiency within data centers does not look to revolutionize their sustainability, but instead to improve upon an already existing infrastructure. This work is not meant to remove the need for renewable resources, but instead to try to alleviate the burdening energy demands during this long transition period away from fossil fuels. As defined by Thomas Seager's Sustainability Spectrum [57], the work introduced in this manuscript lies within the *reliability* phase of sustainability. Nevertheless, it is important to make what energy efficiency enhancements we can in order to minimize the global climate impact (Figure 40.2).

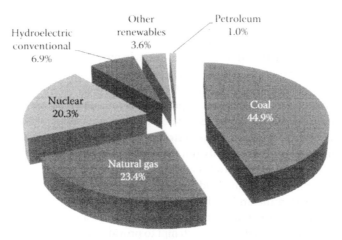

FIGURE 40.1 Sources of energy in the U.S. in 2009.

FIGURE 40.2   Seager's sustainability spectrum.

Cloud computing also makes sense environmentally from a whole inter-level viewpoint. Much of what cloud computing represents is the consolidation of resources from an intra-standpoint to an inter-standpoint, thereby pushing data and services from individual sites to a more centralized system. While there are numerous scaling and privacy issues cloud computing must address, the advantages are clear from an environmental perspective. Typically each individual data center site is relatively small in size and will therefore have more traditional means of cooling. These small data center cooling systems are often just large air conditioning setups that result in a very high data center power usage effectiveness (PUE) [17], resulting in more energy used overall.

Many commercial cloud deployments provide large-scale data center operations. This allows each site to take advantage of the economies of scale [60] which are not available to smaller-sized data centers. These larger cloud centers will be able to implement advanced cooling solutions which keep the PUE low, such as advanced chiller towers or advanced water-cooling setups. Currently, Google has implemented a number of these large data centers and is achieving a PUE value between 1.1 and 1.3, compared to the national average of 2.1 [37]. If the IT industry continues to shift the overall computing resources from many small data centers to a smaller number of larger data centers, the cooling savings alone will have a drastic impact on the environment, just by taking advantage of the basic principals of scaling.

40.2 Related Research

In order to accurately depict the research presented in this article, the topics within cloud computing, grid computing, clusters and, "green" IT will be reviewed.

40.2.1 Cloud Computing

Cloud computing is one of the most explosively expanding technologies in the computing industry today. However, it is important to understand where it came from, in order to figure out where it will be heading in the future. While there is no clear-cut evolutionary path to clouds, many believe the concepts originate from two specific areas: grid computing and Web 2.0.

Grid computing [34,35], in its practical form, represents the concept of connecting two or more spatially and administratively diverse clusters or supercomputers together in a federating manner. The term "grid" was coined in the mid 1990s to represent a large distributed systems infrastructure for advanced scientific and engineering computing problems. Grids aim to enable applications to harness the full potential of resources through coordinated and controlled resource sharing by scalable virtual organizations. While not all of these concepts carry over to the cloud, the control, federation, and dynamic sharing of resources is conceptually the same as in the grid. This is outlined by [36], as grids and clouds are compared at an abstract level and many concepts are remarkably similar. From a scientific perspective, the goals of clouds and grids are also similar. Both systems attempt to provide large amounts of computing power by leveraging a multitude of sites running diverse applications concurrently in symphony. The only significant differences between grids and clouds exist in the implementation details, and the reproductions of them, as outlined later in this section.

The other major component, Web 2.0, is also a relatively new concept in the history of computer science. The term Web 2.0 was originally coined in 1999 in a futuristic prediction by Dracy DiNucci [26]: "The Web we know now, which loads into a browser window in essentially static screenfuls, is only an embryo of the Web to come. The first glimmerings of Web 2.0 are beginning to appear, and we are just starting to see how that embryo might develop. The Web will be understood not as screenfuls of text and graphics but as a transport mechanism, the ether through which interactivity happens. It will [...] appear on your computer screen, [...] on your TV set [...] your car dashboard [...] your cell phone [...] hand-held game machines [...] maybe even your microwave oven." Her vision began to form, as illustrated by the O'Reilly Web 2.0 in 2004 conference, and since then the term has been a pivotal buzz word among the Internet. While many definitions have been provided, Web 2.0 really represents the transition from static HTML to harnessing the Internet and the Web as a platform in of itself.

Web 2.0 provides multiple levels of application services to users across the Internet. In essence, the Web becomes an application suite for users. Data is outsourced to wherever it is wanted, and the users have total control over what they interact with, and spread accordingly. This requires extensive, dynamic, and scalable hosting resources for these applications. This demand provides the user base for much of the commercial cloud computing industry today. Web 2.0 software requires abstracted resources to be allocated and relinquished on the fly, depending on the Web's traffic and service usage at each site. Furthermore, Web 2.0 brought Web Services standards [13] and the Service Oriented Architecture (SOA) [46] which outline the interaction between users and cyberinfrastructure. In summary, Web 2.0 defined the interaction standards and user base, and grid computing defined the underlying infrastructure capabilities.

A cloud computing implementation typically enables users to migrate their data and computation to a remote location with minimal impact on system performance [?]. This provides a number of benefits which could not otherwise be realized. These benefits are as follows:

- *Scalable*: Clouds are designed to deliver as much computing power as any user needs. While in practice the underlying infrastructure is not infinite, the cloud resources are projected to ease the developer's dependence on any specific hardware.
- *Quality of Service (QoS)*: Unlike standard data centers and advanced computing resources, a well-designed cloud can project a much higher QoS than traditionally possible. This is due to the lack of dependence on specific hardware, so any physical machine failures can be mitigated without the prerequisite user awareness.
- *Specialized Environment*: Within a cloud, the user can utilize customized tools and services to meet their needs. This can be done to utilize the latest library, toolkit, or to support legacy code within new infrastructure.
- *Cost Effective*: Users find only the hardware required for each project. This reduces the risk for institutions that potentially want to build a scalable system, thus providing greater flexibility, since the user is only paying for needed infrastructure while maintaining the option to increase services as needed in the future.
- *Simplified Interface*: Whether using a specific application, a set of tools, or Web services, Clouds provide access to a potentially vast amount of computing resources in an easy and user-centric way. We have investigated such an interface within grid systems through the use of the Cyberaide project [64,67].

Many of these features define what cloud computing can be from a user's perspective. However, cloud computing in its physical form has many different meanings and forms. Since clouds are defined by the services they provide and not by applications, an integrated as-a-service paradigm has been defined to illustrate the various levels within a typical cloud, as shown in Figure 40.3.

FIGURE 40.3   View of the layers within a cloud infrastructure.

- *Clients*: A client interacts with a cloud through a predefined, thin layer of abstraction. This layer is responsible for communicating the user requests and displaying data returned in a way that is simple and intuitive for the user. Examples include a Web browser or a thin client application.
- *Software-as-a-Service (SaaS)*: A framework for providing applications or software deployed on the Internet packaged as a unique service for users to consume. By doing so, the burden of running a local application directly on the client's machine is removed. Instead all the application logic and data is managed centrally and to the user through a browser or thin client. Examples include Google Docs, Facebook, or Pandora.
- *Platform-as-a-Service (PaaS)*: A framework for providing a unique computing platform or software stack for applications and services to be developed on. The goal of PaaS is to alleviate many of the burdens of developing complex, scalable software by proving a programming paradigm and tools that make service development and integration a tractable task for many. Examples include Microsoft Azure and Google App Engine.
- *Infrastructure-as-a-Service (IaaS)*: A framework for providing entire computing resources through a service. This typically represents virtualized operating systems, thereby masking the underlying complexity details of the physical infrastructure. This allows users to rent or buy computing resources on demand for their own use without needing to operate or manage physical infrastructure. Examples include Amazon EC2, Eucalyptus, and Nimbus.
- *Physical Hardware*: The underlying set of physical machines and IT equipment that host the various levels of service. These are typically managed at a large scale using virtualization technologies which provide what the QoS users expect. This is the basis for all computing infrastructure.

When all of these layers are combined, a dynamic software stack is created to focus on large-scale deployment of services to users.

40.2.1.1 Virtualization

There are a number of underlying technologies, services, and infrastructure-level configurations that make cloud computing possible. One of the most important technologies is the use of virtualization [?,15]. Virtualization is a way to abstract the hardware and system resources from an operating system. This is typically performed within a cloud environment across a large set of servers using a hypervisor or virtual machine monitor (VMM) which lies in between the hardware and the operating system (OS). From here, one or more virtualized OSs can be started concurrently, as seen in Figure 40.4, leading to one of the key

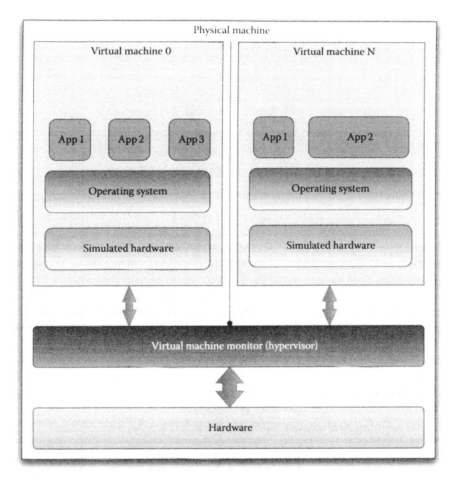

FIGURE 40.4 Virtual machine abstraction.

advantages of cloud computing. This, along with the advent of multicore processing capabilities, allows for consolidation of resources within any data center. It is the cloud's job to exploit this capability to its maximum potential while still maintaining a given QoS.

Virtualization is not specific to cloud computing. IBM originally pioneered the concept in the 1960s with the M44/44X systems. It has only recently been reintroduced for general use on x86 platforms. Today there are a number of clouds that offer IaaS. The Amazon Elastic Compute Cloud (EC2) [12] is probably the most popular and is used extensively in the IT industry. Eucalyptus [54] is becoming popular in both the scientific and industry communities. It provides the same interface as EC2 and allows users to build an EC2-like cloud using their own internal resources. Other scientific cloud-specific projects exist such as OpenNebula [31], In-VIGO [9], and Cluster-on-Demand [23]. They provide their own interpretation of private cloud services within a data center. Using a cloud deployment overlaid on a grid computing system has been explored by the Nimbus project [44] with the Globus Toolkit [33]. All of these clouds leverage the power of virtualization to create an enhanced data center. The virtualization technique of choice for these open platforms has typically been the Xen hypervisor; however, more recently VMWare and the kernel-based Virtual Machine (KVM) have become commonplace.

40.2.1.2 Workload Scheduling

While virtualization provides many key advancements, this technology alone is not sufficient. Rather, a collective scheduling and management for virtual machines is required to piece together a working cloud.

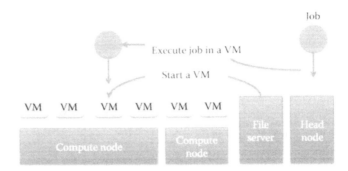

FIGURE 40.5 Example of job scheduling in a virtualized environment.

Let us consider a typical usage for a cloud data center that is used in part to provide computational power for the Large Hadron Collider at CERN [22], a global collaboration from more than 2000 scientists of 182 institutes in 38 nations. Such a system would have a small number of experiments to run. Each experiment would require a very large number of jobs to complete the computation needed for the analysis. Examples of such experiments are the ATLAS [50] and CMS [1] projects, which (combined) require petaflops of computing power on a daily basis. Each job of an experiment is unique, but the application runs are often the same.

Therefore, virtual machines are deployed to execute incoming jobs. There is a file server which provides virtual machine templates. All typical jobs are preconfigured in virtual machine templates. When a job arrives at the head node of the cluster, a correspondent virtual machine is dynamically started on a certain compute node within the cluster to execute the job (see Figure 40.5).

While this is an abstract solution, it is important to keep in mind that these virtual machines create an overhead when compared to running on "bare metal." Current research estimates this as the overhead for CPU-bound operations at 1%–15% depending on the hypervisor; however, more detailed studies are needed to better understand this overhead. While the hypervisor introduces overhead, so does the actual VM image being used. Therefore, it is clear that slimming down the images could yield an increase in overall system efficiency. This provides the motivation for the minimal virtual machine image design discussed in Section 40.5.

40.2.2 Green Information Technology

The past few years have seen an increase in research on developing efficient large computational resources. Supercomputer performance has doubled more than 3000 times in the past 15–20 years, the performance per watt has increased 300 fold, while performance per square foot has only doubled 65 times [24] in the same period of time. This lag in Moore's law over such an extended period of time in computing history has created the need for more efficient management and consolidation of data centers. This can be seen in Figure 40.6 [68].

Much of the recent work in green computing focuses on supercomputers and cluster systems. Currently the fastest supercomputer in the world is the IBM Roadrunner at Los Alamos National Laboratory [16,27], which was fundamentally designed for power efficiency. However, Roadrunner consumes several megawatts of power [58] (not including cooling) and costs millions of dollars to operate every year. The second fastest supercomputer is Jaguar at Oak Ridge National Laboratory. While Jaguar also has a number of power saving features developed by Sandia, Oak Ridge, and Cray [48] such as advanced power metering at the CPU level, 480 V power supplies, and an advanced cooling system developed by Cray. However, the system as a whole still consumes almost 7 MW of power.

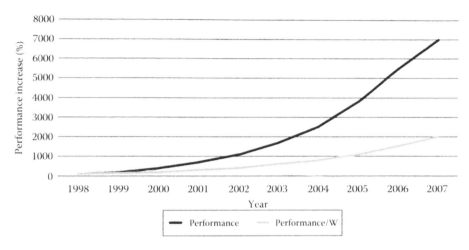

FIGURE 40.6 Performance increases much faster than performance per watt of energy consumed.

40.2.2.1 Dynamic Voltage and Frequency Scaling

One technique being explored is the use of dynamic voltage and frequency scaling (DVFS) within clusters and supercomputers [42,43]. By using DVFS, one can lower the operating frequency and voltage, which results in decreased power consumption of a given computing resource considerably. High-end computing communities, such as cluster computing and supercomputing in large data centers, have applied DVFS techniques to reduce power consumption and achieve high reliability and availability [28,29,38]. A power-aware cluster is defined as a compute cluster where compute nodes support multiple power/performance modes, for example, processors with frequencies that can be turned up or down. This technique was originally used in portable and laptop systems to conserve battery power, and has since migrated to the latest server chipsets. Current technologies exist within the CPU market such as Intel's SpeedStep and AMD's PowerNow! technologies. These dynamically raise and lower both frequency and CPU voltage using ACPI P-states [19]. In [30], DVFS techniques are used to scale down the frequency by 400 MHz while sustaining only a 5% performance loss, resulting in a 20% reduction in power (Figure 40.7).

A power-aware cluster supports multiple power and performance modes, allowing for the creation of an efficient scheduling system that minimizes power consumption of a system while attempting to maximize performance. The scheduler performs the energy performance trade-off within a cluster. Combining

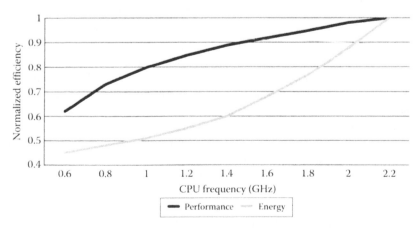

FIGURE 40.7 Possible energy to performance trade-off.

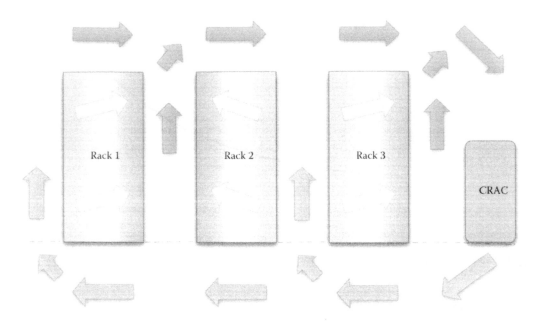

FIGURE 40.8   Data center cooling system.

various power efficiency techniques for data centers with the advanced feature set of clouds could yield drastic results; however, currently no such system exists. This is the premise for much of the work described in Section 40.4.1.

40.2.2.2 Cooling Systems

While there have been numerous reports focused on the green computing aspects of DVFS scheduling, there is also the other side of the coin: cooling solutions. The typical data center cooling layout consists of rows or rack with under-floor cold air distribution. These rows are typically laid back to back, where the hot exhaust air points to each other, such as outlined in Figure 40.8. This forms "cold aisles," where cold air comes from the computer room air conditioning (CRAC) unit under the raised flooring system, and "hot aisles" where the hot air exhausts back around to the CRAC unit.

While the hot and cold aisle cooling approach to a data center improves efficiency compared to a uniform setup, it is far from ideal. With hot and cold air cooling, thermal imbalances are common and interfere with the cooling operations. These imbalances, or "hotspots," can exceed the normal operating temperatures of the servers. This can eventually lead to decreased performance and a high rate of hardware failure. The Arrhenius time-to-fail model [39] describes this, stating that every 10°C increase of temperature leads to a doubling of the system failure rate. Therefore, objectives of thermal-aware workload scheduling are to reduce both the maximum temperature for all compute nodes and the imbalance of the thermal distribution in a data center. In a data center, the thermal distribution and computer node temperatures can be obtained by deploying ambient temperature sensors, on-board sensors [61,63], and with software management architectures like the Data Center Observatory [40].

Many recent data center cooling designs have been created to ease the cooling needs. Many of these designs try to break the standard raised floor hot-cold aisle designs. The first mainstream alternative design was the Sun Blackbox project [5], now known as the Modular Data Center. As seen in Figure 40.9, the design consists of using a trucking cargo holder as a modular server room. By having such a finely controlled system, the overall cooling requirements can be reduced substantially. This was realized by Google, which has since deployed large warehouses of these containers or "pods," reaching a sustained PUE of under 1.3 [8].

FIGURE 40.9 Sun modular datacenter.

40.3 A Green Cloud

There is a pressing need for an efficient yet scalable cloud computing system. This is driven by the ever-increasing demand for greater computational power countered by the continual rise in use expenditures, both economical and environmental. Both business and institutions will be required to meet these needs in a rapidly changing environment in order to survive and flourish in the long term. As such, a systems level design guideline is needed to outline areas of exploration to change efficient cloud data centers from fiction into reality.

40.3.1 Framework

This chapter presents a novel green computing framework which is applied to the cloud paradigm in order to meet the goal of reducing power consumption, as introduced in [69]. The framework is meant to define efficient computing resource management, and green computing technologies can be adapted and applied to cloud systems, but many of the concepts are applicable to various other data center usages. The focus here is on cloud computing data centers for several reasons. First, the cloud is a relatively new concept, one that is able to accept input and be defined most readily. Second, it is a technology that is on the rise with exponential growth, thereby yielding significant gains. Finally, cloud computing's underlying technologies finally allow for the flexibility and precision needed to add in efficiency that makes a difference.

Figure 40.10 illustrates a comprehensive *green cloud framework* for maximizing performance per watt within a cloud. Within the framework, there are two major areas which can lead to widespread improvements in efficiency: virtualization and machine system level designs. The framework tries to outline ways to expand upon the baseline functioning of virtual machines in a cloud environment. This is first done with deriving a more efficient scheduling system for VMs. Section 40.4 addresses the placement of VMs within the cloud infrastructure in a way which maximizes the work capacity while simultaneously minimizing the operating costs of the cloud itself. This is typically achieved by optimizing either power of the server equipment itself or the overall temperature within the data center. Due to the inherent disposability and mobility of stateless VMs within a semi-homogeneous data center, we can leverage

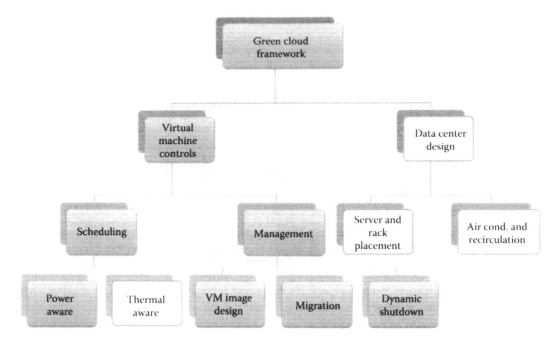

FIGURE 40.10 Green cloud framework. Shaded items represent topics discussed in this paper.

the ability to move and manage the VMs to further improve efficiency. Furthermore, intelligent image management can attempt to control and manipulate the size and placement of VM images in various ways to conserve power and reduce the size of images. Through this, the design of the virtual machine images can also lead to drastic power savings, if architected correctly.

While these operational and run-time chances can have a drastic impact, however, more static data center-level design decisions are also vital for improving energy efficiency. Using more efficient air conditioning units, employing exterior "free" cooling, using completely separated hot and cold isles, or simply picking more efficient power supplies for the servers can lead to incremental but substantial improvements. These best practices can be further enhanced with more radical design, such as a cylindrical or spiral data center design, that brings cool air from the outside in, and exhausts it up through a center chimney and out the top. Also, the excess heat energy of data centers should not go to waste. Instead, the exhaust could be used to heat water or provide ambient heat for surrounding workspaces. Although the potential is great, combining the factors together in such a unified framework and deploying it to a large-scale cloud poses many challenges.

While the last technique may be outside the scope of this manuscript, the integrated components of the green cloud framework in Figure 40.10 provide a sustainable development platform which shows the largest potential impact factor to drastically reduce power requirements within a cloud data center. This framework is not meant to be a solution in and of itself, but rather a set of paths, or guidelines, to reach a solution. As such, only a subset of these tasks are addressed head-on throughout the remaining sections. The hope is that this work will lead to a growing amount of research in this new field to address the growing energy demands within our newest and greatest data centers.

40.4 Scheduling and Management

While supercomputer and cluster scheduling algorithms are designed to schedule individual jobs and not virtual machines, some of the concepts can be translated to the cloud. We have already conducted such research in [65]. In many service-oriented scientific cloud architectures, new VMs are created to perform some work. The idea is similar to sandboxing work within a specialized environment.

40.4.1 DVFS-Enabled Scheduling

As outlined in Section 40.2.2.1, DVFS has proven to be a valuable tool when scheduling work within a cluster. When looking to create a DVFS scheduling system for a cloud data center, there are a few rules of thumb to build a scheduling algorithm which schedules virtual machines in a cluster while minimizing the power consumption:

1. Minimize the processor supply voltage by scaling down the processor frequency.
2. Schedule virtual machines to processing elements with low voltages and try not to scale PEs to high voltages.

Based on the performance model defined earlier, rule 1 is obvious as the power consumption could be reduced when supplied voltages are minimized. Then rule 2 is applied: to schedule virtual machines to processing elements with low voltages and try not to operate PEs with high voltages to support virtual machines.

As shown in Figure 40.11, incoming virtual machine requests arrive at the cluster and are placed in a sorted queue. This system has been modeled, created, and described in [66], but this section will explain in detail this scheduling mechanism.

Algorithm 1 in Appendix 40.A shows the scheduling algorithm for virtual machines in a DVFS-enabled cluster. A scheduling algorithm runs as a demon in a cluster with a predefined schedule interval, *INTERVAL*. During the period of scheduling interval, incoming virtual machines arrive at the scheduler and will be scheduled at the next schedule round. Algorithm 2 in Appendix 40.A is used to schedule incoming virtual machine requests in a certain schedule round defined by the previous algorithm. For each virtual machine (VM), the algorithm checks the processing element (PE) operating point set from low voltage level to high voltage level. The PE with lowest possible voltage level is found; if this PE can fulfill the virtual machine requirement, the VM can be scheduled on this PE. If no PE can schedule the VM, a PE must be selected to operate with higher voltage. This is done using the processor with the highest potential processor speed, which is then incremented to the lowest speed that fulfills the VM requirement, and the VM is then scheduled on that PE. After one schedule interval passes, some virtual machines may

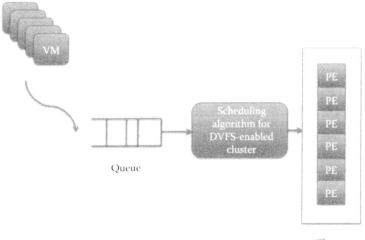

FIGURE 40.11 Working scenario of a DVFS-enabled cluster scheduling.

have finished their execution; thus, Algorithm 3 in Appendix 40.A attempts to reduce a number of PE's supply voltages if they are not fully utilized.

40.4.2 Power-Aware Multicore Scheduling

Currently, there are two competing types of green scheduling systems for supercomputers; power-aware and thermal-aware scheduling. In thermal-aware scheduling [62], jobs are scheduled in a manner that minimizes the overall data center temperature. The goal is not always to conserve the energy used by the servers, but instead to reduce the energy needed to operate the data center cooling systems. In power-aware scheduling [43], jobs are scheduled to nodes in a manner that minimizes the server's total power. The largest operating cost incurred in a cloud data center is in operating the servers. As such, we concentrate on power-aware scheduling in this paper.

Figure 40.12 illustrates the motivation behind power-aware VM scheduling. This graphic documents our recent research findings regarding watts of energy consumed versus the number of processing cores in use. The power consumption curve illustrates that as the number of processing cores increases, the amount of energy used does not increase proportionally. When evaluating using only one processing core, the change in power consumption incurred by using a second processing core is over 20 W. The change from 7 processing cores to all 8 processing cores results in an increase of only 3.5 W.

The impact of this finding is substantial. In a normal round-robin VM scheduling system like the one in Eucalyptus, the load of VMs is distributed evenly to all servers within the data center. While this may be a fair scheduler, in practice it is very inefficient. The result is that each time the scheduler distributes VMs to a processor, the power consumption increases by its greatest potential. In contrast, this research demonstrates that if the scheduler distributes the VMs with the intent to fully utilize all processing cores within each node, the power consumption is decreased dramatically. Therefore, there is a large need for an advanced scheduling algorithm which incorporates the findings in Figure 40.12. To meet this need we propose Algorithm 40.1, a new greedy-based algorithm to minimize power consumption.

Algorithm 40.1 is a VM scheduling algorithm that minimizes power consumption within the data center. This task is accomplished by continually loading each node with as many VMs as possible. In Algorithm 40.1, the pool acts as a collection of nodes and remains static after initialization. While not in the algorithm, the pool can be initialized by a priority-based evaluations system to either maximize performance or further minimize power consumption. At a specified interval t, the algorithm runs through

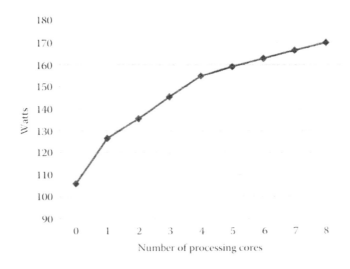

FIGURE 40.12   Power consumption curve of an Intel Core i7 920 CPU.

Algorithm 40.1 Power-based scheduling of VMs

FOR $i = 1$ TO $i \leq |pool|$ DO
  $pe_i =$ num cores in $pool_i$
END FOR

WHILE (*true*)
  FOR $i = 1$ TO $i \leq |queue|$ DO
    $vm = queue_i$
      FOR $j = 1$ TO $j \leq |pool|$ DO
       IF $pe_j \geq 1$ THEN
         IF check capacity *vm* on pe_j THEN
         schedule *vm* on pe_j
         $pe_j - 1$
         END IF
       END IF
      END FOR
  END FOR
  wait for interval t
END WHILE

---

each waiting VM in the queue to be scheduled. The first node in the priority pool is selected and evaluated to see if it has enough virtual cores and capacity available for the new VM. If it does, it is scheduled and the pe_i is decremented by one; and this process is continued until the VM queue is empty. When a VM finishes its execution and terminates, it reports it back to the scheduler and pe_i is increased by one to signify that a core of machine i is freed.

40.4.3 Virtual Machine Management

Another key aspect of a green cloud framework is virtual machine image management. By using virtualization technologies within the cloud, a number of new techniques become possible. Idle physical machines in a cloud can be dynamically shut down and restarted to conserve energy during low-load situations. A similar concept was achieved in grid systems though the use of the Condor Glide-In [49,56] add-on to Condor, which dynamically adds and removes machines from the resource pool. This concept of shutting down unused machines will have no effect on power consumption during peak load as all machines will be running. However, in practice, clouds almost never run at full capacity as this could result in a degradation of the QoS. Therefore, by design, fast dynamic shutdown and start-up of physical machines could have a drastic impact on power consumption, depending on the load of the cloud at any given point in time.

The use of live migration features within cloud systems [25] is a recent concept. Live migration is presently used for proactive fault tolerance by seamlessly moving VMs away from failing hardware to stable hardware without the user noticing a change [53] in a virtualized environment. Live migration can be applied to green computing in order to migrate away machines. VMs can be shifted from low-load to medium-load servers when needed. Low-load servers are subsequently shut down when all VMs have migrated away, thus conserving the energy required to run the low-load idle servers. When using live migration, the user is completely unaware of a change, and there is only a 60–300 ms delay, which is acceptable by most standards.

This process of dynamically allocating and deallocating physical machines is complementary to our scheduling system outlined in Algorithm 40.1. As the scheduling algorithm executes, it will leave a

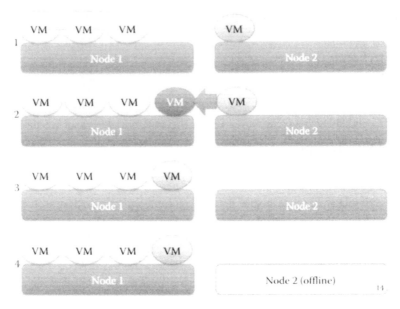

FIGURE 40.13   Virtual machine management dynamic shutdown technique.

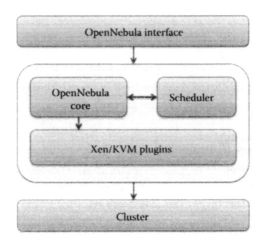

FIGURE 40.14   OpenNebula software architecture.

number of machines idling, potentially for long periods of time. At a predetermined point in time, these idle machines are shut down to conserve energy. When load increases, we use Wake on LAN (WOL) to start them back up. This control can be easily monitored and implemented as a demon running on the cloud head node or scheduler. An illustration of this is presented in Figure 40.13.

40.4.4 Performance Analysis

OpenNebula [31] is an open source distributed virtual machine manager for dynamic allocation of virtual machines in a resource pool. The OpenNebula core components illustrated in Figure 40.14 accept user requirements via the OpenNebula interface, and then place virtual machines in compute nodes within the cluster.

The OpenNebula scheduler is an independent component that provides policies for virtual machine placement. The OpenNebula project was chosen because of this compartmentalized design as it allows

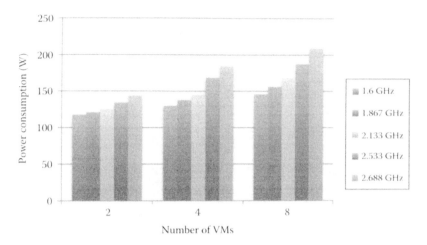

FIGURE 40.15 Power consumption variations for an Intel Nehalem Quad-core processor.

for the integration of our custom scheduling algorithm. The default scheduler provides a scheduling policy based on rank, which allocates compute resources for virtual machines. Scheduling Algorithm 40.1 is implemented by modifying the OpenNebula scheduler to reflect the desired hypothesis that DVFS scheduling leads to a higher performance per unit of energy.

In order to test our design, we created a two-node experimental multicore cluster consisting of Intel Nehalem quad-core processors with Hyperthreading (providing 8 virtual cores). The Nehalem-based CPUs allow for each core to operate on its own independent P-state, thereby maximizing the frequency scaling flexibility. The compute nodes are installed with Ubuntu Server 8.10 with Xen 3.3.4-unstable. The head node consists of a Pentium 4 CPU installed with Ubuntu 8.10, OpenNebula 1.2, and a NFS server to allow compute nodes access to OpenNebula files and VM images. For this experiment, we schedule all virtual machines to the compute nodes and run the nBench [6] Linux Benchmark version 2.2.3 to approximate the system performance. The nBench application is an ideal choice as it is easily compiled in Linux, combines a number of different mathematical applications to evaluate performance, and provides a comparable integer and floating point index that can be used to evaluate overall system performance. The operating frequency of each core can be set to 1.6 Hz, 1.86 GHz, 2.13 GHz, 2.53 GHz, or 2.66 GHz, giving the processor a frequency range of over 1.0 GHz.

Figure 40.15 shows the largest observed power consumption on a WattsUp power meter [7] during the execution of 2, 4, and 8 VMs at each frequency while computing the nBench Linux Benchmark. Here the benchmark effectively simulates a CPU-intensive job running within a VM and provides valuable information on the performance of each VM.

A number of things can be observed from Figure 40.15. First, while scheduling more virtual machines on a node raises power consumption, it seems to consume far less power than operating two separate nodes. Therefore, it seems logical for a scheduler to run as many virtual machines on a node as possible until all available virtual CPUs are taken. Second, when the frequency is dynamically reduced, the difference between running nBench on 2 VMs versus 8 VMs at 1.6 GHz is only 28.3 W. When running the benchmark at 2.668 GHz (the maximum frequency available), this difference grows to 65.2 W, resulting in a larger VM power consumption difference and also a larger overall power consumption of 209 W.

It would be desirable to run each core at its lowest voltage 100% of the time to minimize power consumption; however, one must consider the performance impact of doing so. In Figure 40.16, the average nBench integer calculation index is illustrated with the number of VMs per node and operating frequency dynamically varied for each test.

Figure 40.16 illustrates how the performance degradation due to operating frequency scaling is a linear relationship. This eliminates any question of unexpected slowdowns in performance when running at

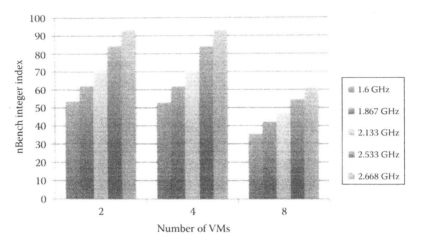

FIGURE 40.16 Performance impact of varying the number of VMs and operating frequency.

frequencies lower than the maximum, such as 1.6 GHz. Another interesting observation is the node's performance running 8 VMs. Due to Intel's Hyperthreading technology, the CPU reports as 8 virtual cores within the host OS (Dom0) even though there are really only 4 cores per node. In our case of running nBench on virtual machines, there appears to be an overall increase in throughput when using 8 VMs instead of just 4. While the performance of each individual VM is only approximately 67% as fast when using 8 VMs instead of 4, there are twice as many VMs contributing to an overall performance improvement of 34%, which is consistent with previous reports [52] of optimal speedups when using Hyperthreading. Therefore, it is even more advisable to schedule as many on one physical node because it maximizes not only power consumption per VM, but also overall system performance.

To evaluate the energy savings of Algorithm 40.1, we consider the following small OpenNebula pool of just four servers. Each server within the pool is a 2.6 GHz Intel Core i7 920 with 12 GB of RAM. We assume each server can hold 8 VMs as it has 8 virtual cores. When idle, they consume 105 W of power and under 100% load they consume 170 W (see Figure 40.12). If we execute the default OpenNebula scheduler to schedule 8 VMs, each server would gain 2 VMs and would consume 138 W with a total pool power consumption of 552 W. However, when Algorithm 40.1 is used, all the VMs are scheduled to the first machine in the pool. This one machine operates at the full 170 W; however, all other machines idle at 105 W, resulting in a pool power consumption of 485 W. Therefore, using our power-based scheduling algorithm, we conserve 12% of the system's power on only four machines on a normal load, as seen in Figure 40.17. If the live migration and shutdown strategy is also deployed, some servers could be dynamically shut down to further conserve energy, leading to further energy savings.

40.5 Virtual Machine Images

While scheduling and management of virtual machines within a private cloud environment is important, one must realize what is actually being scheduled. In a normal cloud environment like the Amazon's EC2 [12], full operating system VMs are scheduled, often to carry out specific tasks in mass. These VM instances contain much more than they need in order to support a wide variety of hardware, software, and varying user tasks. While this is ideal for a desktop-based environment, it leads to wasted time and energy in a server-based solution. A hypervisor provides the same virtualized hardware to each VM, and each VM is typically designed for a specific task. In essence, we want the OS within the VM to act only as a light wrapper which supports a few specific but refined tasks or services, and not an entire desktop/application suite. In order to accomplish this task, we need to concentrate on two areas: VM image size and boot time.

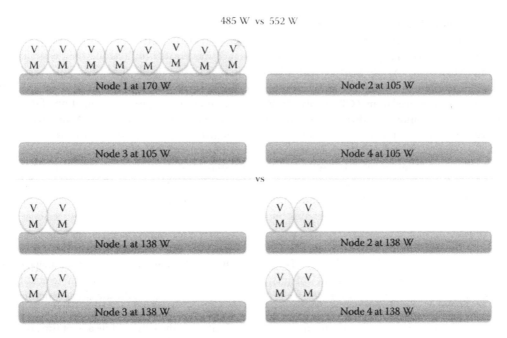

FIGURE 40.17 Illustration of scheduling power savings.

Normal x86 hardware can vary widely, so most modern operating systems are able to detect various hardware and load modules on the fly upon start-up. It is common for bootstrap to spend 15 s running modprobe to load only a single module. This is not an issue with a virtual machine environment since the hardware is standardized and known in advance. The modules in the system and many of the time-consuming probing functions can be reduced upon bootup within a VM environment. In [18], considerable amount of time is saved by changing the IDE delay times for probing new hardware.

Another technique for reducing the boot time is to orchestrate the boot sequence in a more efficient way. Often, many demons and applications are loaded for general use which (in the case of a lightweight VM instance) are not needed and can be removed. This includes stand-alone server applications like window managers and the X11 windowing system. This would also remove the system's disk footprint considerably, saving valuable hard drive space in distributed file systems as well as network traffic when migrating the machines.

Boot time can be further improved by creating a new order which maximizes both the CPU utilization and I/O throughput. The use of bootchart [51] can profile where bootup system inefficiencies occur and to allow for optimization of the boot sequence. Another useful tool is readahead [41]. Readahead profiles the system start-up sequence and uses file pre-fetching techniques to load files into memory before they are requested. Therefore, an application reads directly from system memory and does not have to wait for disk seek time.

40.5.1 Virtual Machine Image Analysis

In order to evaluate the performance of our VM image design, we must create a prototype. There are two paths available to build such a VM OS image. The first is a bottom-up approach where a basic Linux kernel is built upon to reach the minimal feature set needed. This requires developing an entirely new distribution from scratch. While this may be the "cleanest" way, it would require a large development team and is therefore infeasible for this project. The other option involves a top-down approach of taking a common distribution and removing certain components from it, making for a lighter and faster sub-distribution. This route is more practical as it does not require reinventing the wheel, and the option to keep components such as a package management system and a large distribution library are maintained.

Following the second approach, a custom Linux image was created to illustrate the possibility of a fast and lightweight VM OS. Starting with Ubuntu Linux version 9.04, all unnecessary packages were removed, including the Gnome window manager and X11. By removing this multitude of packages, the system image is reduced from 4 GB to only 636 MB. This minimization speeds up migration of the image from one server to another as there is less network traffic during the movement phase. A number of other packages, libraries, and boot-level demons were also removed from the start-up process. At the final stage, the image is a minimal Linux installation with only absolutely necessary components. One thing that was left in was the Synaptic package management system, so if any tools or libraries are needed, it is a trivial process to have them installed on the system. While the package management system does take up some room, it is well worth the extendability it provides to the system. A number of kernel modules were also removed from the 2.6.28-11 kernel to speed up the kernel init and modprobe processes as much as possible.

To test the speed of the custom image, both it and a basic Ubuntu 9.04 installation were moved to a VMWare server with 2.5 GHz Intel Core 2 Duo and 4 GB of RAM. The standard Ubuntu image booted from BIOS in 38 s. With our custom VM image, boot time was reduced dramatically to just 8 s. By comparing the boot charts in Figures 40.18 and 40.19, we can see there is a drastic change in boot time, resulting in a boot time decrease of 30 s. Instead of a large amount of I/O blocking, all disk I/O is done at once toward the beginning, allowing for much higher utilization of the CPU. While a boot time of 8 s is a considerable improvement, we can do better. The kernel still takes a full 2 s to load; with some additional improvements, a second or more could possibly be saved.

40.5.1.1 Minimal VM Discussion

Consider a VM which is started on a machine that requires 250 W of power. Spending 30 s on booting up the VM results in 2.08 Wh or 0.002 kWh of energy used. While this saving of 0.002 kWh or 30 s does not

FIGURE 40.18   Bootup chart of the default Ubuntu Linux VM image.

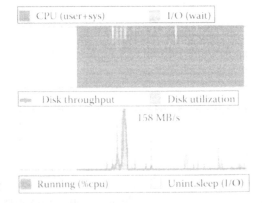

FIGURE 40.19   Bootup chart of Minimal Linux VM image.

seem like much, its effects are actually quite significant. In just a small private cloud system, it is common for 100 VMs to be created every hour, depending on system load and utilization. As such, over 1750 kWh will be wasted per year. Thus, these changes in the VM image may lead to hundreds or thousands of dollars in savings. Furthermore, the power savings realized through using lightweight VM images on a 10 MW facility where thousands of VMs are started every minute equate to tens or hundreds of thousands of dollars a year.

40.6 Conclusion

As the prevalence of cloud computing continues to rise, the need for power-saving mechanisms within the cloud also increases. In this paper we have presented a novel green cloud framework for improving system efficiency in a data center. To demonstrate the potential of our framework, we have presented new energy-efficient scheduling, VM system image, and image management components that explore new ways to conserve power. Through our research presented in this paper, we have found new ways to save vast amounts of energy while minimally impacting performance.

Acknowledgment

This document was developed with support from the National Science Foundation (NSF) under Grant No. 0910812 to Indiana University for FutureGrid: An Experimental, High Performance Grid Test-bed. Any opinions, findings, and conclusions or recommendations expressed in this material are those of the author(s) and do not necessarily reflect the views of the NSF.

40.A Appendix

40.A.1 DVFS Algorithms

Algorithm 1 Scheduling VMs on DVFS-enable cluster

$F_1 = F_2... = F_J = \emptyset$

FOR $k = 1$ TO K DO
  $pe_k.s^{pe} = pe_k.s^a = s_{min}$
  $pe_k.v^{pe} = pe_k.v^a = s_{min}$
  $F_1 = F_1 \cup \{pe_k\}$
  $pe_k.\nabla s = s_{max}$
END FOR

  $t = 0$

WHILE ($\neg$ finished) DO
  reduce current power profiles with Algorithm 40.A.1
  schedule the set of incoming virtual machine requests
with Algorithm 40.A.1
  $t = t+ INTERVAL$
END WHILE

Algorithm 2 Scheduling VMs in an interval of Algorithm 40.6

FOR $i = 1$ TO $i \leq I$ DO

  FOR $j = 1$ TO $j \leq J$ DO

    find $pe_n \in F_j$ which has the max $pe_n.s^a$

    IF $pe_n.s^a \geq vm_i.s^r$ THEN

      schedule vm_i on pe_n

      $pe_n.s^a = pe_n.s^a - vm_i.s^r$

      $pe_n.\nabla s = pe_n.\nabla s - vm_i^r$

      go to schedule vm_{i+1}

    END IF

  END FOR

  IF vm_i is not scheduled THEN

    find pe_n, which has the max $pe_n.\nabla s$

    IF $pe_n.\nabla s \geq vm_i.s^r$ THEN

      $j = pe_n.op^{pe}$

      move pe_n from F_j to F_k, op_k is the lowest possible voltage level that can schedule vm_i, $j < k \leq J$.

      $pe_n.s^{pe} = op_k.s^{op}$

      $pe_n.v^{pe} = op_k.v^{op}$

      $pe_n.op^{pe} = k$

      schedule vm_i on pe_n

      $pe_n.s^a = pe_n.s^a + (op_k.s^{op} - op_j.s^{op}) - vm_i.s^r$

      $pe_n.\nabla s = pe_n.s^a + (s_{max} - op_k.s^{op}) - vm_i.s^r$

    ELSE

      vm_i cannot be scheduled now;

    END IF

  END IF

END FOR

Algorithm 3 Level down VM voltage profiles in a scheduling round

FOR $i = 1$ to I DO

  IF $t - vm_i.t^r \geq vm_i.\Delta t$ THEN

    set vm_i finished its execution on $pe_n \in F_j$

    $pe_n.s^a = pe_n.s^a + vm_i.s^r$

    $pe_n.\nabla s = pe_n.\nabla s + vm_i.s^r$

  ENDIF

FOR $j = J$ TO 2 DO

  FOR $pe_n \in F_j$ DO

    level down pe_n to the operating point set F_k with lowest possible voltage, $1 \leq k < j$, if $op_k.s^{op}$ can support all current virtual machines on pe_n

  ENDFOR

References

1. "CMS," Web page. [Online]. Available: http://cms.cern.ch/
2. G. von Laszewski, G.C. Fox, F. Wang, A.J. Younge, A. Kulshrestha, G.G. Pike, W. Smith et al. Design of the FutureGrid Experiment Management Framework. In *GCE2010 at SC10, New Orleans, 2010. IEEE.*
3. K. Jackson, L. Ramakrishnan, K. Muriki, S. Canon, S. Cholia, J. Shalf, H. Wasserman, and N. Wright. Performance analysis of high performance computing applications on the Amazon Web Services Cloud. In *2nd IEEE International Conference on Cloud Computing Technology and Science.* IEEE, pp. 159–168, 2010.
4. Report to Congress on server and data center energy efficiency — Public Law 109–431. *Tech Report*, July 2007.
5. M. Waldrop. Data center in a box. *Scientific American Magazine.* 297(2):90–93, 2007.
6. Unix/Linux nBench. Webpage, May 2008. [Online]. Available: http://www.tux.org/~mayer/linux/bmark.html
7. Wattsup power meters. Website, 2009. [Online]. Available: https://www.wattsupmeters.com/secure/index.php
8. Google Data Centers: An overview. Web page, 2010. [Online]. Available: http://www.google.com/corporate/green/datacenters/measuring.html
9. S. Adabala, V. Chadha, P. Chawla, R. Figueiredo, J. Fortes, I. Krsul, A. Matsunaga, M. Tsugawa, J. Zhang, Mi. Zhao, L. Zhu, and X. Zhu. From virtualized resources to virtual computing grids: The In-VIGO system. *Future Generation Computer Systems*, 21(6):896–909, 2005.
10. U.S. Energy Information Administration. Net generation by energy source: Total (all sectors), March 2010.
11. B. Alexander. Web 2.0: A new wave of innovation for teaching and learning? *Learning*, 41(2):32–44, 2006.
12. Amazon. Elastic Compute Cloud. [Online]. Available: http://aws.amazon.com/ec2/
13. A. Arkin, S. Askary, S. Fordin, W. Jekeli, K. Kawaguchi, D. Orchard, S.P., K. Riemer et al. Web services choreography interface, June 2002.
14. M. Armbrust, A. Fox, R. Griffith, A.D. Joseph, R. Katz, A. Konwinski, G. Lee, D. Patterson, A. Rabkin, I. Stoica, and M. Zaharia. Above the clouds: A Berkeley view of cloud computing. Technical report, University of California, Berkeley, CA, February 2009.
15. P. Barham, B. Dragovic, K. Fraser, S. Hand, T.L. Harris, A. Ho, R. Neugebauer, I. Pratt, and A. Warfield. Xen and the art of virtualization. In *Proceedings of the 19th ACM Symposium on Operating Systems Principles*, New York, pp. 164–177, October 2003.
16. K.J. Barker, K. Davis, A. Hoisie, D.J. Kerbyson, M. Lang, S. Pakin, and J.C. Sancho. Entering the petaflop era: The architecture and performance of Roadrunner. In *Proceedings of the 2008 ACM/IEEE Conference on Supercomputing*, IEEE Press, Piscataway, NJ, 2008.
17. C. Belady. The green grid data center efficiency metrics: PUE and DCIE. Technical report, The Green Grid, Beaverton, OR, February 2007.
18. T.R. Bird. Methods to improve bootup time in Linux. In *Proceedings of the Ottawa Linux Symposium*, Ottawa, Omtario, Canada 2004.
19. D. Bodas. Data center power management and benefits to modular computing. In *Intel Developer Forum*, 2003. Available at http://www.intel.com/idf/us/spr2003/presentations/SO3US-MODS137_OS.pdf
20. R. Buyya, C.S. Yeo, and S. Venugopal. Market-oriented cloud computing: Vision, hype, and reality for delivering it services as computing utilities. In *Proceedings of the Tenth IEEE International Conference on High Performance Computing and Communications (HPCC-08)*, IEEE CS Press, Los Alamitos, CA, pp. 5–13, 2008.

21. C. Catlett. The philosophy of TeraGrid: Building an open, extensible, distributed terascale facility. In *Second IEEE/ACM International Symposium on Cluster Computing and the Grid, 2002*, p. 8, 2002.

22. CERN. LHC Computing Grid Project. Web page, Dec. 2003. [Online]. Available: http://lcg.web.cern.ch/LCG/

23. J.S. Chase, D.E. Irwin, L.E. Grit, J.D. Moore, and S.E. Sprenkle. Dynamic virtual clusters in a grid site manager. In *Proceedings of the 12th IEEE International Symposium on High Performance Distributed Computing*, pp. 90–100, 2003.

24. W.C. Feng and K.W. Cameron. The Green500 list: Encouraging sustainable supercomputing. *IEEE Computer*, 40(12):50–55, 2007.

25. C. Clark, K. Fraser, S. Hand, J.G. Hansen, E. Jul, C. Limpach, I. Pratt, and A. Warfield. Live migration of virtual machines. In *Proceedings of the Second ACM/USENIX Symposium on Networked Systems Design and Implementation (NSDI)*, Boston, MA, May 2005.

26. D. DiNucci. Fragmented future. *AllBusiness-Champions of Small Business*, 1999.

27. J. Dongarra, H. Meuer, and E. Strohmaier. Top 500 supercomputers. Web site, November 2008. Available: http://ns2.supersys.org/static/lists/1993/06/top500_199306.pdf

28. W. Feng and K.W. Cameron. The Green500 list: Encouraging sustainable supercomputing. *IEEE Computer Society*, 40(12), 2007.

29. W. Feng, A. Ching, C.H. Hsu, and V. Tech. Green Supercomputing in a desktop box. In *Proceedings of the 21st IEEE International Parallel and Distributed Processing Symposium, 2007. IPDPS 2007*, Long Beach, CA, pp. 1–8, 2007.

30. W. Feng, X. Feng, and R. Ge. Green supercomputing comes of age. *IT Professional*, 10(1):17–23, 2008.

31. J. Fontan, T. Vazquez, L. Gonzalez, R.S. Montero, and I.M. Llorente. OpenNEbula: The open source virtual machine manager for cluster computing. In *Open Source Grid and Cluster Software Conference*, San Francisco, CA, May 2008.

32. W. Forrest. How to cut data centre carbon emissions? Web site, December 2008. [Online]. Available: http://www.computerweekly.com/Articles/2008/12/05/233748/how-to-cut-data-centre-carbon-emissions.htm

33. I. Foster and C. Kesselman. Globus: A metacomputing infrastructure toolkit. *International Journal of Supercomputer Applications*, 11(2):115–128, 1997. ftp://ftp.globus.org/pub/globus/papers/globus.pdf.

34. I. Foster, C. Kesselman et al. The physiology of the grid: An open grid services architecture for distributed systems integration. Technical report, Argonne National Laboratory, Chicago, IL, January 2002.

35. I. Foster, C. Kesselman, and S. Tuecke. The anatomy of the grid: Enabling scalable virtual organizations. *International Journal of Supercomputer Applications*, 15(3), 2001.

36. I. Foster, Y. Zhao, I. Raicu, and S. Lu. Cloud computing and grid computing 360-degree compared. In *Grid Computing Environments Workshop, 2008. GCE'08*, Austin, TX, pp. 1–10, 2008.

37. Google. Data center efficiency measurements. Web page. [Online]. Available: http://www.google.com/corporate/green/datacenters/measuring.html

38. I. Gorton, P. Greenfield, A. Szalay, and R. Williams. Data-intensive computing in the 21st century. *IEEE Computer*, 41(4):30–32, 2008.

39. P.W. Hale. Acceleration and time to fail. *Quality and Reliability Engineering International*, 2(4):259–262, 1986.

40. E. Hoke, J. Sun, J.D. Strunk, G.R. Ganger, and C. Faloutsos. InteMon: Continuous mining of sensor data in large-scale self-infrastructures. *Operating Systems Review*, 40(3):38–44, 2006.

41. H. Hoyer and K. Zak. Readahead. Web page. [Online]. Available: https://fedorahosted.org/readahead/

42. C.H. Hsu and W.C. Feng. A feasibility analysis of power awareness in commodity-based high-performance clusters. In *Proceedings of IEEE International Conference on Cluster Computing*, IEEE Computer Society, Los Alamitos, CA, pp. 1–10, 2005.

43. C. Hsu and W. Feng. A power-aware run-time system for high-performance computing. In *Proceedings of the 2005 ACM/IEEE conference on Supercomputing*. IEEE Computer Society, Washington, DC, 2005.

44. K. Keahey, I. Foster, T. Freeman, X. Zhang, and D. Galron. Virtual workspaces in the grid. *Lecture Notes in Computer Science*, 3648:421–431, 2005.

45. J.G. Koomey. Estimating total power consumption by servers in the US and the world. Final report. February, 15, 2007.

46. D. Krafzig, K. Banke, and D. Slama. *Enterprise SOA: Service-Oriented Architecture Best Practices (The Coad Series)*. Prentice Hall PTR, Upper Saddle River, NJ, 2004.

47. T.S. Kuhn. *The Structure of Scientific Revolutions*. University of Chicago Press, Chicago, IL, 1970.

48. J.H. Laros III, K.T. Pedretti, S.M. Kelly, J.P. Vandyke, K.B. Ferreira, C.T. Vaughan, and M. Swan. Topics on measuring real power usage on high performance computing platforms. In *IEEE Cluster*, New Orleans, LA, 2009.

49. M.J. Litzkow, M. Livny, and M.W. Mutka. Condor—A hunter of idle workstations. In *Proceedings of the Eighth International Conference on Distributed Computing Systems (ICDCS)*, IEEE Computer Society, San Jose, CA, pp. 104–111, June 1988.

50. J. Luo, A. Song, Y. Zhu, X. Wang, T. Ma, Z. Wu, Y. Xu, and L. Ge. Grid supporting platform for AMS data processing. *Parallel and Distributed Processing and Applications–ISPA 2005 Workshops*, LCNS 3759, 276, 2005.

51. Z. Mahkovec. Bootchart. Webpage, 2005. [Online]. Available: http://www.bootchart.org/

52. D.T. Marr, F. Binns, D.L. Hill, G. Hinton, D.A. Koufaty, J.A. Miller, and M. Upton. Hyper-threading technology architecture and microarchitecture. *Intel Technology Journal*, 6(1):4–15, 2002.

53. A.B. Nagarajan, F. Mueller, C. Engelmann, and S.L. Scott. Proactive fault tolerance for hpc with xen virtualization. In *ICS '07: Proceedings of the 21st Annual International Conference on Supercomputing*, New York, pp. 23–32, 2007. ACM.

54. D. Nurmi, R. Wolski, C. Grzegorczyk, G. Obertelli, S. Soman, L. Youseff, and D. Zagorodnov. The Eucalyptus open-source cloud-computing system. In *Proceedings of Cloud Computing and Its Applications*, Chicago, IL, 2008.

55. R. Pordes, D. Petravick, B. Kramer, D. Olson, M. Livny, A. Roy, P. Avery, K. Blackburn, T. Wenaus, F. Würthwein. The open science grid. In *Journal of Physics: Conference Series*, 78:012057, Institute of Physics Publishing, 2007.

56. S. Sarkar and I. Sfiligoi. GlideCNAF: A purely condor glide-in based CDF analysis farm. Technical report, CDF/DOC/COMP UPG/PUBLIC/7630P, 2005.

57. T.P. Seager. The sustainability spectrum and the sciences of sustainability. *Business Strategy and the Environment*, 17(7):444–453, 2008.

58. S. Sharma, C.-H. Hsu, and W. C. Feng. Making a case for a Green500 list. In *IEEE International Parallel and Distributed Processing Symposium (IPDPS 2006)/ Workshop on High Performance—Power Aware Computing*, Rhodes, Greece, 2006.

59. T. Sterling, Beowulf cluster computing with Linux. The MIT Press, 2001.

60. G.J. Stigler. Economies of scale, *Journal of Law & Economics*, 1:54–71, 1958.

61. Q. Tang, S.K.S. Gupta, and G. Varsamopoulos. Thermal-aware task scheduling for data centers through minimizing heat recirculation. In *CLUSTER'07 Proceedings of the 2007 IEEE International Conference on Cluster Computing*, pp. 129–138, 2007.

62. Q. Tang, S.K.S. Gupta, and G. Varsamopoulos. Energy-efficient thermal-aware task scheduling for homogeneous high-performance computing data centers: A cyber-physical approach. *IEEE Transactions on Parallel and Distributed Systems*, 19(11):1458–1472, 2008.

63. Q. Tang, T. Mukherjee, S. K.S. Gupta, and P. Cayton. Sensor-based fast thermal evaluation model for energy efficient high-performance datacenters. In *Proceedings of the Fourth International Conference on Intelligent Sensing and Information Processing*, Bangalore, India, pp. 203–208, October 2006.

64. G.V. Laszewski, F. Wang, A. Younge, X. He, Z.H. Guo, and M. Pierce. Cyberaide JavaScript: A JavaScript commodity grid kit. In *Grid Computing Environments Workshop GCE08 at SC'08*, Austin, TX, November 16, 2008. IEEE.

65. G.V. Laszewski, L. Wang, A.J. Younge, and X. He. Power-aware scheduling of virtual machines in DVFS-enabled clusters. In *IEEE Cluster 2009*, New Orleans, LA, August 2009. IEEE.

66. G.V Laszewski, L. Wang, A.J. Younge, and X. He. Power-aware scheduling of virtual machines in DVFS-enabled clusters. In *IEEE Cluster 2009*, New Orleans, LA, August 31–September 4, 2009. IEEE.

67. G.V. Laszewski, A. Younge, X. He, K. Mahinthakumar, and L. Wang. Experiment and workflow management using cyberaide shell. In *Fourth International Workshop on Workflow Systems in e-Science (WSES 09) in Conjunction with Nineth IEEE International Symposium on Cluster Computing and the Grid*, Bangalore, India, 2009. IEEE.

68. M. Wagner. The efficiency challenge: Balancing total cost of ownership with real estate demands. Technical report, Cherokee International, 2008.

69. A.J. Younge, G.V. Laszewski, L. Wang, S. Lopez-Alarcon, and W. Carithers. Efficient resource management for cloud computing environments. In *WIPGC in International Green Computing Conference*. IEEE, August 2010.

70. L. Wang, G.V. Laszecoski, A.J. Younge, X. He, M. Kunze, and J. Tao. Cloud computing: A perspective study, *New Generation Computing*, Vol. 28, pp. 63–69, March 2010 [online].

71. VMware ESX server [online], http://www.vmware.com/de/products/vi/esx

41

Environmentally Opportunistic Computing

Paul Brenner
University of Notre Dame

Douglas Thain
University of Notre Dame

Aimee Buccellato
University of Notre Dame

David B. Go
University of Notre Dame

41.1 Introduction

Energy utilization by high-performance computing and information communications technology (HPC/ICT) is a critical resource management issue. In the United States, billions of dollars are spent annually on power and cool data systems. The 2007 U.S. Environmental Protection Agency "Report to Congress on Server and Data Center Efficiency" estimated that the United States spent $4.5 billion on electrical power to operate and cool HPC and ICT servers in 2006 with the same report forecasting that our national ICT electrical energy expenditure will nearly double—ballooning to $7.4 billion in 2011. Current energy demand for HPC/ICT is already 3% of U.S. electricity consumption and places considerable pressure on the domestic power grid: the peak load from HPC/ICT is estimated at 7 GW or the equivalent output of 15 base load power plants [1].

Recognizing that power resources for data centers are finite, several professional entities within the technology industry have begun to explore this problem including the High-Performance Buildings for High Tech Industries Team at Lawrence Berkeley National Laboratory [2], the ASHRAE Technical Committee 9.9 for Mission Critical Facilities, Technology Spaces, and Electronic Equipment [3,4], the Uptime Institute [5], and the Green Grid [6]. At the same time, efforts by corporations, universities, and government labs to reduce their environmental footprint and more effectively manage their energy consumption have resulted in the development of novel waste heat exhaust and free cooling applications, such as the installation of the Barcelona Supercomputing Center, MareNostrum, in an eighteenth century Gothic masonry church [7], and novel waste heat recirculation applications, such as a centralized data center in Winnipeg that uses recirculated waste heat to heat the editorial offices of a newspaper directly above [8]. Similar centralized data centers in Israel [9] and Paris [10] use recaptured waste heat to condition adjacent office spaces and an on-site arboretum, respectively.

The ICT industry must continually evolve with the rapid development of individual but interrelated components which form the basis of internal infrastructure and end-user interfaces. Simultaneous changes in hardware, software, and networking technology are simply one example of the daily dynamic systems considerations. This high degree of variability and uncertainty makes it difficult to justify long-term infrastructure investments based on speculative total cost of ownership (TCO) analyses. It is not surprising therefore that major energy efficiency improvements to ICT data center facilities are often delayed until they can be coupled with renovations or new construction where improved reliability and capacity are key components of the funding justification. It is also worth noting that energy efficiency may be a smaller component of a total energy cost savings justification when a new facility location is selected based on proximity to less expensive power.

Given the technical- and market-based challenges to more energy efficient ICT facilities, it is promising to see publicly reported power usage effectiveness (PUE) values continuing to shrink toward 1.0. With hardware vendors and large ICT market players continuing to expand the acceptable ranges for free cooling, corporations and professional societies partnering to improve efficiency through cooperatives such as the Green Grid, and the U.S. federal government leading multiple initiatives through the DOE (FEMP Data Center Energy Efficiency) and EPA, it is plausible to expect continued (albeit slow) progress toward lower PUE values approaching 1.0. This begs the question our research intends to answer "Is a PUE of 1.0 the best possible outcome?"

41.2 Environmentally Opportunistic Computing Methodologies

Environmentally opportunistic computing (EOC) is the philosophy that ICT data centers do not need to be isolated entities, but can be integrated with existing buildings and facilities to synergistically share energy needs. Waste heat from high-output ICT can be harvested by the facility (waste heat recovery), and the facility's existing plumbing and air movement can be used as cooling for the ICT (free cooling). To that end, EOC uses *distributed* ICT infrastructure to create heat where it is already needed, to exploit cooling where it is already available, to utilize energy when and where it is least expensive, and to minimize the overall energy consumption of an organization. EOC engages sustainable computing at the macroscale, taking advantage of current scheduler, operating system, and hardware level efficiency [11–16] improvements as well as consolidation efforts in the virtualized data center. The aggressive growth of users—and the capability demanded by those users—must necessarily be met with new, integrated design paradigms, and EOC capitalizes on the dynamic mobility of virtualized services to exploit energy volatility for cost savings and lower environmental impact.

The focus of EOC research is to develop models, methods of delivery, and building/system design integrations that reach beyond current waste heat utilization applications and minimum energy standards to optimize the consumption of computational waste heat in the built environment. Transforming current approaches to ICT data center deployment on a wide scale via EOC requires the development of a systematic method for assessing, balancing, and effectively integrating various interrelated "market" forces (Table 41.1) related to the generation and efficient consumption of computer heat.

At the building scale, the efficient consumption of computer waste heat must be closely coordinated with building heating, ventilation, and air conditioning (HVAC) systems, whether these are existing technologies or new recovery and distribution systems designed specifically for waste heat recovery and free cooling. A sensor–control relationship must be established between these systems, the hardware they monitor, and the local input and output temperatures necessitated by the hardware and demanded by the building occupants, respectively. The controls network must mediate not only the dynamic relationship between source and target but also the variation in source and target interaction due to governing outside factors such as seasonal variations. In the colder winter months, the computational heat source can provide necessary thermal energy whereas the relationship inverts during the hot summer months when the facility can provide reasonably cool exhaust/makeup to the computational components.

TABLE 41.1   Relevant Market Forces for Integrating HPC/ICT into the Built Environment

1. User demand for computational capability
   a. Iterative examination of utilization patterns for various applications (science, business, entertainment, education, etc.)
   b. Iterative correlation of utilization characteristics with developing software, hardware, and network capabilities
2. Computational capability mobility and associated security concerns
   a. Evolution and adoption of grid/cloud computing and virtualization technology
   b. Security algorithms and implementations to allow sensitive/classified information transfer
3. Hardware thermal and environmental limits (temperature, humidity, particulate, etc.)
4. Facility concerns
   a. Integration with existing or novel active HVAC and/or passive systems
   b. General thermal performance variables (building materials, orientation, size, location of openings, etc.)
5. Facility occupant demands/concerns
   a. Thermal control (minimum user expectations and current standards and guidelines)
   b. Indoor air/environmental quality and perception of heat source (radiant computer heat)
6. Temperature variability (indoor/outdoor; day/night; seasonal)
7. Return on investment, TCO, and carbon reduction cost benefits/avoidance

The development of efficiency standards and increased expectations with respect to building occupant comfort require that the optimized integration of computational waste heat in a facility or group of facilities take into account the prevailing thermal comfort standards, like ASHRAE Standard 55-2004 Thermal Comfort Conditions for Human Occupancy [17] which specifies "the combinations of indoor space environment and personal factors that will produce thermal environmental conditions acceptable to 80% or more of the occupants within a space"; and more recent provisions for enhanced controllability of systems by building occupants, like the USGBC's LEED rating system Environmental Quality Credit 6.2 Controllability of Systems [18] which calls for the provision of "individual comfort controls for a minimum of 50% of the building occupants to enable adjustments to suit individual task needs and preferences." Comfort system control may be achieved as long as the building occupants have control over at least one of the primary indoor space environment criteria designated in ASHRAE Standard 55-2004: air temperature, radiant temperature, humidity, and air speed (USGBC 2007), all of which are critical considerations for the utilization and optimization of waste heat in a user-occupied facility.

41.3 EOC in Practice

The authors have evolved toward the EOC concept based a number of related smaller scale research prototypes. Our first model framework called grid heating (GH) [19] specifically focused on utilization and control of server exhaust heat within individual controlled human work centers. We were able to successfully demonstrate the dynamic, energy-based migration of ICT services in response to environmental stimuli [20]. The GH work proposed larger scale container-based possibilities as shown in Figure 41.1a. We then investigated CPU core level energy utilization characteristics for benchmark grid loads to shape our macroscale migration policies [21], finding dominant efficiency benefits of hibernation states over voltage scaling and disabling individual cores.

Most recently, the authors have constructed a heterogeneous, geographically distributed, multi-institutional grid infrastructure for executing service migration in production ICT environments and

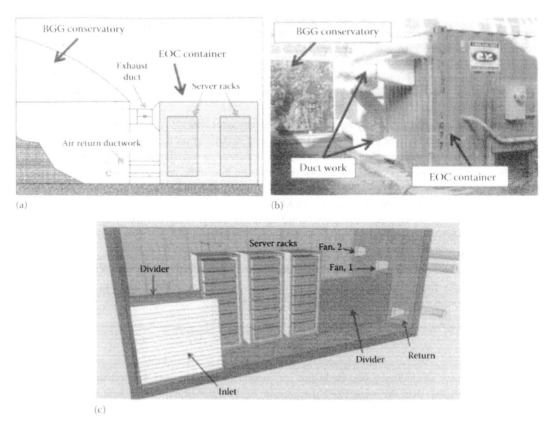

FIGURE 41.1   (a) Layout of prototype EOC container integrated into Greenhouse facility. (b) Photograph of the GC prototype at the Greenhouse. (c) Schematic of prototype EOC container.

recovering waste heat into an existing facility. The University of Notre Dame Center for Research Computing (CRC), the City of South Bend (IN), and the South Bend Botanical Society have collaborated on a prototype building-integrated distributed data center at the South Bend Botanical Conservatory and Greenhouse (Greenhouse) called the Green Cloud (GC) Project [22]. Currently, the ICT infrastructure is based on the University of Notre Dame Condor [23] pool serving a wide variety of high-throughput research computing needs. Jobs can migrate from personal workstations to the traditional data center to dedicated Condor servers at the Greenhouse, where the authors have designed and deployed a sustainable distributed data center (SDDC) container-based prototype. This SDDC serves as a dedicated resource hub for the GC prototype and is shown in Figure 41.1b and c.

The GC hardware resources in the SDDC are fully integrated into the general access Notre Dame Condor pool appearing no different than other resources to end-user jobs. The differentiating factor lies in the environmentally aware controls system set in place for job management and scheduling. The controls system currently has two primary components: Condor and xCAT [24]. The Condor component handles the entire workload management and each server's response to the workload based on environment (such as system temperature). The xCAT component handles real-time monitoring of system vitals through interface with the hardware's service processor Intelligent Platform Management Interface (IPMI). The interface between the two is handled with short Python scripts. It is rational to consider a direct interface between Condor and the hardware level diagnostics; however, the robust existing capabilities of xCAT in this regard for a variety of hardware models has made using the two separate software (Condor and xCAT), the most readily viable option. (xCAT is also our standard cluster management tool for server installation and administration.)

The SDDC was designed to minimize capital cost while still providing a suitably secure facility for use outdoors in a publicly accessible venue. The SDDC container is a standard 20 ft long by 8 ft wide shipping container retrofitted with the following additions: a 40 kW capacity power panel with 208 V power supplies to each rack, lighting, internally insulated walls, man and cargo door access, ventilation louvers, small fans, and ductwork connecting the SDDC to the Greenhouse; for a total cost under $20,000. Exterior power infrastructure including the transformer, underground conduit, panel, and meter was coordinated by the local power utility (American Electric Power) and the City of South Bend. The slab foundation was provided by the City of South Bend. The high-bandwidth network connectivity critical to viable scaling is possible via 1 Gb fiber network connectivity to the Notre Dame campus on the St. Joseph County MetroNet backbone.

From the outset, the SDDC was designed for operation utilizing only direct free cooling via outdoor or Greenhouse air; that is, no specialized air conditioning system was integrated in the container. It was also determined that hardware performance and mean time to failure will be evaluated when pushing systems beyond ASHRAE and industry specified limits for temperature and particulate. By allowing the hardware in the SDDC to endure a larger window of thermal fluctuation, we are working to provide variable exhaust heat densities for delivery to the Greenhouse. System level temperatures are provided by the hardware IPMI and validated by occasional infrared camera measurements. The system has multiple inlet, outlet, and fan options, which will allow for hands-on education of mechanical engineering undergraduates studying heat transfer mechanisms.

The eBay Corporation graciously provided over 100 servers for use in this rigorous prototype environment. Per our agreement with eBay, we are not able to provide the reader with specific details on the server models utilized. We can however summarize that they are commercially available multicore systems. The servers began accepting jobs from the Notre Dame Condor pool in December, 2009. Since the SDDC came online, seasonal temperature variations and continued tuning have allowed the numbers of concurrently running jobs to vary from 0 to 250. Dynamic, real-time machine utilization is posted publicly on the GC Website. As the afternoon temperatures warm up in the SDDC, machines idle or migrate jobs to keep their core temperatures below specified limits; the jobs return in the evening when the environment is more suitable. In the same temporal cycle, the Greenhouse will become a priority service location in the evening as a power tariff under negotiation with the local utility will provide much lower costs at night.

During moderate-temperature months, external air ($\sim$50°F/10°C) is introduced into the container through a single 54 in. $\times$ 48 in. (1.4 m $\times$ 1.2 m) louver, heated by the hardware, and expelled into the conservatory. Conversely, during cold-temperature months, when external air is too cold ($<$50°F/10°C) to appreciably heat for benefit to the conservatory, a return vent has been ducted to the conservatory to draw air directly from the conservatory into the container, heat it from the hardware, and then return it directly into the conservatory. Air is driven by a set of three axial fans through two ducts into the Greenhouse. The fans deliver a total volume flow rate of approximately 1260 cfm (3.6 L/min) at a speed of approximately 26.9 ft/s (8.2 m/s) through one duct and 18.4 ft/s (5.6 m/s) through the other. For operation during summer months, when the conservatory does not require additional heating, the ductwork is disconnected and the container simply uses external air cooling for the hardware.

41.4 EOC Control

One of the largest challenges in the development of the GC prototype has been heat management. As opposed to a traditional data center, machines in the GC pool group are additionally managed by an environmentally aware GC Manager (GCM) controls system. This is necessary because the SDDC is not fitted with its own HVAC equipment and relies solely on external air cooling by either the outside air through the louvers or Greenhouse air through a return vent, during the hot and cold seasons, respectively. The primary role of the GCM is to maintain each machine within its safe operating temperatures as stated by their manufacturers by shutting it down if it exceeds them to prevent damage. At the same time, the GCM attempts to maximize the number of machines available for scientific computations, therefore maximizing the temperature of the hot-aisle air that is used for Greenhouse heating.

The GCM interfaces both Condor and xCAT. xCAT provides access, by means of IPMI, to the functionality of the hardware's built-in service processors: power state control and measurements of intake, RAM and CPU temperatures, fan speeds and voltages. The measurements are used by the GCM to decide whether or not the machine is operating within a safe temperature range. The Condor component handles all of the scientific workload management: deploying jobs on running servers, evicting jobs from machines meant to be shut down, and monitoring the work state of each core available in the GC. The GCM posts new Condor configurations to the machines whenever any actions are required.

The GCM is written in Python and provides rule-based control of the servers running in the GC based upon xCAT data as well as online measurements of the cold-/hot-aisle temperatures in the SDDC container gathered from the two APC sensors. The rules are applied every 2 min to each machine individually, deciding what action the machine should take under the current conditions: start (machine is started, Condor starts running jobs), suspend (Condor jobs are suspended, machine is running idly), and hibernate (Condor jobs are evicted and migrated to a different machine, machine shuts down). Apart from system control, the GCM also provides detailed logging of the transient conditions. Each log entry is stored separately and consists of air temperature measurements in the SDDC and individual machine values. The logs are saved to provide reference points after adjustments to GCM rules or the physical setup of the prototype are made.

To provide ease in interpretation of the GCM logs, an AJAX-based GC Viewer has been created. This online tool provides a near real-time view of the rack-space with color gradient representation of temperatures and information about core utilization and machine states. It also allows users to choose any data point in the measurement period and run a slideshow-like presentation of the changes to the machine states and temperatures.

As previously mentioned, the SDDC prototype container is fitted with air temperature sensors, whose readings are stored in log files and easily viewed using the GC Viewer. Studying the exact distribution of temperatures in the GC provided us with much needed feedback on how to improve the management logic of the GCM. Experiments covered within this chapter were conducted during the months of June, July, and August 2010. Given that these months are typically the hottest on average year round in South Bend, IN, they represent the worst-case free cooling situation of the GC, which only improve during the remainder of the year. The outside temperatures referenced in the following are based on measurements from weather station MC7428 of Weather Underground, which is nearest to the SDDC prototype. The first trials were designed to run the GC machines within vendor-specified environment temperature ranges to prevent machine wear due to overheating. In the basic rule-set, seen in Algorithm 1, whenever a machine's intake temperature exceeded the stop temperature (106°F) value, it was hibernated. The machine was only restarted after its intake temperature dropped below a starting point (99°F) to prevent excessive power cycling.

Algorithm 41.1 Basic Rule = Set for the GCM

```
if   $mytemp<=$start_temp : start;
if   $mytemp,$sleep_temp : continue;
if   $mytemp>$sleep_temp : hibernate
```

With the basic rule-set, we had 25–30 of 60 machines running Condor jobs during July afternoons with outside temperatures ranging from 82°F to 95°F. Given that the machines were cooled by the ambient air and there was no artificial cooling, this result could be reasonable. However, viewing the situation in GC Viewer provided clues on how this could be improved. As expected, the coolest machines were at the bottom of the racks, and the coolest rack of all was the one placed nearest the intake vent. On all three racks, the top 3–5 machines were always in hibernation due to their excessive intake air temperatures, except for cold early mornings (5:00 a.m.), which effectively rendered them useless. This indicates that the initial physical arrangement of the SDDC air flow is suboptimal for rack cooling using ambient air and could be improved.

Additionally, a very important observation was made. There usually was a large intake temperature difference (8°F–15°F) between machines that were on and running jobs and the ones that were hibernated. The difference was even sharper between machines directly neighboring each other (vertically) in a rack. The issue was because the intake fans do not spin when a machine is hibernated, and therefore, there was insufficient air flow for the machine to cool down below the starting temperature, possibly even causing recirculation of hot air from the hot aisle to the cold one. As a result, during the early hours of the night, there were hibernated machines as hot as 115°F at the top of the racks while machines at the bottom of the racks that were running jobs were registering 75°F intake temperatures.

The previous observation prompted further work on the GCM, which included real-time monitoring of the hot-/cold-aisle temperatures and providing information about the placement of the machine inside the rack. With these being exposed in the rule engine, a new spatially aware rule-set, shown in Algorithm 2, was introduced. The rules regarding hibernation were split into two, depending on the temperature of the cold aisle. When the cold aisle was hot (above 85°F), the behavior stayed the same as in the basic approach. In colder cases, the behavior changed. Rules would suspend the machine from performing Condor jobs instead of hibernating it after exceeding the sleeping temperature (106°F). This would keep the internal server fans spinning, facilitating more rapid cooling of the machine below the starting temperature (99°F). The machine would only be hibernated (turned off) if it exceeded an upper operating temperature (109°F). This rule prevented newly started machines from rehibernating more quickly than they could cool down.

The biggest difference came because of the introduction of the last rule. It would force hibernated machines to wake up, even if they reported modest overheating, in order for the fans to spin up and cool the machine down. However, this behavior was only applied if the average intake temperature of the

Algorithm 41.2 Spatially Aware Rule-Set for the GCM

```
if   $mytem<=$start_temp : start;
if   $mytemp<$sleep_tcmp : continue;
if   $cold_aisle>85 and $mytemp>temp=
     $sleep_temp : hibernate;

if   $scold_aisle <=85 and $mytemp>=
     $sleep_temp and $mystate= ="on"
     :suspend;
if   $scold_aisle <=85 and $mytemp>=
     $danter_temp : hibernate;

if   $cold_aisle <=85 and $mystate!="on"
     and $mytemp<=$danter+temp+6 and
     $neightemp<=$start+temp : start;
```

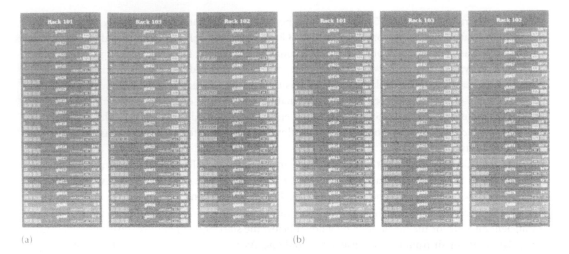

(a)   (b)

FIGURE 41.2 Comparison of rule-set data points in GC Viewer. Note the white/black boxes representing hiberna-tion/running power states, respectively. Only the top 17 machines are shown for each rack, due to space limitations. (a) [left three columns] Spatially aware rule-set (August 1, 2010, 20:00 p.m.): running: 41, hibernated: 19, cold aisle: 82°F, hot aisle: 111°F. (b) [right three columns] Basic rule-set (August 2, 2010, 20:00 p.m.): running: 33, hibernated: 27, cold aisle: 84 F, hot aisle: 109°F.

two machines placed directly beneath it was below the starting point temperature (99°F). This prevented rapid reheating due to natural convection from its nearest neighbors.

The impact of the spatially aware rule-set can be seen in Figure 41.2. During two consecutive days of similar weather, we ran both rule-sets and measured their impact on the state of the GC during the evening (8 p.m., 81°F outside temperature in both cases). While using the basic rule-set, only 33 machines were running and 27 were hibernated. Using the spatially aware rule-set, the number of running machines rose to 41, meaning 36 more processor cores were available to run computations. Moreover, in the case of the spatially aware rule-set, the hot-/cold-aisle temperature difference rose as well, resulting in a 13.7% increase, from 9.72 to 11.29 kW, in available waste heat for the Greenhouse. We observed that even during the hottest parts of the day, we could run 2–3 more machines using the spatially aware rule-set.

While the controls algorithm are important for the reality of suboptimal container and environmental conditions, it should be noted that the availability of outside air free cooling (below 95°) has the potential to provide nearly year round server operation (no suspension, etc.).

41.5 EOC Thermal Measurements

Thermal measurements were conducted on the SDDC prototype during the summer months of June, July, and August of 2010 when the container was fully operational, but not actively heating the conservatory. The thermal measurements consisted of the constant monitoring of various local air temperatures throughout the container as well as server temperatures and server loads. In this way, local temperature and heat recovery could be estimated and directly correlated to server usage and activity. The prototype container was analyzed as a single, pseudo-steady state control volume as shown in Figure 41.3 where q_{waste} is the heat generated by the servers (as well as other electrical equipment), q_{in} is the energy advected into the container through the louvers, q_{loss} accounts for any thermal loss through the insulated container walls and doorways, and q_{out} is the heat leaving the container to effectively heat the Greenhouse.

In principle, the waste heat delivered to the Greenhouse is simply the difference between the exhausted heat q_{out} and the incoming heat not generated by the servers q_{in}. Because the configuration of the servers

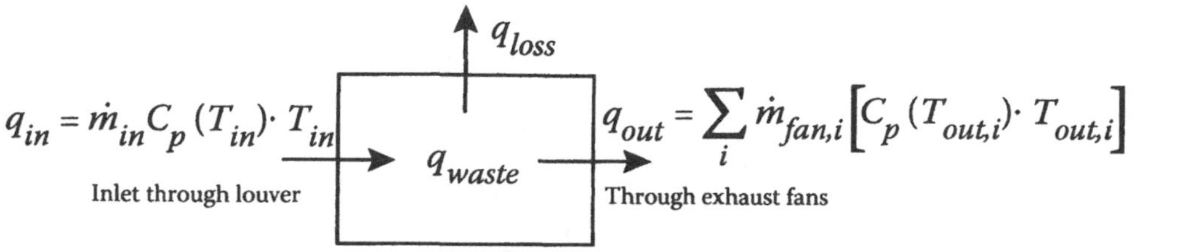

$$q_{in} = \dot{m}_{in} C_p (T_{in}) \cdot T_{in}$$

Inlet through louver

$$q_{out} = \sum_i \dot{m}_{fan,i} \left[C_p (T_{out,i}) \cdot T_{out,i} \right]$$

Through exhaust fans

FIGURE 41.3 Schematic of heat flow through the prototype container.

and hot/cold aisles was not optimal, the true waste heat delivered could not be easily measured. Therefore, for this preliminary study, the amount of waste heat available for recovery was estimated as the amount of heat picked up across the servers minus any heat losses.

$$q_{waste} = \dot{m} \left[c_p(T_{ha}) T_{ha} - c_p(T_{ca}) T_{ca} \right] - q_{loss}, \tag{41.1}$$

where

$c_p(T)$ is the specific heat of the air at the local temperature

T_{ca} and T_{ha} are the temperature of the cold-aisle upstream of the servers and the hot-aisle downstream of the servers, respectively

It is difficult to exactly quantify the heat loss q_{loss} through the insulated container walls, doorways, and louver. For this preliminary analysis of the prototype data, the heat loss was estimated using a 1D conduction analysis through the insulated walls with thermal conductivity of 0.0058 Btu/(h ft°F) (0.01 W/m K), by assuming the wall exteriors were at ambient and the wall interiors were either at the hot-aisle or cold-aisle temperatures. The mass flow rate was determined by applying mass conservation and calculating the total flow rate passing through the two exhaust fans:

$$\dot{m} = \sum_{i=1,2} \dot{m}_{fan,i} = \sum_{i=1,2} \rho \left(T_{out,i} \right) U_{avg,i} A_{duct,i}, \tag{41.2}$$

where

i indicates the two exhaust fan ducts

ρ is the local air density

A_{duct} is the cross-sectional area of the exhaust duct for each fan

$U_{avg,i}$ is the average flow speed at the fan exit

The exit flow speed was measured one time (the fans operated at constant speed) using a handheld hot wire anemometer with integrated thermocouple (Extech Model 407123) with an accuracy of approximately ±3.0% + 0.3 m/s. The velocimeter was placed at the exit of the fan duct, perpendicular to the flow, to record flow speeds for 20 s, and the time-averaged speed was calculated. To obtain the average exit flow speed, the same measurement was conducted over a number of locations across the duct exit (Figure 41.4), and the average flow speed was calculated as

$$U_{avg} = \frac{1}{A_{duct}} \int_{A_{duct}} U(r, \theta) \, r \, dr \, d\theta. \tag{41.3}$$

The integration was conducted numerically using trapezoidal rule, and the uncertainty in the mass flow rate was estimated to be ±8.1%.

Air temperatures in the SDDC were recorded with four temperature/humidity sensors (APC Model AP9512THBLK) connected to two networked control boxes (APC Model AP9319). One sensor was placed just upstream of each exhaust fan ($T_{out,i}$), one sensor was placed in the cool aisle (T_{ca}), and one

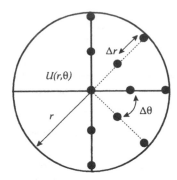

FIGURE 41.4 Schematic of measurements to determine the average outlet velocity. The radii of ducts 1 and 2 were $r = 5$ in. and $r = 4$ in., respectively. The spacing between measurements was $\Delta\theta = \pi/4$ and $\Delta r = 2.5$ in. for duct 1 and $\Delta r = 2$ in. for duct 2. Symmetry was assumed and the flow was only measured in two quadrants.

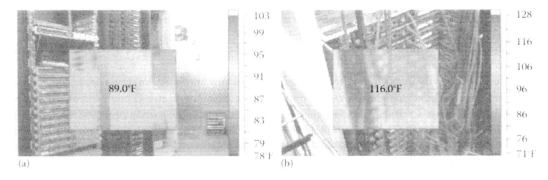

FIGURE 41.5 Illustrative infrared thermal maps of the server (a) inlet and (b) outlet temperatures.

was placed in the hot aisle (T_{ha}). Temperatures were recorded in real time at a rate of four readings per hour. At present, the prototype is not configured to enable temporal temperature measurements in the ducts downstream of the fans, which would be the best way to accurately measure heat recovery. The temperature at the inlet of the louver T_{in} was taken from weather measurements from the Indiana State Climate Office (ISCO 2010). The heat recovery as a function of time $q_{rec}(t)$ was estimated using Equation 41.1 along with the sensor readings, and mass flow rate was estimated using Equation 41.2. In addition to occasional spot checks with a handheld thermocouple for assurance that the sensors were correct, the temperature readings were also validated by comparisons with the temperatures of the HPC hardware as well as occasional infrared camera measurements (Figure 41.5). Hardware temperatures (T_{HPC}) were recorded from the hardware's internal temperature sensors using the IPMI.

Figure 41.6a shows a representative plot of the temperature measurements from the four sensors, the estimated fan downstream temperatures, the inlet temperature, and the temperatures from one of the servers over a 48 h period. The plot illustrates three significant points. The temperature of the hot aisle is significantly hotter than that of the cold aisle, exemplifying the vast amount of waste heat that is generated in data centers, and, though the single hardware temperature (T_{HPC}) varies significantly because of dynamic computational loads, overall the temperatures are fairly constant because the total number of active servers stays fairly constant. Finally, the average server inlet temperatures range from $\sim 70°F$ to $100°F$ ($21°C–38°C$), which exceeds current recommended HPC hardware operating ranges. One aspect of the EOC philosophy is that ITC can be operated beyond current limits, and the data demonstrates server operation at temperatures greatly exceeding standards.

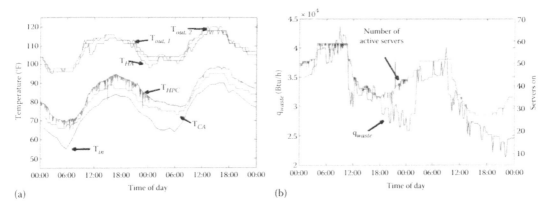

FIGURE 41.6 (a) Representative measured temperatures and (b) available waste heat over a 48 h period in July 2010.

Figure 41.6b shows the amount of waste heat available for recovery for the same 48 h period. On average, nearly 33.6×10^3 Btu/h (9.39 kW) was extracted from the data servers for this period of time, for a total energy recovery of 1.61×10^6 Btu (450.7 kW h). Though this value is limited by approximations for the heat loss and simple bulk temperature measurements, it was consistent with the energy consumed by the container according to energy bills during the same period. At an estimated cost of $0.10/kW h, the measured heat recovery corresponds to a $45.07 energy savings for this time period, and, when extrapolated to a month, approximately $676, or ~4.5% of the average monthly expenditures for the Greenhouse during a winter month (an average ~$15,000 during the months of December through February from 2003 to 2006). In these preliminary studies, the container was limited to a maximum of 60 servers operating at any time in part due to the operational layout but primarily because the warm summer air necessarily increased T_{ca}. With improved configuration and in cooler months, this prototype container should be able to operate at a nearly constant heat recovery of 102,364 Btu (30 kW) corresponding to ~15% savings in the Greenhouse's monthly energy consumption during the winter. Further, the installation of additional containers will provide incremental increases to the total savings accordingly.

One of the primary challenges of successfully integrating HPC into built structures is balancing the market forces illustrated previously in Table 41.1. Successful energy recovery from an EOC container requires that heat be delivered to its partner facility when it is required and in a predictable manner. With EOC, the availability of heat depends on the computing users' need for computational power at that given moment. This basic issue is illustrated in Figure 41.7 when a preset pilot test was conducted during which the computational load on the servers was intentionally varied intermittently—the servers were initially idle for a 27 h period followed by a period of normal loading capacity for 12 h, idle for 12 h, active again for 12 h, and then idle for 12 h. As the pilot test demonstrated, the temperature rise not only reduces dramatically when the HPC hardware is idle, but there is also a transient recovery period when the hardware is active but the temperature rise follows more slowly. The thermal time constant of this system is related to not only the heat capacity of the air, but of the entire set of server components and infrastructure (racks, etc.) as well, which serve as heat sinks during any heat up time. For this prototype, the time constant, when the temperature rise reached 95% of maximum, was estimated to be 53 min. Similarly, during times when the servers were idle, the EOC container took approximately 44 min to reach its ambient condition. In an application where EOC containers are distributed across multiple facilities, control algorithms will be needed to balance the demands of both the compute and building users to balance performance for each set of customers, and such time constants will play an integral role.

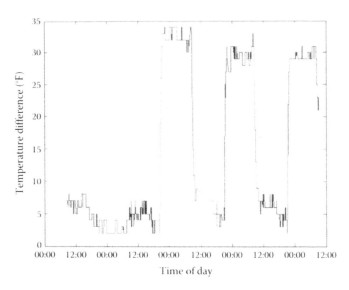

FIGURE 41.7 Temperature measurements for a pilot test over a 72 h period where the servers were left inactive for a 27 h period followed by the servers being operated and then idled for alternating 12 h periods.

41.6 Related Work

It is important to note the evolving commercial applications in grid, utility, and cloud computing [25,26] that will directly benefit from this technology. As the computational infrastructure configuration and locality are obscured from the end user, the flexibility to distribute and configure grows, allowing for additional economic and environmental optimizations. Along these lines, the growing acceptance of virtualization [27,28] in commercial applications will also allow greater flexibility in the design and deployment of EOC-based solutions.

While these grid frameworks are evolving, a large body of work has studied the problem of managing energy, heat, and load in large centralized data centers. Schmidt et al. [29,30] provide a good overview of the mechanical issues of cooling units, heat sinks, fans, and so forth. Server management techniques can also be applied to reduce energy costs. For example, inactive servers can be shut down, or loads migrated as more energy efficient hardware become idle/available. Chase [31] and Bradley [32] describe techniques for balancing performance, cost, and energy in this situation. To avoid hot spots, it is necessary to map the relation between components and heat [33], and then shape loads so as to evenly distribute the heat. Further, large institutions such as the University of Illinois and NCSA are taking a holistic look at their entire campus utilities infrastructure to efficiently operate their data center. Despite the new efficiency benefits, Patel et al. [34] report that a typical data center still consumes about as much energy for cooling as it does for productive work. The advances toward seven more efficient traditional infrastructures and EOC frameworks will serve in tandem to provide greater computational capability while reducing economic and environmental costs.

41.7 Conclusions

In this work, we have introduced the principles of EOC and demonstrated its efficacy with the GC prototype that is integrated with a City of South Bend botanical garden and greenhouse. The prototype successfully conducts significant amounts of workload from the University of Notre Dame Condor pool despite geographical separation and operating only with outside air cooling. It has become a viable tool for analysis and research into the management of ICT energy reutilization, server states, and load distribution. To that end, EOC provides economic and environmental benefits to improve the sustainability of information and communication technology infrastructure.

Our GC prototype serves as a successful demonstration that ICT resources can be integrated with the dominate energy footprint of existing facilities and dynamically controlled to balance process throughput, thermal energy transfer, and available cooling via process management and migration. Apart from not using energy-consuming air conditioning cooling at a centralized data center, the GC further improves energy efficiency by harvesting the waste hot air coming from the working servers for the adjacent Greenhouse facility. The success of this technique makes the EOC concept even more attractive for sustainable cloud computing frameworks.

EOC is a sustainable computing technology that complements existing efficiency improvements at the application, operating system and hardware levels. With the growing enterprise utilization of cloud computing and virtualized services, EOC becomes more viable in across the range of ICT services. The measured average and maximum throughput continues to improve as we tune our GCM controls system and modify the physical layout of the prototype. This initial success will stimulate continued project growth to the economic and environmental benefit of both our organization and our community partners.

Acknowledgments

The authors would like to recognize contributions to this work from multiple organizations and individuals. The City of South Bend and South Bend Botanical Society have been close partners contributing funds and facility expertise. The eBay Corporation generously donated over 100 servers to the Green Cloud prototype. Dr. Jaroslaw (Jarek) Nabrzyski and the Center for Research Computing operations group contributed essential financial resources and system administration expertise. Our undergraduate and summer interns Ryan Jansen, Eric Ward, and Michal Witkowski were key contributors to the infrastructure development and operations. Partial support for this work was provided by a Department of Energy grant for the Northwest Indiana Computational Grid.

References

1. U.S. EPA. August 2007. Report to Congress on server and data center energy efficiency—public law 109–431. Technical Report. Washington, DC: U.S. Environmental Protection Agency.
2. Blazek, M., Mills, E., Naughton, W., Tschudi, P., Sartor, D., Seese, R., and Shamshoian, G. 2007. The business case for energy management in high-tech industries. *Journal of Energy Efficiency*, 1:1–16.
3. ASHRAE Technical Committee 9.9. 2008a. Best practices for datacom facility energy efficiency. Atlanta, GA: American Society of Heating Refrigeration, and Air-Conditioning Engineers.
4. ASHRAE Technical Committee 9.9. 2008b. High density data centers—Case studies and best practices. Atlanta, GA: American Society of Heating Refrigeration, and Air-Conditioning Engineers.
5. Brill, K.G. 2008. Special report: Energy efficiency strategies survey results. Technical Report. Santa Clara, CA: The Uptime Institute.
6. Green Grid: 7 × 24 Exchange International, ASHRAE, Silicon Valley Leadership Group, U.S. DOE Save Energy Now Program, Energy Star Program: U.S. EPA, U.S. GBC, Uptime Institute. 2010. Recommendations for measuring and reporting overall data center efficiency. Version 1: Measuring PUE at Dedicated Data Centers.
7. Barcelona Supercomputing Center—Centro Nacional Supercomputación. 2006. Annual Report 2005. Barcelona, Spain.
8. Fontecchio, M. 2008. Data center news: Companies reuse data center waste heat to improve energy efficiency. Winnipeg, Manitoba, Canada: Search Data Center.

9. Alger, D. 2010. *Grow a Greener Data Center*. Indianapolis, IN: Cisco Press.
10. Miller, R. February 2010. *Data Centers Heat Offices, Greenhouses, Pools*. Lawrenceville, NJ: Data Center Knowledge.
11. Ahmad, I., Ranka, S., and Khan, S.U. Using game theory for scheduling tasks on multi-core processors for simultaneous optimization of performance and energy. In *The Next Generation Software (NGS) Workshop 2008*, Melbourne, Victoria, Australia, April 2008.
12. Gunaratne, C., Christensen, K., and Nordman, B. 2005. Managing energy consumption costs in desktop PCs and LAN switches with proxying, split TCP connections, and scaling of link speed. *International Journal of Network Management*, 15(5):297–310.
13. Jejurikar, R. and Gupta, R. Dynamic voltage scaling for systemwide energy minimization in real-time embedded systems. In *Proceedings of the 2004 International Symposium on Low Power Electronics and Design (ISLPED'04)*, Newport, CA, 2004.
14. Sharma, S., Hsu, C.H., and Feng, W.-C. Making a case for a Green500 list. In *IEEE International Parallel and Distributed Processing Symposium 2006, Workshop on High Performance Power Aware Computing*, Rhodes Island, Greece, 2006.
15. Eilam, T., Appleby, K., Breh, J., Breiter, G., Daur, H., Fakhouri, S.A., Hunt, G.D.H., Lu, T., Miller, S.D., Mummert, L.B., Pershing, J.A., and Wagner, H. 2004. Using a utility computing framework to develop utility systems. *IBM Systems Journal*, 43:97–120.
16. Figueiredo, R., Dinda, P., and Fortes, J. 2005. Resource virtualization renaissance. *Computer*, 38:28–31.
17. ASHRAE. 2004. *ANSI/ ASHRAE Standard 55-2004, Thermal Comfort Conditions for Human Occupancy*. Atlanta, GA: American Society of Heating, Air-Conditioning and Refrigeration Engineers, Inc.
18. USGBC. 2007. *LEED for New Construction & Major Renovation Reference Guide*, Version 2.2. Indoor Environmental Quality: EQ Credit 6.2, Controllability of Systems: Thermal Comfort. Washington, DC: USGBC.
19. Brenner, P., Thain, D., and Latimer, D. Grid heating clusters: Transforming cooling constraints into thermal benefits. In *Uptime Institute—IT Lean, Clean, and Green Symposium*, New York, 2009.
20. Brenner, P., Thain, D., and Latimer, D. April 2008. Grid heating: Transforming cooling constraints into thermal benefits. Technical Report 2008-30, Computer Science and Engineering Department, University of Notre Dame, Notre Dame, IN.
21. Lammie, M., Brenner, P., and Thain, D. Scheduling grid workloads on multicore clusters to minimize energy and maximize performance. In *10th IEEE/ACM International Conference on Grid Computing*, Banff, Alberta, Canada, 2009.
22. Brenner, P., Jansen, R., Go, D., and Thain, D. Environmentally opportunistic computing transforming the data center for economic and environmental sustainability. In *First International Green Computing Conference, Technically Co-Sponsored by IEEE Computer Society*, Chicago, FL, 2010.
23. Litzkow, M., Livny, M., and Mutka, M. Condor—A hunter of idle workstations. In *Eighth International Conference of Distributed Computing Systems*, San Jose, CA, June 1988.
24. Ford, E. 1999. xCAT extreme cloud administration toolkit. IBM internal software development. Open Sourced 2007.
25. Raghavan, B., Vishwanath, K., Ramabhadran, S., Yocum, K., and Snoeren, A.C. 2007. Cloud control with distributed rate limiting. In *SIGCOMM'07: Proceedings of the 2007 Conference on Applications, Technologies, Architectures, and Protocols for Computer Communications*, Kyoto, Japan, pp. 337–348, New York: ACM.
26. Mitchell, W. August 2007. Data center in a BOX. *Scientific American*, 297:90–93.
27. Barham, P., Dragovic, B., Fraser, K., Hand, S., Harris, T., Ho, A., Neugebauer, R., Pratt, I., and Warfield, A. 2003. Xen and the art of virtualization. In *SOSP'03: Proceedings of the Nineteenth ACM Symposium on Operating Systems Principles*, Bolton Landing, NY, pp. 164–177, New York: ACM.

28. Uhlig, R., Neiger, G., Rodgers, D., Santoni, A., Martins, F., Anderson, A., Bennett, S., Kagi, A., Leung, F., and Smith, L. 2005. Intel virtualization technology. *Computer*, 38:48–56.

29. Schmidt, R., Cruz, E., and Iyengar, M. July 2005. Challenges of data center thermal management. *IBM Journal of Research and Development*, 49(4–5):709–724.

30. Schmidt, R., and Iyengar, M. January 2007. Best practices for data center thermal and energy management—Review of literature. *ASHRAE Transactions*, 113(1):206.

31. Chase, J., Anderson, D., Thakar, P., Vahdat, A., and Doyle, R. Managing energy and server resources in hosting centers. In *Symposium on Operating Systems Principles*, Banff, Alberta, Canada, 2001.

32. Bradley, D., Harper, R., and Hunter, S. 2003. Workload based power management for parallel computer systems. *IBM Journal of Research and Development*, 47:703–718.

33. Kang, J., and Ranka, S. Assignment algorithm for energy minimization on parallel machines. In *2009 International Conference on Parallel Processing Workshops*, Vienna, Austria, 2009.

34. Patel, C.D., Bash, C.E., Sharma, R., and Beitelmal, M. Smart cooling of data centers. In *Proceedings of IPACK*, Maui, HI, July 2003.

42

Energy-Efficient Data Transfers in Large-Scale Distributed Systems

Anne-Cécile Orgerie
*Ecole Normale Supérieure de
Lyon*
University of Lyon

Laurent Lefèvre
INRIA
University of Lyon

42.1 Introduction

At the age of petascale machines and cloud computing, large-scale distributed systems need an ever-increasing amount of energy. Due to the growth of such distributed systems in terms of computing and storage resources, the networking resources are more and more requested.

In-advance reservation mechanisms are widely used in large-scale distributed systems [14,45,57] since they guarantee users a certain quality of service, including the respect of deadlines and specific hardware and software constraints, in an infrastructure-as-a-service way.

Bandwidth provisioning is feasible for network operators for several years thanks to protocols such as Multiprotocol Label Switching (MPLS) [50] and Reservation Protocol (RSVP) [61]. However, for end users with no knowledge of network traffic, this task is impossible without collaboration with the other nodes.

On the other hand, as networks become increasingly essential, their electric consumption reaches unprecedented peaks [3]. Up to now, the main concern to design network equipments and protocols was performance only; energy consumption was not taken into account. With the costly growth in network electricity demand, it is high time to consider energy as a main priority for network design.

To this end, we propose a new complete and energy-efficient bulk data transfer (BDT) framework including scheduling algorithms which provide an adaptive and predictive management of the advance bandwidth reservations (ABR). This model is called HERMES: High-level Energy-awaRe Model for bandwidth reservation in End-to-end networkS.

The rest of the chapter is organized as follows. Section 42.2 summarizes the background knowledge on data center, grid and cloud networks, on bulk data transfers and bandwidth provisioning, and on green wired networking. Section 42.3 details the architecture of HERMES and its working. In Section 42.4, we present a new network simulator: BoNeS (Bookable Network Simulator) and we discuss experimental results of our model using this simulator. Finally, Section 42.5 concludes the chapter and propose directions for future work.

42.2 Background and Framework Design Choices

This section presents some background knowledge on data center, grid and cloud networks, on bulk data transfers and bandwidth provisioning, and on green wired networking. We motivate the associated choices made for the design of the framework components.

42.2.1 Data Center, Grid, and Cloud Networks

The need for big computing facilities is always increasing due to new killer applications. Providing these services with high quality and reliability rely on robust networks which are specially designed for data centers, grids and clouds.

Previous work consider data center network and traffic in order to develop energy-efficient data center task schedulers [1,42]. The common scheme is as follows: a centralized scheduler performs the traffic flow optimization inside the data center network [1]. We have opted for a decentralized approach to gain scalability. The authors of [42] propose a traffic-aware virtual machine placement. We have chosen to de-correlate task scheduling and traffic scheduling in order to optimize only the traffic scheduling. Indeed, these two problems are not equivalent since traffic is not exclusively generated by the computing tasks. For example, in a cloud environment, virtual machine deployment causes transfers from the server containing the images to the deployed nodes. The networking performance of the famous cloud Amazon EC2 is evaluated in [58]. The authors show that virtualization and processor sharing on server hosts lead to unstable network characteristics from the application point of view.

Three-tier fat trees are among the typical network topologies for today's data centers [1,29,34]. We evaluated our framework on this three-tier architecture, but it is independent of the network architecture. Such networks are designed for peak workload and run well below capacity most of the time following periodic patterns [7,29,35]. The periodicity of the traffic makes its predictability more easy and thus, the traffic can be scheduled.

42.2.2 Bulk Data Transfer and Bandwidth Provisioning

Typical bulk data transfer applications include peer-to-peer protocols [28,56] and content delivery network facilities [13] with media servers that require timely transfers of large amounts of data among these different servers [52]. They can even be used on the Internet with water-filling techniques [36] that are compared in terms of performance and cost to the FedEx courier service. We are focusing on the specific traffic appearing inside the data center, grid and cloud networks.

42.2.2.1 Bandwidth Allocation and Routing Algorithms

To provision bandwidth for BDT, the two main problems are to allocate bandwidth in time and in space. These two issues are solved by bandwidth scheduling and path computation algorithms, respectively.

Two basic provisioning modes are commonly distinguished [24,33,51]: (1) on-demand mode: a connection request is made when needed, and it is then accepted or denied depending on the current bandwidth availability; arriving requests are queued until their allocation; (2) in-advance mode: a connection request is granted for future time slots based on bandwidth-allocation schedules; arriving requests are scheduled in the future as soon as they arrive (system of agendas).

Two approaches can also be taken to provision bandwidth over the network [46]: the centralized and the distributed approaches. However, the centralized approach has issues with scalability [62] while the distributed approach has issues with the processing time of requests. The network-management system is often called *bandwidth broker* [10,62], and is in charge of the reservation requests' admission control.

Many problems related to bandwidth allocation and path computation are NP-complete. Lin and Wu [39] describe two basic scheduling problems: fixed path with variable bandwidth and variable path with variable bandwidth with a view to minimize the transfer end time of a given data size. They prove that both problems are NP-complete and they propose greedy heuristic algorithms to solve them. In [38], they consider two other problems dealing with multiple data transfers. The bandwidth allocation and path computation algorithms are mainly inspired from the Dijkstra and Bellman–Ford algorithms [33,38,51].

Our model is based on a decentralized bandwidth provisioning scheme and relies on existing routing protocols. It is built as an energy-efficient overlay on the existing architecture of data centers, grids, and clouds.

42.2.2.2 Network Protocols

The usual network protocols are not adapted to dedicated networks since they are designed to work in a best-effort mode with congestion, failures, and resource competition. Dedicated networks have different characteristics, among which high speed, reliability, and high delay-bandwidth product. It has been known for a long time that TCP is inefficient in this kind of environment [32]. New protocols using UDP for data transfer and TCP for control have been developed, such as reliable blast UDP [27] or SABUL [22]. Other emerging solutions include a better adaptivity to maximize the data rate according to the receiver's capacity and to maximize the goodput by minimizing synchronous, latency-bound communication (Adaptive UDP [18], FRTP: Fixed Rate Transport Protocol [64]). These protocols are implemented on top of UDP as application-level processes.

Furthermore, specific reservation protocols have been developed for a long time, like for example RSVP (ReSerVation Protocol) [61] where resources are reserved across a network for integrated services in QoS-oriented networks. RSVP also allows protocols to be designed on top of it to complete its functionality [52].

Yet, none of these protocols is energy-aware. So, we develop our own reservation protocol using, in particular, disruptive-tolerant network (DTN) techniques and minimizing the number of control messages exchanged in order to save energy.

42.2.2.3 Advance-Reservation Algorithms

The on-demand mode can be seen as a special case of in-advance mode [33]. Thus, focusing on advance reservations does not restrict our scope. The idea of making advance reservations of network resources is not recent [48]. The main issue is the unpredictability of the routing behavior. However, with the emergence of the MPLS (Multi-Protocol Label Switching) [50] standard with traffic engineering and explicit routing features, it becomes possible to disconnect the reservation management from the network layer, thus leading to an easier interoperability for the ABR management systems.

For advance reservation, different BDT scheduling techniques can be used: online scheduling where requests are processed as soon as they arrive or periodic batch scheduling where they are scheduled with a certain periodicity [37]. Different time models can also be used: continuous time models [33,37] and discrete models [10,47] with fixed time slots (slices) during which the resource allocations are similar.

Several other issues related to ABRs have been explored: fault tolerance [11], rerouting strategies [12], load-balancing strategies [60], time-shift reservations [46], etc.

For the moment, none of the proposed solutions takes into account the network's energy consumption as a major issue that should influence the design of each algorithm related to the network's management, from scheduling to routing.

42.2.3 Green Wired Networking

Gupta and Singh [25] have shown that transmitting data through wired networks takes more energy (in bits per joule) than transmitting data through wireless networks. Energy is indeed one of the main concerns for wireless networks while, for now, it is not the case for wired networks since they are not battery-constrained. However, the energy issue is becoming more and more present in wired networks because of the need to maintain network connectivity at all times [17]. The ever-increasing demand in energy can still be greatly reduced. Studies have indeed shown for a few years that network links, and especially edge links, are lightly utilized [17,44]. This fact has led researchers to propose several approaches to take advantage of link underutilization in order to save energy.

We have classified these approaches in several categories: the *optimization* approach, which is based on improvements of the hardware components such as routers and network interface controllers (NICs); the *shutdown* approach, which takes advantage of the idle periods to switch off the network components such as switches and the ports of routers; the *slowdown* approach, which puts the network components in low power modes during underutilization periods; and the *coordination* approach, which advocates a network-wide power management and global solutions including energy-efficient routing, for example.

42.2.3.1 Energy Consumption

Before being able to save energy with new technologies and mechanisms, researchers and network designers need to know how energy is consumed in network equipment. This preliminary analysis is key to understand how energy can be saved and to design energy models of network equipment that will be used to validate new hardware components and new algorithms.

Several models have been proposed for the different network components [2,30,59] and for the whole Internet [3,4]. Based on real energy measurements, they allow researchers to validate their new frameworks and algorithms. To our knowledge, no study has been made on the specific energy consumption of data center, grid or cloud networks.

42.2.3.2 The Optimization Approach: Hardware Improvements

The first way to reduce the energy consumption of a component is to increase its energy efficiency. This is why network-equipment manufacturers are proposing more and more *green* routers and switches [2], for example. D-Link,* Cisco,† and Netgear‡ are among the manufacturers proposing new *green* functionalities in their products, such as the adaptation of the transmission's power to the cable length, the adaptation of the power to the load, power off buttons, and more energy-efficient power supply. These new products come with green initiatives (GreenTouch,§ GreenStar Network,¶ ECR initiative,‖ etc.) and study groups (such as the IEEE 802.3 Energy Efficient Ethernet Study Group**) which aim to standardize and enforce new regulations in terms of energy consumption for network equipment.

---

* http://dlinkgreen.com/energyefficiency.asp
† http://www.cisco.com/en/US/products/ps10195/index.html
‡ http://www.netgear.com/NETGEARGreen/GreenProducts/GreenRoutersGateways.aspx
§ http://www.greentouch.org/
¶ http://www.greenstarnetwork.com/
‖ http://www.ecrinitiative.org/
** http://grouper.ieee.org/groups/802/3/eee_study/index.html

42.2.3.3 The Shutdown Approach: Sleeping

Network links, and especially edge links, are lightly utilized [17,44]. For this reason, researchers have proposed switching off (sleeping mode) the network equipment when it is not used [16,25]. This technique raises several problems: connectivity loss, long resynchronization time, and the fact that constantly switching on and off can be more energy consuming than doing nothing. New mechanisms have been designed to settle these issues: proxying techniques to keep the connectivity alive [43] and new mechanisms to quickly resynchronize both ends of a link [26], for example.

Our model uses an on/off algorithm to reduce the energy consumption of ports, NICs and switches when they are not used. The shutdown approach takes advantage of idle periods in network traffic. Energy savings can also be made during low-demand periods with the slowdown approach.

42.2.3.4 The Slowdown Approach: Rate Adaptation

Gunaratne et al. [23] have shown that there is a negligible difference in power consumption whether an Ethernet link is idle or fully utilized. This is mainly due to the fact that when there is no data to transmit on the link, idle bit patterns are still continuously transmitted in order to keep both NICs synchronized. However, the Gigabit Ethernet specifications include backward compatibility with previous specifications and are thus also designed to operate at 10, 100 Mb/s, and 1 Gb/s. Moreover, when NICs and switches operate at lower data rates, they consume less energy [23]. This observation has led several research teams to propose methods to dynamically adjust the link's data rates to the load [5,53], based on the same principle as dynamic voltage frequency scaling (DVFS) techniques for CPUs. This technique can only be used with mechanisms to quickly switch the data rate of an Ethernet link [9].

As we are using bandwidth reservation mechanisms, the bandwidth used is always known and thus we take advantage of slowdown techniques when the links are not used at full capacity.

42.2.3.5 The Coordination Approach: Network-Wide Management and Global Solutions

The optimization, shutdown, and slowdown approaches are focused on specific network components. However, energy-efficiency improvements have to be made at wider scales as well. For example, routing algorithms [15,49] and network protocols [8,31] can be improved to save energy.

Coordinated power-management schemes benefit from previously cited techniques, such as the on/off and adapting-rate techniques, and they take decisions at a wider scale. Thus, they should make greater energy savings. For example, in low-demand scenarios with network redundancy, entire network paths can be switched off, and the traffic is routed on other paths [54,55].

Combining shutdown and slowdown techniques at the scale of the network, we have developed a coordinated model responsible for managing the networks with end-to-end bandwidth reservations in an energy-efficient way.

42.3 HERMES

Bandwidth brokers have been designed for grid infrastructures in order to guarantee reliability and robustness to grid applications that require strong quality of service [45]. Yet, these infrastructures are only focused on performance and not energy saving. Our model is based on an essential trade-off between performance (number of accepted requests) and energy consumption. This model is called HERMES.

42.3.1 HERMES Architecture

The HERMES framework is not only designed for networks of large-scale distributed systems such as data centers, grids, and clouds; it can also be deployed in any kind of dedicated network, e.g., backup networks, e-science networks, dedicated networks.

In the context of grids and clouds data transfers, HERMES is used to manage the entire network. Three traffic characteristics of such systems make them the perfect candidate for HERMES:

1. The traffic stays mainly inside this network: traffic coming from and going outside represents about 20% [21];
2. Packet arrivals exhibit ON/OFF patterns [6,7];
3. The network runs well below capacity most of the time [6,7].

The first characteristic allows HERMES to have end-to-end bandwidth reservations in order to have complete control on the network. The second characteristic fosters on/off and traffic aggregation algorithms for energy-saving purposes. The third characteristic guarantees that energy savings are feasible by using shutdown and slowdown techniques.

To achieve energy efficiency, HERMES combines several techniques:

- Unused network components put into sleep mode
- Energy optimization of the reservation scheduling through reservation aggregation
- Minimization of the control messages required by the infrastructure
- Usage of DTN to manage the infrastructure
- Network-usage prediction to avoid too frequent on/off cycles

The principle of HERMES is that each data transfer between any two nodes of the managed infrastructure (e.g., data center, grid or cloud) should first be submitted to the reservation system of HERMES. Then, HERMES schedules it, informs the sender about the transfer scheduling, and guarantees that the transfer will occur without congestion.

The amount of bandwidth required by the traffic induced by the applications running on the computing nodes is booked when the task is submitted to the infrastructure's task scheduler. At the same time, the data transfers required to launch the application (e.g., virtual machine migration, image deployment) are also submitted to the reservation system of HERMES. Computing tasks can also submit data transfer requests when they are running.

Each network component (router, switch, bridge, network interface card) has two agendas per port: for both ways (in and out). An *agenda* stores all the future reservations concerning its one-way link. This information is sometimes called the book-ahead interval [10]. Figure 42.1 presents an example of such an agenda. A *free bandwidth portion* is always kept on each link for management messages and for the ACKs. This portion can be either a fixed amount of bandwidth or a fraction of the link's capacity.

Furthermore, each network component also has an agenda stating the on and off periods and the switching stages between on and off. This global agenda is in fact the combination of all equipment's per-port agendas: When no port is used for a certain amount of time (not too small), the network equipment

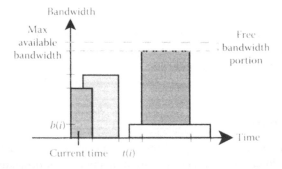

FIGURE 42.1 Agenda example.

can be switched off. Usage-prediction algorithms are used to avoid switching the equipment off if it is going to be useful in a near future. The prediction algorithms will be described later.

Our model uses a continuous time model and not a discrete model with fixed time slots during which the resource allocations are similar [10,47]. Indeed, as explained in [33], the storage of agendas is more flexible and less space-consuming with the continuous time model. Thus, to store the per-port agendas, we use the time-bandwidth list structure used in several previous works [38,40,51].

Each port maintains its reservation status using a *time-bandwidth list* (TB list) which is formed by $(t[i], b[i])$ tuples, where $t[i]$ is a time and $b[i]$ is a bandwidth. These tuples are sorted in increasing order of $t[i]$. Thus $b[i]$ denotes the available bandwidth of the concerned port during the time period $(t[i], t[i+1])$. If $(t[i], b[i])$ is the last tuple, then it means that a bandwidth of $b[i]$ is available from $t[i]$ to ∞. Each $t[i]$ is called an *event* in the agenda.

42.3.2 The Reservation Process

The reservation process is as follows:

1. A user submits a reservation request (specifying at least the data volume and the required deadline) to the network-management system (which will be detailed later).
2. The advance-reservation environment launches the negotiation phase including admission control, reservation scheduling, and optimization policies.
3. The notification is sent to the user when his/her request is accepted or rejected, and when it is scheduled.
4. The reservation starts at the scheduled start time and ends at the scheduled end time, which occurs before the user-submitted deadline.

The reservation process is based on gateways. Figure 42.2 presents the HERMES architecture as seen by the end users or servers: They are linked to a gateway and know no more about the network. The format of the reservation requests is comprehensive and sent to a daemon running on the gateway.

Each request should, at least, contain a volume of data to transfer and a deadline (the transfer should be finished before this deadline). Other options can be added: the maximum transmission rate, a bandwidth profile (maximum available bandwidth over time for applications that generate data online with various rates), an earliest start time (by default, it is the submission time).

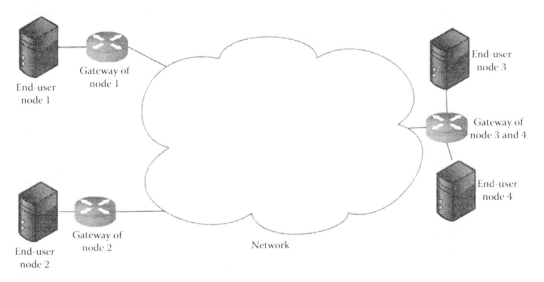

FIGURE 42.2 End-user view of HERMES architecture.

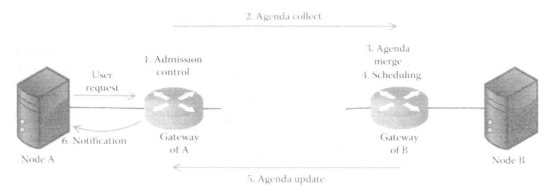

FIGURE 42.3   Reservation process.

A global view of this reservation process is presented on Figure 42.3. When a gateway receives a reservation request, the first operation to execute is admission control. The validity of the request is checked. Then, each request requires collecting the agendas of all the equipment (ports and routers) along the network paths between the source and the destination.

In order to do this agenda collection, the agendas of the possible shortest paths will be sent to the gateway of the receiver. The sender gateway will send a particular management message. The first node to receive it adds its own availability agenda to this message and send it to the two next nodes which are the nearest to the destination. If the network topology is, for example, a simple tree with no redundant link, only one path is available and thus, the message is sent only to the next node. The agendas of the ports used to transmit these messages are also included in the message.

Each node includes the required agenda it has and pass the message to the next nodes or the next node if there are no alternative. At the end, the destination gateway rebuilds the end-to-end paths. This algorithm limits the number of nodes that are involved in the reservation since each node only transmits the request to the one or two next nodes toward the requested destination, and not to all of its neighbors. This limitation does not compromise the energy efficiency of HERMES, because one expects that long paths (using numerous nodes) will be more energy consuming than short ones. Moreover, this limitation considerably reduces the computing time since it does not consider all the paths of the network.

Thus, the receiver gateway ends up with all the required agendas. It merges the corresponding availability agendas of the nodes to obtain one availability agenda per end-to-end path. Once they are computed, the end-to-end paths are put into cache in order to avoid doing this computation again. The speed of the merging operation depends linearly on the total number of events in the considered agendas for each end-to-end path. This is fast since the agendas have been truncated to get only the part between the submission time and the deadline.

42.3.3 Energy-Efficient Scheduling Algorithm

The end-to-end availability agenda is scanned using the HERMES scheduling algorithm (Algorithm 42.1) to find the solution consuming the least energy. At each time, the solution tries to use as much bandwidth as it can to reduce the reservation's duration, and thus its cost. We estimate the energy consumption of each possible solution (i.e., place in the agenda), and we compare each solution to pick the least consuming one.

42.3.4 Resource Management: On/Off Algorithm, Slowdown Techniques, Usage of DTN, and Prediction Algorithms

At the end of a transfer between two nodes, if one port is idle for a consequent time (over a certain threshold), the port is switched off and if all the ports of a router are switched off, then the router itself

Algorithm 42.1 Scheduling algorithm

if *the availability agenda of the path is empty* **then**
| Put the reservation in the middle of the remaining period before the deadline, if possible. Otherwise, put it
| now ($+\epsilon$ for the request processing time).
else
| **if** *there is no event before the deadline* **then**
| | Put the reservation in the middle of the remaining period before the deadline if possible. Otherwise,
| | put it as soon as possible.
| **else**
| | **foreach** *event in the availability agenda of the path and while it occurs before deadline* **do**
| | | Try to place the reservation after and before the event. Memorize the possible places (no collision
| | | with other reservations and end-before deadline).
| | **end**
| | **if** *there is no possible place* **then**
| | | **if** *the reservation can be put before the deadline* **then**
| | | | Put the reservation now ($+\epsilon$ for the quest's processing time).
| | | **else**
| | | | **if** *some events were not possible because of the deadline constraint* **then**
| | | | | **if** *the reservation can be put now (some bandwidth is available) without respecting the*
| | | | | *deadline* **then**
| | | | | | Propose this solution to the user.
| | | | | **else**
| | | | | | **foreach** *of these remaining events while no solution has been found* **do**
| | | | | | | Try to place the reservation after the event without respecting the deadline. Store
| | | | | | | the earliest possible place (no collision with other reservations) to propose it to the
| | | | | | | user.
| | | | | | **end**
| | | | | **end**
| | | | **end**
| | | **end**
| | **else**
| | | **foreach** *possible place* **do**
| | | | Estimate the energy consumption of the transfer using each equipment's energy-cost functions.
| | | **end**
| | | **if** *there is one less energy-consuming solution* **then**
| | | | Take that place!
| | | **else**
| | | | Take the earliest place among the less energy-consuming ones.
| | | **end**
| | **end**
| **end**
end

can be switched off if it should stay idle for a certain time. In addition, to avoid unnecessary on/off cycles, prediction algorithms are used to predict the next utilization of a link. The process happening at the end of a reservation is presented in Algorithm 42.2.

This algorithm is in fact distributed and executed at the end of a transfer by each port independently of one another. The prediction algorithms rely on recent history (past agenda) of the port. They are based on average values of past inactivity period durations and feedbacks that are average values of differences between past predictions and the past corresponding events in the agenda.

In addition to these shutdown techniques, HERMES uses adaptive link rate (ALR) during the transfers to dynamically adjust the transmission rate of each port to the bandwidth used. As each transfer is

Algorithm 42.2 At the end of a reservation

foreach *port used by this transfer* **do**

    if *there is a reservation in the port's agenda starting in less than T_s seconds* **then**

       | Leave the port powered on (at a lower transmitting rate)

    else

       | Predict the next utilization of this port **if** *the predicted usage is in less than T_s seconds* **then**

          | Let the port powered on (at a lower transmitting rate)

       else

          | Switch the port off (sleeping mode) **if** *this port was the last powered-on port* **then**

             | Switch the router off (sleeping mode)

          end

       end

    end

end

scheduled, the traffic at any time is known, and thus, the complex queue threshold mechanism of ALR used to change the transmission rate is useless here.

The reservation process works only if the necessary ports and routers are on when the agenda collection is done. Indeed, when they are not used, the network equipment (individual ports or entire routers) is put into sleep mode. To solve this issue, DTN (disruption-tolerant networking) [20] techniques are used. Indeed, DTN is perfectly adapted to this type of scenario where parts of the network are not always available without any guarantee of end-to-end connectivity at any time.

The idea is to add a kind of TTL (time-to-live) in seconds to each end-user request: When the TTL expires, if the request has not reached the receiver gateway and has not come back, then all the sleeping nodes of the path are awakened and the agenda collection is performed. While the TTL is not expired, the agenda-collection message moves forward along the path until meeting a sleeping node. Then, as long as the TTL has not expired, the message waits in the previous node for the sleeping node to wake up, and when it wakes up (wake-up detection managed by the DTN protocol), the message is sent to it and continues its way. Thus, hop by hop, the agenda-collection message moves toward the receiver gateway.

The gateways are always fully powered on to ensure high availability and reactivity for the overall system. The gateways are able to wake up the nodes they are linked to. So, each sleeping node needs just one awake component (or two if it is connected to a gateway) linked to its manager to be remotely awakened, and not one component per port (i.e., per outgoing link).

Compared to a centralized resource management, our approach uses more control messages. However, the size of targeted networks is quite limited, and the number of messages depends on the number of hops in the reservations path. Moreover, a free bandwidth portion on each link is kept for these messages.

As a result, the overhead due to control messages is negligible. In terms of computational cost, for each request, our approach computes several scheduling possibilities. But, we have limited this number to the number of events already put into the agendas and going from the submission time to the deadline, so this number is quite limited too. Moreover, in the worst case, if the reservation is not possible (agendas too busy), our approach has a better complexity than an algorithm that would check all the possible dates. The major advantage compared to a centralized mechanism is the scalability, and thus the reactivity of the whole infrastructure.

42.3.5 Discussion

The proposed network management optimizes the energy consumption of the overall architecture at any time. However, we have not yet studied the energy optimization of transfers themselves.

Indeed, we have assumed that at any time, the most energy-efficient behavior is to use as much bandwidth as possible (from source to destination). However, we have not proved that this algorithm leads to the minimum energy consumption.

Let's consider this example: Node A wants to send 200 Mb of data to node B and nodes A and B are directly linked by a 1 Gb/s link. Our algorithm will schedule the transfer and set the bandwidth at 1 Gb/s (minus the free bandwidth portion). If we assume that the free bandwidth portion is negligible, it takes 0.2 s to transmit 200 Mb of data at 1 Gb/s. Thus, this transfer will consume $E_{transfer}$ with $P_{EthernetCard}(Node\ A, 1\ Gb/s)$ that denotes the power consumed by node A when it transmits data at 1 Gb/s:

$$E_{transfer} = E_{EthernetCard}(Node\ A, 1\ Gb/s, 0.2\ s) + E_{EthernetCard}(Node\ B, 1\ Gb/s, 0.2\ s)$$

$$= P_{EthernetCard}(Node\ A, 1\ Gb/s) \times 0.2 + P_{EthernetCard}(Node\ B, 1\ Gb/s) \times 0.2$$

However, another solution could be to adjust the Ethernet card to work at 100 Mb/s, and thus, it does not use the full capacity and it takes more time. In that case, the transfer consumes

$$E'_{transfer} = E_{EthernetCard}(Node\ A, 100\ Mb/s, 2\ s) + E_{EthernetCard}(Node\ B, 100\ Mb/s, 2\ s)$$

$$= P_{EthernetCard}(Node\ A, 100\ Mb/s) \times 2 + P_{EthernetCard}(Node\ B, 100\ Mb/s) \times 2$$

If we assume that the NICs are identical and thus have the same power consumption $P_{EthernetCard}(100\ Mb/s)$ and $P_{EthernetCard}(1\ Gb/s)$ depending on the rate, then the second solution uses less energy to transfer the data if and only if

$$P_{EthernetCard}(1\ Gb/s) > 10 \times P_{EthernetCard}(100\ Mb/s)$$

If we use the figures provided in [63], we have $P_{EthernetCard}(100\ Mb/s) = 0.4\ W$, and $P_{EthernetCard}(1\ Gb/s) = 3.6\ W$ for a NIC. In that case, our scenario is the most energy efficient with a consumption equal to 0.72 J (and 0.8 J for the second scenario). However, here, we only considered the energy used to transfer data and not the overall energy of the infrastructure during a certain period of time. These two energy consumptions thus do not represent the same period of time (0.2 and 2 s). To compare them over an identical time period, we should add to $E_{transfer}$ the cost of staying off during 1.8 s.

We have not taken into account the energy required to switch the NICs on at the beginning and to switch them off at the end of the transfer since these energy costs are identical in both scenarios.

This remark shows that our algorithm should be compared with other solutions and that the optimal solution is hard to find, even in scenarios with fixed routing. This situation is the result of the non-proportionality between energy and usage: Cost functions are linear by steps and not just linear.

42.4 A Test Case

42.4.1 BoNeS: Bookable Network Simulator

To validate our model, we have designed a network simulator written in Python which is called BoNeS: Bookable Network Simulator (more than 4300 code lines). It takes as input a network-description file (topology and router and link capacities) and some network-traffic characteristics (e.g., statistical distribution of interarrival submissions, distribution of the reservation durations, source and destination nodes, distribution of the deadlines and TTLs). It generates the network and an ABR traffic according to the characteristics given in input. It then simulates, with this traffic and topology, different scheduling algorithms and compares them in terms of both performance and energy consumption.

To generate the routes, we use the Dijkstra algorithm at the beginning of the simulation to compute the shortest route between any pair of source and destination nodes. Then, another routing algorithm is used to compute a 2-shortest path: one of the second shortest paths which is the most different from the shortest path in terms of used links. So, for each request, we first check if the first route (shortest path) can be used. If it cannot, we use the other route if it exists. This mechanism allows us to accept requests even if the primary route between sources and destinations is not available due to some failures or to other reservations. This ensures fault tolerance of the system.

Currently, the simulator runs five different schedulings on the generated traffic and network:

- *First*: The reservation is scheduled at the earliest possible place.
- *First green*: The reservation is aggregated with the first possible reservation already accepted (before deadline), or scheduled at the earliest possible place.
- *Last*: The reservation is scheduled at the latest possible place (before deadline).
- *Last green*: The reservation is aggregated with the latest possible reservation already accepted (before deadline).
- *Green*: This scheduling is the implementation of our framework: the energy consumption is estimated for each possible allocation, and the least consuming one is chosen.

Our simulator provides the energy consumption for these five schedulings combined with our on/off algorithm where resources are switched off when they are not used. The simulator also computes the energy consumption of the *first* scheduling without any on/off algorithm but with ALR (as it could be the case presently), this case is called *no off*. The generated network traffic consists in requests with

- Submission times distributed according to a log-normal distribution
- Data volumes generated with a negative exponential distribution
- Sources and destinations chosen randomly (equiprobability)
- Times between submission times and deadlines generated with a Poisson distribution

The probability distributions of the different traffic's characteristics are parameters and can be changed. The distributions presented here are the ones that we have used in the following for the experiments. They have been inspired by the results presented in [19].

The energy consumption of a networking component (i.e., switch or router) depends on the type of equipment, the number of ports, the port transmission rates (with ALR), and the employed cabling solutions [41]. So, for each router, we have modeled the energy consumption with two values for the chassis power ($P_{chassis}$) depending if it is on or off. We take several values for the port power P_{port}: one for when it is off, one for when it is idle (working at the lower transmission rate), and one for each possible transmission rate.

So, for example, here, for a 1 Gbps router, the values used in the simulations are presented in Table 42.1.

The values are in watts and per chassis and per port. As we only focus on networking equipment, we do not take into account the energy consumption of servers, we only take into account the energy consumption of their Ethernet card. Each 1 Gbps Ethernet card is assumed to consume 10 W when idle, and 15 W at full capacity.

TABLE 42.1　Power Parameters Used for a 1 Gbps per-Link Router

Component	State	Power (W)
Chassis	ON	150
	OFF	10
Port	1 Gbps	5
	100 Mbps	3
	Idle, 10 Mbps	1

42.4.2 Typical Three-Tier Network Architecture

The topology used to evaluate HERMES with BoNeS is described on Figure 42.4. The core tier comprises four layer-3 switches. The aggregation tier is responsible for routing with its 8 switches. Finally, each access tier switch is directly connected to 40 servers with 1 Gbps links. Two data servers, which contain for example the images or virtual machines to deploy on the nodes, are directly connected at the core network.

So, this topology comprises 482 servers (including 2 data servers), 24 routers and 552 links.

Two kinds of traffic will be simulated on this network:

- The transfers among the nodes themselves; this traffic is induced by the user's applications
- The transfers between the data servers and the computing nodes

42.4.3 Evaluation

In order to evaluate HERMES, we have simulated the previous three-tier architecture with BoNeS and using different workloads. For each experiment, the simulation has been launched 70 times with requests generated as explained previously. Each simulation represents the behavior of the network during 2 h of real time.

Table 42.2 shows the results obtained with a workload of 30% on the links (i.e., links are used at their full capacity during 30% of the time). This workload has been obtained by simulating 20,000 requests between the computing servers among themselves and 5,000 requests between the data servers and the computing servers.

The *last green* scheduling is the less energy consuming one. Yet, it accepts 6.6% requests less than the *green* scheduling. This represents 1650 requests. The *first* scheduling has almost the same percentage of

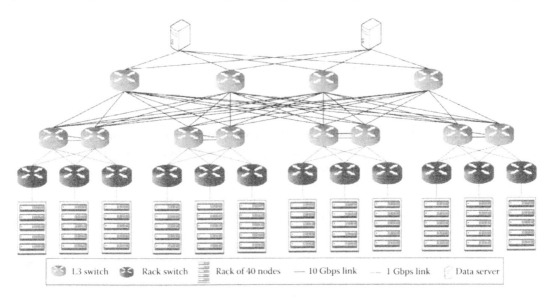

FIGURE 42.4 Typical three-tier fat-tree architecture.

TABLE 42.2 Energy Consumption in Wh for the Different Schedulings with 30% Workload

Scheduling	First	First Green	Last	Last Green	Green	No Off
Average	14,642	13,477	14,667	9,868	14,275	26,860
Standard deviation	392	311.7	346	233	340	430
Accepted requests (%)	86.73	85.36	78.69	80.12	86.72	86.73

TABLE 42.3 Energy Consumption in Wh for the Different Schedulings with 50% Workload

Scheduling	First	First Green	Last	Last Green	Green	No Off
Average	16,093	15,791	16,130	11,748	15,395	28,528
Standard deviation	305	336	297	254	322	464
Accepted requests (%)	74.81	74.19	70.07	71.16	74.87	74.81

TABLE 42.4 Results Summary
Comparing *Green* and *No Off* Results

Load	30%	50%
Energy saved	47%	46%

accepted requests, yet it consumes 367 Wh more than the *green* scheduling. This is just for a 2 h simulation. Thus, the *green* scheduling is the best trade-off between energy savings and request's acceptance rate.

Table 42.3 shows the results obtained with a workload of 50% on the links. This workload has been obtained by simulating 30,000 requests between the computing servers among themselves and 8,000 requests between the data servers and the computing servers.

By comparing Tables 42.2 and 42.3, one can notice that increasing the link's workload does not affect the performance of the schedulings in terms of both energy savings and accepted requests. For a 50% workload, the *green* scheduling is still the best option.

Table 42.4 presents the energy saved with the green scheduling compared to the energy used when no switch off is allowed (*no off* case) for the two workloads.

In both cases, energy savings that could be made using HERMES are almost half of the energy consumed in the case of current infrastructures.

42.5 Conclusion and Future Work

This chapter presents a detailed state of the art on large-scale distributed systems (data center, grid and cloud) networks, on bulk data transfer and bandwidth provisioning, and green wired networking.

This state of the art is then used to design HERMES for bandwidth reservation in end-to-end networks. This framework ensures energy efficiency through energy-aware scheduling with reservation aggregation, and on/off mechanisms for resource management with usage prediction to switch off unused resources.

Using our own simulator, BoNeS (Bookable Network Simulator), we present an evaluation of HERMES on a realistic typical three-tier network architecture. This evaluation gives really encouraging results: almost half of the energy used by a current data center, grid and cloud networks could be saved using HERMES.

Our future work will focus on exploring more scheduling algorithms, and in particular, off-line algorithms that lead to better optimizations. Such algorithms can be used by discretizing the time in short intervals, and by launching the scheduling at the end of each interval. It lowers the reactivity of the overall infrastructure, but should reach better scheduling in terms of both energy consumption and performance.

References

1. M. Al-Fares, S. Radhakrishnan, B. Raghavan, N. Huang, and A. Vahdat. Hedera: Dynamic flow scheduling for data center networks. In *Symposium on Networked Systems Design and Implementation (NSDI)*, pp. 281–296, San Jose, CA, 2010.

2. G. Ananthanarayanan and R. Katz. Greening the switch. Technical report, EECS Department, University of California, Berkeley, CA, 2008.

3. M. Baldi and Y. Ofek. Time for a "greener" Internet. In *GreenComm'09: International Workshop on Green Communications*, Dresden, Germany, 2009.

4. J. Baliga, K. Hinton, and R. Tucker. Energy consumption of the Internet. In *COIN-ACOFT 2007: Joint International Conference on Optical Internet, and Australian Conference on Optical Fibre Technology*, pp. 1–3, Melbourne, Australia, 2007.

5. M. Bennett, K. Christensen, and B. Nordman. Improving the energy efficiency of Ethernet: Adaptive link rate proposal, *Ethernet Alliance*, Version 1.0, Austin, TX, 2006.

6. T. Benson, A. Akella, and D. Maltz. Network traffic characteristics of data centers in the wild. In *Conference on Internet Measurement, IMC '10*, pp. 267–280, Melbourne, Australia, 2010.

7. T. Benson, A. Anand, A. Akella, and M. Zhang. Understanding data center traffic characteristics. In *ACM Workshop on Research on Enterprise Networking, WREN '09*, pp. 65–72, Barcelona, Spain, 2009.

8. J. Blackburn and K. Christensen. A simulation study of a new green bittorrent. In *GreenCom'09: First International Workshop on Green Communications*, Dresden, Germany, 2009.

9. F. Blanquicet and K. Christensen. An initial performance evaluation of rapid PHY selection (RPS) for energy efficient Ethernet. In *Conference on Local Computer Networks*, Dublin, Ireland 2007.

10. L.-O. Burchard. Networks with advance reservations: Applications, architecture, and performance. *Journal of Network and Systems Management*, 13(4):429–449, 2005.

11. L.-O. Burchard and M. Droste-Franke. Fault tolerance in networks with an advance reservation service. In *11th International Workshop on Quality of Service (IWQoS 2003)*, pp. 215–228, Monterey, CA, 2003.

12. L.-O. Burchard, B. Linnert, and J. Schneider. Rerouting strategies for networks with advance reservations. In *International Conference on e-Science and Grid Computing (E-Science)*, Melbourne, Australia, 2005.

13. J. W. Byers, J. Considine, M. Mitzenmacher, and S. Rost. Informed content delivery across adaptive overlay networks. *IEEE/ACM Transactions on Networking*, 12(5):767–780, 2004.

14. C. Castillo, G. N. Rouskas, and K. Harfoush. Efficient resource management using advance reservations for heterogeneous grids. In *International Symposium on Parallel and Distributed Processing (IPDPS)*, pp. 1–12, Miami, FL, 2008.

15. J. Chabarek, J. Sommers, P. Barford, C. Estan, D. Tsiang, and S. Wright. Power awareness in network design and routing. In *INFOCOM 2008*, pp. 457–465, Phoenix, AZ, 2008.

16. L. Chiaraviglio, M. Mellia, and F. Neri. Energy-aware networks: Reducing power consumption by switching off network elements. In *FEDERICA-Phosphorus Tutorial and Workshop (TNC)*, Bruges, Belgium, 2008.

17. K. Christensen, C. Gunaratne, B. Nordman, and A. George. The next frontier for communications networks: Power management. *Computer Communications*, 27(18):1758–1770, 2004.

18. B. Eckart, X. He, and Q. Wu. Performance adaptive UDP for high-speed bulk data transfer over dedicated links. In *IEEE International Symposium on Parallel and Distributed Processing (IPDPS 2008)*, pp. 1–10, Miami, FL, 2008.

19. D. Ersoz, M. S. Yousif, and C. R. Das. Characterizing network traffic in a cluster-based, multi-tier data center. In *International Conference on Distributed Computing Systems (ICDCS)*, Toronto, Ontario, Canada, 2007.

20. S. Farrell, V. Cahill, D. Geraghty, I. Humphreys, and P. McDonald. When TCP breaks: Delay- and disruption-tolerant networking. *IEEE Internet Computing*, 10(4):72–78, 2006.

21. A. Greenberg et al. VL2: A scalable and flexible data center network. In *ACM SIGCOMM Conference on Data Communication, SIGCOMM '09*, pp. 51–62, Barcelona, Spain, 2009.

22. Y. Gu and R. Grossman. SABUL: A transport protocol for grid computing. *Journal of Grid Computing*, 1(4):377–386, 2003.

23. C. Gunaratne, K. Christensen, and B. Nordman. Managing energy consumption costs in desktop PCs and LAN switches with proxying, split TCP connections, and scaling of link speed. *International Journal of Network Management*, 15(5):297–310, 2005.

24. C. Guok, J. Lee, and K. Berket. Improving the bulk data transfer experience. *International Journal of Internet Protocol Technology*, 3(1):46–53, 2008.

25. M. Gupta and S. Singh. Greening of the Internet. In *SIGCOMM '03*, pp. 19–26, Karlsruhe, Germany, 2003.

26. M. Gupta and S. Singh. Dynamic Ethernet link shutdown for energy conservation on Ethernet links. In *International Conference on Communications (ICC '07)*, pp. 6156–6161, Glasgow, Scotland, 2007.

27. E. He, J. Leigh, O. Yu, and T. Defanti. Reliable blast UDP : Predictable high performance bulk data transfer. In *IEEE International Conference on Cluster Computing*, pp. 317–324, Chicago, IL, 2002.

28. M. Hefeeda, A. Habib, D. Xu, B. Bhargava, and B. Botev. CollectCast: A peer-to-peer service for media streaming. *Multimedia Systems*, 11(1):68–81, 2005.

29. B. Heller, S. Seetharaman, P. Mahadevan, Y. Yiakoumis, P. Sharma, S. Banerjee, and N. McKeown. ElasticTree: Saving energy in data center networks. In *USENIX Conference on Networked Systems Design and Implementation, NSDI '10*, San Jose, CA, 2010.

30. H. Hlavacs, G. Da Costa, and J.-M. Pierson. Energy consumption of residential and professional switches. In *International Conference on Computational Science and Engineering (CSE '09)*, vol. 1, pp. 240–246, Vancouver, British Columbia, Canada, 2009.

31. L. Irish and K. Christensen. A "Green TCP/IP" to reduce electricity consumed by computers. In *Proceedings of IEEE Southeastcon*, pp. 302–305, Orlando, FL, 1998.

32. V. Jacobson, R. Braden, and D. Borman. *TCP Extensions for High Performance*. Network Working Group, RFC 1323, 1992.

33. E.-S. Jung, Y. Li, S. Ranka, and S. Sahni. An evaluation of in-advance bandwidth scheduling algorithms for connection-oriented networks. In *International Symposium on Parallel Architectures, Algorithms, and Networks (I-SPAN 2008)*, pp. 133–138, Sydney, New South Wales, Australia, 2008.

34. S. Kandula, J. Padhye, and P. Bahl. Flyways to de-congest data center networks. In *ACM Workshop on Hot Topics in Networks (HotNets)*, New York, 2009.

35. S. Kandula, S. Sengupta, A. Greenberg, P. Patel, and R. Chaiken. The nature of data center traffic: Measurements & analysis. In *Conference on Internet Measurement Conference, IMC '09*, pp. 202–208, Chicago, IL, 2009.

36. N. Laoutaris, G. Smaragdakis, P. Rodriguez, and R. Sundaram. Delay tolerant bulk data transfers on the internet. In *SIGMETRICS '09*, pp. 229–238, Seattle, WA, 2009.

37. Y. Li, S. Ranka, and S. Sahni. In-advance path reservation for file transfers In e-science applications. In *IEEE Symposium on Computers and Communications (ISCC 2009)*, pp. 176–181, Sousse, Tunisia, 2009.

38. Y. Lin and Q. Wu. On design of bandwidth scheduling algorithms for multiple data transfers in dedicated networks. In *ANCS '08: Symposium on Architectures for Networking and Communications Systems*, pp. 151–160, San Jose, CA, 2008.

39. Y. Lin and Q. Wu. Path computation with variable bandwidth for bulk data transfer in high-performance networks. In *High-Speed Networks Workshop (HSN 2009)*, Rio de Janeiro, Brazil, 2009.

40. Y. Lin, Q. Wu, N. Rao, and M. Zhu. On design of scheduling algorithms for advance bandwidth reservation in dedicated networks. In *IEEE INFOCOM Workshops 2008*, pp. 1–6, Phoenix, AZ, 2008.

41. P. Mahadevan, P. Sharma, S. Banerjee, and P. Ranganathan. A power benchmarking framework for network devices. *Networking 2009*, 5550:795–808, 2009.

42. X. Meng, V. Pappas, and L. Zhang. Improving the scalability of data center networks with traffic-aware virtual machine placement. In *INFOCOM*, pp. 1–9, San Diego, CA, 2010.

43. B. Nordman and K. Christensen. Proxying: The next step in reducing IT energy use. *Computer*, 43(1):91–93, 2010.

44. A. Odlyzko. Data networks are lightly utilized, and will stay that way. *Review of Network Economics*, 2:210–237, 2003.

45. C. Palansuriya, M. Buchli, K. Kavoussanakis, A. Patil, C. Tziouvaras, A. Trew, A. Simpson, and R. Baxter. End-to-end bandwidth allocation and reservation for grid applications. In *Conference on Broadband Communications, Networks and Systems (BROADNETS)*, pp. 1–9, San Jose, CA, 2006.

46. A. Patel, Y. Zhu, Q. She, and J. Jue. Routing and scheduling for time-shift advance reservation. In *Conference on Computer Communications and Networks (ICCCN 2009)*, pp. 1–6, San Francisco, CA, 2009.

47. K. Rajah, S. Ranka, and Y. Xia. Advance reservations and scheduling for bulk transfers in research networks. *Transactions on Parallel and Distributed Systems*, 20(11):1682–1697, 2009.

48. W. Reinhardt. Advance reservation of network resources for multimedia applications. In *International Workshop on Multimedia (IWACA 1994)*, pp. 23–33, Heidelberg, Germany, 1994.

49. J. Restrepo, C. Gruber, and C. Machuca. Energy profile aware routing. In *IEEE International Conference on Communications (ICC Workshops 2009)*, pp. 1–5, Dresden, Germany, 2009.

50. E. Rosen, A. Viswanathan, and R. Callon. *Multiprotocol Label Switching Architecture*. Network Working Group, RFC 3031, 2001.

51. S. Sahni, N. Rao, S. Ranka, Y. Li, E.-S. Jung, and N. Kamath. Bandwidth scheduling and path computation algorithms for connection-oriented networks. In *Sixth International Conference on Networking (ICN 2007)*, Sainte-Luce, Martinique, France, 2007.

52. A. Schill, S. Kühn, and F. Breiter. Design and evaluation of an advance reservation protocol on top of RSVP. In *International Conference on Broadband Communications (BC '98)*, pp. 23–40, Stuttgart, Germany, 1998.

53. L. Shang, L.-S. Peh, and N. Jha. Dynamic voltage scaling with links for power optimization of interconnection networks. In *International Symposium on High Performance Computer Architecture (HPCA '03)*, Anaheim, CA, 2003.

54. L. Shang, L.-S. Peh, and N. Jha. PowerHerd: A distributed scheme for dynamically satisfying peak-power constraints in interconnection networks. *IEEE Transactions on Computer-Aided Design of Integrated Circuits and Systems*, 25(1):92–110, 2006.

55. G. Shen and R. Tucker. Energy-minimized design for IP over WDM networks. *Journal of Optical Communications and Networking*, 1(1):176–186, 2009.

56. R. Sherwood, R. Braud, and B. Bhattacharjee. Slurpie: A cooperative bulk data transfer protocol. In *INFOCOM 2004*, vol. 2, pp. 941–951, Hong Kong, China, 2004.

57. B. Sotomayor, R. S. Montero, I. M. Llorente, and I. Foster. Resource leasing and the art of suspending virtual machines. In *Conference on High Performance Computing and Communications (HPCC)*, pp. 59–68, Seoul, Korea, 2009.

58. G. Wang and T. S. E. Ng. The impact of virtualization on network performance of amazon EC2 data center. In *INFOCOM 2010*, San Diego, CA, 2010.

59. H.-S. Wang, L.-S. Peh, and S. Malik. A power model for routers: Modeling Alpha 21364 and InfiniBand routers. *Symposium on High-Performance Interconnects*, 0:21, Stanford, CA, 2002.

60. C. Xie, F. Xu, N. Ghani, E. Chaniotakis, C. Guok, and T. Lehman. Load-balancing for advance reservation connection rerouting. *IEEE Communications Letters*, 14(6):578–580, 2010.

61. L. Zhang, S. Deering, D. Estrin, S. Shenker, and D. Zappala. RSVP: A new resource ReSerVation protocol. *IEEE Network*, 7:8–18, 1993.

62. Z.-L. Zhang, Z. Duan, and Y. Hou. On scalable design of bandwidth brokers. *IEICE Transactions on Communications*, 8:2011–2025, 2001.

63. B. Zhang, K. Sabhanatarajan, A. Gordon-Ross, and A. George. Real-time performance analysis of adaptive link rate. In *IEEE Conference on Local Computer Networks (LCN)*, pp. 282–288, Montreal, Canada, 2008.

64. X. Zheng, A. Mudambi, and M. Veeraraghavan. FRTP: Fixed rate transport protocol—A modified version of SABUL for end-to-end circuits. In *First Workshop on Provisioning and Transport for Hybrid Networks (PATHNets, BroadNets Workshop)*, San Jose, CA, 2004.

43

Overview of Data Centers Energy Efficiency Evolution

Lennart Johnsson
University of Houston
Royal Institute of
Technology

43.1 Introduction

Energy efficiency in computation has become the prime concern for infrastructure providers for environmental and cost reasons. Both concerns have driven and continue to drive energy efficiency in design of data centers and computer systems and component technologies. The concerns also have impacted the way data centers and systems are operated with dynamic management of major system components increasingly being introduced, and they have also impacted the selection of energy sources for major data centers and their location.

There is a huge difference in greenhouse gas emissions by different energy sources. A study regarding emissions related to electricity generation by the U.K. Parliamentary Office of Science and Technology carried out in 2006 [1] resulted in the life-cycle assessment of CO_2 and other greenhouse gas emissions shown in Table 43.1. The emissions are expressed as gram CO_2 equivalents (gCO_2eq) per kWh. This measure accounts for the warming effects of CO_2 and other greenhouse gases. As can be seen from Table 43.1, the range in gCO_2eq/kWh between the energy sources with the lowest and highest emissions is more than a factor of 200. Life-cycle assessment includes greenhouse gas emissions for all stages related to electricity generation including plant construction; operation; maintenance and decommissioning; and fuel extraction, transport, and processing.

Unfortunately, much of the world's electricity generation is based on coal. According to the 2010 Key World Energy Statistics by the Internal Energy Agency (IEA) [2] and the Pew Center on Global Climate Change Climate TechBook's Electricity Generation Overview [3], about 41% of electricity is based on

TABLE 43.1 Life-Cycle Assessment of Greenhouse Gas Emissions Expressed as Grams of CO_2 Equivalents per kWh (gCO_2eq/kWh) with 2006 Technologies

Coal	Oil	Gas	Biomass	Solar PV	Marine	Hydro	Wind	Nuclear
>1000	~650	~500	25–93	35–58	25–50	5–30	~5	~5

Source: UK Parliamentary Office of Science and Technology, Carbon footprint of electricity generation, Postnote 268, October 2006. http://www.parliament.uk/documents/post/postpn268.pdf

TABLE 43.2 World Electric Energy Generation by Energy Source, 2008

Coal	Oil	Gas	Hydro	Nuclear	Other
41%	6%	20%	16%	15%	2%

Source: International Energy Agency, 2010 key world energy statistics. http://www.iea.org/textbase/nppdf/free/2010/key_stats_2010.pdf

TABLE 43.3 Relative Contribution to Greenhouse Gas Emissions for World Electricity Generation by Energy Source, 2008

Coal	Oil	Gas	Other
73%	7%	19%	1%

coal, about 20% on natural gas, and about 6% on oil. Thus, about 67% of all electric energy comes from the three sources that generate the most gCO_2eq/kWh. Hydroelectric energy accounts for 16% of total world electric energy generation and nuclear energy for 15%. The world electricity generation by energy source in 2008 is summarized in Table 43.2.

The result is that electric energy generation from fossil fuel energy sources accounts for almost all greenhouse gas emissions for electric energy generation, with about 73% of CO_2 emissions for electricity generation due to coal as a source of energy, about 19% due to natural gas and 7% due to oil. Table 43.3 summarizes the greenhouse gas emissions for world electricity generation by energy source from Tables 43.1 and 43.2.

The need for electric energy is expected to more than double by 2050, and dramatic changes in electric energy production are necessary to meet climate change targets. In the IEA Blue Scenario [4], it is estimated that limiting the temperature rise to 2°C by 2050 will require a 50% reduction in greenhouse gas emissions compared to 2005. To achieve this level of reduction in greenhouse gas emissions due to electricity generation, the gCO_2eq/kWh emissions must be reduced by 85% compared to 2008, which requires a drastic change in energy source for electric energy as shown in Figure 43.1. All coal generation of electric energy will need to use carbon capture and storage (CCS) techniques and coal reduced to account for about 12% of all electric energy generation or about 5 PWh of an estimated total 40 PWh. Natural gas is projected to account for about 10% and natural gas with CCS another 2%. Oil as an energy source for electric energy is projected to be very small and hence fossil fuel reduced to account for about 27% of all electric energy generation. Nuclear power based electric energy is projected to account for about as much as fossil fuel based electric energy, or about 25% with electric energy from renewable sources accounting for close to half of all electric energy.

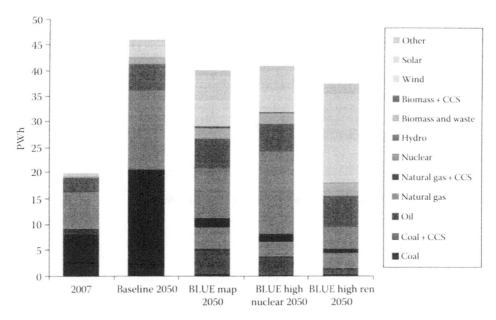

FIGURE 43.1 Electric energy by source. Decarbonizing the electricity sector to limit temperature rise to about 2°C by 2050. (From International Energy Agency, Energy technology perspectives—Scenarios and strategies to 2050, July 2010. http://www.iea.org/techno/etp/)

43.2 Data Centers

43.2.1 Energy

The contribution to greenhouse gas emissions of the information and communications technology (ICT) sector, though small, is growing faster than the overall growth in emissions. The Smart2020 report [5] estimated that total emissions from all sources will increase by about 30% from 2002 to 2020, while the ICT sector emissions (including PCs) will grow by 180% during the same period. However, the report also estimated that by 2020, the ICT sector will contribute to a reduction in emissions in other sectors more than fivefold its own emissions. The benefit on the overall energy consumption due to the ICT sector was also studied by the American Council for an Energy-Efficient Economy. This study found that, for the United States, for every kWh consumed by the IT industry about 10 kWh is saved in other parts of the economy [6]. Though the economies are different in different countries, in most modern economies IT should be part of the solution for more energy-efficient and environmentally friendly economies.

The growth in greenhouse gas emissions by data centers is predicted by the Smart2020 report to grow even faster than the overall emissions by the ICT sector, or by 240% from 2002 to 2020.

The capital and operation costs of energy for operating and cooling computer systems have increased very rapidly over the last 15–20 years. Though there is a wide range in power consumption, for servers used for high-performance computing (HPC) systems servers based on the x86 architecture have come to dominate the HPC systems market. Therefore, the energy consumption of this type of server can be used for a view of typical HPC center's energy efficiency and cost evolution. The energy consumption for a typical x86-based server has increased during the last decade from somewhat less than 100 W on average to about 250 W [7] an almost threefold increase while server costs have remained fairly constant or even decreased slightly [7]. In 2007, Belady [8] estimated that in 2008 energy cost for operating and cooling a standard x86 server would equal the cost of the server, while already in 2004 the capital expense

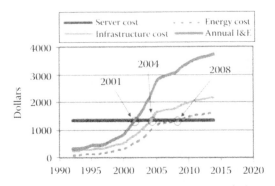

FIGURE 43.2   Evolution of U.S. power and cooling costs for a standard x86 server.

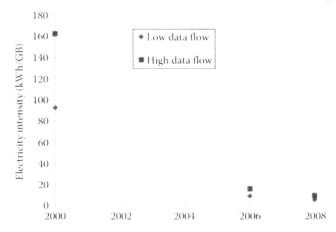

FIGURE 43.3   Electricity intensity of the Internet. (From Weber, C.L., et al., The energy and climate change impacts of different music delivery methods, August 17, 2009. Intel.
http://download.intel.com/pressroom/pdf/cdsvsdownloadsrelease.pdf)

for power and cooling equaled the cost of the server. The combined capital and operating cost for power and cooling equaled the server cost already in 2001 according to Belady, and by now, 2010, lifetime power and cooling costs amounts to more than double the server cost as shown in Figure 43.2.

Though it is difficult to find good data on the cost of energy and transport losses delivered to a data center from different sources, and energy requirements for information transport between a data center and its users, it is a generally held opinion that it is economically advantageous to locate data centers close to an electric power source, preferably an inexpensive, clean, and renewable source of energy, or in a location in which cooling can be realized at low cost. Some insight into the cost of data transport to and from a data center can be gained from the studies that have estimated the energy efficiency of the Internet. According to [9], the Internet in 2008 required about 7 kWh/GB of traffic with an efficiency improvement of 30%/year (Figure 43.3). However, about two-thirds of this energy consumption is estimated to be due to servers and storage systems [10], and only about one-third due to data transport, as seen from Table 43.4.

The cost of electricity from different sources may vary significantly by location, but for the United States, the predictions made by the U.S. Department of Energy's Energy Information Administration (EIA) gives an indication of expected energy costs by source. Table 43.5 shows predictions for 2016 [11].

TABLE 43.4 Internet and Phone System Direct Energy Use

Equipment Type	2000 Electricity Use (TWh/Year)	2006 Electricity Use (TWh/Year)
Internet[a]	19.3	42.3
Servers[b]	11.6	24.5
Data storage[c]	1.5	4.4
Hubs[d]	1.6	3.5
Routers[e]	1.1	2.4
LAN switches[d]	3.3	7.2
WAN switches[d]	0.2	0.3
Telephone systems[a]	3.8	2.5
Transmission[e]	1.8	1.2
Public phone network[e]	1.0	0.7
Private branch Exchanges (PBX)[e]	1.0	0.7
Total	23.1	44.9

[a]These estimates do not include energy use for ventilation, cooling, and auxiliary equipment.

[b]From EPA [29]. Includes energy use from all types of servers.

[c]Year 2000 value from Roth [12], year 2006 value scaled by growth factor for Enterprise Storage Devices from EPA [29].

[d]Year 2000 value from Roth [12], year 2006 value scaled by growth factor for network equipment from EPA [29].

[e]Year 2000 data from Roth [12], year 2006 value scaled by growth in total phone system data traffic. The estimated decline of energy use in transmission equipment for voice traffic may be offset somewhat by increasing energy use of co-located transmission equipment to carry data traffic.

Of energy sources with low environmental impact, hydro, biomass, geothermal, and nuclear are estimated to be very cost competitive.

An example of industry locating a major data center close to a clean, renewable electric energy source is Google's data center at The Dalles, Oregon. Google, which publicly stresses both energy efficiency of its infrastructure as well as high environmental standards, located one of its major U.S. data centers in The Dalles [13,14], Oregon, next to the Columbia River and close to a 2 GW [15] hydroelectric power plant. In Europe, Google is currently building a data center in Hamina, Finland, [16,17] that will use Baltic Sea water for cooling, enabling "free" cooling, that is, cooling without chillers, and some wind power from a wind power farm being built next to the data center. Furthermore, in Finland, nuclear, hydro, wind, biomass, and peat account for over 60% of electric energy generation whereas coal, oil, and natural gas only account for less than 25% [18]. Another example of the use of cool "natural" water for "free" cooling is the Swiss National Supercomputing Centre's (CSCS) new data center [19] under construction in Lugano, Switzerland, that will use water from Lake Lugano through 2.8 km long pipes for a 16 MW data center design. Another example of the use of free cooling is Stanford University's planned new data center that is estimated to save $3 million/year compared to their current data center that use chillers [20] Other examples of data center locations enabling low cost cooling through "free" cooling by using outside air and eliminating chillers are Microsoft's data center in Dublin, Ireland [21] and Google's data center in Belgium [22,23]. To reduce its environmental impact, Google has also made major investments in renewable energy, such as wind, investing in two wind farms in North Dakota with a total of 170 MW capacity [24], purchasing 114 MW over a 20-year period from Idaho wind farms [25,26], and investing in wind power transmission infrastructure [27]. Other companies operating large data centers also consider both cost and environmental impact in locating their data centers.

TABLE 43.5   Estimated Levelized Cost of New Generation Resources, 2016

Plant Type	Capacity Factor (%)	U.S. Average Levelized Costs (2008 $/MWh) for Plants Entering Service in 2016				
		Levelized Capital Cost	Fixed O&M	Variable O&M (Including Fuel)	Transmission Investment	Total System Levelized Cost
Conventional coal	85	69.2	3.8	23.9	3.6	100.4
Advanced coal	85	81.2	5.3	20.4	3.6	110.5
Advanced coal with CCS	85	92.6	6.3	26.4	3.9	129.3
Natural gas-fired						
Conventional combined cycle	87	22.9	1.7	54.9	3.6	83.1
Advanced combined cycle	87	22.4	1.6	51.7	3.6	79.3
Advanced CC with CCS	87	43.8	2.7	63.0	3.8	113.3
Conventional combustion turbine	30	41.1	4.7	82.9	10.8	139.5
Advanced combustion turbine	30	38.5	4.1	70.0	10.8	123.5
Advanced nuclear	90	94.9	11.7	9.4	3.0	119.0
Wind	34.4	130.5	10.4	0.0	8.4	149.3
Wind—Offshore	39.3	159.9	23.8	0.0	7.4	191.1
Solar PV	21.7	376.8	6.4	0.0	13.0	396.1
Solar thermal	31.2	224.4	21.8	0.0	10.4	256.6
Geothermal	90	88.0	22.9	0.0	4.8	115.7
Biomass	83	73.3	9.1	24.9	3.8	111.0
Hydro	51.4	103.7	3.5	7.1	5.7	119.9

Source: Energy Information Administration, *Annual Energy Outlook 2010*, December 2009, DOE/EIA-0383.

43.2.2 Energy Efficiency Measures

The energy efficiency of data centers has been of great concern for major centers for close to a decade, and significant improvements have been made in their design and operation. Low cost of the infrastructure and its operation is a major competitive advantage for Web and Internet companies such as Amazon, Google, Microsoft, and Yahoo, resulting in limited openness about center efficiencies and, in particular, how this efficiency is achieved. However, in recent years, the secrecy has decreased.

A typical distribution of energy consumption in a traditional data center is illustrated in Figure 43.4 [28]. Some specific results are reported in the 2007 EPA report to the U.S. Congress [29]. From Figure 43.4, it is apparent that high energy efficiency requires elimination of chillers and reduced losses in power conversion. To measure the effectiveness of energy use, the Green Grid introduced two measures for data centers: power usage effectiveness (PUE) and its inverse datacenter infrastructure effectiveness (DCiE) [30] that are illustrated in Figure 43.5.

State-of-the-art data centers today claim a PUE of about 1.2 [31–33], which in the case of [32] and [33] refer to chillerless data centers, which most likely also is the case for [31]. To reach efficiencies at the reported level, it is clear that significant other improvements have been made.

43.2.3 Cooling

In regard to cooling, a key aspect is controlled airflow to minimize or completely prevent mixing of cold air entering servers and hot air leaving servers. Air cooling dominates today's data centers and is an implicit assumption in Figures 43.4 and 43.5. A typical approach is to arrange computer racks such that the airflow forms alternating hot and cold aisles as illustrated in Figure 43.6 [34]. The hot/cold aisle arrangement can be combined with enclosures to further assure separation of hot and cold air as shown in Figure 43.7 [35]. This form of arrangement is often combined with in-row cooling for high heat densities as illustrated in Figure 43.8 [36]. The management of airflow for effective cooling is a strong contributor to the energy

PUE: Power usage effectiveness
DCE: Data center efficiency

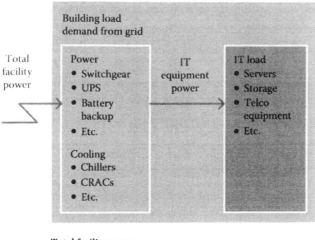

$$PUE = \frac{Total\ facility\ power}{IT\ equipment\ power}$$

$$DCE = \frac{1}{PUE} = \frac{IT\ equipment\ power}{Total\ facility\ power}$$

FIGURE 43.4 Electricity use in a traditional data center. (From The Green Grid, Guidelines for energy-efficient data centers, February 16, 2007.
http://www.thegreengrid.org/~/media/WhitePapers/Green_Grid_Guidelines_WP.ashx?lang=en)

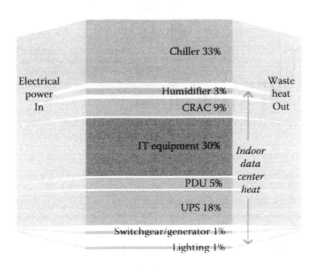

FIGURE 43.5 PUE and DCiE. (From The Green Grid, Green Grid data center power efficiency metrics: PUE and DCiE. White Paper no 6, 2008. http://www.thegreengrid.org/~/media/WhitePapers/White_Paper_6_-_ PUE_and_DCiE_Eff_Metrics_30_December_2008.pdf?lang=en)

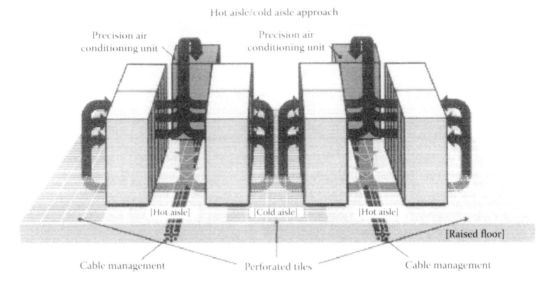

FIGURE 43.6   Hot-aisle/cold-aisle arrangement of racks.

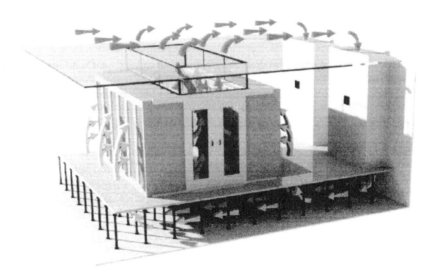

FIGURE 43.7   Hot-aisle enclosure.

effectiveness of containerized data centers as illustrated in Figure 43.9 [37]. Practically all major vendors now have some form of containerized data center [38]. The idea of modularized (large) data centers originates from a need for rapid and cost effective deployment of large data centers with Google filing for a patent on containerized data centers in 2003 [39], Figure 43.10, and Microsoft discussing containerized centers in 2007 and 2008 [40,41] and showing one of their containerized centers in 2009 [42]. For an overview of data center trends see [43].

   Not all servers have a front-to-back airflow as assumed earlier. For server designs that have a sideway flow, such as the IBM Blue Gene (BG), arrangements as in Figure 43.11 can be made to prevent mixing of cold and hot air [44]. Transverse flow is also planned for the next-generation Cray systems that have a transverse flow across an entire rack row with temperature restoring water coils for each rack and blowers for each pair of racks to maintain temperature and speed of the transverse airflow as shown in

FIGURE 43.8 Hot-aisle enclosure with in-row cooling.

FIGURE 43.9 Data center in a container.

Figure 43.12 [45]. The claim is that despite being open this design will bring the PUE down to less than 1.05. A 6 kW blower cabinet serves two 100+ kW cabinets.

Currently, Cray systems are designed for a vertical airflow [46], as shown in Figure 43.13. Vertical airflow is also used by some other vendors having high-density solutions.

As evident from Figures 43.12 and 43.13 and the in-row cooling units in Figure 43.8, liquid cooling has moved closer to the racks, or even into the racks. An alternative to in-row cooling or top mounted cooling is cooled rack doors that restore the air temperature, mostly, to that close to the inlet temperature. The idea is illustrated in Figure 43.14 [47]. Several vendors have doors of this type with cooling capacity of up to 40 kW or more depending on water temperatures and water flow. With water cooling at the rack level, claims are made that less than 2% energy is required for cooling of servers.

The evolution of cooling solutions is not only a reflection of a need for more energy-efficient and environmentally friendly solutions, but also a consequence of increased heat densities. Traditional data center designs had raised floors with open airflow and computer room air conditioning

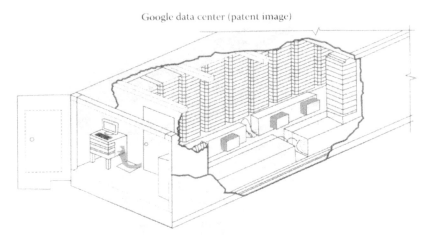

FIGURE 43.10 Google's container data center concept. (From US Patent Application, Google. Modular Data Center, December 30, 2003. http://patft.uspto.gov/netacgi/nph-Parser?Sect1=PTO1&Sect2=HITOFF&d=PALL& p=1&u=%2Fnetahtml%2FPTO%2Fsrchnum.htm&r=1&f=G&l=50&s1=7,278,273.PN.&OS=PN/7,278,273&RS=PN/ 7,278,273)

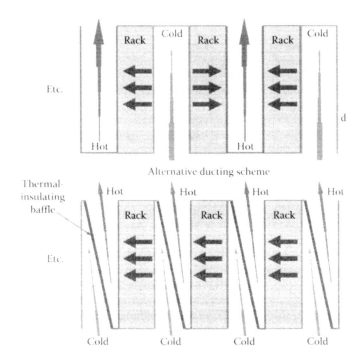

FIGURE 43.11 Hot/cold air separation typical for an IBM BG system. (From Takken, T., Blue Gene/L power, packaging and cooling, February 2004.
http://www.physik.uni-regensburg.de/studium/uebungen/ws0405/scomp/BGWS_03_PowerPackagingCooling.pdf)

(CRAC) units along the walls as illustrated in Figure 43.15 [48]. This design was appropriate when heat densities were low. According to American Society of Heating, Refrigerating and Air-Conditioning Engineers (ASHRAE) [49], 20 years ago densities of about $3\,kW/m^2$ was common for centers dominated by computer servers [50], but recently announced products [51–53] result in heat densities of about 20 times that, as shown in Figure 43.16. The increased heat densities are due in part to increased component

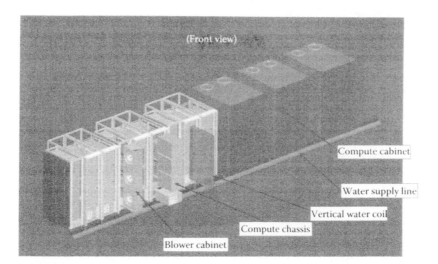

FIGURE 43.12 Airflow and cooling of the Cray Cascade system. (From Pel, V., Energy efficient aspects in Cray supercomputers, September 2010.
http://www.ena-hpc.org/2010/talks/EnA-HPC2010-Pel-Energy_Efficiency_Ascpects_in_Cray_Supercomputers.pdf)

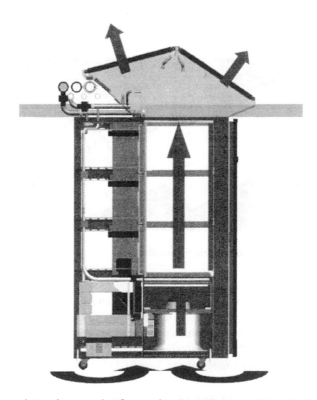

FIGURE 43.13 Cooling solution for vertical airflow used in Cray XT systems. (From Graham, D. and Laatsch, M., Meeting the demands of computer cooling with superior efficiency, 2009.
http://www.cray.com/Assets/PDF/products/xt/whitepaper_ecophlex.pdf)

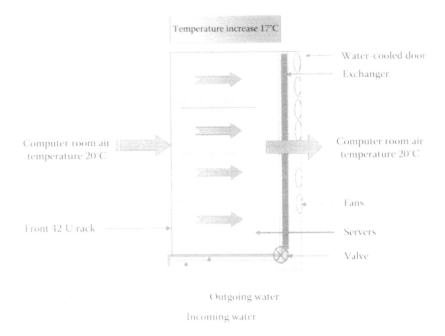

FIGURE 43.14   Cooled rear door of computer rack. (From Bull Cabinet Door, October, 2010. http://www.bull.com/extreme-computing/download/S-HPCwaterdoor-en2.pdf)

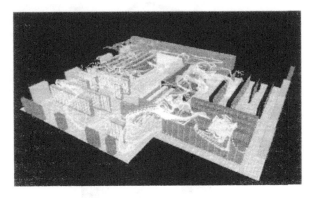

FIGURE 43.15   Typical open air cooled data center floor plan.

heat densities, and in part to improved cooling techniques enabling increased packing densities and consequent increased heat densities. Figure 43.17 illustrates the CPU heat density trends that dominated through the early part of the last decade, at which point heat densities forced a cap on power dissipation so that for the last several years CPUs have largely been designed for a nonincreasing maximum power dissipation. In fact, in recent years a range of x86-based CPUs have been introduced for lower power dissipation and clock frequencies.

The high heat densities of some components has lead to the introduction of component liquid cooling techniques, such as used, for instance, in IBM's Power7 based servers shown in Figures 43.18 and 43.19 [52]. Recently, liquid cooling in the form of liquid enclosed blades [54] has also been introduced as shown in Figure 43.20, or even entirely liquid filled racks, Figure 43.21 [55].

Operating temperatures have an impact on the energy consumption of data centers, though the relationship is not simple and component temperatures may affect reliability and longevity. ASHRAE has

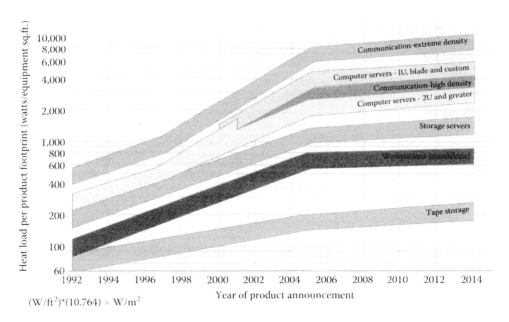

FIGURE 43.16 Evolution of heat densities in data centers. (From ASHRAE, Datacom equipment power trends and cooling applications, 2005. http://esdc.pnl.gov/SC07_BOF/SC07BOF_McCann.pdf)

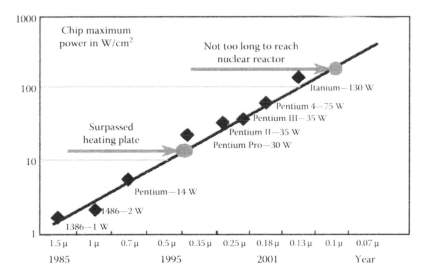

FIGURE 43.17 CPU heat density evolution. (From Shekhar Borkar, Intel circa 2001.)

made thorough studies and recommendations for server inlet temperatures. The first set of recommendations of 25°C were made in 2004 then revised to 27°C in 2008 [56]. It has been claimed [57] that for every degree in increased set point, an energy savings of 4% can be realized. In [58], it was shown that a reduction in cooling energy requirements of as much as 30% can result from raising the inlet temperature from 18°C to 27°C, but that the total net energy savings maybe about 10% due to increased energy consumption for other systems in the data center including the computer system itself, see Figure 43.22. As the temperature set point is raised, the load on the server fans increases to assure that component temperatures stay below the target values. According to ASHRAE when set points increase over 25°C,

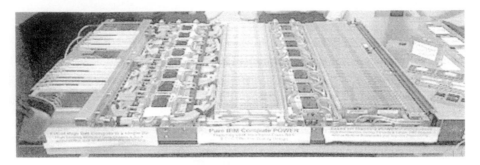

FIGURE 43.18 An IBM Power7 water cooled 8 Tflop/s server unit (2U rack units high) with 8 multi-chip-modules with a total of 256 cores.

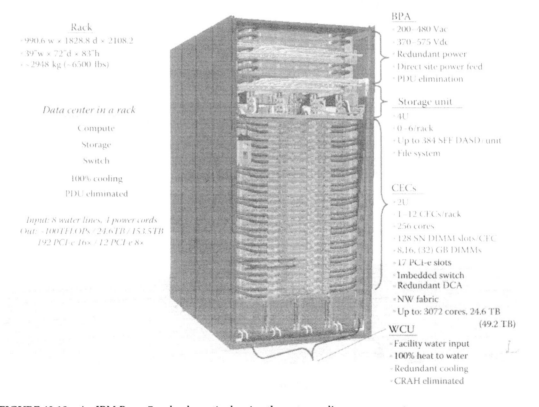

FIGURE 43.19 An IBM Power7 rack schematic showing the water cooling arrangement.

the fan energy consumption increases significantly. The study reported in [58] also noticed a significant increase in server energy consumption with increased temperature values even with constant fan speeds under high loads. It is interesting to note that according to a recent study [59] many data centers operate at significantly lower temperatures than ASHRAE recommends, see Figure 43.23.

Increased temperatures do not only reduce required cooling energy but also increase the potential for energy reuse. Typical return water from CRAC units is not warm enough for a variety of needs and hence may in fact represent more of a problem than an asset. However, direct cooling of components as explored by IBM in a research project [60] would enable using cooling water with an inlet temperature as high as 60°C. The idea is illustrated in Figure 43.24.

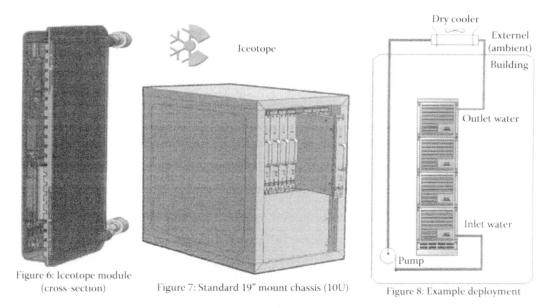

Iceotope
Dry cooler
Externel
(ambient)
Building
Outlet water
Inlet water
Pump
Figure 6: Iceotope module
(cross-section)
Figure 7: Standard 19" mount chassis (10U)
Figure 8: Example deployment

FIGURE 43.20 Liquid submerged and enclosed blades. (From Liquid cooling for servers. http://www.boston.co.uk/technical/2009/12/liquid-cooling-for-servers.aspx)

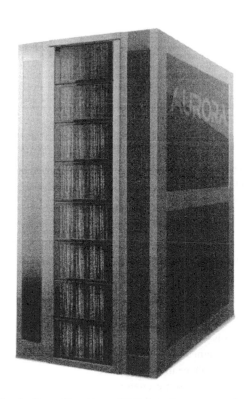

FIGURE 43.21 Liquid cooled rack. (From Data Center HPC. http://www.eurotech.com/en/hpc)

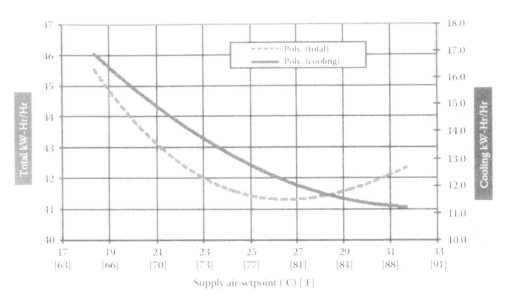

FIGURE 43.22 Energy consumption as a function of air inlet temperature set point. (From Bean, J.H. and Moss, D.L. Energy impact of increased server inlet temperature. Optimal Temperature Operation. http://i.dell.com/sites/content/business/solutions/whitepapers/en/Documents/dci-energy-impact-of-increased-inlet-temp.pdf)

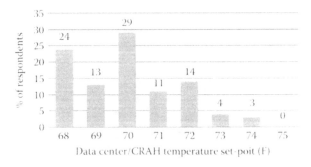

FIGURE 43.23 Data center inlet temperature set point as reported in Patterson et al. (2009). (From Patterson, M.K. et al., Energy efficiency through the integration of information and communications technology management and facilities controls, July, 2009. IPACK. http://download.intel.com/pressroom/archive/reference/IPACK2009.pdf)

The Green Grid is defining a new metric to account for energy reuse by introducing an energy reuse factor, ERF, which is the fraction of energy used for the IT equipment that is being reused [61].

43.2.4 Power Supply

The typical data center power supply structure is illustrated in Figure 43.25, which excludes a typical substation in which power supply voltage is stepped down to 480 or 400 V 3-phase from several kV. However, common servers are designed to work with 12 V DC internally. Thus, conversion from AC to DC as well as further reduction in voltage must take place.

Since every conversion step implies some energy loss, reducing the number of conversions has a potential to increase energy efficiency. One such way would be to use DC for power distribution in the data center instead of the current practice of using AC for distribution with conversion to DC at the server. Another way would be to convert from AC to DC at the rack level and use DC for distribution within the rack. Most studies on the potential gains from using DC for power distribution conclude that DC

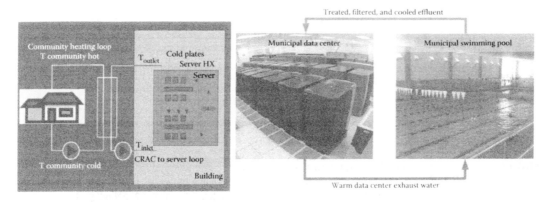

FIGURE 43.24 Data center energy reuse schematic.

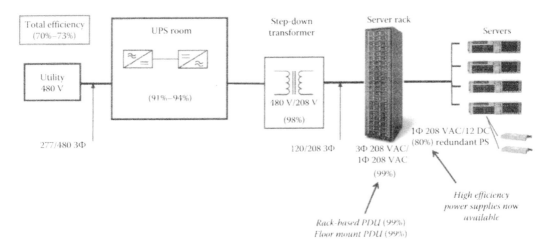

FIGURE 43.25 Data center power infrastructure.

distribution would yield higher efficiency though there is disagreement about the potential gain [62–65]. The estimated efficiency of a well-designed DC power distribution system is about 88%, Figure 43.26, including UPS (uninterruptible power supplies), wiring losses, and PSUs (power supply units) [62]. At this level of efficiency, DC distribution is estimated to offer an efficiency advantage over a well-designed and operated AC system of 2%–5%, which is likely to be too small for DC to become the dominating data center power distribution method since such a conversion would require major investments and adoption of technologies that today do not have a broad market.

UPS is used by many data centers to assure high availability. The most common UPS equipment uses batteries to supply power to the data center in case of a loss of external power. Thus, a conversion from AC to DC is required to keep batteries charged and a conversion from DC to AC required for the distribution of power in the data center when AC is used for this task. The efficiency of this double conversion has improved to up to 98% from a typical of around 80% several years ago. At this level of efficiency, battery-based UPS solutions are comparable to flywheels from an efficiency point of view, Figure 43.27 [66]. Even though the efficiency of state-of-the art UPS is very high, some data centers do not use UPS, mostly for cost reasons. For instance, Google is reported to use batteries directly on their servers instead of UPS [67].

The 480 or 400 V 3-phase power used for distribution in the data center is in most parts of the world routed directly to server racks, whereas in the United States, it is stepped down to a lower voltage in a power distribution unit (PDU) incurring some losses [62].

	UPS		Distribution wiring		IT power supply		Overall efficiency
DC	96.0%	×	99.5%	×	91.75%	=	87.64%
AC	96.2%	×	99.5%	×	90.25%	=	86.39%

FIGURE 43.26 Comparison of the energy efficiency of DC and AC power distribution in the data center. (From Rasmussen, N. and Spitaels, J., A quantitative comparison of high efficiency AC vs DC power distribution for data centers. White Paper 127. APC. http://www.apcmedia.com/salestools/NRAN-76TTJY_R2_EN.pdf)

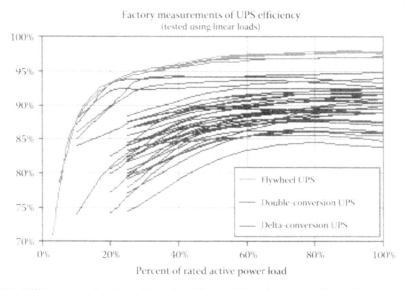

FIGURE 43.27 Efficiency as a function of load for different UPS technologies. (From Ton, M. and Fortenbury, B., High performance buildings: Data Centers, uninterruptible power supplies (UPS), December 2005. LBNL. Ecos Consulting. EPRI Solutions. http://hightech.lbl.gov/documents/UPS/Final_UPS_Report.pdf)

Server PSUs convert the AC used for distribution to DC used in the servers and also steps down the voltage to 12 V. The PSU efficiency has increased significantly in recent years from below 80%–90% or better for a broad range of loads with peak efficiencies in excess of 94% for high-quality PSUs [68,69] as seen in Figure 43.28.

Most of the inefficiencies in the power supply chain within the data center have been eliminated in recent years and remaining inefficiencies are small. It is worth noting that at least one conversion is necessary since AC is used for electricity generation and distribution and DC used for electronics in servers. It is also the case that conversion is necessary for difference in voltage levels, and for "isolating" servers from the energy source.

43.2.5 Data Center Infrastructure Efficiency Summary

Over the last several years, data center design and operation have lead to an exceptional improvement in energy efficiency largely through improved cooling techniques and through improvements in the efficiencies of UPS and PSUs. Data center efficiency has also improved by locating new data centers where

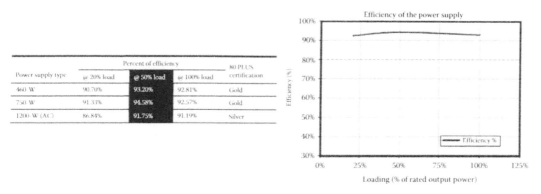

Power supply type	Percent of efficiency			80 PLUS certification
	@ 20% load	@ 50% load	@ 100% load	
460 W	90.70%	93.20%	92.81%	Gold
750 W	91.33%	94.58%	92.57%	Gold
1200 W (AC)	86.84%	91.75%	91.19%	Silver

FIGURE 43.28 PSU efficiencies from two vendors as certified by 80-Plus. (From 80-Plus. http://www.plugload-solutions.com/80PlusPowerSupplies.aspx)

"free" cooling can be used and chillers eliminated. The reuse of energy consumed in the data center can contribute to a significant reduction in overall energy use and emissions and can be an important consideration for data centers in areas where hot water can be an effective energy source. Increasing data center operating temperatures and in particular use of direct cooling technologies enabling significantly raised outlet water temperatures is of great interest and pursued by industry. In the planning of new data centers, these issues should be considered together in a comprehensive way, as was done in the planning of the Computational Research and Theory Facility (CRTF) at University of California Berkeley [71].

With state-of-the-art data center PUEs of 1.2 or less, there clearly are very limited energy efficiency gains possible from improved data center design and operation. Significant additional gains must come from energy reuse, and improved energy efficiency of the IT equipment and its use. Though power consumption of data centers generally has increased substantially, the energy efficiency of computer systems measured in terms of work per energy unit has improved considerably for decades, largely due to Moore's law, but also due to numerous innovations in many areas, including management and operations. Next, we will review some of these changes.

43.3 HPC System Energy Efficiency

The exponentially improved performance of computers, usually referred to as Moore's Law [72], is well known. The technology evolution has for the last few decades resulted in a halving of feature sizes about every 54 months, a doubling of transistors per processor every 21 months, that is, more rapidly than what reduced feature sizes would predict, and a doubling in performance as measured by MIPS (million instructions per second) about every 20 months. For larger HPC systems, the performance improvement as measured by the Linpack benchmark [73] has been even more rapid. From the plot in Figure 43.29 of the history of systems on the Top500 list [74], the list of the 500 most powerful computer systems in the world as measured by the Linpack benchmark, it can be deduced that the performance has doubled on average every 13.64 months for the number 1 system whereas for the number 500 system the doubling time on average is 12.90 months.

A number of studies have been carried out to attempt to assess the improved energy efficiency of computers. Figure 43.30 shows the findings by Koomey et al. reported in [75] that extend the study by Nordhaus reported in [76]. The results indicate a doubling in the energy efficiency of computation about every 18.84 months. Note that the Top500 list measures performance of systems by the Linpack benchmark while Koomey et al. [75] and Nordhaus [76] use a composite synthetic measure that includes both elements of the Standard Performance Evaluation Corporation (SPEC) benchmarks [77] and other benchmarks as well as theoretical performance [76]. But, the rate of improvement in terms of energy

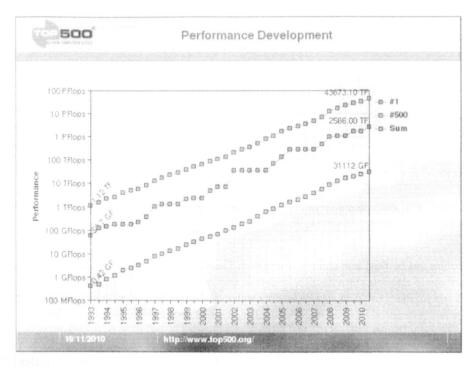

FIGURE 43.29   Evolution of systems on the Top500 list.

efficiency should still be relevant for HPC systems since Nordhaus in [76] provide a (fixed) relation between floating-point performance and computations per second used in [76].

The difference in floating-point performance growth rate and improvement in energy efficiency of computation is a good indicator of the growth in energy consumption of HPC systems, which according to these observations would amount to about 20%/year, or about a factor of 6 over a decade. This is higher than the EPA projected growth rate of 14%/year on average for the 2000–2006 period, or 17%/year average for volume servers in its report to Congress [29], Figure 43.31, but in line with the findings of the Uptime Institute [78–80]. The growth rate of about 20%/year on average is also consistent with our experience at packet data channel (PDC) at Tammsvik Konferens och Herrgird (KTH) [81] where we have had to expand the infrastructure from less than 400 kVA in 2003 to 2 MW in 2010. The growth factor of 6 over a decade in power for HPC systems is also consistent with the average power consumption for the Top50 systems on the Top500 list as presented in [82] for June 2000, about 230 kW, and in [83] for June 2010, 1401 kW. In connection with procurements in 2007/2008, we estimated that the capital and operating cost of the infrastructure for the lifetime of the procured systems would be about 1.5 times the cost of the hardware, which is in line with the predictions of Belady [8]. The reason for this very significant change is rapidly increasing energy costs, and the increased energy consumption of servers, as discussed earlier. For Sweden, the electricity cost has increased about 7%/year on average over the last 25 years while the United States has experienced a lower growth rate. As can be seen from Figure 43.32 [84], the United States price evolution has been highly variable and averages between 4% and 5% over the last 50 years for different consumer sectors. For the last 5 years, the price increase has been about 6% on average.

43.3.1 System Architecture from an Energy Perspective

Koomey [75] and Nordhaus [76] have both reported a rapid improvement in the overall energy efficiency of computation. To understand both the past improvements in energy efficiency and future possibilities

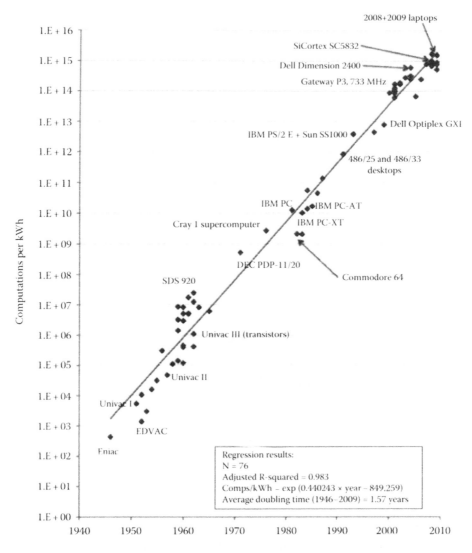

FIGURE 43.30 Energy efficiency of computers over time.

and challenges for HPC system energy efficiency improvements, it is helpful both to understand the relative energy consumption of different parts of a system and the physics that governs the energy consumption of complementary metal-oxide semiconductor (CMOS) technology, the dominating technology today for CPUs and memory. It is also useful to understand the power management techniques introduced by vendors.

The energy consumption per transistor has improved by a factor of about 1 million over 30 years according to [85], as shown in Figure 43.33, which corresponds to a halving of energy consumption about every 18 months in line with the observations in [75].

As reported in [86], we recently designed a four socket blade server targeting energy efficiency based on 6-core high-efficiency CPUs resulting in a power distribution in the design stage as shown in Table 43.6 for a chassis of 10 blades. Since our design emphasized energy efficiency, our nodes are diskless, which removes one source of energy consumption. For the estimates in Table 43.6, four DIMMs per CPU socket are assumed. In [87,88], subsystem power consumption is given for a two-socket server, but only two

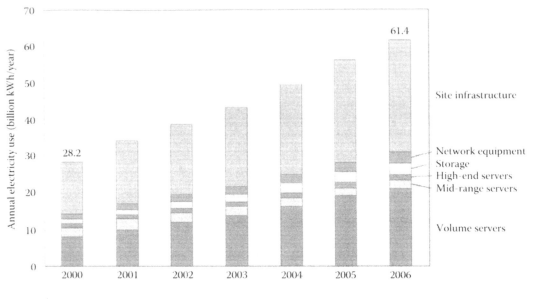

End Use Component	2000		2006		2000–2006
	Electricity Use (Billion kWh)	% Total	Electricity Use (Billion kWh)	% Total	Electricity Use CAGR
Site infrastructure	14.1	50%	30.7	50%	14%
Network equipment	1.4	5%	3.0	5%	14%
Storage	1.1	4%	3.2	5%	20%
High-end servers	1.1	4%	1.5	2%	5%
Mid-range servers	2.5	9%	2.2	4%	−2%
Volume servers	8.0	29%	20.9	34%	17%
Total	28.2		61.4		14%

FIGURE 43.31 Data center energy growth according to the EPA 2007 report to the U.S. Congress. (From U.S. Environmental Protection Agency, Energy star program, August 2, 2007. Report to Congress on server and data center energy efficiency. Public Law 109-431. http://www.energystar.gov/ia/partners/prod_development/downloads/EPA_Datacenter_Report_Congress_Final1.pdf)

DIMMs per socket is assumed and the CPU power consumption seems exceptionally low given that high-efficiency (HE) x86 CPUs typically have a peak power rating of about 80 W and high-performance x86 CPUs have a peak power rating of 130–140 W. The measured peak power consumption for a chassis of the servers we designed is about 4650 W, or about 92% of the estimated peak power consumption. About 2/3rds of the peak power is consumed by CPUs and memory. The power distribution among subsystems for our design is fairly typical for current HPC servers. Though gains in energy efficiency are possible by reducing energy consumed by PSUs, fans, interconnect, and motherboards, major improvements must address the energy efficiency of CPUs and memory.

The energy consumption of memory and CPUs depends on the feature sizes of the technology being used as indicated by Figure 43.33, but for any given feature size it also depends on operating voltage and frequency. For CMOS, the relationship between power, voltage, and frequency is

$$P = c_1 V^2 f + c_2 V + c_3 + O(V^4),$$

where

$c_1, c_2,$ and c_3 are constants
V is the supply voltage
f is the operating frequency

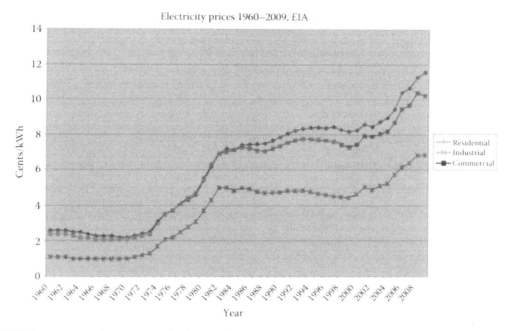

FIGURE 43.32 U.S. electricity costs for the period 1960–2009. (From U.S. Energy Information Administration.)

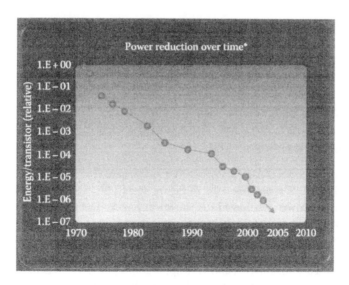

FIGURE 43.33 Evolution of energy consumption per transistor. (From Eco-technology: Delivering efficiency & innovation, in *Behavior, Energy & Climate Change Conference*, November 16–19, Sacramento, CA. http://piee.stanford.edu/cgi-bin/docs/behavior/becc/2008/presentations/18-4C-01-Eco-Technology_-_Delivering_Efficiency_and_Innovation.pdf)

The first term represents dynamic power and is dominant in today's CMOS; the second and third terms represent leakage and board power while the last term captures fan power. With the first term dominating, the power needed scales with the square of the voltage and the clock frequency. But, it is also the case that the frequency is fairly proportional to the voltage setting for normal operating conditions. Hence, in fact, the power is related to f^3. This relationship is exploited both in terms of

TABLE 43.6   Power Ratings for Subsystems in a 10-Blade Chassis

Subsystem	Power (W)	Percentage
CPUs	2880	56.8
Memory	800	15.8
PSU	355	7.0
Fans	350	6.9
Motherboards	300	5.9
HT3 links	120	2.4
IB HCAs	100	2.0
IB switch	100	2.0
GigE switch	40	0.8
CMM	20	0.4
Total	5065	100.0

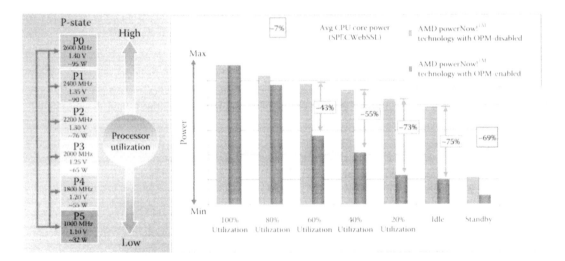

FIGURE 43.34   Energy gain as a function of CPU speed.

controlling standard x86 CPUs, as illustrated in Figure 43.34 [89], and in design points for different CPUs. A typical load to power relationship is shown in Figure 43.35 [88] in which the disk power draw is independent of load and memory power consumption also is fairly constant, except for the idle case. CPU power more than triples from idle to full load. But, idle power is nevertheless about half of power under full load. This is of great concern from an energy efficiency perspective for many usage scenarios, in particular for Internet and Web applications [90]. HPC systems often have workloads and queuing system assuring a sustained high load and hence idle power is not of major concern for HPC.

Another good illustration of the relationship between voltage, frequency, performance, and power is shown in Figure 43.36 showing the characteristics of Intel's 80-core experimental CPU [91].

From these observations, we conclude that minimizing execution time by maximizing execution rate may in fact be very energy inefficient because power needs increase more rapidly than execution time decrease. This is the premise on which multi-core chips are based, as seen in current CPUs that for a fixed power envelope and technology tend to have slower cores the more cores there are, as illustrated for the AMD Magny-Cours chips in Table 43.7. We also conclude that managing the state of the cores as a function of workload is important for overall energy efficiency. Further, we

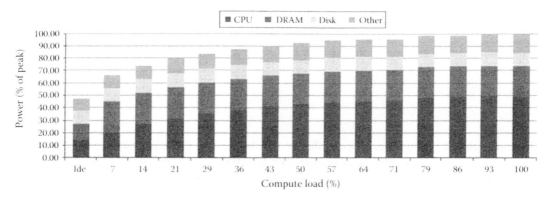

FIGURE 43.35 Power consumption as a function of load on a typical server. (From Barroso, L.A. and Hölzle, U., The data center as a computer: An introduction to the design of warehouse-scale machines, Morgan & Claypool Publishers, 2009. http://www.morganclaypool.com/doi/pdf/10.2200/S00193ED1V01Y200905CAC006)

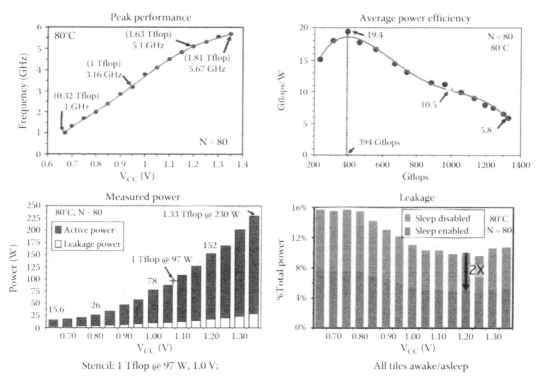

FIGURE 43.36 Power and performance relationship for the Intel Polaris research chip. (From Vangal, S. et al., An 80-tile 1.28 Tflops network-on-chip in 65 nm CMOS, in *IEEE Solid-States Circuits Conference*, San Francisco, CA, pp. 98–99, February 11–15, 2007. http://ieeexplore.ieee.org/xpl/freeabs_all.jsp?arnumber=4242283)

observe that the Intel 80-core experimental CPU operates in the same power range as a standard Intel CPU, which for 65 nm technology that was used for the 80-core chip had up to 4 cores. (Clovertown) [92], highlighting that the cores on the experimental chip are much simpler and smaller, 100 million transistors for 80 cores versus 582 million for the 4-core Clovertown [93]. Yet, for the Linpack benchmark, the experimental chip achieves in excess of 1 TF compared to about 38 GF for the

TABLE 43.7 Sample AMD Multi-Core CPUs

Cores	Clock (GHz)	ADP Power (W)
12	2.2	80
8	2.4	80
12	1.7	65
8	2.0	65

Clovertown chip. This illustrates the point that in regards to energy efficiency there is a possibility that simpler, lower power cores may be of great interest for HPC, as also discussed in [94] where it was shown that for some scientific application only 80 out of 300 x86 assembly language instructions were needed.

43.3.2 Multi-Core CPUs

The heat density of standard CPUs, as illustrated in Figure 43.17, forced commodity CPU vendors to seek new ways to exploit the continually increased capabilities offered by decreased feature sizes ("Moore's law"). The exponential improvement is expected to continue through this decade [95]. The industry's solution to exploit increased capability without increased power consumption was multi-core CPUs.

Technology demonstration systems based on dual-core AMD CPUs [96,97] and dual-core PowerPC CPUs [98,99] appeared in 2004. AMD, Intel [100], and IBM all delivered dual-core microprocessor CPUs for production systems in 2005. For more complex processors, IBM had already introduced dual-core CPUs in 2001 [101] for their Power4 processors. Today, AMD offers CPUs with up to 12 cores with frequencies up to 2.3 GHz and a maximum power dissipation of about 137 W while Intel offers CPUs with up to 8 cores. Intel's 6-core CPUs have a maximum power dissipation of 130 W and a maximum clock frequency in turbo mode of 3.6 GHz.

Specialized CPUs, such as graphics processing units (GPUs), today typically have hundreds of cores with, for example, the nVidia Fermi GPU having 512 stream processor cores [102] with a maximum power consumption of 225 W and a peak theoretical double-precision performance of 515 GF [103] and the AMD FireStream 9370 having 1600 stream processor cores [104] with a maximum power consumption of 225 W and a theoretical peak double-precision performance of 528 GF. Though the power consumption of GPUs is about twice that of x86 architecture CPUs, or more, the peak double-precision performance/W is about three times higher than that of the x86 CPUs.

Recently, in the quest for energy-efficient servers, there has been an increased interest in processors used in the embedded and mobile markets, such as the ARM processors [105] that are widely used in the mobile market, the Intel Atom processor [106], and digital signal processors (DSPs), such as the Texas Instruments TMS320C6678 [107] capable of 40 GF in double precision at about 10 W, which is still significantly less than the ClearSpeed CX700 floating-point processor [108] that has about the same power consumption but a theoretical peak of 96 GF.

The possible improvement in energy efficiency of conventional CPUs is well demonstrated by the CPU designed for the IBM BG/Q for which little information is publicly available. However, from [74], it can be deduced that the CPU has an impressive energy efficiency with a theoretical peak performance of 204.8 GF at 1.6 GHz and an estimated power draw of about 50 W. The processor characteristics are summarized in Table 43.8. Power and theoretical peak performance data in the table are in some cases estimates, in other cases from public specifications. The intent is only to show relative qualities.

From Table 43.8, we note that in terms of theoretical peak double-precision floating-point performance, the CPUs designed for mobile markets have a performance/W comparable to the x86-based CPUs that are designed with floating-point intensive applications in mind. The mobile CPUs, however, are expected to have advantages over x86-based CPUs for applications not dominated by floating-point operations and have more evolved power management features than x86-based systems. The

TABLE 43.8 Estimates of Theoretical Performance/W for Some Processor Alternatives

ARM Coretx-9			ATOM			AMD 12-Core			Intel 6-Core			ATI 9370		
Cores	W	GF/W	Cores	W	GF/W	Cores	W	GF/W	Cores	W	GF/W	Cores	W	GF/W
4	~2	~0.5	2	2+	~0.5	12	115	~0.9	6	130	~0.6	1600	225	~2.3

nVidia Fermi			TMS320C6678			IBM BQC			ClearSpeed CX700		
Cores	W	GF/W	Cores	W	GF/W	Cores	W	GF/W	Cores	W	GF/W
512	225	~2.3	8	10	~4	16	~50	~4	192	10	~10

benefits of such features are not captured in Table 43.8. The focus on low power in the design of mobile CPUs and their energy management features are the foundation for the current interest in servers based on low-power processors [109,110], such as ARM [111–113] and ATOM [114]. Servers accelerated with GPUs have received a great deal of interest in recent years as their performance has increased dramatically, in particular in terms of double-precision floating-point, and programmability improved. However, integration into servers still is via the I/O bus (PCI Express), which can degrade possible application performance gains substantially. The ClearSpeed accelerator faces the same integration issues, though it fared better than GPUs in a study carried out in porting some benchmarks to accelerated systems [115]. The IBM BQC processor for which not much information is available at this time has an impressive energy efficiency and is likely to offer good performance and not require much effort in porting codes used on clusters, unlike porting of codes to accelerator-based systems.

43.3.3 Energy-Efficient HPC Systems

Though energy efficiency at the CPU and server level has been a key consideration for component and system vendors for a good part of the last decade, the number of whole system design efforts focusing on energy efficiency has been few. However, a good example of the industry's efforts at energy-efficient HPC systems design is IBM's BG series starting with the BG/L introduced in 2004 after a 5 year development effort, followed by the BG/P in 2007 and to be followed by the BG/Q in 2011 [116,117]. The BG/L was based on the dual-core PowerPC CPU. The BG/L set a record not only in terms of performance assuming the no. 1 position in the November 2004 Top550 list, but also in terms of energy efficiency. In the inaugural Green500 list in November 2007 [118] of the most energy-efficient systems on the Top500 list, BG/L systems held positions 6 through 26 with positions 1 through 5 held by the second generation BG system, the BG/P introduced the same year. The BG/Q to be delivered in 2011 holds the number 1 position in the most recent Green500 list [119] with an efficiency of 1684 MF/W. The second most energy-efficient system on the November 2010 Top500 list used SPARC64 VIII CPUs with no accelerator and achieved an energy efficiency about half of the BG/Q, 829 MF/W, and position 4 on the list. The third most energy-efficient system without acceleration used Intel 6-core CPUs and achieved 400 MF/W, about a quarter of the BG/Q, and position 18. The most energy-efficient GPU-accelerated system achieved an efficiency of 958 MF/W and the no. 2 position, while the most energy-efficient system using the Cell Broadband Engine [120,121] for acceleration [122] achieved 773 MF/W and assumed position 5. On the June 2008 Green500 list, cell-accelerated IBM systems occupied the three top positions with an efficiency of 488 MF/W. The top 10 positions on the November 2010 Green500 list are shown in Figure 43.37.

Another design targeting energy efficiency was SiCortex's MIPS-based systems (the company closed during the Spring of 2009), that used relatively slow cores, initially 500 MHz, later 700 MHz [123] [similar to the BG/L (700 MHz) and BG/P (850 MHz at ~7 W/core)]. Another interesting system design is the proposed Lawrence Berkley National Laboratories Green Flash [124,125] proposed architecture based on Tensilica Xtensa CPUs (650 MHz at ~0.7 W/core) [126,127]. A comparison of the expected energy

Green500 Rank	MFLOPS/W	Site[a]	Computer[a]	Total Power (kW)
1	1684.20	IBM Thomas J. Watson Research Center	NNSA/SC Blue Gene/Q Prototype	38.80
2	958.35	GSIC Center, Tokyo Institute of Technology	HP Proliant SL390s G7 Xeon 6C X5670, Nvidia GPU, Linux/Windows	1243.80
3	933.06	NCSA	Hybrid Cluster Core i3 2.93GHz Dual Core, NVIDIA C2050, Infiniband	36.00
4	828.67	RIKEN Advanced Institute for Computational Science	K computer, SPARC64 VIIIfx 2.0 GHz, Tofu interconnect	57.96
5	773.38	Forschungszentrum Juelich (FZJ)	QPACE SFB TR Cluster, powerXCell 8i, 3.2 GHz, 3D-Torus	57.54
5	773.38	Universitaet Regensburg	QPACE SFB TR Cluster, PowerXCell 8i, 3.2 GHz, 3D-Torus	57.54
5	773.38	Universitaet Wuppertal	QPACE SFB TR Cluster, PowerXCell 8i, 3.2 GHz, 3D-Torus	57.54
8	740.78	Universitaet Frankfurt	Supermicro Cluster, QC Opteron 2.1 GHz, ATI Radeon GPU, Infiniband	385.00
9	677.12	Georgia Institute of Technology	HP Proliant SL390s G7 Xeon 6C X5660 2.8GHz, nVidia Fermi, Infiniband QDR	94.40
10	636.36	National Institute for Environmental Studies	GOSAT Research Computation Facility, nvidia	117.15

FIGURE 43.37 The 10 most energy-efficient systems on the Top500 list, November 2010. (From The Green500 list, November 2010. http://www.green500.org/lists/2010/11/top/list.php?from=1&to=100)

TABLE 43.9 Comparisons of Expected Power Consumption of a 200 PF System Based on Different CPUs

Processor	Clock (GHz)	Peak/Core (GF)	Cores/Socket	Sockets (k)	Cores (M)	Power (MW)	Cost 2008
AMD Opteron	2.8	5.6	2	890	1.7	179	$1B+
IBM BG/P	0.850	3.4	4	740	3.0	20	$1B+
Green Flash Tensilica Xtensa	0.650	2.7	32	120	4.0	3	$75 M

Source: Simon, H., The greening of HPC—Will power consumption become the limiting factor for future growth in HPC?, October 10, 2008. http://www.hpcuserforum.com/presentations/Germany/EnergyandComputing_Stgt.pdf

efficiency with x86- and PowerPC-based CPUs is shown in Table 43.9. As seen from the table, a factor of seven improved efficiency over using the PowerPC's used in the BG/P and a factor of 60 over using an AMD Opteron processor is expected.

43.3.4 Energy Efficiency HPC Systems Summary

Whereas for over a decade there was convergence in the design of HPC systems driven by the cost-effectiveness of commodity technologies, the increased energy consumption and increased cost of energy has brought about a divergence in HPC architecture. Some form of acceleration is likely to become common, but in the near term, it is hampered by the relatively poor integration into systems by the fact that accelerators typically use I/O bus technology for communication to their hosts. However, that might change in the not too distant future. Intel recently released a development platform with 32 vector cores per chip [129,130] with plans to release a follow-up product with 50 or more cores. It will also be interesting to follow the adoption (or not) of low-energy CPUs for HPC systems for which some (many) features of today's x86 CPUs are not needed as well as how the opportunities for dynamic/application-dependent power management of components and subsystems will evolve.

The challenges in building systems for the next major HPC performance target, Exa-scale systems, that is, systems capable of 10^{18} operations per second are significant not only in terms of the level of concurrency applications need to exhibit, but also how systems with tens of billions of threads will be managed, and how the energy issues will be resolved. Without a significant change in technology and possibly architecture, Exa-scale systems toward the end of the decade have been estimated to consume 70 MW [131] to 130 MW [132]. In estimates by Intel [133], the majority of the power consumption is expected to be due to memory and processor interconnection network. In Intel's prediction an improvement in the CPU energy efficiency to 100 GF/W by 2018 is assumed resulting in 10 MW of power for the CPUs of a system with a peak performance of 1 EF. This level of efficiency represents a 25-fold improvement over the IBM BQC efficiency. This level of improvement is in line with the historical trend [75], though significant technical challenges must be successfully addressed for the past trend to continue. For memory, 40 MW is estimated for a 1 EF peak system assuming memory is integrated with the processors in a memory in processor (MIP) [134,135] architecture. For interconnect, a power requirement of 5 MW/Gbps is estimated with a target of 0.1 byte/F 50 MW would be required for the interconnect [133]. For the interconnection network, this represents an improvement of about a factor of 10 compared to state-of-the-art today. Though a high-speed network clearly is needed, the extent to which network performance impacts application performance is subject to debate and also application as well as architecture dependent. In [136], it is shown that the nominally highest performing interconnection technology is not necessarily the most efficient, whereas in [137], the use of a low-latency high-bandwidth interconnection network is shown to improve the energy efficiency by close to a factor of 2, as shown in Figure 43.38. However, in this study, no detailed energy measurements are reported and energy consumption assumed independent of the interconnection network used and proportional to run-time.

Energy consumption of an Exa-scale system is perceived as one of the most serious challenges in realizing such a system [138]. Because of the great challenges in regard to power consumption for future high-end systems and no clear pathway, approaches taken in markets with traditional high emphasis on energy efficiency are now being considered also for HPC. In [139], the impact of using approaches used in the embedded market including instruction set simplification and alternate memory system designs is discussed. The potential benefits of new architectures on the energy efficiency of computations, also using the embedded market as a starting-point, is illustrated in [140] in which it is highlighted that a typical laptop CPU requires about 4000 times as much energy as an AISC for many operations,

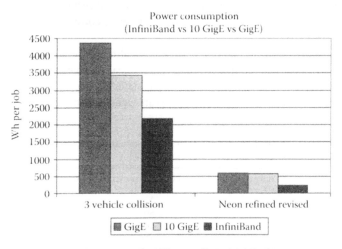

FIGURE 43.38 Comparison of energy efficiency for different interconnection networks for LS-DYNA on a 24-node cluster. (From Shainer, G. et al., LS-DYNA productivity and power-aware simulations in cluster environments. http://www.hpcadvisorycouncil.com/pdf/7th%20European%20LS-DYNA%20Users%20Conference.pdf)

whereas a DSP may require about 250 times the energy of an application-specific integrated circuit (ASIC) for comparable operations. The architecture proposed in [140] is claimed to only consume energy that is a small multiple of an ASIC design.

43.4 System Operations

Data center design and operations have evolved to a point where the inefficiencies in the infrastructure are relatively small and most potential energy gains from a facilities point of view are to be made by energy reuse. The component and platform industries have energy efficiency and environmental impact as one of their foremost concerns, a concern that has driven architecture and system design for a few years. One aspect of this concern is the trend to increased ability to adjust the operating state of systems according to the workload resulting in technologies such as Intel's SpeedStep [141] and AMD's PowerNow [142]. Underlying these technologies is the ability to control the CPU operating conditions wholly or in part. The power management has been structured into power management through power planes with all parts in a power plane having a common power feed and voltage. Within a power domain, operating conditions can still be controlled through what is commonly known as performance states. The Advanced Configuration and Power Interface (ACPI) [143] standard refers to these two aspects as C-states (processor power states) and P-states (processor performance states), respectively. The processor power states are characterized as one operating state (labeled C0) and a range of sleep states in which instructions are not executed. Depending on the parts being powered down, restoring execution state will take different amounts of time. Sleep states do not include a state requiring reboot. Table 43.10 [144] illustrates the times to active state from a few sleep states on Intel mobile CPUs.

With the emergence of multi-core processors, the number of power domains on a chip have been increasing, with Intel in its current generation six-core chips having a separate power domain for each core with the memory controller having its own power domain as well [145]. Within a power domain, CPU manufacturers have chosen to implement different number of C-states, a number that tend to evolve with product generations. The significance of the sleep states in regards to energy consumption in idle mode is illustrated by Figure 43.39 showing the power consumption in a typical active state. In a sleep state allowing for a very quick recovery to active state, the power for logic and local clocks can be eliminated reducing the power consumption to typically less than 50% of active power. CPU states of "deeper sleep" may imply shutting down clock distribution and sections of logic for reduced leakage currents and hence enabling an inactive core to reduce its power consumption to a small fraction of its active power, Figure 43.40.

With cores in active state, the performance is controlled by altering the clock frequency of the cores with different cores possibly operating at different frequencies, a feature enabled on recent CPUs from both AMD and Intel. AMD currently supports five different P-states from a low frequency of 800 MHz up to the maximum for the product [146,147]. Intel supports an even larger number of P-states. The increased

TABLE 43.10 Time to Active State from Sleep State for Some Mobile Intel CPUs

C-State	Typical Worst Case Exit Latency Time (μs)
C1	1
C3	80
C6	104
C7	109

Source: Intel., Intel® 64 and IA-32 architectures optimization reference manual, September, 2010. http://www.intel.com/Assets/PDF/manual/248966.pdf

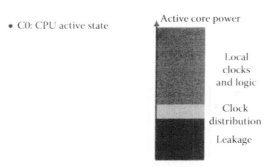

- C0: CPU active state

FIGURE 43.39 Typical power consumption for a core in active state. (From Shrout, R., Inside the Nehalem: Intel's new Core i7 microarchitecture, PC perspective, August 25, 2008. http://swfan.com/reviews/processors/inside-nehalem-intels-new-core-i7-microarchitecture/nehalem-power-management?page=199)

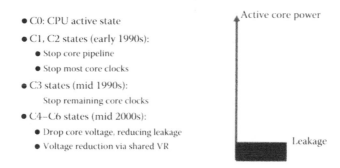

- C0: CPU active state
- C1, C2 states (early 1990s):
  - Stop core pipeline
  - Stop most core clocks
- C3 states (mid 1990s):
   Stop remaining core clocks
- C4–C6 states (mid 2000s):
  - Drop core voltage, reducing leakage
  - Voltage reduction via shared VR

FIGURE 43.40 Reduced power consumption using "deep" sleep power states on Intel CPUs. (From Shrout, R., Inside the Nehalem: Intel's new Core i7 microarchitecture, PC perspective, August 25, 2008. http://swfan.com/reviews/processors/inside-nehalem-intels-new-core-i7-microarchitecture/nehalem-power-management?page=199)

ability to control the power state and operating conditions on an increasing number of components on a chip also increases the complexity of control of the CPU leading to the introduction of a separate power control unit (PCU) on recent CPUs [145], Figure 43.41. Controlling the power states of the caches represent its own set of challenges with AMD using its SmartFetch technology [142] to store L1 and L2 cache content in the L3 cache to enable powering down L1 and L2 caches. A similar approach is used by Intel on their Nehalem CPUs, but on its Westmere generation CPUs, Intel is reported to use a special static random-access memory (SRAM) for saving cache content enabling all three levels of cache to be powered down [148]. A summary of the power management features on the current generation AMD Opteron CPUs can be found at [147].

To stimulate research into multi-core chip technology including power management Intel has produced the single-chip cloud computer (SCC) [149–151] that has 24 dual-core processors (total 48 cores) in six power domains with one additional power domain for the on-chip interconnection network and routers, and another power domain for the remaining parts of the chip, that is, eight power domains in total. Each dual-core processor has its own frequency control, but the cores on a processor do not have individual frequency control. Memory, I/O, and on-chip networks have their own independent frequency control. In all, there are 8 power domains and 28 frequency domains on the SCC, Figure 43.42. Figure 43.43 shows the power consumption of the chip under light and high load [150]. The effectiveness of the core power management is apparent with the cores in low-load state consuming less than 10%

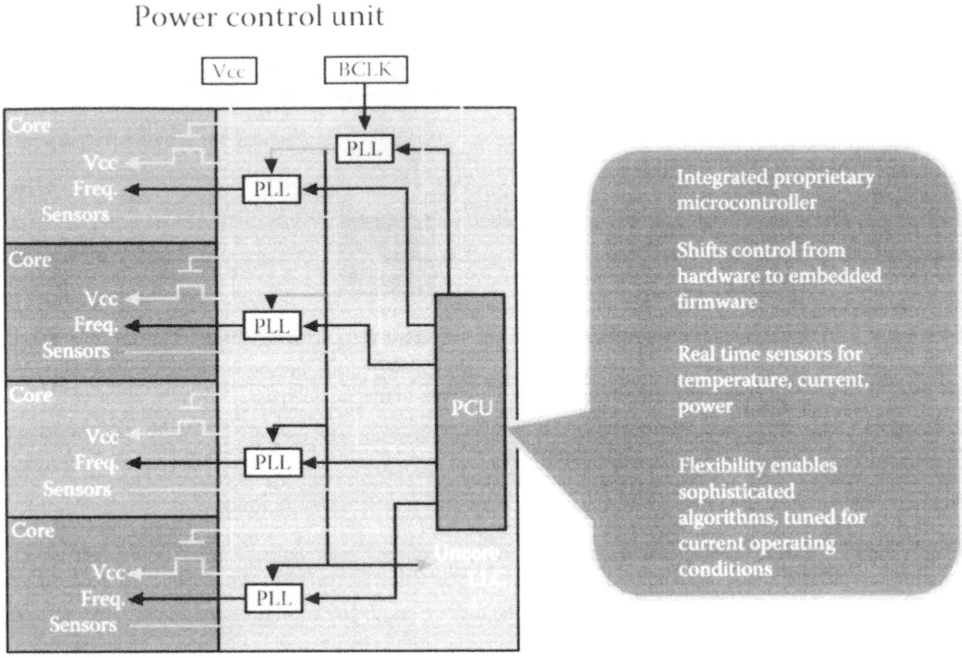

Power control unit

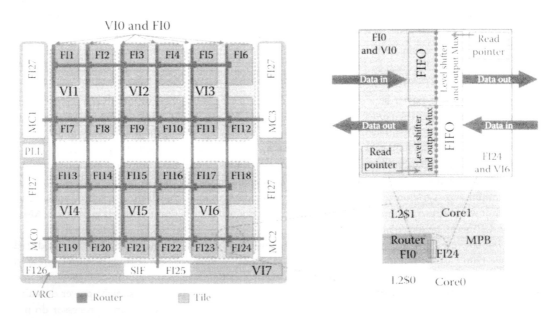

FIGURE 43.41 PCU schematic for recent Intel CPUs. (From Shrout, R., Inside the Nehalem: Intel's new Core i7 microarchitecture, PC perspective, August 25, 2008. http://swfan.com/reviews/processors/inside-nehalem-intels-new-core-i7-microarchitecture/nehalem-power-management?page=199)

FIGURE 43.42 Power and frequency domains on the SCC. (From Intel, Single-chip cloud computer—An experimental many-core processor from Intel Labs, *Intel Labs Single-Chip Cloud Computer Symposium*, March 16, 2010. http://communities.intel.com/servlet/JiveServlet/downloadBody/5075-102-1-8132/SCC_Sympossium_Mar162010_GML_final.pdf)

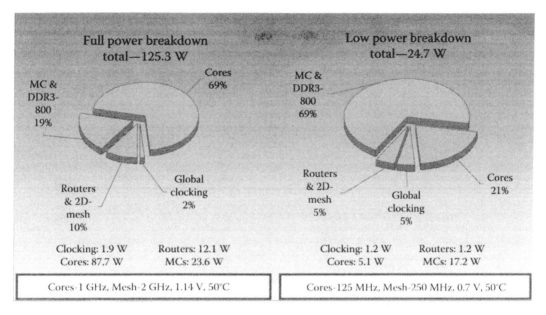

FIGURE 43.43 Power consumption of the Intel SCC under high and low load. (From Intel, Single-chip cloud computer—An experimental many-core processor from Intel Labs, *Intel Labs Single-Chip Cloud Computer Symposium*, March 16, 2010. http://communities.intel.com/servlet/JiveServlet/downloadBody/5075-102-1-8132/SCC_Sympossium_Mar162010_GML_final.pdf)

TABLE 43.11 Execution Time and Energy Consumption for a Sparse-Matrix Vector Multiplication on an Intel T7700 Laptop CPU

Core Frequency (GHz)	$T_{exec}(s)$	Energy (J)
0.8	6.74	50.21
1.2	4.53	57.21
1.6	4.50	85.09
2.0	4.46	116.99
2.4	4.45	155.75

Source: Keller, V., Optimal application-oriented resource brokering in a high performance computing grid, PhD thesis, EPFL, 2008. http://library.epfl.ch/theses/?nr=4221

of their high-load state. But Figure 43.43 also shows the validity of the concerns about memory power consumption of about 20% of the power in the high-load state but about 70% of the power in the low-load state. The power consumption by the memory only declines by about 20% from high to low load.

The interest in dynamic voltage and frequency scaling (DVFS) to gain energy efficiency is relatively recent for the HPC market but is common practice in the mobile market. For HPC applications, a 20%–25% gain in power consumption for a 3%–5% slowdown was reported in [152] using an automatic run-time procedure adjusting frequency to load. The tests were carried out on AMD CPUs. Similar predictions were made in [153]. Another test was carried out in [154] on an Intel Mobile CPU showing a potential energy saving of about a factor of 2.7 for about a 2% slowdown, Table 43.11.

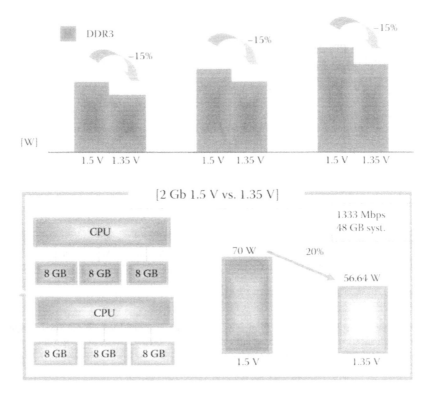

FIGURE 43.44 The energy benefit of low-voltage DDR3 memory. (From Kadivar, S.R., New memory technology: Evolving toward greener solutions. http://www.samsung.com/us/business/semiconductor/downloads/greeningPresentation.pdf)

In summary, load-related power management can result in significant energy savings. CPUs for the HPC market have an increasingly rich set of control possibilities to adjust CPU behavior to the application demands. How to exploit these features has only received modest interest from the HPC research community. Initiatives such as Intel's MIC [129] and SSC [149] will hopefully change that. A big problem though is the relatively large power consumption of memory and the still limited ability to control its power consumption. Memory is produced using the same basic technology as CPUs and hence the power and frequency scaling is similar. Low-voltage DDR3 memory that operates at 1.35 V instead of standard 1.5 V is estimated to reduce memory power consumption by about 15%–20% [155], Figure 43.44. Though the use of low-voltage dynamic random-access memory (DRAM) will improve energy efficiency a different memory architecture, or technologies will be necessary [131] as well as effective memory management from an energy perspective [90]. The difference in energy consumption by memory depending on its integration into the systems may be a factor of as much as 50 [139]. Embedded DRAM [156], eDRAM, used in many mobile devices but also in the IBM Power7 [157], may offer an energy savings of a factor of 2–4 or more [158]. An exciting development that seems to make good advances toward commercial reality [159] is that of the memristor [160,161] that has the potential to reduce memory power consumption, increase memory density up to 1 Tbit/cm^2, and speed by up to an order of magnitude.

The processor interconnection network, though currently not a dominating energy consumer for HPC systems, is predicted to become one. Today, like memory, interconnection networks are always on and the energy consumption fairly independent of load, Figure 43.45 [162]. However, the concern about energy efficiency of networks has also caused the network community to engage in work toward management of

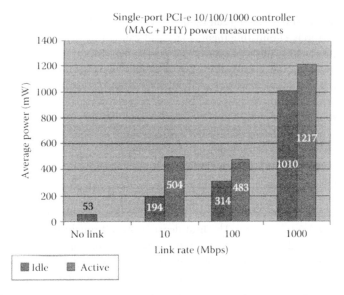

FIGURE 43.45 Ethernet power consumption in idle and active state. (From Hays, R., Active/idle toggling with 0BASE-x for energy efficient Ethernet, IEEE 802.3az Task Force, November 2007. http://grouper.ieee.org/groups/802/3/az/public/nov07/hays_1_1107.pdf)

networks from an energy perspective seeking to make energy consumption related to usage either through rate adjustment or through introducing sleep modes [163].

43.5 Software

Finally, software at various levels has a big role in improving delivered energy efficiency. Some specific examples of the efficiency of scientific applications measured as percentage of peak floating-point performance, typically in the 5%–30% or so range, are found in [164,165]. A highly optimized newly developed code to achieve scalability received a Gordon Bell Award for performance at the SC10 conference with an efficiency of 34% [166]. If the application is memory bandwidth limited this may be acceptable from an energy efficiency point of view, in particular if the CPUs can be controlled to operate with reduced power. However, we suspect that in most cases poor scalability, poor match between chosen algorithms and the architecture, or simply codes not written for efficiency of resource use or energy efficiency may be the source of poor efficiency. Rewriting codes with energy efficiency in mind may result in a significant payoff. As an example, improved software resulted in a server efficiency gain of 29% in 2009 at Akamai Technologies [167]. But as discussed in [165], new execution models may be required for significantly improved energy efficiency in addition to choosing algorithms based on their possible implementations in energy-efficient codes.

43.6 Summary

Data center energy demands have been rising much more rapidly than overall electric energy demands. In fact, the demands has risen with about 20%/year on average for the last decade causing many existing data centers to either have to refurbish existing facilities or acquire, refurbish or build new data centers. The rapidly increased energy demands and associated cooling have made energy-related capital and operating infrastructure cost exceed that of the IT equipment and led to large improvement in data center energy efficiency. The use of free cooling can significantly improve the energy efficiency by reducing the need for chillers, or entirely eliminating them, a fact that can both affect the location as well as design and operation of data centers. The efficiency of the power distribution system in data centers has also improved significantly with high-quality server power supplies having an efficiency in excess of 90% for a broad range of loads and a peak efficiency of about 95%. High-quality uninterruptible power supplies now reaches efficiencies of about 98% for load levels of 50% or more. But, in the case of HPC systems many centers do restrict the use of UPS to critical servers and networks and do not cover HPC systems by UPS. The improvements in data center design and operation, including raised inlet temperatures, has led to 80% or more of the energy being used by the IT systems.

Further improvement in energy efficiency will largely need to come from energy reuse and improved energy efficiency of the computer systems themselves. Energy reuse is a consideration in many data centers, in particular if there are nearby needs for energy, such as heating of buildings or for use in industrial processes. For reuse of energy in the form of hot water from data centers, the warmer the water is the more useful it tends to be. The heat densities have brought liquid cooling into rack rows in the form of liquid-cooled rack doors, or liquid cooling coils on top of racks or between racks, or direct liquid cooling of components or complete servers. Direct cooling has the possibility of generating the highest water temperatures of the liquid cooling options and hence the highest quality energy for reuse.

The energy efficiency of CMOS continues to improve rapidly, but since not all logic is actively used all the time, it is important to develop and use techniques to seek to make energy consumption related to the work carried out. Over the last few years, CPU designs have included abilities to control power states as well as performance through control of clock frequencies as means of making energy consumption increasingly related to workload. An increasing number of power domains on chips allow for independent control of many chip areas and an increasing number of power states enable different levels of (deep) sleep with corresponding savings in energy consumption. The control of clock rates, DVFS, enables optimization of energy consumption for workloads in that for some workloads, power consumption decreases more rapidly than execution time increases with reduced clock frequency and hence a lower clock rate would be beneficial from an energy point of view. At this time, the operating system makes use of some of these features, firmware makes use of some others, and few are accessible from applications. In addition to improved energy efficiency through controlling the CPU operating state, savings can also be made through simplified instruction sets reducing the complexity of the CPUs. Many instructions in the x86 instruction set are not used in a given application with a range of scientific applications using less than 30% of the instruction set.

The control of CPUs state from an energy perspective is the most advanced at this time. However, in recent years efforts have also been made to introduce load-related control of interconnection networks, in particular for Ethernet, for which both sleep mode as well as load-related data rates have been proposed. In the first standard for an energy-efficient Ethernet that was approved on September 30, 2010, 100 Mbps and 1 Gbps Ethernet chips are to transition into sleep mode when idle whereas for 10 Gbps chips a transition to lower data rates should take place under light or no load. The reason for not specifying a sleep mode for 10 Gbps is the potentially long time to return to active mode from sleep mode. The estimated saving across all uses of Ethernet had these energy-efficient features of the standard been in place is 5 TWh/year [168].

Memory is the second largest consumer of energy in most computer systems today and is expected to become the largest energy consuming subsystem in large computer systems of the future. At this time, there are no dynamic control features for memory and the power consumption in idle mode is only up to 20% lower than in active mode. Dynamic control of memory, or significantly reduced power consumption for memory through the use of new technology is required for achieving significantly improved energy efficiency of computer systems. Energy efficiency has been the focus for a long time in the mobile device market, including a focus on energy-efficient memory systems. Significant gains in energy efficiency are possible by a different memory system architecture bringing it closer to the CPU [140]. With improved tools for generating chip designs, the advantages of domain-specific designs may outweigh the increased costs due to limited volumes.

The largest potential in increased energy efficiency is in increased utilization of the hardware with many applications achieving efficiencies in the 5%–30% range based on fraction of peak floating-point capability. Though this measure is questionable since the CPU is a diminishing part of both the system capital cost and energy use, it is likely in today's systems that the fraction of peak memory bandwidth or network bandwidth is no higher. Hence, the opportunities for improved energy efficiency by improved algorithms and software, new architectures, and control of operating conditions are significant and need to be pursued vigorously.

References

1. U.K. Parliamentary Office of Science and Technology. October 2006. Carbon footprint of electricity generation. Postnote 268. http://www.parliament.uk/documents/post/postpn268.pdf
2. International Energy Agency. 2010. Key world energy statistics. http://www.iea.org/textbase/nppdf/free/2010/key_stats_2010.pdf
3. Pew Center on Global Climate Change. Electricity overview. http://www.pewclimate.org/technology/overview/electricity
4. International Energy Agency. July 2010. Energy technology perspectives—Scenarios and strategies to 2050. http://www.iea.org/techno/etp/
5. The Climate Group. 2008. SMART 2020: Enabling the low carbon economy in the information age. http://www.smart2020.org/_assets/files/02_Smart2020Report.pdf
6. J. A. Laitner and K. Ehrhardt-Martinez. February 2008. Information and communication technologies: The power of productivity. Report E081. American Council for an Energy Efficient Economy. http://www.aceee.org/pubs/e081.pdf
7. J. G. Koomey, C. Belady, M. Patterson, A. Santos, and K.-D. Lange. August 17, 2009. Assessing trends over time in performance, costs, and energy use for servers. Intel. http://www3.intel.com/assets/pdf/general/servertrendsreleasecomplete-v25.pdf
8. C. Belady. February 1, 2007. In the data center, power and cooling costs more than the IT equipment it supports. Electronics cooling. http://www.electronics-cooling.com/2007/02/in-the-data-center-power-and-cooling-costs-more-than-the-it-equipment-it-supports
9. C. L. Weber, J. G. Koomey, and H. S. Matthews. August 17, 2009. The energy and climate change impacts of different music delivery methods. Intel. http://download.intel.com/pressroom/pdf/cdsvsdownloadsrelease.pdf
10. C. Taylor and J. G. Koomey. February 14, 2008. Estimating energy use and greenhouse gas emissions of Internet advertising. Working Paper. http://evanmills.lbl.gov/commentary/docs/carbonemissions.pdf
11. Energy Information Administration, Department of Energy. 2016 levelized cost of new generation resources from the annual energy outlook 2010. http://www.eia.doe.gov/oiaf/aeo/pdf/2016levelized_costs_aeo2010.pdf
12. K. W. Roth, F. Goldstein, and J. Kleinman. 2002. *Energy Consumption by Office and Telecommunications Equipment in Commercial Buildings. Vol. I: Energy Consumption Baseline.* Reference No. 72895-00. Prepared by Arthur D. Little for U.S. Department of Energy, Washington, DC. http://www.eren.doe.gov/buildings/documents
13. Google. The Dalles, OR. Data Center. http://www.google.com/datacenter/thedalles/index.html
14. J. Markoff, S. Hansell. June 14, 2006. Hiding in plain sight, Google seeks more power. *New York Times.* http://www.nytimes.com/2006/06/14/technology/14search.html?_r=1
15. The Dalles Dam. http://en.wikipedia.org/wiki/The_Dalles_Dam#Specifications
16. Google. Hamina Data Center. http://www.google.com/datacenter/hamina

17. Revealed: Google's new mega data center in Finland. September 15, 2010. http://royal.pingdom.
    com/2010/09/15/googles-mega-data-center-in-finland
18. Energy year 2008—Electricity. January 22, 2009. Energiateollisuus. http://www.energia.fi/en/news/
    energy%20year%202008%20electricity.html
19. CSCS. Swiss National Supercomputing Centre. http://www.cscs.ch/491.0.html
20. Green Data Center Blog. January 9, 2009. Stanford Universities green data center efforts.
    http://www.greenm3.com/2009/01/stanford-universities-green-data-center-efforts.html
21. R. Miller. June 29, 2009. Microsoft to open two massive data centers. Data center knowl-
    edge. http://www.datacenterknowledge.com/archives/2009/06/29/microsoft-to-open-two-massive-
    data-centers
22. R. Miller. July 15, 2009. Google's chiller-less data center. Data center knowledge. http://www.
    datacenterknowledge.com/archives/2009/07/15/googles-chiller-less-data-center
23. C. Metz. July 16, 2009. Google data center born without chillers, Belgian free cooling. *The Register.*
    http://www.theregister.co.uk/2009/07/16/google_chillerless_data_center
24. M. G. Richard. March 5, 2010. Google move! Google invests $38.8 million in two North Dakota
    wind farms. http://www.treehugger.com/files/2010/05/google-invests-in-two-wind-farms-north-
    dakota-169-megawatts.php
25. Going green at Google. http://www.google.com/intl/en/corporate/green/114megawatt.html
26. K. Fehrenbacher. July 20, 2010. Google buys wind power, first deal for "Google energy." GigaOM.
    http://gigaom.com/cleantech/google-buys-wind-power-first-deal-for-google-energy
27. J. Eilperin. October 12, 2010. Google helps finance "superhighway" for wind power. http://www.
    washingtonpost.com/wp-dyn/content/article/2010/10/12/AR2010101202271.html
28. The Green Grid. February 16, 2007. Guidelines for energy-efficient data centers. http://www.
    thegreengrid.org/~/media/WhitePapers/Green_Grid_Guidelines_WP.ashx?lang=en
29. U.S. Environmental Protection Agency, Energy Star Program. August 2, 2007. Report to Congress
    on server and data center energy efficiency. Public Law 109-431. http://www.energystar.gov/ia/
    partners/prod_development/downloads/EPA_Datacenter_Report_Congress_Final1.pdf
30. The Green Grid. 2008. Green grid data center power efficiency metrics: PUE and DCiE.
    White Paper No 6. http://www.thegreengrid.org/~/media/WhitePapers/White_Paper_6_-_PUE_
    and_DCiE_Eff_Metrics_30_December_2008.ashx?lang=en
31. Google. Data center efficiency measurements. http://www.google.com/corporate/green/datacenters/
    measuring.html
32. HP's Wynyard data center and its unique cooling setup. http://www.datacenterdynamics.com/ME2/
    dirmod.asp?sid=&nm=&type=news&mod=News&mid=9A02E3B96F2A415ABC72CB5F516
    B4C10&tier=3&nid=A351A7B4E4DF49D3AA5DFD7ED8701B17
33. R. Miller. October 20, 2008. Microsoft: PUE of 1.22 for data center containers. http://www.
    datacenterknowledge.com/archives/2008/10/20/microsoft-pue-of-122-for-data-center-containers
34. S. Madera. January 29, 2007. How do I cool high-density racks? http://searchdatacenter.techtarget.
    com/feature/How-do-I-cool-high-density-racks
35. Solutions—Hot air return. http://cold-aisle-containment.co.uk/solutions/hot-air-return/index.html
36. M. Stansberry and M. Fontecchio. June 29, 2009. Hot-aisle vs. cold-aisle containment. Liebert
    and APC Face off. http://searchdatacenter.techtarget.com/news/1360462/Hot-aisle-vs-cold-aisle-
    containment-Liebert-and-APC-face-off
37. SGI. Next generation data center infrastructure. ICE Cube Modular Data Center. Overview and
    FEATURES. http://www.sgi.com/pdfs/4172.pdf
38. S. Christensen. June 15, 2009. Data center containers. http://www.datacentermap.com/blog/
    datacenter-container-55.html
39. U.S. Patent Application. December 30, 2003. Google. Modular data center. http://patft.
    uspto.gov/netacgi/nph-Parser?Sect1=PTO1&Sect2=HITOFF&d=PALL&p=1&u=%2Fnetahtml%

2FPTO%2Fsrchnum.htm&r=1&f=G&l=50&s1=7,278,273.PN.&OS=PN/7,278,273&RS=PN/7, 278,273

40. C. Thucker. October 25, 2007. Microsoft research. Rethinking data centers. http://netseminar. stanford.edu/seminars/10_25_07.ppt

41. R. Miller. December 2, 2008. Microsoft goes all-in on container data centers. Data center knowledge. http://www.datacenterknowledge.com/archives/2008/12/02/microsoft-goes-all-in-on-container-data-centers

42. R. Miller. October 1, 2009. Microsoft Chicago: The road ahead. http://www.datacenterknowledge. com/inside-microsofts-chicago-data-center/microsoft-chicago-the-road-ahead

43. R. H. Katz. February 2009. Tech titans building boom. *IEEE Spectrum*. http://spectrum.ieee.org/ green-tech/buildings/tech-titans-building-boom

44. T. Takken. February 2004. Blue gene/L power, packaging and cooling. http://www.physik.uni-regensburg.de/studium/uebungen/ws0405/scomp/BGWS_03_PowerPackagingCooling.pdf

45. V. Pel. September 2010. Energy efficient aspects in Cray supercomputers. http://www.ena-hpc.org/ 2010/talks/EnA-HPC2010-Pel-Energy_Efficiency_Ascpects_in_Cray_Supercomputers.pdf

46. D. Graham and M. Laatsch. 2009. Meeting the demands of computer cooling with superior efficiency. http://www.cray.com/Assets/PDF/products/xt/whitepaper_ecophlex.pdf

47. Bull Cabinet Door. October 2010. http://www.bull.com/extreme-computing/download/S-HPCwaterdoor-en2.pdf

48. M. Michelotto. Workshop infrastructure tier 2. http://www.infn.it/CCR/riunioni/presentazioni/ pres_ott_05/michelotto_summary_workshop.pdf

49. American Society of Heating, Refrigerating and Air-Conditioning Engineers. ASHRAE. www.ashrae.org

50. ASHRAE. 2005. Datacom equipment power trends and cooling applications. http://esdc.pnl.gov/ SC07_BOF/SC07BOF_McCann.pdf

51. IBM. February 2010. IBM unveils new Power7 systems to manage increasingly data-intensive services. http://www-03.ibm.com/press/us/en/pressrelease/29315.wss

52. M. Snir. Blue waters. Power7 for scientific computing. http://www.ncsa.illinois.edu/BlueWaters/ pdfs/snir-power7.pdf

53. M. Feldman. June 10, 2010. Cray sets sight on cascade supercomputer. Exascale milestone. HPC Wire. http://www.hpcwire.com/features/Cray-Sets-Sights-On-Cascade-Supercomputer-Exascale-Milestone-96066794.html

54. Liquid cooling for servers. http://www.boston.co.uk/technical/2009/12/liquid-cooling-for-servers. aspx

55. Data Center HPC. http://www.eurotech.com/en/hpc

56. 2008 ASHRAE environmental guidelines for datacom equipment. http://tc99.ashraetcs.org/ documents/ASHRAE_Extended_Environmental_Envelope_Final_Aug_1_2008.pdf

57. R. Miller. September 24, 2007. Data center cooling set-points debated. Data center knowledge. http://www.datacenterknowledge.com/archives/2007/09/24/data-center-cooling-set-points-debated

58. J. H. Bean and D. L. Moss. Energy impact of increased server inlet temperature. Optimal temperature operation. http://i.dell.com/sites/content/business/solutions/whitepapers/en/Documents/dci-energy-impact-of-increased-inlet-temp.pdf

59. M. K. Patterson, M. Meakins, D. Nasont, P. Pusuluri, W. Tschudi, G. Bell, R. Schmidt, K. Schneebeli, T. Brey, M. McGraw, W. Vinson, and J. Gloeckner. July, 2009. Energy efficiency through the integration of information and communications technology management and facilities controls. IPACK. http://download.intel.com/pressroom/archive/reference/IPACK2009.pdf

60. T. Brunschwiler, B. Smith, E. Ruetsche, and B. Michel. May 2009. Toward zero-emission data centers through direct reuse of thermal energy. *IBM Journal of Research and Development*. 53(3),

11:1–11:13.   http://ieeexplore.ieee.org/Xplore/login.jsp?url=http%3A%2F%2Fieeexplore.ieee.org%2Fiel5%2F5288520%2F5429013%2F05429024.pdf%3Farnumber%3D5429024&authDecision=-203

61. White Paper 29. September 22, 2010. ERE: A metric for measuring the benefit of reuse of energy from a data center. The Green Grid. http://www.thegreengrid.com/~/media/WhitePapers/ERE_WP_101510_v2.ashx?lang=en

62. N. Rasmussen and J. Spitaels. A quantitative comparison of high efficiency AC vs. DC power distribution for data centers. White Paper 127. APC. http://www.apcmedia.com/salestools/NRAN-76TTJY_R2_EN.pdf

63. M. Ton, B. Fortenbery, and W. Tschudi. March 2008. DC power for improved data center efficiency. LBNL. Ecos Consulting. EPRI. http://hightech.lbl.gov/documents/DATA_CENTERS/DCDemoFinalReport.pdf

64. Lawrence Berkeley National Laboratory. April 2, 2008. The Green Grid peer review of "DC power for improved data center efficiency". White Paper 12. http://www.thegreengrid.org/~/media/WhitePapers/WhitePaper12LBNLPeerReview050908.ashx?lang=en,http://www.apcmedia.com/salestools/SADE-5TNRLG_R6_EN.pdf

65. N. Rasmussen. 2007. AC vs DC power distribution for data centers. White Paper 63. http://www.apcmedia.com/salestools/SADE-5TNRLG_R6_EN.pdf

66. M. Ton and B. Fortenbury. December 2005. High performance buildings: Data centers, uninterruptible power supplies (UPS). LBNL. Ecos Consulting. EPRI Solutions. http://hightech.lbl.gov/documents/UPS/Final_UPS_Report.pdf

67. U.S. Patent Office Application. June 1, 2007. Data center uninterruptible power distribution architecture.   http://appft1.uspto.gov/netacgi/nph-Parser?Sect1=PTO1&Sect2=HITOFF&d=PG01&p=1&u=%2Fnetahtml%2FPTO%2Fsrchnum.html&r=1&f=G&l=50&s1=%2220080030078%22.PGNR.&OS=DN/20080030078&RS=DN/20080030078

68. HP. August 2010. Technologies in the HP Blade system c7000 enclosure. http://h20000.www2.hp.com/bc/docs/support/SupportManual/c00816246/c00816246.pdf

69. EPRI. 80 Plus verification and testing report. http://www.supermicro.com/products/powersupply/80PLUS/80PLUS_PWS-920P-1R.pdf

70. 80-Plus. http://www.plugloadsolutions.com/80PlusPowerSupplies.aspx

71. D. Sartor and M. Wilson. 2010. Money for research, not energy bills: Finding energy and cost savings in high-performance computer facility designs. *Computing in Science & Engineering.* 12(6), 11–22. IEEE Computer Society. http://ieeexplore.ieee.org/xpls/abs_all.jsp?arnumber=5624675&tag=1

72. G. E. Moore. April 1965. Cramming more components onto integrated circuits. *Electronics.* 38(8), 114–117. ftp://download.intel.com/research/silicon/moorespaper.pdf

73. A. Petitet, R. C. Whaley, J. Dongarra, and A. Cleary. HPL—A portable implementation of the high-performance Linpack benchmark for distributed-memory computers. http://www.netlib.org/benchmark/hpl

74. Top500 Supercomputer sites. www.top500.org

75. J. G. Koomey, S. Berard, M. Sanchez, and H. Wong. August 17, 2009. Assessing in the trends in the electrical efficiency of computation over time. Intel. http://download.intel.com/pressroom/pdf/computertrendsrelease.pdf

76. W. D. Nordhaus. March 2007. Two centuries of productivity growth in computing. *Journal of Economic History.* 67(1): 128–159. http://nordhaus.econ.yale.edu/nordhaus_computers_jeh_2007.pdf

77. Standards Performance Evaluation Corporation. http://www.spec.org

78. K. G. Brill. 2008. Findings on data center energy consumption growth may already exceed EPA's prediction through 2010! The Uptime Institute. http://uptimeinstitute.org/content/view/155/147

79. B. Schultz. March 12, 2008. Data-center energy consumption: Worse than we thought? *NetworkWorld.* http://www.networkworld.com/community/node/25913

80. K. G. Brill. 2007. The invisible crises in the data center: The economic meltdown of Moore's law. The Uptime Institute. http://www.uptimeinstitute.org/wp_pdf/(TUI3008)Moore'sLawWP_080107.pdf
81. PDC—Center for High Performance Computing. www.pdc.kth.se
82. J. Shalf, S. Kamil, E. Strohmaier, and D. Bailey. November 14, 2007. Power efficiency and the top 500. SC2007. Reno, Nevada. http://www.nersc.gov/projects/SDSA/reports/uploaded/Top500PowerNov14SC07.pdf
83. H. Simon. October 6, 2010. Exascale challenges for the computational science community. Oklahoma Supercomputing Symposium. http://symposium2010.oscer.ou.edu/oksupercompsymp2010_talk_simon_20101006.pdf
84. Annual Energy Review 2009. August 19, 2010. http://www.eia.doe.gov/aer/pdf/aer.pdf
85. *Behavior, Energy & Climate Change Conference.* November 16–19. Eco-technology: Delivering efficiency & innovation. Sacramento, CA. http://piee.stanford.edu/cgi-bin/docs/behavior/becc/2008/presentations/18-4C-01-Eco-Technology_-_Delivering_Efficiency_and_Innovation.pdf
86. L. Johnsson, D. Ahlin, and J. Wang. August 15–18, 2010. The SNIC/KTH PRACE prototype: Achieving high energy efficiency with commodity technology without acceleration. *The International Conference on Green Computing.* IEEE Computer Society. http://www.computer.org/portal/web/csdl/doi/10.1109/GREENCOMP.2010.5598259
87. X. Fan, W.-D. Weber, and L. Barroso. June 2007. Power provisioning for a warehouse-sized computer. *ISCA'07.* San Diego, CA, pp. 13–23. ACM. http://portal.acm.org/citation.cfm?id=1250665&CFID=948065&CFTOKEN=70566076
88. L. A. Barroso and U. Hölzle. 2009. The data center as a computer: An introduction to the design of warehouse-scale machines. Morgan & Claypool Publishers. http://www.morganclaypool.com/doi/pdf/10.2200/S00193ED1V01Y200905CAC006
89. B. Kerby. May 23–25, 2006. Delivering performance while conserving power. *Windows Hardware Engineering Conference,* 2006. Seattle, WA. http://www.microsoft.com/whdc/winhec/pres06.mspx
90. L. A. Barroso and U. Holzle. December 2007. The case for energy-proportional computing. *IEEE Computer.* 40(12), 33–37. http://ieeexplore.ieee.org/xpls/abs_all.jsp?arnumber=4404806&tag=1
91. S. Vangal, J. Howard, G. Ruhl, S. Dighe, H. Wilson, J. Tschanz, D. Finan, P. Iyer, A. Singh, T. Jacob, S. Jain, S. Venkataraman, Y. Hoskote, and N. Borkar. February 11–15, 2007. An 80-tile 1.28 Tflops network-on-chip in 65 nm CMOS. *IEEE Solid-States Circuits Conference,* San Francisco, CA, pp. 98–99. http://ieeexplore.ieee.org/xpl/freeabs_all.jsp?arnumber=4242283
92. Intel. http://ark.intel.com/ProductCollection.aspx?codeName=23349
93. Intel CPU X5355 specification. http://ark.intel.com/Product.aspx?id=28035
94. J. Shalf, D. Donofrio, L. Oliker, and M. Wehner. April 2009. Green flash: Application driven system design for power efficient HPC. *Salishan Conference on High-Speed Computing.* Salishan, Gleneden Beach, OR. http://www.nersc.gov/projects/SDSA/reports/uploaded/SalishanGreenFlash_Shalf.pdf
95. International Technology Roadmap for Semiconductors. http://www.itrs.net
96. A. L. Shimpi. June 14, 2004. AMD Begins tape-out of first dual-core Opteron. AnandTech. http://www.anandtech.com/show/1351
97. IBM. September 9, 2004. IBM unveils second-generation AMD Opteron-based server with support for dual-core processor specification. http://www-03.ibm.com/press/us/en/pressrelease/7285.wss
98. E. Bangeman. September 28, 2004. Dual-core PowerPC processor announced. *Ars Technica.* http://arstechnica.com/old/content/2004/09/4246.ars
99. T. Smith. February 2, 2004. First 65 nm IBM PowerPC chip to be dual-core. *The Register.* http://www.theregister.co.uk/2004/02/02/first_65nm_ibm_powerpc_chip
100. Intel. April 18, 2005. Intel dual-core processor based platforms. http://www.intel.com/pressroom/kits/pentiumee/index.htm
101. IBM Power4. http://en.wikipedia.org/wiki/POWER4

102. nVidia. September 1, 2009. nVidia's next generation CUDA computer architecture. White Paper. http://www.nvidia.com/content/PDF/fermi_white_papers/NVIDIA_Fermi_Compute_ Architecture_Whitepaper.pdf

103. nVidia Tesla. http://en.wikipedia.org/wiki/Nvidia_Tesla

104. AMD FireStream. http://en.wikipedia.org/wiki/AMD_FireStream

105. ARM. ARM—The architecture for the digital world. http://www.arm.com/products/processors/ index.php

106. Intel. Intel atom processor. http://www.intel.com/technology/atom

107. Texas Instruments. November 9, 2010. Texas Instruments, TMS320C6678 multi-core fixed and floating-point digital signal processor. http://focus.ti.com/docs/prod/folders/print/tms320c6678. html

108. ClearSpeed. September 2010. ClearSpeed CSX700 floating-point processor datasheet. http://www. clearspeed.com/products/documents/CSX700_Datasheet_Rev1D.pdf

109. J. Niccolai. October 7, 2010. Will web-scale servers find a role in the enterprise? *NetworkWorld.* http://www.networkworld.com/news/2010/100710-web-scale-servers.html?pid=165

110. J. Brodkin. October 25, 2010. 25 New IT companies to watch. *NetworkWorld.* http://www. networkworld.com/news/2010/102210-25-tech-startups.html?page=1

111. Data Center Performance, Cell phone power. http://www.calxeda.com

112. Marvell adopts quad-core ARM for low power servers. November 16, 2010. http://www.eetindia.co. in/ART_8800626016_1800001_NT_90257de7.HTM

113. Marvell. Marvell, Armada XP. http://www.marvell.com/products/processors/embedded/armada_xp

114. SeaMicro, The SM 10000 high density, low power server. http://www.seamicro.com/?q=node/38

115. Deliverable D8.3.2 partnership for advanced computing in Europe. June 25, 2010. Final Technical Report and Architecture Proposal. http://www.prace-project.eu/documents/public-deliverables/ d8-3-2-extended.pdf

116. IBM. IBM System Blue Gene. http://www-03.ibm.com/systems/deepcomputing/solutions/bluegene

117. Blue Gene. http://en.wikipedia.org/wiki/Blue_Gene

118. The Green500 list—November 2007. http://www.green500.org/lists/2007/11/top/list.php?from=1& to=100

119. The Green500 list—November 2010. http://www.green500.org/lists/2010/11/top/list.php?from=1& to=100

120. Cell (Microprocessor). http://en.wikipedia.org/wiki/Cell_(microprocessor)

121. IBM. The Cell Project at IBM Research. http://www.research.ibm.com/cell

122. Juelich. February 9–10, 2009. Network specification and software data structures for the eQPACE Architecture. http://www.fz-juelich.de/jsc/juice/eQPACE_Meeting

123. SiCortex. http://en.wikipedia.org/wiki/SiCortex

124. M. Wehner, L. Oliker, and J. Shalf. May 2008. Towards ultra-high resolution models of climate and weather. *International Journal of High Performance Computing Applications.* 22(2), 49–165. http://portal.acm.org/citation.cfm?id=1361723

125. D. Donofrio, L. Oliker, J. Shalf, M. F. Wehner, C. Rowen, J. Krueger, S. Kamil, and M. Mohiyuddin. November 2009. Energy-efficient computing for extreme-scale science. *IEEE Computer.* 42(11), 62–71. http://www.tensilica.com/products/xtensa-customizable.htm

126. Tensilica. Xtensa LX3—Customizable DPU. http://www.tensilica.com/uploads/pdf/LX3.pdf

127. T. R. Halfhill. November 30, 2009. Tensilica Tweaks Xtensa, Xtensa LX2 and Xtensa 8 cores boost performance, tweaks power. Microprocessor Report. In-Stat. http://www.tensilica.com/ uploads/pdf/Tensilica-LX3_Reprint.pdf

128. H. Simon. October 10, 2008. The Greening of HPC—Will power consumption become the limiting factor for future growth in HPC? http://www.hpcuserforum.com/presentations/Germany/ EnergyandComputing_Stgt.pdf

129. Intel. May 31, 2010. Intel unveils new product plans for high-performance computing. http://www.intel.com/pressroom/archive/releases/20100531comp.htm

130. J. Hruska. November 17, 2010. Intel demos Knights Ferry development platform, Tesla scores with Amazon. Hot Hardware. http://hothardware.com/News/Intel-Demos-Knights-Ferry-Development-Platform-Tesla-Scores-With-Amazon/

131. DARPA Information Processing Techniques Office and Air Force Research Laboratory. September 28, 2008. ExaScale computing study: Technology challenges in achieving ExaScale systems. http://www.er.doe.gov/ascr/Research/CS/DARPA%20exascale%20-%20hardware%20(2008).pdf

132. H. Simon, T. Zacharia, and R. Stevens. Spring 2007 (DoE E^3 Report). Modeling and simulation at the exascale for energy and the environment. Office of Science, The U.S. Department of Energy. http://www.er.doe.gov/ascr/ProgramDocuments/Docs/TownHall.pdf

133. B. Camp. April 21–24, 200. Exa-scale ambitions: What, me worry? *The Salishan Conference on High-Speed Computing.* Salishan. Gleneden Beach, OR. http://www.lanl.gov/orgs/hpc/salishan/pdfs/Salishan%20slides/Camp2.pdf

134. P. M. Kogge, T. Sunaga, E. Retter et al. March 1995. Combined DRAM and logic chip for massively parallel applications. *16th IEEE Conference on Advanced Research in VLSI,* Raleigh, NC. pp. 4–16. IEEE Computer Society Press # PR07047. http://www.cse.nd.edu/~kogge/reports.html

135. P. M. Kogg. September 1996. Processor-in-memory (PIM) based architectures for petaflops potential massively parallel processing. Final Report CSE TR-962. Notre Dame. http://www.cse.nd.edu/~kogge/reports.html

136. T. Hoeffler. 2010. Software and hardware techniques for power-efficient HPC networking. *Computing in Science & Engineering.* 12(6), 30–37. IEEE Computer Society. http://ieeexplore.ieee.org/xpls/abs_all.jsp?arnumber=5506076&tag=1

137. G. Shainer, T. Liu, J. Liberman, J. Layton, O. Celebioglu, S. A. Schultz, J. Mora, and D. Cownie. LS-DYNA Productivity and power-aware simulations in cluster environments. http://www.hpcadvisorycouncil.com/pdf/7th%20European%20LS-DYNA%20Users%20Conference.pdf

138. P. M. Kogge. September 22–23, 2009. Exa-scale computing embedded style. *13th Annual Workshop. High Performance Embedded Computing,* Lexington, MA. http://www.ll.mit.edu/HPEC/agendas/proc09/Day1/S1_0955_Kogge_presentation.ppt

139. D. W. Jensen and A. F. Ridrigues. 2010. Embedded systems and exa-scale computing. *Computing in Science & Engineering.* 12(6), 20–29. IEEE Computer Society. http://ieeexplore.ieee.org/xpls/abs_all.jsp?arnumber=5506076&tag=1

140. W. J. Dally, J. Balfour, D. Black-Shaffer, J. Chen, R. C. Harting, V. Parikh, J. Park, and D. Sheffield. 2008. Efficient embedded computing. *IEEE Computer.* 41(7), 27–32. http://ieeexplore.ieee.org/xpls/abs_all.jsp?arnumber=4563875&tag=1

141. Intel. August 3, 2009. Using enhanced Intel SpeedStep® features in HPC clusters. http://software.intel.com/en-us/articles/using-enhanced-intel-speedstep-features-in-hpc-clusters/?wapkw=(speedstep)

142. S. Troia. March 29, 2010. You down with AMD-P? http://blogs.amd.com/developer/2010/03/29/you-down-with-amd-p

143. Advanced configuration and power interface specification. http://www.acpi.info/DOWNLOADS/ACPIspec40a.pdf

144. Intel. September 2010. Intel® 64 and IA-32 architectures optimization reference manual. http://www.intel.com/Assets/PDF/manual/248966.pdf

145. R. Shrout. August 25, 2008. Inside the Nehalem: Intel's new Core i7 microarchitecture. PC perspective. http://swfan.com/reviews/processors/inside-nehalem-intels-new-core-i7-microarchitecture/nehalem-power-management?page=199

146. AMD. April 22, 2010. BIOS and kernel developer's guide (BKDG) for AMD family 10h processors. http://support.amd.com/us/Processor_TechDocs/31116.pdf

147. AMD. AMD Opteron series platform features and benefits. http://www.amd.com/us/products/server/processors/6000-series-platform/pages/6000-series-features.aspx

148. A. L. Shimpi. February 3, 2010. New Westmere details emerge: Power efficiency and 4/6 core plans. AnandTech. http://www.anandtech.com/show/2930

149. Intel. Single-chip cloud computer. http://techresearch.intel.com/ProjectDetails.aspx?Id=1

150. Intel. March 16, 2010. Single-chip cloud computer—An experimental many-core processor from Intel Labs. *Intel Labs Single-chip Cloud Computer Symposium*, http://communities.intel.com/servlet/JiveServlet/downloadBody/5075-102-1-8132/SCC_Sympossium_Mar162010_GML_final.pdf

151. Intel Labs. 2010. The SSC platform overview. http://techresearch.intel.com/spaw2/uploads/files/SCC_Platform_Overview.pdf

152. C. H. Hsu and W.-C. Feng. 2005. A power-aware run-time system for high-performance computing. *Proceedings of the 2005 ACM/IEEE Conference on Supercomputing.* http://portal.acm.org/citation.cfm?id=1105766

153. Y. Ding, K. Malkowski, P. Raghavan, and M. Kandemir. April 14–18, 2008. Towards energy efficient scaling of scientific codes. *IEEE International Symposium on Parallel and Distributed Processing*, pp. 1–8. Miami, FL. http://www.cse.psu.edu/~yding/hppac_final.pdf

154. V. Keller. 2008. Optimal application-oriented resource brokering in a high performance computing grid. PhD thesis. EPFL. http://library.epfl.ch/theses/?nr=4221

155. S. R. Kadivar. New memory technology: Evolving toward greener solutions. http://www.samsung.com/us/business/semiconductor/downloads/greeningPresentation.pdf

156. Physorg.com. February 14, 2007. IBM reveals breakthrough eDRAM memory technology. http://www.physorg.com/news90661936.html

157. J. Stokes. September 1, 2009. IBM's 8-core Power7: Twice the muscle, half the transistors. *Ars Technica.* http://arstechnica.com/hardware/news/2009/09/ibms-8-core-power7-twice-the-muscle-half-the-transistors.ars

158. N. When and S. Hei. February 23–26, 1998. Embedded DRAM architectural trade-offs. *Conference on Design, Automation and Test in Europe*, pp. 704–708. Paris, France. http://citeseerx.ist.psu.edu/viewdoc/download?doi=10.1.1.59.7390&rep=rep1&type=pdf

159. E. Bauley. August 31, 2010. HP and Hynix—Bringing the memristor to market in next-generation memory. http://h30507.www3.hp.com/t5/blogs/blogarticleprintpage/blog-id/datacentral/article-id/67

160. D. B. Strukov, G. S. Snider, D. R. Stewart, and R. S. Williams. May 1, 2008. The missing memristor found. *Nature.* 453, 80–83. http://www.nature.com/nature/journal/v453/n7191/full/nature06932.html

161. R. C. Johnson. April 30, 2008. "Missing link" memristor created: Rewrite the textbooks. *EETimes.* http://www.eetimes.com/General/DisplayPrintViewContent?contentItemId=4076910

162. R. Hays. November 2007. Active/idle toggling with 0BASE-x for energy efficient Ethernet. *IEEE 802.3az Task Force.* http://grouper.ieee.org/groups/802/3/az/public/nov07/hays_1_1107.pdf

163. P. Patel-Predd. May 2008. Energy-efficient Ethernet. *IEEE Spectrum.* http://spectrum.ieee.org/computing/networks/energyefficient-ethernet

164. L. Oliker, A. Canning, J. Carter, C. Iancu, M. Lijewski, S. Kamil, J. Shalf, H. Shan, E. Strohmaier, S. Ethier, and T. Goodale. May 26–30, 2007. Scientific application performance on candidate petascale platforms. *IEEE International Parallel & Distributed Processing Symposium.* Long Beach, CA. http://crd.lbl.gov/~oliker/papers/ipdps07.pdf

165. R. Murphy, T. Sterling, and C. Dekate. 2010. Advanced architectures and execution models to support green computing. *Computing in Science & Engineering.* 12(6), 8–47. IEEE Computer Society. http://ieeexplore.ieee.org/xpls/abs_all.jsp?arnumber=5506076&tag=1

166. A. Rahimiari, I. Lashuk, S. Veerapaneni, A. Chandramowlishwaran, D. Malhotra, L. Moon, R. Sampath, A. Shringarpure, J. Vetter, R. Vuduc, D. Zorin, and G. Biros. 2010. Petascale direct

numerical simulation of blood flow on 200K cores and heterogeneous architectures. *Proceedings of the 2010 ACM/IEEE International Conference on High Performance Computing, Networking, Storage and Analysis*, pp. 1–11. New Orleans, LA. http://portal.acm.org/citation.cfm?id=1884648

167. Akamai Technologies. Akamai environmental sustainability initiative. http://www.akamai.com/dl/sustainability/Environmental_Sustainability.pdf

168. Lightwave. October 5, 2010. IEEE ratifies new 802.3az standard to reduce network energy footprint. http://www.lightwaveonline.com/education/news/IEEE-ratifies-new-8023az-standard-to-reduce-network-energy-footprint--104392129.html?cmpid=EnlDirectOctober72010

44

Evaluating Performance, Power, and Cooling in High-Performance Computing (HPC) Data Centers

Jeffrey J. Evans
Purdue University

Sandeep Gupta
Arizona State University

Karen L. Karavanic
Portland State University

Andres Marquez
Pacific Northwest National Laboratory

Georgios Varsamopoulos
Arizona State University

44.1 Introduction

Large-scale high-performance computing (HPC) systems have historically been designed and deployed to assist the scientific community with discovery. These systems have evolved from strictly proprietary monolithic designs to include commodity-like networks-of-workstations (NOWs) called "clusters." Clusters outwardly resemble machines found in traditional information technology (IT) data centers; however, HPC data centers tend to be extremely dense, and almost always incorporate state-of-the-art interconnection and storage technologies to counteract speed-of-light latencies associated with data movement. Therefore, HPC systems are generally more complex than commercial IT systems, requiring new mechanisms to handle increased reliability risk. These factors, high density and increased reliability risk, invariably translate into increased power requirements, presenting new challenges and opportunities in the areas of power and thermal management.

While HPC systems' computational capacity continues to grow beyond petascale (10^{15} floating point operations per second), so does the need for energy to achieve this performance, both for power and cooling. Under these conditions, the operational objectives involving sustained systemic performance and energy efficiency can become conflicted. Users of HPC resources (scientists) wish for faster computational throughput on larger problems while policy makers have mandated operational reductions in energy consumption. Unfortunately, the problems associated with minimizing energy consumption in scientific HPC data centers while maximizing the throughput of work are not well understood. This is further complicated by the rates of change in hardware, software, and heating, ventilating, and air conditioning (HVAC) components. It is not unusual for typical HPC hardware to be replaced every 3 years while system software components are often maintained, updated, or replaced almost weekly.

When comparing HPC systems to traditional IT data centers, the recommended energy saving and power budget-capping techniques such as Uptime Institute's Best Practices have limited value in an HPC environment. For example, there is almost no opportunity to deactivate idling HPC compute and storage resources, since these systems almost always have jobs (executable programs) waiting in a queue while other jobs run on the machine. Similarly, resource virtualization provides little to no direct energy benefit for compute and storage resources.

Therefore, the fundamental challenge to successful "green" HPC computing is understanding, then balancing the notion of energy efficiency (work performed per joule of energy consumed) with system performance (work performed per unit of time) and application performance (time to solution). One of the challenges involves the determination of precise models and metrics for describing energy consumption, the work produced by the machine, and their interrelationships. Other challenges involve the collection of measurements on the system, what information to collect, where it is collected, and how much, without unduly affecting normal operation. Once these data are collected, it must be analyzed and placed in proper perspective in terms of true energy savings in meaningful units, and impact on system and application performance.

The development of models and interfacing of systems software, such as operating systems, job schedulers, and room controls continues to evolve through experimentation with smaller systems instrumented to collect environmental information. Our understanding of user-application behavior, particularly large scientific parallel applications, and their interactions with machine subsystems and other applications continues to be underdeveloped. As a result, standardized interfaces between the machine, its subsystems, and its operating environment (the room, HVAC control system, users) do not yet exist.

The remaining sections of this chapter explore current research focused on developing our understanding of the interrelationships involved with HPC performance and energy management. Section 44.2 explores data center instrumentation, measurement, and performance analysis techniques, followed by a section focusing on work in data center thermal management and resource allocation. This is followed by an exploration of emerging techniques to identify application behavioral attributes that can provide clues and advice to HPC resource and energy management systems for the purpose of balancing HPC performance and energy efficiency.

44.2 HPC Data Center Instrumentation, Data Collection, and Performance Analysis

Recent studies conclude that without widespread adoption of improved energy efficiency practices, total energy consumption is expected to double in the 5 year period from 2006 to 2011 [1,2]. At the same time, HPC researchers have taken on the challenge to achieve an exascale class computer, a level of performance that would require an unachievable amount of power using today's technologies. In response to this critical situation, we see a greatly increased focus on the trade-offs between execution time and power and cooling costs. In HPC, application developers require answers to a number of important questions. *Can we keep the execution time the same yet decrease the power requirements? Can we decrease the power requirements with only a moderate increase in execution time? Is cooling cost affected by the physical location of the compute nodes used? By the number of cores used per node?* Obtaining useful answers to questions such as these requires a holistic analysis approach that integrates previously disparate approaches for measuring power, cooling, and application performance.

44.2.1 Energy Smart Data Center

An example of the state-of-the-art in data center power and cooling research is the Energy Smart Data Center (ESDC) [3–5]. The ESDC program was established at Pacific Northwest National Laboratory (PNNL) with funding from the U.S. National Nuclear Security Administration (NNSA). The focal point of the ESDC program is the ESDC Test Bed (ESDC-TB), which is a highly instrumented data center devoted to research in energy efficiency and an eight-rack combined air-cooled, liquid-cooled (spray-cooled) IBM x3550, 9.58 TFlop, cluster called NW-ICE. NW-ICE is a dedicated research cluster that is being used to make significant advances in power-aware computing, data center productivity, and data center energy efficiency. NW-ICE, which utilizes both air and liquid cooling, constitutes the IT equipment and infrastructure under test (ITUT). NW-ICE is housed in the ESDC-TB, within PNNL's Environmental Molecular Sciences Laboratory (EMSL). EMSL is a mixed-use facility that also houses the Molecular Sciences Computing Facility's (MSCF's) main clusters. The combination of ESDC-TB and EMSL presents a very industry-relevant problem of data center productivity and energy efficiency in a mixed-use facility.

A depiction of EMSL's condenser and chilled water loops is shown in Figure 44.1. The figure shows the general mechanical layout for the facility and highlights the key mechanical subsystems such as the chiller plant, cooling towers, condenser pumps, and chilled water pumps. Figure 44.1 also shows the key instrumentation points, the general plumbing layout, and in particular, the manner in which chilled water is delivered to NW-ICE via the "building secondary chilled water system." A depiction of the electrical distribution system for EMSL is shown in Figure 44.2. The figure shows the general electrical layout for the facility and highlights the key electrical subsystems such as the uninterruptible power supplies (UPSs).

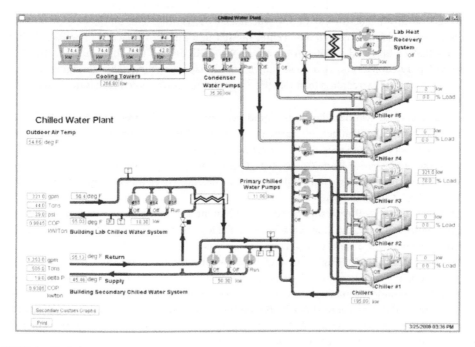

FIGURE 44.1   EMSL layout.

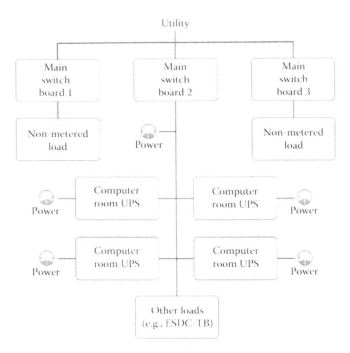

FIGURE 44.2   EMSL electrical distribution system.

Figure 44.2 also shows the key instrumentation points and the manner in which NW-ICE is electrically connected to the facility via the "Main Switch Board 2."

The ESDC-TB is a state-of-the-art 700 ft² data center located within PNNL's mixed-use EMSL facility. The ESDC-TB shares power distribution units, chillers, chilled water pumps, condenser water pumps, cooling towers, and other utilities as in any true mixed-use facility. The uniqueness of the ESDC-TB is that it is a true research-dedicated data center housed in a mixed-use facility. The ESDC-TB houses an eight-rack 9.58TFlops IBM x3550 cluster called NW-ICE. Five of the eight racks in the cluster are liquid cooled (spray cooled) and two racks utilize traditional air cooling. This configuration facilitates comparison between the two cooling technologies. One additional air-cooled rack is dedicated to network hardware.

Water is delivered to two heat liquid-to-liquid heat exchangers housed within the ESDC-TB. The purpose of these two heat exchangers is to provide independent control over the temperature of the water delivered to the racks within the ESDC-TB. These two heat exchangers allow the water delivered to each liquid-cooled rack to be varied from 7°C to 30°C. One of the key objectives is to demonstrate the ability to cool NW-ICE with condenser (non-chilled) water. The condenser water temperature is expected to range from 18°C to 30°C in the worst case.

PNNL has developed a real-time software tool, Fundamental Research in Energy-Efficient Data centers (FRED), to monitor, analyze, and store data from the ESDC-TB and EMSL facility instrumentation. This instrumentation measures a variety of parameters such as chilled water flow rates, temperatures, and electrical power. FRED's underlying technology is derived from PNNL's experience in developing power plant, distribution, and facility monitoring and diagnostic systems for applications ranging from nuclear power generation to building management and to public housing (Jarrell and Bond 2006).

FRED (Figure 44.3) consists of the ESDC-TB monitoring system, the EMSL facility monitoring system, a data collector, a central database, and a Web-based graphical user interface (GUI) client. The ESDC-TB monitoring system derives from PNNL's Decision Support for Operations and Maintenance (DSOM) software, an advanced, flexible diagnostic monitoring application for energy supply and demand systems.

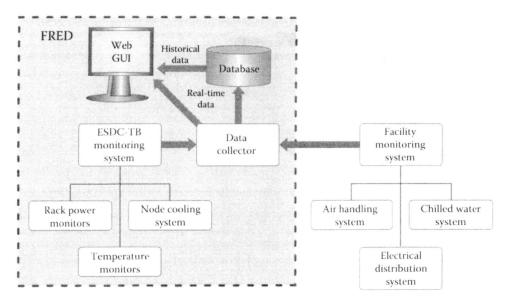

FIGURE 44.3 Fundamental research in energy-efficient data centers (FRED).

DSOM is currently deployed in several U.S. military installations and a large public housing project in New York City.

The ESDC-TB monitoring system interfaces to auxiliary data acquisition systems that collect data specific to NW-ICE. A network of power monitors measure the electrical power consumption of individual NW-ICE racks. Multichannel acquisition units measure temperatures from nearly 700 thermocouples installed throughout NW-ICE. Most of these thermocouples are mounted at the front or rear of NW-ICE nodes to measure inlet and outlet air temperatures. Remaining thermocouples are mounted inside three nodes (top, middle, and bottom) within each NW-ICE rack to measure contact temperatures on dual inline memory modules (DIMMs), application-specific integrated circuits (ASICs) and central processing units (CPUs). In addition, two racks have thermocouples on all CPUs. Finally, the ESDC-TB monitoring system acquires the chilled water flow inlet and outlet temperatures from the node cooling system as well as various system statuses. The ESDC-TB monitoring system also implements versatile calculation and decision engines for deriving high-level, real-time metrics and for performing real-time system diagnostics.

FRED's facility monitoring systems collect real-time parametric data from the air handling system, chilled water system, and electrical distribution system. FRED's data collector module fuses data from the ESDC-TB and facility monitoring systems. The data collector then routes data to one or more Web-based GUIs for real-time display, as well as to a central database for long-term storage and analysis.

The FRED GUI presents real-time data within the context of a graphical depiction of the ESDC-TB. Information is organized as a hierarchy of graphical displays providing greater, more focused detail as the user navigates to lower tiered displays. This display provides an overview of the condition of the ESDC-TB via real-time indication of various parameters including chilled water temperatures, power consumption, and energy transfer rate from node cooling systems to the building chilled water system. The GUI also displays current node air inlet and outlet temperatures with color coding that provides a quick indication of heat generation (red indicates higher temperatures at air outlets and violet indicates lower temperatures at air inlets). From the main display, the user may display details of a specific NW-ICE rack. Figure 44.4 depicts an individual NW-ICE rack. Nodes in this rack are equipped with liquid cooling and heat is removed from the node by the thermal management unit (TMU) to the building chilled water system. Three nodes in this rack are instrumented with additional thermocouples to monitor the temperatures

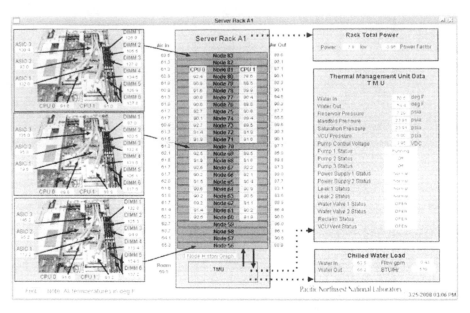

FIGURE 44.4   FRED user display.

of the two CPUs, six DIMMS, and three ASICs. From the main display, users may also view specific data related to the building chilled water system. From this display, the user can view real-time data related to the cooling towers. Real-time tower status information reflects fan and spray pump activity for each of the four towers. Individual chillers are characterized in real-time by their electrical and mechanical operation points.

44.2.2 Environment-Aware Performance Analysis

In collaboration with ESDC, researchers at Portland State University are developing a holistic approach to parallel performance analysis that integrates traditional application-oriented performance data with measurements of the physical run time environment. This new integrated approach, called environment-aware parallel performance analysis, breaks down longstanding barriers between views of different levels of an application run time: application, system, and physical environment.

Traditional application-level performance measurement and analysis tools focus on execution time, and in some cases on system software behavior, but ignore power and cooling. Tools developed to diagnose parallel application performance [6–14] provide information about the causes of application performance in terms of the application itself, but excludes the status or behavior of the run time environment in their analyses. Operating System (OS) measurement tools such as NWPerf [15] are designed to report metrics of interest to system developers and administrators on a global view of system behavior. Unfortunately, such tools do not correlate changes to particular applications.

To accomplish our goal of integrated performance analysis, we must combine measurements from different types of tools. The main challenge for data collection is to measure at the different levels (application, hardware, system, and room) while maintaining an acceptable level of perturbation and solving scaling challenges that arise as the data volume increases. Our design views a running application from four different perspectives: application, hardware, system, and room. Metrics specific to each layer are defined and measured through a variety of instrumentation. Our measurement infrastructure is detailed as follows:

Room Layer: Temperature and Cooling. At the node level, we collect temperature data for CPUs, memory modules, and ASICs. Two air temperature measurements are taken for all nodes in each rack: the *air in temperature* measures the air entering the node and the *air exchange temperature* measures the air exiting the node. The air exiting the node is typically warmer than the entry air because it has been warmed by the computing equipment. At the rack level, we collect temperature and cooling data for either liquid spray cooling or more traditional air cooling. Each spray-cooled rack has a TMU, which contains pumps, a liquid-to-liquid heat exchanger, and control electronics. The TMU supplies the rack with cool single-phase fluid to distribute to the spray modules mounted in the rack. The spray modules spray the coolant; it heats up and then vaporizes, creating a two-phase fluid that returns to the TMU. The heat exchanger condenses and cools the fluid, preparing it for redistribution to the rack. The *reservoir temperature* is the temperature of the liquid in the TMU reservoir—where the liquid is cooled after returning from the rack. The TMU liquid-to-liquid heat exchanger is supplied with facility water; the *water inlet temperature* is the incoming temperature of this water. The *water outlet temperature* is the outgoing temperature of this water (on its return to the facility water supply).

Room Layer: Power. We collect *power*, measured in kilowatts, at regular intervals, at the rack level.

Hardware Layer: We run scripts at the beginning of each run to collect information about the underlying hardware and software. This metadata includes system version number, node type, processor core count, etc.

System Layer: Our design includes a variety of measurements of system-level behavior such as interrupts and processor speed. System-level data may be collected using hardware counters and timers, or in software. Our design anticipates run time processing for these data.

Application Layer: Our design for the application layer is not limited to one particular measurement approach. A variety of software tools exist to provide a traditional application layer view of performance as measured in time, for example, the Message Passing Interface (MPI) profiling library. Our MPI instrumentation library reports per process *MPI function call counts* for the whole execution, per process *total inclusive time* spent in each MPI function for the whole execution, and for each function, it reports the inclusive time spent in that function for the process with the *maximum inclusive time* in that MPI function call. The resulting application-level data generated at run time is output into one file per MPI task.

Postmortem processing stores the variety of data collected for each of the three layers in a performance database tool called PerfTrack [10,16,17]. PerfTrack is a tool for storing, exploring, and analyzing application performance data. Developers collect and store a description of each platform, each build, and each application run. Then they can load performance data collected from any of a wide variety of different measurement tools that can be queried by selecting a wide range of system and run characteristics. The PerfTrack GUI includes functionality for browsing and selecting data and simple visualizations. The integration of metadata describing the run time environment with the performance data allows us to retrieve whatever data are stored, from the different levels of the application and run time environment. We show a query over room layer data in Figure 44.5. The tool user selects particular metrics, locations, and nodes, as appropriate, to view subsets of the available data. Our example shows a list of selection parameters in the right-hand side and the number of performance results matching those parameters.

Figure 44.6 shows results from a case study that investigates the relationship between the physical mapping of an MPI application onto the racks in the machine room and the resulting energy consumption. The application room map is a display of nodes organized by rack or cabinet. The coloring of the nodes indicates usage—uncolored (white) nodes are in use and shaded nodes are not in use—by a particular application. In the right-hand side of each node box is a gauge that is either empty (indicating zero), partly shaded, or completely shaded. This indicates the percentage of the node in use. The figure shows two configurations that used 100% (high density) and 25% (low density) of the eight processors in each node. We ran an MPI application in each configuration and collected and stored data from the different

FIGURE 44.5   Screenshot of PerfTrack GUI selection of rack and node resources and a mix of room and system metrics.

layers into PerfTrack. Although the low-density approach required more power, it ran more quickly, with almost a 2X Speedup compared to the high-density version. We show selected results for power, temperature, and execution time. This infrastructure allows further study of the trade-offs between lower power (high density) and faster execution time (low density).

44.3 Models and Metrics for Data Centers

The energy consumption and the sustainability of computing systems has been an increasing concern [18]. In this section, we describe our research in the following two important areas:

1. Quantitatively modeling and relating performance, consumption to the workload: This area includes efforts on developing models that describe the interaction of computing components and models that estimate the performance given the incoming workload.
2. Predicting the workload: This area involves all the efforts that relate to predictability of the workload, various patterns and properties (e.g., self-similarity), and deriving predictive models of workload.

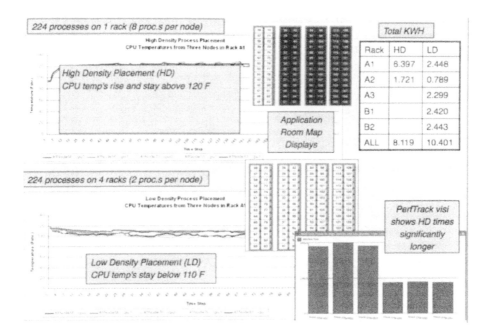

FIGURE 44.6 Environment-aware performance analysis.

44.3.1 Reference System Model and Layout

The assumed system layout and system model can be divided into the following parts: (i) the layout of a single-computing unit (or computing node); (ii) the layout of a data center, which is essentially a collection of computing nodes in a certain arrangement; (iii) the organization of the managing software; and (iv) the nature of the workload. A computing node consists of an enclosing frame and the motherboard, which hosts the bulk of the electronic and integrated circuits. The frame also encloses peripheral devices such as a hard drive. The frame features openings at the front and at the back to serve as "air inlets" and "air outlets," respectively. Systems like the one in Figure 44.7a, an IBM xSeries 336, can pack equipment consuming several hundreds of watts. The air inlets and outlets allow ambient (cool) air to enter the frame and pick up the heat produced.

Figure 44.7b depicts a computational fluid dynamics (CFD) representation of a data center. The picture clearly shows the four rows of equipment resting on the raised floor plenum. Data centers also feature lowered ceilings for cooling air circulation (not shown in the figure), with the computing equipment organized in rows of 42U racks arranged in an aisle-based layout, with alternating cold aisles and hot aisles. The figure also shows several streams of air going from the air outlets of the servers to the inlet of the *computer room air conditioners* (CRACs), also known as the HVACs. They supply cool air into the data center through perforated tiles on the raised floor. The computing equipment is usually in blade-server form, organized in 7U chassis.

Although data centers can exhibit several varieties of row arrangement or other physical layouts, they are classified more upon the workload they serve. This is mainly because the nature of the workload determines the density and energy consumption of data centers. High-performance data centers service workload that is characterized by batch arrival and CPU-intense computation. They run massive, highly parallelized scientific applications, for example, weather analysis and prediction simulators. Maximization of the performance, in terms of CPU throughput, is of utmost importance.

On the other hand, there are data centers that service less CPU-intensive workload that is characterized by a fine-grained, flow-like arrival of small IO-oriented jobs, which are referred to as *transactions*. These transactions are most frequently part of human-interactive processes, such as bank transactions

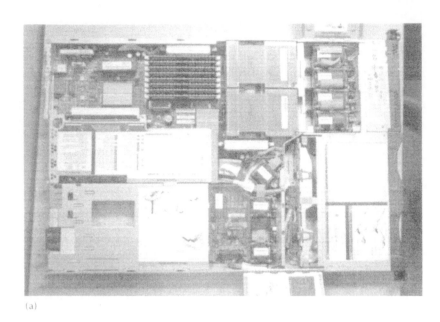

Test recirculation

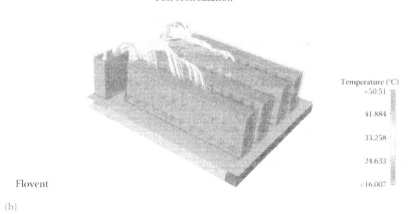

Flovent

(b)

FIGURE 44.7   Layout and operation of (a) a computing unit and (b) a data center.

originating from ATMs and database lookups or file retrievals initiated from a Web browser. For that reason, quality of service in terms of end-to-end delay is the cornerstone criterion of performance in these types of data centers (Figure 44.8).

The requirement of ensuring low delays implies that the server systems should not be overly utilized. Consequently, HPC data centers may seem more noisy (higher speed fans) and perhaps a bit hotter than

FIGURE 44.8   Examples of workload over time from the context of the (a) system and (b) CPU.

their transactional counterparts. In the context of this chapter, the phrase "data centers" will refer to HPC centers. Section 44.3.2 will talk more about a data center's energy consumption.

44.3.2 Energy and Power Consumption Modeling

There are basically two types of power measurements that can be taken, and they reflect whether the power is measured at each server system, or "globally" as the consumption of the entire data center.

Power at the plug is the usual way of measuring a computing system's power consumption. There are two implications with this measuring method:

1. This measuring takes into account the consumption of the power units.
2. Blade servers, which are a dominant server technology, share one power unit per *chassis of blade servers*. Servers on one chassis may run different software, which means that distinguishing the power consumption between the blades is a difficult process.

Power at the meter is the usual way of measuring a data center's power consumption. It is the power as measured at the point of delivery from the power grid. Implications of this measuring method include:

- Power conversion and distribution losses are part of the measured value.
- Power consumption of non-computing equipment, for example, lighting or cooling equipment is also included in this measurement.

A general way of representing a power model of a system is expressing the relationship between the serviced workload and the power consumed servicing it, as shown in Figure 44.9.

Despite the elegance of this definition, there is the fundamental question of what is "workload intensity." Workload intensity, however, translates to different measures depending on the system context. A common power model of computing equipment is the power consumed over the system utilization.

Since a system contains many components, for example, peripherals, hard drives, memory, and CPU, system utilization may be a multidimensional quantity. Hence, the conventional way of representing system utilization with a scalar value is to use CPU utilization level (Figure 44.10).

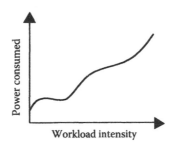

FIGURE 44.9 Relationship between workload intensity and power consumption.

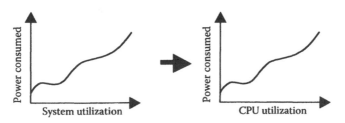

FIGURE 44.10 Power consumption as a function of utilization.

Power versus utilization is the standard way of profiling the power consumption of a system. SPECPower is a family of power-profiling benchmarks. The latest version is SPECTpower_ssj2008 (http://www.spec.org/power_ssj2008/), and several power-profiling results have been published (http://www.spec.org/power_ssj2008/results/).

Figure 44.11 gives three sample graphs taken from the Web site. Note the inverted axes.

Until recently (ca. 2006), systems showed a near-linear power consumption with respect to utilization. This has been witnessed by several research efforts [19–21]. Understandably, the dominant algebraic model of power consumption has been a linear one:

$$\text{Power} = a \times \text{Utilization} + b.$$

For example, the leftmost system in Figure 44.11 shows a near-linear power consumption.

However, the power curve of contemporary computing systems is not a straight line. The graphs in Figure 44.11b and c demonstrate these properties.

Another characteristic of the newer power curves is that systems tend to have a much lower idle power consumption with respect to their peak consumption. This characteristic goes hand-in-hand with the concept of *energy-proportional computing* [22,23], which states that computing systems should consume power in proportion to the amount of computation they perform. Achieving energy proportionality in computing systems has several challenges that are the system level [23] and the need to address energy wastage [24].

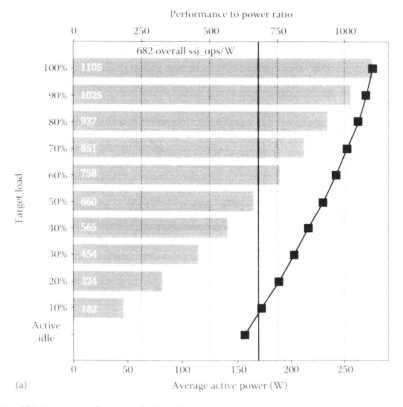

FIGURE 44.11   SPECpower results examples (2008).

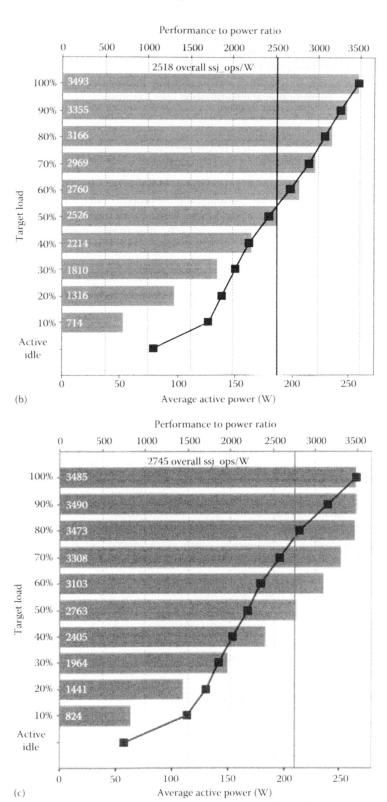

FIGURE 44.11 (continued)

In a recent study on energy proportionality, two metrics are defined to characterize energy proportionality of a system: a metric that measures how low the system's idle consumption and another metric that measures the linearity of the power curve.

For measuring the idle level, the proposed metric is the idle-to-peak power ratio (IPR), which is defined as the ratio of the power consumption at 0% utilization over the power consumption at 100% utilization (see Figure 44.3):

$$\text{IPR} = \frac{P_{\text{idle}}}{P_{\text{peak}}}.$$

Lower IPR values imply that the power curve is relatively closer to the origin.

For the "linearity" aspect of the power consumption, the proposed metric is the linear deviation ratio (LDR), which is defined as the maximum ratio of the actual power consumption's difference from the hypothetical linear power consumption P_{idle} at 0% to P_{peak} at 100% utilization, over the hypothetical linear consumption:

$$\text{LDR} = \max_u \frac{P(u) - (P_{\text{peak}} - P_{\text{idle}})u + P_{\text{idle}}}{(P_{\text{peak}} - P_{\text{idle}})u + P_{\text{idle}}}.$$

Lower LDR values denote a more linear system.

Although the focal point in data centers are the computing servers and their performance (either in terms of computational performance or in terms of energy performance), the data centers contain energy systems that are integral parts of the infrastructure. Namely, *power distribution units* and *cooling systems* are important components for the operation of a data center. Power units consume energy in direct relation to the power drawn by the computing systems, that is, the load. Cooling systems, on the other hand, deserve a closer look, because their cooling performance greatly affects the power consumption of a data center. It is not unreasonable, under certain load conditions, for half of a conventional data center's power consumption to come from the cooling systems (CRACs).

There are two aspects that characterize the performance of a cooling system: the *coefficient-of-performance* curve, and the *return-vs.-supplied air temperature* curve. The coefficient of performance is the ratio of the work consumed by the CRAC in order to remove heat from the data center room. The coefficient of performance is bounded by the Carnot efficiency (CoP = $|T_{\text{in}} - T_{\text{out}}|/T_{\text{in}}$). Measurements done by HP Labs, the CoP of a profiled system demonstrates a quadratic behavior.

Data centers use air-cooling technologies for nearly every powered component. The ambient air may not be well controlled (however, recent aisle containment technologies are very effective in that respect) and the hot air that leaves the equipment's air outlets may not reach the CRAC, but it may enter a cold aisle, mix with the cool air, and reduce the efficiency of cooling. Previous research has shown that if the air recirculation patterns do not significantly change, then the amount of heat that is recirculated between each pair of the equipment follows a linear dependency with respect to the heat produced [19,21]. Specifically, a matrix of coefficients (*heat recirculation matrix*) can be derived to denote what portions of heat between each pair of nodes is being recirculated.

44.3.3 Predictive Workload Models

Figure 44.12 shows that the conversion of electricity into released heat is driven by the incoming workload. In plain terms, the incoming workload is the proverbial gas pedal to the data center's engine. A smart and flexible data center is one that has knowledge of its internals and can reason about and control its own behavior. Such a data center can switch modes in order to achieve the best energy efficiency for the given workload.

The challenge of this approach lies in the fact that the process of realizing the volume of the workload and switching to efficient internal modes could incur a considerable delay, during which the data center

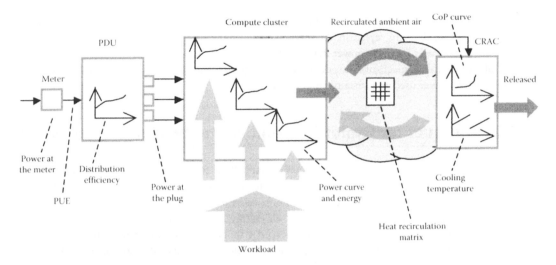

FIGURE 44.12   System power conversion.

may exhibit extremely poor performance, or it may have the ambient temperature exceed the servers' *redline*, that is, the maximum operating temperature.

Research should focus on the predictability of workload, so that the data center does not switch to an internal state that may prove inappropriate in the subsequent decision epoch.

Determining what metrics characterize the sustainability of a data center is still evolving [25]. Most of the extant research on sustainability of data centers rely on energy efficiency metrics; in essence, according to the bulk of the research work, a sustainable data center is an energy-efficient data center.

The three most used energy efficiency metrics are the following:

1. *Energy efficiency of a system:* This is usually defined as *(useful) work over energy* and as *work rate over power*. Usual units used are cycles per joule, or instructions per joule, and Million Instructions Per Second (MIPS) per watt, respectively. As the units suggest, energy efficiency is how much energy was spent to execute a program.
2. *Power efficiency of a system:* This is usually defined as *work over power* or *power over work*. Usual units used are watts per cycle or watts per instruction.
3. *Power usage effectiveness (PUE):* This is the ratio of the power at the plug over the power at the meter: PUE = power_at_the_plug/power_for_IT.

Recently, a new metric has been introduced to capture the effectiveness of data center energy reuse behavior, called *energy reuse effectiveness (ERE)*. This metric is defined as follows:

$$\frac{\text{Power\_at\_the\_plug} - \text{Reused\_power}}{\text{Power\_for\_IT}}.$$

Note that there is a distinct difference between the energy and power efficiency metrics on the one hand, and PUE and ERE metrics on the other: the first two metrics take the computational performance into consideration, whereas PUE and ERE are independent of the data center's computational performance.

While energy efficiency is a key aspect of a sustainable process, it cannot be a metric of sustainability. It is easy to see that a process or a product may be extremely energy efficient, but it can still be unsustainable; that is, it may be efficient yet unaffordable. This "affordability" can be viewed as (i) peak power affordability and (ii) long-term viability. "Peak affordability" can be easily translated to a cap on the peak power. Any system that can exceed the power cap is not sustainable. This can be clearly relevant to large-scale data centers, because they can be significant contributors to black out risk. Long-term viability is the affordability in the long run. Long-term viability directly relates to using renewable sources and never draining them before they can replenish. A tighter aspect of the long-term viability with energy sources is the use of "dirty power," that is, power that is not "green." A typical example is electricity from coal-powered plants. In this respect, one can define the average percentage of green power over the entire (green plus dirty) power used as a sustainability metric.

44.4 Techniques for Exploiting Application Behavioral Attributes

U.S. Executive Order (EO) 13423 was issued in January 2007 to strengthen federal environmental, energy, and transportation management. It outlines goals for agencies to improve energy efficiency and reduce greenhouse gas emissions through reduction in energy intensity by 3% annually or 30% by the end of fiscal year 2015 relative to the baseline of an energy agencies use in 2003. The order further defines energy intensity as "energy consumption per square foot of building space, including industrial or laboratory facilities" [26]. Applying the definition of energy intensity and its reduction goals to future HPC facilities presents significant challenges. HPC systems are being designed to consume significantly more energy per square/cubic foot, not less. New HPC node designs, however, are incorporating techniques for reducing energy consumption. Systemically though, HPC systems are built and used with the objectives of maximum sustained performance and utilization. Achieving maximum sustained performance continues to be elusive. This is at least in part due to the unintended consequences of subsystem interactions. The consequences of subsystem interactions not only degrade system performance, they also do nothing to improve power consumption and in fact, contribute to increased power consumption by extending one or more application's run time for the same amount of work performed (i.e., watts over time).

Work done to date in areas such as dynamic HVAC control [27–37], disk and compute node, (and core) power awareness [38–40], and HPC application sensitivity assessment [41–47] have extended and deepened each knowledge base, yet gaps in our understanding of subtle but holistic problems still remain. Some of these gaps include the following:

1. Conflicting subsystem operational objectives: HPC subsystems, like interconnection systems, schedulers, etc., are developed in isolation. This means their designs generally do not account for the interactions of other subsystems. Often it is the case that these subsystems have different, sometimes conflicting, operational objectives. For example, the HPC job scheduler strives to utilize every resource available, regardless of its physical location in the HPC system. Network elements strive to deliver information as quickly as possible, with no concern for systemic ramifications in the event that information needs to be rerouted due to congestion.

2. Subsystem interactions trigger autonomous subsystem behavior: Subsystems sometimes detect downgraded performance or failure, which can trigger significant changes to subsystem behavior, which in turn may have serious ramifications to other subsystems. An example of this might be where an application detects degraded node performance, triggering a process migration that unintentionally creates a congestion point in the network.

3. Cost of state transition: There is a cost involved to change a subsystem's behavior, such as a network reroute or application process migration. These actions can lead to unintended consequences for the rest of the system, like creating new congestion points in the network.

4. Centralized control of semiautonomous subsystems: The operational objectives of HPC centralized resource management and scheduling subsystems include the optimization of the system for fair, near 100% utilization. These objectives have traditionally been independent (for the most part) of systemic run time or energy performance considerations. In other words, energy efficiency has generally not been a serious consideration in HPC systems.

5. The emergence of multicore processing elements (general purpose and application specific, such as Graphics Processing Units [GPUs]) has shifted the concept of HPC interprocess communication

away from a well-defined distributed memory model, where a node contains one single-core processor and therefore a single-memory hierarchy. Today, it is more common to see a hybrid model where each node is equipped with more than one multicore processor, each with its own memory hierarchy, where some caches are local to each core, while others are shared between cores.

One emerging area of research focuses on relating "application behavioral attributes" to energy management. The idea originated as a theory to improve HPC systemic performance consistency by assessing a parallel application's run time sensitivity to network performance degradation in clusters made up of single-core, single-processor compute nodes [45–47]. This idea is being extended to include the notion that a more complete understanding of a parallel application's behavior under benign conditions and those where certain subsystems interact can serve to provide guidance to HPC system and data center level controls to help optimize energy efficiency. The main idea is to identify application behavioral attributes that can affect run time performance. The concept of an application's sensitivity to network performance degradation is presented in [46]. Extending this concept further, it is somewhat intuitive to consider that if an application can be sensitive to network performance degradation, it is possible that the same application might contribute to network performance degradation, being disruptive to other applications. This concept is introduced in [48] while also expanding the notion of network performance to communication performance to take into account multicore, multiprocessor nodes in HPC systems.

One of the first things to address is whether the concepts studied and identified in [44–46] still hold for multicore, multiprocessor nodes. In [49], communication performance between cores, processors, and nodes is revisited along with the concept of application sensitivity to communication performance degradation. Figure 44.13 illustrates the idea of run time sensitivity of the NAS MG Class C benchmark when the communication subsystem is perturbed using a tool called parallel application communication emulator (PACE) [44] and an add-on used to evaluate parallel application run time sensitivity called PARSE [45]. Here, the processes used by the application under test (AUT) were spread evenly across two quad-core dual processor nodes for the baseline trace. The loaded case then interlaced the AUT processes with PACE processes communicating with each other 95% of the time. While extreme, Figure 44.13 illustrates the negative run time impact of communication subsystem performance degradation. In terms of energy, the "loaded" trace in Figure 44.13 shows that the run time was increased by approximately 30%, meaning 30% more energy was needed to perform the same amount of work. Moreover, it is visually obvious that the AUT run time becomes less predictable as well.

Parallel applications exhibit a run time behavioral attribute of sensitivity to communication subsystem performance variability (often called degradation). Conceptually, parallel applications may also exhibit the inverse to sensitivity, or its propensity to be "disruptive" to the communication subsystem thereby causing extended run time (and therefore wasted energy) for other applications. By using the same tools (PACE and PARSE [45,46]) in a way so that PACE might be sensitive to the communication degradation imposed by the AUT and using similar statistical comparisons, the run time disruptiveness of an AUT can be quantified, as shown in Figure 44.14. The important thing to note in Figure 44.14 is not necessarily the absolute values of sensitivity and disruptiveness, but the fact that the range is remarkably large, over two orders of magnitude.

The variables used to determine sensitivity and disruptiveness may also provide interesting clues into managing HPC system energy. These variables are ratios of expected run times and their coefficients of variation under benign and perturbed operating conditions. By examining Figure 44.13, it becomes clearer to appreciate that if the system can locate application processes using sensitivity and disruptiveness values as inputs then it may be possible for application's to approach their baseline run time more often, wasting less (saving) energy, and requiring less effort in more exotic low (core) level energy-management techniques. For applications that behave like NAS MG Class C, the goals set forth in EO 13423 can be easily met if reducing "wasted" energy is considered equal in importance to finding reduction opportunities.

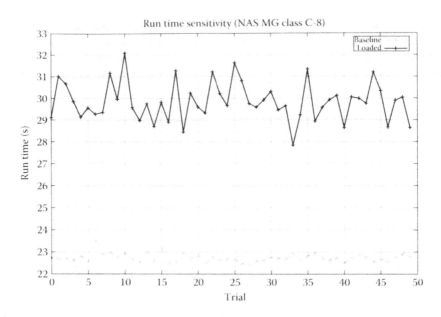

FIGURE 44.13   Run time effect of communication subsystem perturbation.

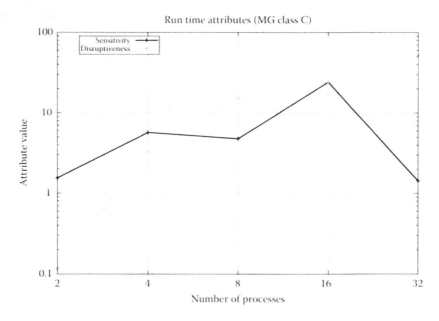

FIGURE 44.14   Run time attributes of NAS MG Class C.

44.5 Summary

Although there is some overlap in research and solutions for power and thermal management between HPC and more traditional IT data centers, HPC applications introduce some unique challenges that require new approaches. In this chapter, we have discussed three important aspects: direct measurement and performance analysis; modeling of energy and power consumption, and of workloads; and application behavioral attributes. Work continues on these challenging issues, as the current push to achieve exascale computing is intensifying the focus on power and cooling efficiency for HPC data centers.

Acknowledgments

PNNL work has been funded in part by DE-AC05-76RL01830 (59193, 55430). Gupta's and Varsamopoulos's work at ASU has been funded in part by NSF CRI grant #0855527 and NSF CNS grant #0834797. Evans' work has been funded in part by U.S. Department of Energy (DOE) grant DE-SC0004596.

References

1. J. Koomey. *Estimating Total Power Consumption by Servers in the U.S. and the World.* Analytics Press, Oakland, CA, February 15, 2007.
2. A. Marquez, L. Sego, K. Fox, D. Sisk, D. Hatley, D. Johnson, S. Elbert, and M. Khaleel. Towards real-world HPC energy efficiency and productivity metrics in a fully instrumented datacenter. In *SC'08: 2008 ACM/IEEE Conference on Supercomputing,* Austin, TX, November 2008.
3. The Energy Smart Data Center. http://esdc.pnl.gov/ (accessed Nov 2010).
4. T. Cader, V. Sorrell, L. Westra, and A. Marquez. Liquid cooling in data centers. Presented at the *ASHRAE Winter Conference,* Chicago, IL, January 2009.
5. T. Cader, L. Westra, A. Marquez, H. McAllister, and K. Regimbal. Performance of a rack of liquid-cooled servers. *ASHRAE Transactions,* 113(1):101–114, 2007.
6. L. Adhianto, S. Banerjee, M. Fagan, M. Krentel, G. Marin, J. Mellor-Crummey, and N. Tallent. HPC toolkit: Tools for performance analysis of optimized parallel programs. *Concurrency and Computation: Practice and Experience,* 22(6):685–701, April 2010.
7. I.-H. Chung, R. Walkup, H. Wen, and H. Yu. MPI performance analysis tools on Blue Gene/L. In *Supercomputing,* IBM Thomas J. Watson Research Center, New York, November 2006.
8. Cray. Cray performance analysis tool, http://docs.cray.com (accessed Nov 2010).
9. M. Geimer, F. Wolf, B. J. Wylie, E. Ábrahm, D. Becker, and B. Mohr. The SCALASCA performance toolset architecture. *Concurrency and Computation: Practice and Experience,* 22(6):702–719, April 2010.
10. K. L. Karavanic, J. May, K. Mohror, B. Miller, K. Huck, R. Knapp, and B. Pugh. Integrating database technology with comparison-based parallel performance diagnosis: The Perftrack performance experiment management tool. In *SC'05: Proceedings of the 2005 ACM/IEEE Conference on Supercomputing,* IEEE Computer Society, Washington, DC, p. 39, 2005.
11. R. Kufrin. Measuring and improving application performance with PerfSuite. *Linux Journal,* 135(4), July 2005.
12. B. P. Miller, M. D. Callaghan, J. M. Cargille, J. K. Hollingsworth, R. B. Irvin, K. L. Karavanic, K. Kunchithapadam, and T. Newhall. The Paradyn parallel performance measurement tools. *IEEE Computer,* 28(11):37–46, November 1995.
13. B. Mohr and F. Wolf. Kojak—A tool set for automatic performance analysis of parallel programs. In *Euro-Par,* August 2003.
14. S. S. Shende and A. D. Malony. The tau parallel performance system. *International Journal of High Performance Computing Applications,* 20(2):287–311, 2006.
15. R. Mooney, K. P. Schmidt, and R. S. Studham. NWPerf: A system wide performance monitoring tool for large Linux Clusters. In *CLUSTER'04 Proceedings of the 2004 IEEE International Conference on Cluster Computing,* San Diego, CA, USA, Sep 20–23, pp. 379–389, 2004.
16. R. L. Knapp, K. L. Karavanic, and A. Marquez. Integrating power and cooling data into parallel performance analysis. In *Second International Workshop on Green Computing,* San Diego, CA, September 2010, ICPP.

17. R. L. Knapp, K. Mohror, A. Amauba, K. L. Karavanic, A. Neben, T. Conerly, and J. May. Perftrack: Scalable application performance diagnosis for Linux clusters. In *Linux Clusters Institute 2007 Conference on High Performance Computing*, 2007.

18. NewsLink, Spring 08. *Tackling Todays's Data Center Energy Efficiency Challenges*, 2008.

19. Q. Tang, T. Mukherjee, S. K. S. Gupta, and P. Cayton, Sensor-based fast thermal evaluation model for energy efficient high-performance datacenters. In *International Conference Intelligent Sensing and Information Processing* (ICISIP 2006), December 2006.

20. T. Heath, A. P. Centeno, P. George, L. Ramos, and Y. Jaluria. Mercury and Freon: Temperature emulation and management for server systems. In *ASPLOS-XII: Proceedings of the 12th International Conference on Architectural Support for Programming Languages and Operating Systems*, ACM Press, New York, pp. 106–116, 2006.

21. Q. Tang, S. K. S. Gupta, and G. Varsamopoulos. Energy-efficient thermal-aware task scheduling for homogeneous high-performance computing data centers: A cyber-physical approach. *IEEE Transactions on Parallel and Distributed Systems, Special Issue on Power-Aware Parallel and Distributed Systems (TPDS/PAPADS)*, 19(11):1458–1472, November 2008.

22. L. A. Barroso and U. Hölzle. The case for energy-proportional computing. *IEEE Computer*, 40(12):33–37, 2007.

23. K. W. Cameron. The challenges of energy-proportional computing. *IEEE Computer*, 43:82–83, 2010.

24. P. Ranganathan. Recipe for efficiency: Principles of power-aware computing. *Communications of the ACM*, 53(4):60–67, 2010.

25. S. K. S. Gupta, T. Mukherjee, G. Varsamopoulos, and A. Banerjee. Research directions in energy-sustainable cyber-physical systems. *Journal on Sustainable Computing: Informatics and Systems*, 1(1):57–74, (2010), doi:10.1016/j.suscom.2010.10.003.

26. U.S. National Archives and Records Administration. Executive order 13423, January 2007. [Online]. Available: edocket.access.gpo.gov/2007/pdf/07-374.pdf

27. I. Cohen, M. Goldszmidt, T. Kelly, and J. Symons. Correlating instrumentation data to system states: A building block for automated diagnosis and control. In *Sixth Symposium on Operating Systems Design and Implementation (OSDI'04)*, pp. 231–244, 2004.

28. J. Moore, R. Sharma, R. Shih, J. Chase, R. Patel, and P. Ranganathan. Going beyond CPUS: The potential of temperature-aware solutions for the data center. In *Proceedings of the International Symposium on Computer Architecture (ISCA), Workshop on Temperature-Aware Computer Systems (TACS-1)*, 2004.

29. Q. Tang, S. K. S. Gupta, and G. Varsamopoulos. Thermal-aware task scheduling for data centers through minimizing heat recirculation. In *IEEE Cluster*, 2007.

30. Q. Tang, S. K. S. Gupta, and G. Varsamopoulos. Energy-efficient thermal-aware task scheduling for homogeneous high-performance computing data centers: A cyber-physical approach. *IEEE Transactions on Parallel and Distributed Systems*, 19(11):1458–1472, November 2008.

31. G. Varsamopoulos, A. Banerjee, and S. K. Gupta. Energy efficiency of thermal-aware job scheduling algorithms under various cooling models. In *International Conference on Contemporary Computing (IC3)*, 2009, pp. 568–580.

32. W. Watts, M. Koplow, A. Redfern, and P. Wright. Application of multizone HVAC control using wireless sensor networks and actuating vent registers. Energy Systems Laboratory, Texas A&M University, Technical Report, 2007.

33. F. Hu and J. J. Evans. Power and environment aware control of Beowulf clusters. *Cluster Computing*, 12(3):299–308, September 2009.

34. Q. Tang, T. Mukherjee, S. K. S. Gupta, and P. Cayton. Sensor-based fast thermal evaluation model for energy efficient high-performance datacenters. In *Proceedings of the International Conference on Intelligent Sensing & Information Processing (ICISIP2006)*, pp. 203–208, December 2006.

35. J. Moore and J. Chase. Data center workload monitoring, analysis, and emulation. In *Eighth Workshop on Computer Architecture Evaluation Using Commercial Workloads*, 2005.

36. T. Mukherjee, G. Varsamopoulos, S. K. S. Gupta, and S. Rungta. Measurement based power profiling of data center equipment. In *Workshop on Green Computing (in Conjunction with CLUSTER 2007)*, 2007.

37. T. Mukherjee, A. Banerjee, G. Varsamopoulos, S. K. S. Gupta, and S. Rungta. Spatio-temporal thermal-aware job scheduling to minimize energy consumption in virtualized heterogeneous data centers. *Computer Networks*, 53(17):1389–1286, 2009.

38. Q. Zhu, F. M. David, C. F. Devaraj, Z. Li, Y. Zhou, and P. Cao. Reducing energy consumption of disk storage using power-aware cache management. In *Proceedings of the 10th International Symposium on High Performance Computer Architecture (HPCA-10)*, pp. 118–129, February 2004.

39. C. Isci, A. Buyuktosunoglu, C.-Y. Cher, P. Bose, and M. Martonosi. An analysis of efficient multi-core global power management policies: Maximizing performance for a given power budget. In *Proceedings of the 39th Annual IEEE/ACM International Symposium on Microarchitecture*, pp. 347–358, 2006.

40. D. H. Albonesi, R. Balasubramonian, S. G. Dropsho, S. Dwarkadas, E. G. Friedman, M. C. Huang, V. Kursun, et al. Dynamically tuning processor resources with adaptive processing. *Computer*, 36(12):49–58, 2003.

41. J. J. Evans, C. S. Hood, and W. D. Gropp. Exploring the relationship between parallel application runtime variability and network performance in clusters. In *Workshop on High Speed Local Networks (HSLN) from the Proceedings of the 28th IEEE Conference on Local Computer Networks (LCN)*, pp. 538–547, October 2003.

42. J. J. Evans, S. Baik, J. Kroculick, and C. S. Hood. Network adaptability in clusters and grids. In *Proceedings from the Conference on Advances in Internet Technologies and Applications (CAITA)*. p. CDROM, IPSI, July 2004.

43. J. J. Evans and C. S. Hood. Network performance variability in NOW clusters. In *Proceedings of the 5th IEEE International Symposium on Cluster Computing and the Grid, (CCGrid05) (CDROM)*, May 2005.

44. J. J. Evans and C. S. Hood. Application communication emulation for performance management of NOW clusters. In *Proceedings of the nineth IFIP/IEEE International Symposium on Integrated Network Management*, May 2005, p. CDROM.

45. J. J. Evans and C. S. Hood. PARSE: A tool for parallel application run time sensitivity evaluation. In *Proceedings of the 12th International Conference on Parallel and Distributed Systems (ICPADS)*, pp. 475–484, July 2006.

46. J. J. Evans and C. S. Hood. A network performance sensitivity metric for parallel applications. In *Proceedings of the Fifth International Symposium on Parallel and Distributed Processing and Applications (ISPA07) (Best Paper)*, pp. 920–932, August 2007.

47. J. J. Evans and C. S. Hood. A network performance sensitivity metric for parallel applications. In *International Journal of High-Performance Computing and Networking*, invited paper. Vol. 7, No. 1, pp. 8–18, 2011.

48. J. J. Evans. On performance and energy management in high performance computing systems. In *39th International Conference on Parallel Processing, GreenCom Workshop* (invited paper), pp. 445–452, September 2010.

49. P. P. Veeraraghavan and J. J. Evans. Parallel application communication performance on multi-core high performance computing systems. In *Proceedings of the IASTED International Conference Parallel and Distributed Computing and Systems (PDCS 2010)*, pp. 9–16, November 2010.

Green
Applications

45

GreenGPS-Assisted Vehicular Navigation

Tarek F. Abdelzaher
*University of Illinois at
Urbana–Champaign*

45.1 Introduction

In this chapter, we investigate the potential of green navigation services based on experiences with an implemented green navigation service, called *GreenGPS*, that gives drivers the most fuel-efficient route for their vehicle as opposed to the shortest or fastest route. The material in this chapter is largely based on the original GreenGPS manuscript, published in Mobisys 2010 [16].

Green navigation is possible thanks to the *on-board diagnostic* (OBD-II) interface, standardized in all vehicles that have been sold in the United States after 1996. The OBD-II is a diagnostic system that monitors the health of the automobile using sensors that measure approximately 100 different engine parameters. Examples of monitored measurements include fuel consumption, engine RPM, coolant temperature, vehicle speed, and engine idle time. A comprehensive list of measured parameters can be obtained from standard specifications as well as manufacturers of OBD-II scanners [3]. Several commercial OBD-II scanner tools are available [2–5] that can read and record these sensor values. Apart from such scanners, remote diagnostic systems such as GM's OnStar, BMW's ConnectedDrive, and Lexus Link are capable of monitoring the car's engine parameters from a remote location (e.g., home of driver of the car).

Green navigation utilizes a vehicle's OBD-II system to enable collection of fuel consumption data that can be then used to compute fuel-efficient routes. Compared to traditional mapping and navigation tools, such as Google maps [19] and MapQuest [26], which provide either the fastest or the shortest route between two points, green navigation collects the necessary information to compute and answer queries on the *most fuel-efficient route*. The most fuel-efficient route between two points may be different from the shortest and fastest routes. For example, a fastest route that uses a freeway may consume more fuel than the most fuel-efficient route because fuel consumption increases nonlinearly with speed or because it is longer. Similarly, the shortest route that traverses busy city streets may be suboptimal because of downtown traffic.

In this study, data was collected by individual vehicles and processed to construct generalized prediction models that estimate fuel consumption of arbitrary vehicles on arbitrary routes. We first demonstrate how to develop a fuel-saving navigation service that relies on data collection by individual vehicles. Second, we provide a brief experimental evaluation of the system, where users are shown to save 6% on average over the shortest route and 13% over the fastest.

The motivation for green navigation does not need elaboration. Users of the service might be driven by benefits such as savings on fuel or reducing CO_2 emissions and the carbon footprint. With the increase in the use of Bluetooth devices (e.g., cell phones) and in-vehicle Wi-Fi, such a service can be easily supported by inexpensive OBD-II-to-Bluetooth or OBD-II-to-WiFi adaptors that can upload OBD-II measurements opportunistically, for example, to applications running on the driver's cell phone [30]. It can also be supported by scanning tools that read and store OBD-II measurements on storage media such as SD cards. At the time of writing, OBD-II Bluetooth adaptors, such as the ELM327 Bluetooth OBD-II Wireless Transceiver Dongle, are available for approximately $50, together with software that interfaces them to phones and handhelds.

Green navigation services are expected to support two types of users; members and nonmembers. Members are those who own OBD-II adaptors or scanning tools and contribute their data to the service repository from the OBD-II sensors described earlier. They may have accounts to store their vehicle's data and benefit from more accurate estimates of route fuel efficiency, customized to the performance of their individual vehicles.

Nonmembers can use the service to query for fuel-efficient routes as well. Since the service does not have measurements from their specific vehicles, it may answer queries based on the average estimated performance for their vehicle's make, model, and year (or some subset thereof, as available).

To accommodate nonmembers, it is desired to demonstrate how sparse samples of high-dimensional spaces (i.e., measurements made by members) can be generalized to develop models of complex nonlinear phenomena, where one size (i.e., model) does not fit all. Namely, we develop prediction models that enable us to extrapolate from fuel-efficiency data of some vehicles on some streets to the fuel consumption of arbitrary vehicles on arbitrary streets.

Models, described in this chapter, abstract vehicles and routes by a set of parameters such that fuel efficiency can be computed simply by plugging in the parameters of the right car and route. Using Dijkstra's algorithm, the minimum-fuel route can then be computed. An experimental study is performed over the course of 3 months using 16 different cars with different drivers (and a total of over 1000 miles driven) to determine the accuracy of prediction models. It is shown that a prediction accuracy of 1% is attainable.

The rest of this chapter is divided into nine sections. Section 45.2 presents a feasibility study that investigates the amount of fuel savings that can be achieved by using green navigation by following fuel-efficient routes. The details of the proof-of-concept system are described in Section 45.3. Models for estimating fuel consumption are presented in Section 45.4. Implementation details and evaluation results are presented in Sections 45.5 and 45.6, respectively. Section 45.7 discusses the results and lessons learned. Section 45.8 reviews related work. Finally, we conclude with directions for future work in Section 45.9.

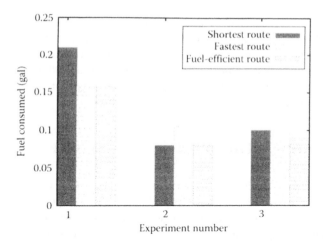

FIGURE 45.1 Figure showing fuel consumption for multiple routes between multiple-selected landmarks for different cars and drivers.

45.2 Feasibility Study

In this section, we present a feasibility study that provides the reader with a proof of concept estimate of fuel savings that can be achieved by driving on the most fuel-efficient routes.

We compute fuel consumption between landmarks in Urbana-Champaign for three different cars (and drivers) and compare these values across multiple routes between the same pairs of landmarks. The landmarks chosen were frequently visited destinations such as the work place of the authors, a major shopping center, and a football stadium. Three landmarks were initially chosen. The shortest and fastest routes were obtained using MapQuest [26].* In Figure 45.1, we plot the fuel consumption for the shortest route, the fastest route, and the route that consumes the least fuel (as computed from our models) for the aforementioned landmarks.

We observe, from Figure 45.1, that in the first experiment, the fastest route is also the most fuel-efficient route. In the second experiment, the shortest route consumes the least amount of fuel. In the third experiment, the most fuel-efficient route is different from both the shortest and the fastest routes. We conclude from the aforementioned observations that simply choosing the shortest or the fastest route will not always be fuel-optimal.

For example, if the user always chooses the fastest route, their extra fuel consumption compared to taking the optimal route is 0%, 24%, and 10% for the three landmarks, respectively (an average of about 11% overhead). Similarly, if the user always chooses the shortest route, their average extra fuel consumption is about 11.5%. Hence, following the fuel-optimal route can translate, at the current national average gas price,† into savings of at least 30 cents per gallon, which is not bad for "cash back."

The aforementioned results are only a proof of concept. They simply show that there may exist situations where using a fuel-optimal route can save money. A more extensive study of route models and savings is presented in the evaluation section.

To estimate the amount of savings that can be achieved on a global scale, we provide back of the envelope calculations based on data from the Environmental Protection Agency (EPA) [12]. An estimated 200 million light vehicles (passenger cars and light trucks) are on the road in the United States. Each of

---

* Google Maps provides only the shortest route. MapQuest provides both fastest and shortest routes. Hence, we use MapQuest to get route information.
† At the time of writing this chapter, the national average gas price was $2.76.

them is driven, on an average, 12,000 miles in a year. The average mile-per-gallon (mpg) rating for light vehicles is 20.3 mpg. Even if 5% of these vehicles adopted green navigation and 10%, fuel savings were achieved on only a quarter of the routes traveled by each of these vehicles, the amount of overall fuel savings is nearly 177 million gallons of fuel $((12,000/20.3)*0.3*(0.05*200M)*0.1)$. This translates into nearly half a billion dollars in savings at the pump (based on the current national average pump prices for a gallon of gasoline). This justifies investigating the service further.

45.3 Implementing a Green Navigation Service

To study the benefits of green navigation, the authors developed a service, called GreenGPS [16]. It is similar to a regular map application, such as Google Maps [19] or MapQuest [26]. Google Maps and MapQuest provide the shortest or fastest routes between two points, whereas GreenGPS computes the most fuel-efficient route. A snapshot of the Web-based GreenGPS's user interface is shown in Figure 45.2 along with the most fuel-efficient route between two points for a user with a Toyota Celica, 2001. In the following sections, we will discuss the concept of green navigation, then describe a proof-of-concept study that evaluates its efficacy.

45.3.1 Concept of Green Navigation

As with regular navigation services, individuals who want to compute the most fuel-efficient route between two points enter the source and destination address via the interface. Members of the service (i.e., those individuals who contributed their vehicle's data) can register their vehicles and upload OBD-II data. Hence, the service can compute a route specifically for the registered vehicle. Other users may enter their

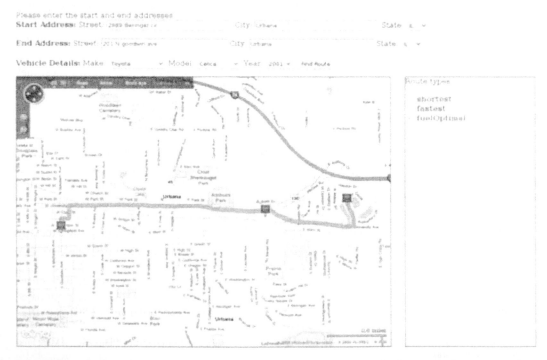

FIGURE 45.2   Figure showing the user interface of GreenGPS with the most fuel-efficient route between two points on the map for a Toyota Celica, 2001.

vehicle's make, model, and year of manufacture. Since different vehicles have different fuel consumption characteristics, these car details are used to compute the most fuel-efficient route for the given vehicle brand. The advantage for the users who contribute data is that the system provides better estimates of the most fuel-efficient routes to these individuals, thus allowing them to have higher savings.

It is impractical to assume that service members will measure all city streets and cover all vehicle types. Instead, measurements of members are used to calibrate generalized fuel-efficiency *prediction models*. These models, discussed in Section 45.4, show that the fuel consumption on an arbitrary street can be predicted accurately from set of *static* street parameters (e.g., the number of traffic lights and the number of stop signs) and a set of *dynamic* street parameters (such as the average speed on the street or the average congestion level), plus of course the vehicle parameters (e.g., weight and frontal area). It is the mathematical model describing the relation between these general parameters and fuel-efficiency that gets estimated from participant data. Hence, the larger and more diverse is the set of participants, the better the generalized model.

For most streets, static street parameters can be readily obtained from traffic databases. For example, the number of traffic lights and the number of stop signs on streets can be obtained from the red light database [20]. Dynamically changing parameters such as the congestion levels or average speed are more tricky to obtain. In larger cities, real-time traffic monitoring services can supply these parameters [35]. Many GPS device vendors, such as TomTom, also collect and provide congestion information. Finally, participatory sensing applications, such as Traffic Analyzer [17] and CarTel [24], have been described in prior literature that have the potential to provide congestion and speed data.

Speed information is obtained from the collected data using the hardware described in the next section. The speed data are aggregated for different city blocks, based on the GPS data. Thus, given a street name (or the latitude/longitude of a location), the navigation service provides the average speed information for the corresponding block.

45.3.2 Data Collection Framework

In our feasibility study, we utilized a participatory sensing framework developed in our prior work, called PoolView [17]. PoolView facilitates developing data collection applications. It provides a client-side interface for data upload and delivers all data to a central server called the *aggregation server*, that is application-specific. We implemented a service, called GreenGPS, by writing our aggregation server for PoolView. An individual who wants to share their OBD-II sensor data can thus download the client side software of PoolView, and use it to upload their data to the GreenGPS aggregation server. The aggregation server uses these data to calibrate models that relate street and vehicle parameters to fuel efficiency and offers the GreenGPS query interface for fuel-efficient routes.

Individuals who wish to contribute OBD-II data to GreenGPS install, in their vehicle, a commercial OBD-II scanner along with a GPS unit. In our deployment, we use one such off-the-shelf device for data collection purposes, called DashDyno [3], shown in Figure 45.3. The DashDyno's OBD-II scanner is interfaced to a Garmin eTrex Legend GPS [18] to get location data. The DashDyno records trip data (including Garmin's GPS location) on an SD card that the user later uploads it to the GreenGPS server.

A total of 16 parameters are obtained from the car and the GPS, the most important of them being instantaneous vehicle speed, total fuel consumption, rate of fuel consumption, latitude, longitude, and time.

45.4 Generalizing from Sparse Data

Green navigation relies on fuel-efficiency measurements by individual vehicles. It is therefore a participatory sensing application, where collection of sensory data is outsourced to the crowd (crowd-sourcing). In this section, we demonstrate the foundations of one of the key mechanisms in participatory sensing that

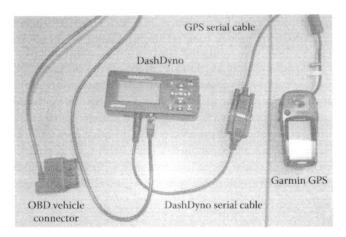

FIGURE 45.3 Hardware used for data collection.

are tolerant to conditions of sparse deployment; namely, the generalization from sparse multidimensional data. Such generalization is complicated by the fact that, in high-dimensional data sets, one size does not fit all. Hence, for example, developing a single regression model to represent all data is highly suboptimal. In the case of green navigation, the lack of widespread availability of OBD-II scanner tools suggests that the data contributed by members will be a sparse sampling of routes and cars. Hence, we aim to use data collected by a smaller population to build models capable of predicting the fuel consumption characteristics of a larger population. Admittedly, the conditions of sparse deployment are typically temporary, making the aforementioned contribution short-lived in nature. Nevertheless, it solves a key problem at a critical phase of most newly deployed systems, which makes it important. Before we explain the details of the generalization mechanism, we will provide a brief description of our data collection for the purpose of developing models.

45.4.1 Data Collection

The vision for green navigation, when fully deployed, is to collect data from everyday users, which can be employed to update and refine predictions of fuel consumption when such users (or others with similar vehicles) embark on new itineraries. We conducted a limited proof-of-concept study involving 16 users (with different cars) over the course of 3 months. A total of over 1000 miles were driven by our users to construct the initial models. Figure 45.4 shows a partial map of the paths on which data were collected. The details of the car make, model, year, and the number of miles of data collected for each car are summarized in Table 45.1.

In the aforementioned experiments, each user was asked to drive among a specific set of landmarks in the city. We split each drive into smaller *segments* to capture the variation in the fuel consumption on individual streets. We determined that 1 mile constitutes a good segment size. These segments were the "samples" used to capture the variables affecting fuel consumption and develop initial prediction models.

45.4.2 Derivation of Model Structure

The first part of data generalization is to derive a model structure. The structure describes how various parameters are related, but does not evaluate the various constants and proportionality coefficients. In this case, we derive the structure of fuel prediction models from physical analysis. The analysis is straightforward but is included for completeness.

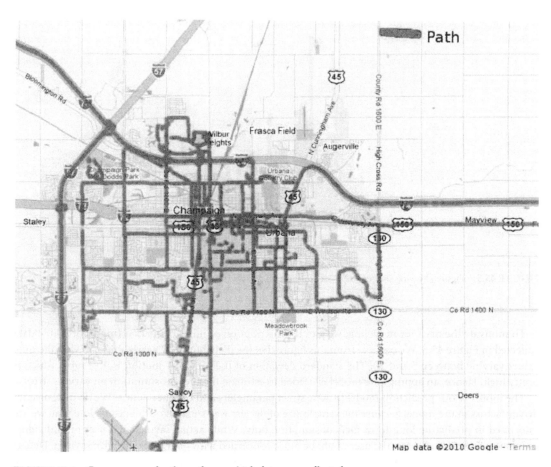

FIGURE 45.4 Coverage map for the paths on which data were collected.

TABLE 45.1 Table Summarizing the Cars Used and the Amount of Data Collected

Car Make	Car Model	Car Year	Miles Driven
Ford	Taurus	2001	135
Toyota	Solara	2001	45
BMW	325i	2006	70
Toyota	Prius	2004	140
Ford	Taurus	2001	136
Ford	Focus	2009	95
Toyota	Corolla	2009	45
Honda	Accord	2003	102
Ford	Contour	1999	22
Honda	Accord	2001	18
Pontiac	Grand Prix	1997	25
Honda	Civic	2002	11
Chevrolet	Prizm	1998	16
Ford	Taurus	2001	10
Mazda	626	2001	9
Toyota	Celica	2001	120
Hyundai	Santa Fe	2008	22

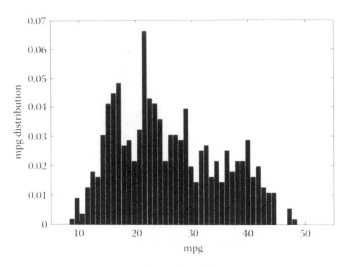

FIGURE 45.5   Figure showing the real mpg distribution for all the 16 users.

To motivate the need for modeling, we plot the distribution of miles per gallon (mpg) for all the data collected in Figure 45.5. We observe from this figure that the distribution is nearly uniform with the mpg values varying between 5 and 50. The standard deviation of the mpg distribution is 9.12 mpg, which is pretty high. Hence, an appropriate model is needed to estimate the fuel consumption on various streets.

The inputs to our prediction model include street parameters and car parameters. We do not consider driver factors in the model because the sample size of drivers was small in our dataset. Note that, we are interested in predicting long-term fuel consumption only. While actual savings of a user on individual commutes to work may vary, the user might be more concerned with their net long-term savings. Hence, it is important to capture only the statistical averages of fuel consumption. As long as the errors have near-zero mean, the savings are accurate in the long term. As a given user drives more segments, a value of interest is the total end-to-end prediction error that results (which improves over time as the individual positive and negative segment errors cancel out). We call that end-to-end error the *cumulative error*. It is useful to normalize that error to the total distance driven. We call the result *cumulative percentage error*. It represents how far we are off in our estimate of total fuel consumption.

We derive the model structure for fuel consumption from the basic principles of physics. Many such models exist in prior literature [6,14,36], which simplifies the task. We divide the parameters that affect fuel consumption into (i) *static street parameters*, namely, numbers of stop signs (*ST*), numbers of traffic lights (*TL*), distance traveled (Δ*d*) and slope (θ), (ii) *dynamic street parameters*, namely, average speed ($\bar{v}$), and *car specific parameters*, namely, weight of the car (*m*), and car frontal area (*A*).

The approximated fuel consumption model as a function of the aforementioned parameters is derived in Appendix 45.A and is given as follows (where *gpm* is the inverse of *mpg* and the unit of measure is gallons per mile):

$$gpm = k_1 m\bar{v}^2 \frac{(ST + vTL)}{\Delta d} + k_2 m \frac{\bar{v}^2}{\Delta d}$$
$$+ k_3 mcos(\theta) + k_4 A\bar{v}^2 + k_5 msin(\theta) \qquad (45.1)$$

In the next section, we show that the coefficients of our model, $k_1, k_2, k_3, k_4,$ and k_5, differ among different vehicles making it harder to generalize from vehicles we have data for to those we do not.

TABLE 45.2   Table Summarizing the Cumulative Percentage Errors for the Individual Car Models and the Generalized Case When All the Data (Except One Car) Are Used to Obtain the Model

Car Make	Car Model	Car Year	Individual Model Cumulative Error %	General Model Cumulative Error %
Hyundai	Santa Fe	2008	2.75	27.06
Honda	Accord	2003	1.81	19.34
Ford	Contour	1999	1.43	88.91
Ford	Focus	2009	0.05	29.35
Ford	Taurus	2001	0.27	15.09
Toyota	Corolla	2009	0.21	85.2
Ford	Taurus	2001	0.33	1.98

45.4.3 Model Evaluation: One Size Fits All?

Regression analysis is a standard technique for estimating coefficients of models with known structure. In this section, we demonstrate that a single regression model is a bad fit for our data. Said differently, while a regression model that accurately predicts fuel consumption can be found for each car from data of that one car, the model found from the collective data pool of all cars is not a good predictor for any single vehicle. Hence, in a sparse data set (where data are not available for all cars), it is not trivial to generalize. We illustrate that challenge by first evaluating the performance of car models obtained from their own data (which is good), then comparing it to the trivial generalization approach: one that finds a single model based on all car data then uses it to predict fuel consumption of other cars. A solution to the challenge is presented in the next section.

One should add that while the generalization challenge is common to many participatory sensing applications, the evaluation reported in the following is not intended to be a definitive study on vehicular fuel consumption. For example, we evaluate fuel consumption in Urbana-Champaign only, which is quite flat. Hence, $\theta = 0$ is a good approximation. (We therefore set the last term, $k_5 m sin(\theta)$, of our physical model to zero, so k_5 is no longer needed.) Furthermore, the city is rarely congested. Moreover, the range of cars used in the study is rather skewed toward sedans, and hence not representative of the diversity of cars on the streets. Fortunately, even this rather homogenous data set is sufficient to show that generalization is hard.

We evaluate the accuracy of models derived from vehicle data using a cross validation approach. We choose a random data point (i.e., a given 1-mile *segment* of a street driven by some car) to predict fuel consumption for. We then use other points to train a model. We distinguish models based on other segments of the same car from models based on data from other cars in predicting the fuel consumption of the one segment. The fourth and fifth columns of Table 45.2 summarize the resulting errors, respectively, for a fraction of the used cars.

We observe from Table 45.2 that the cumulative percentage error for individual car models are quite good. Most of them are below 2%. On the other hand, when we predict one car's consumption using data from other cars, the errors are quite high. This suggests the existence of nontrivial bias in error that does not cancel out by aggregation. In the next section, we propose a way to mitigate this problem based on grouping cars into clusters, such that prediction can be done based on other *similar* cars by some metric of similarity.

45.4.4 Model Clustering

The preceding text suggests a need for better generalization over vehicle data. Different car types behave differently. Even though the model is parameterized by factors such as car weight and frontal area, they are not enough to account for differences among cars. This is a common problem in high-dimensional data sets collected in participatory sensing applications. The question becomes, if we cannot generalize over the whole set, can we generalize over a subset of dimensions?

A solution is borrowed from the general literature on data cubes [21]. Data cubes are structures for online analytical processing (OLAP) that are widely used for multidimensional data analysis. They group data using multiple attributes and extract similarities within each group. For example, previous work showed how to efficiently construct regression models for various subsets of data [9]. The data cube framework can thus help compute the optimal generalization hierarchy in that it can help generalize data based on those dimensions that results in the minimum modeling error.

We consider three major attributes (data dimensions) of a given car: *make, year,* and *class*. Based on these three attributes, data can be grouped in eight ways. At one extreme, all cars may be grouped together, thus producing a single regression model (which we have shown is not acceptable). At the other extreme, cars can be partitioned into clusters based on their (make, year, class) tuple. A separate model is derived for each cluster. Therefore, a 2001 compact Ford is modeled differently from a 2001 mid-size Ford, a 2002 compact Ford or a 2001 compact Toyota.

Between these two extremes, to find out which clustering scheme gives the best accuracy, we obtain the cumulative percentage error for each scheme. The results, summarized in Figure 45.6, show that different generalizations have different quality. These generalizations are somewhat better than using all car data lumped together. While our data set is too small to make general conclusions (from only 16 cars), as more data are collected in a deployed participatory sensing application (e.g., say deployment reaches 100s of cars), progressively better generalizations can be attained.

To use results of Figure 45.6, one would build models for each pair of make and year (the lowest error clustering scheme). If a car is encountered for which we do not have data on its (make, year) cluster, we go one level up and use (make) clusters or (year) clusters as generalizations for the (make, year) cluster. If there are no models corresponding to either make or year of a given car, we have no recourse but to go one level up and use the model computed from all data. Figure 45.7 depicts the generalization process among various model clusters.

We evaluate the performance of our model clustering technique by measuring how accurately an individual car can be modeled using the data from cars with similar make or year. Specifically, we

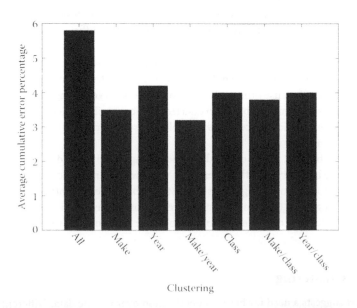

FIGURE 45.6 Mean error percentage of the models obtained from various clustering approaches.

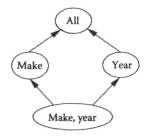

FIGURE 45.7 Model generalization from fine-grained clusters.

TABLE 45.3 Table Showing the Mean Error Percentage for Each Individual Car When Model Clustering Is Used

Car Make	Car Model	Car Year	Cumulative Error %
Hyundai	Santa Fe	2008	0.73
Honda	Accord	2003	1.01
Ford	Contour	1999	1.42
Ford	Focus	2009	2.7
Ford	Taurus	2001	3.38
Toyota	Corolla	2009	1.28

construct the model cluster while removing data of a certain car type. We use the model cluster to estimate the fuel consumption for a given car. The resulting mean error percentage in estimating a segment is presented in Table 45.3. Later, in the evaluation segment, we show that on longer paths, individual segment errors cancel out, leading to a more accurate end-to-end prediction.

To put the aforementioned results in perspective, the reader is reminded that the nature of the landscape in Urbana-Champaign limits our study in that we do not have data on hilly terrain. The study would have been more interesting if conducted on uneven grounds, where changes in incline modulate fuel consumption. Another limitation of our modeling approach arises from the class of cars for which data has been collected. We observe from Table 45.1 that the majority of the cars are sedans (with the exception of one SUV). The generalization tree shown in Figure 45.7 is specific to the data set collected. The point of this section is to illustrate an approach to improve prediction in the temporary but important conditions of sparse deployment. Ultimately, when all cars have their own OBD-II readers supplying data to drivers' cell-phones, we shall not need the generalization scheme described earlier.

45.5 Implementation of Green Navigation

The green navigation service is based on a server that combines provides fuel-efficient route computation. The various modules that are part of the GreenGPS implementation are depicted in Figure 45.8. The server maintains the map of a given area as an *OpenStreetMap* (*OSM*) [29]. OSM is the equivalent of Wikipedia for maps, where data are collected from various free sources (such as the US TIGER database [37], Landsat 7 [27], and user contributed GPS data) and an editable street map of the given area is created in an XML format. The OSM map is essentially a directed graph, which is composed of three basic object types, *nodes*, *ways*, and *relations*. A node has fixed coordinates and expresses points of interest (e.g., junction of roads, Marriott hotel). A way is an ordered list of nodes with tags to specify the meaning of the way, e.g., a road, a river, a park. A relation models the relationship between objects, where each member of the relation

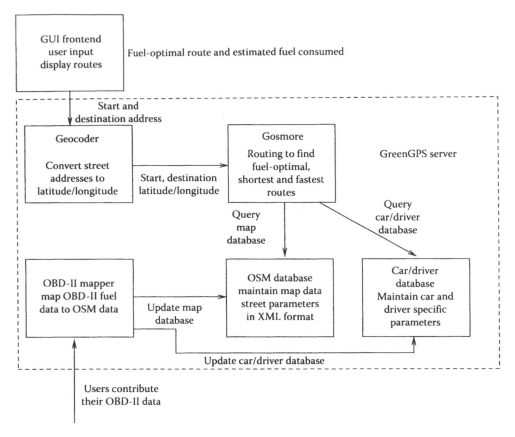

FIGURE 45.8 Figure depicting the various modules of GreenGPS.

has a specific role. Relations are used in specifying routes (e.g., bus routes, cycle routes), enforcing traffic (e.g., one way routes).

The service maintains the street variables affecting fuel consumption as additional parameters in the OSM map. This information is stored as a tag/value pair in the way object, where tag is a street parameter and value is the corresponding value of the parameter. Further, the car and driver specific parameters are maintained in a separate database. The model to compute fuel consumption on a given way (for a given car and driver) queries these databases and computes the fuel consumption on the way.

The OBD-II data shared by individuals is used to compute regression models that predict the fuel consumption on specific streets given the car details (e.g., make, model, age) and driver behavior. The regression variables which are street specific are stored in the OSM map database, whereas the car and driver specific variables are stored in a similar database.

45.5.1 Model Clustering Implementation

The service implements the model clustering technique described in Section 45.4.4 using *Data Cubes* [21].

We implement a three-dimensional (make, class, year) regression cube [9] in C++. Each one mile segment is organized as a row in a database where *five* of its attributes are the values of physical model parameters (see Section 45.4.2) and are used for regression. Three other attributes (make, class, year) are used for grouping. After computing the regression models for all clusters (i.e., materializing the cube), search for a specific triple of (make, class, year) is done consecutively within the (make, year) cluster, the (make) cluster, and the (year) cluster. The first regression model that matches the query is used for prediction.

45.5.2 Routing

Routing is achieved by customizing the open source routing software, Gosmore [28]. Gosmore is a C++ based implementation of a generic routing algorithm that provides shortest and fastest routes between two arbitrary end-points. Gosmore uses OSM XML map data for doing routing. Gosmore's routing algorithm is a heuristic that by default computes the shortest route. This routing algorithm can be thought of as a weighted Dijkstra's algorithm on the OSM map, where the nodes of the graph are OSM nodes and the edges of the graph are OSM ways and the weights of the edges are the lengths (distance) of the ways. The fastest route is computed by multiplying the distance by an inverse speed factor (thus giving lower weights to faster ways). Our fuel-optimal routing algorithm multiplies the distance by an inverse mpg metric that results in lower weights for fuel-optimal ways.

45.5.3 Other Implementation Issues

Street address inputs provided by the user are translated into latitude/longitude pairs using the open source geocoding perl module, Geo::Coder::US. This module is used for geocoding U.S. addresses only. Geocoding is the process of finding corresponding latitude/longitude data given a street address, intersection, or zipcode.

The GUI frontend to display the fuel-optimal route (shown in Figure 45.2) utilizes Microsoft Bing maps. Routes are color coded and rendered as *polylines* on Bing maps. For example, the fuel-optimal route is a "green" color polyline.

When a query is posed to the server for the fuel-optimal route between the start address and destination address, the addresses are first geocoded into their corresponding latitude and longitude pairs using the geocoder module. The latitude and longitude pairs of the start and destination addresses are then fed to the routing module which computes the fuel-optimal route (along with the shortest and fastest routes) using the OSM XML database and the prediction models of fuel consumption on streets (computed from the OBD-II sensor data contributed by users). The computed routes are then displayed on the Bing maps based GUI frontend.

45.6 Evaluation

We evaluate the performance of our implemented green navigation service in two stages. First, we evaluate the performance of our model by using it to predict the end-to-end fuel consumption for long routes. Second, we evaluate the potential fuel savings of an individual using the service.

45.6.1 Model Accuracy

We first evaluate the accuracy of our prediction model in estimating fuel consumption on long routes. These routes are continuous sequences of segments that individuals drove. Only six cars are used in this experiment* because the data from the rest of the cars did not include multiple paths (and hence we would not be able to path-based cross-validation, where data collected on one path is used to predict fuel consumption on another). For cross-validation, we remove the data points associated with a given path

---

* Ford Focus, 2009; Ford Taurus, 2001; Toyota Corolla, 2009; Ford Taurus, 2001; Honda Accord, 2001; and Ford Taurus, 2001.

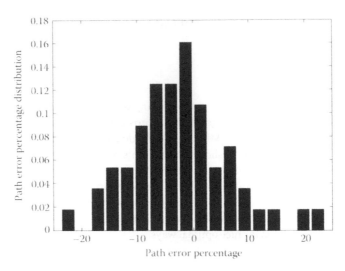

FIGURE 45.9 Distribution of path error percentages when training is done using individual cars.

and obtain a model for the car, then obtain the error in predicting fuel consumption for this path based on the computed model. We repeat the aforementioned process for all the paths.

The entire path error distribution corresponding to the aforementioned experiment when prediction for each car is used based on data of the same car (on other paths) is shown in Figure 45.9. We observe that the path error distribution is nearly normal. We conduct a similar experiment to derive the path error distribution that is achieved by employing clustering such that fuel consumption of cars is predicted from that of other cars in the nearest cluster. To experiment with prediction accuracy of clusters, we remove the data points for each car (as if that car was not known to the system) and cluster the rest of data points, as described in Section 45.4.4, based on make, year, and both. Fuel consumption of the removed car is then predicted using the nearest cluster. Namely, we first check if a cluster exists with the same car make and year. If no such cluster exists, we check for a cluster of the same make or the same year, respectively. Finally, a model based on all car data is used if all the previous steps fail. The prediction error for each path is computed as before and the distribution is presented in Figure 45.10. Again, a normal distribution of the path errors is observed.

In order to understand how path errors vary with path lengths, we bin the paths based on their length and compute the average of the absolute path errors as a function of path length. We repeat this experiment for the case where models are derived for each car individually and the case where models are derived for clusters and the nearest cluster is used. We plot the mean of the absolute path errors for varying path lengths in Figure 45.11. We observe from Figure 45.11 that the error decreases with increasing path length for both the individual and cluster-based models. As expected, models based on the owner's car do better than models based on the nearest cluster, but the cumulative error continues to decrease with distance driven, which is what we want. We have not explored if this holds true when the commutes have large dynamics in speeds, such as in larger cities. The current data set is limited in that it was collected in a fairly quiet town.

From the perspective of building participatory sensing applications, the preceding text suggests the importance of finding models that do not have *biased error*. Since the models often try to predict aggregate or long-term behavior (such as long-term exposure to pollutants, annual cost of energy consumption, eventual weight-loss on a given diet, etc.), if the error in day-by-day predictions is normally distributed with zero mean, the long-term estimates will remain accurate. Hence, rather than worrying about exact models, GreenGPS attempts to find *unbiased* models, which is easier.

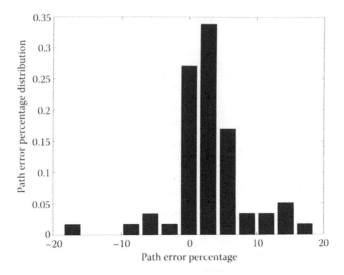

FIGURE 45.10 Distribution of path error percentages for the clustering approach.

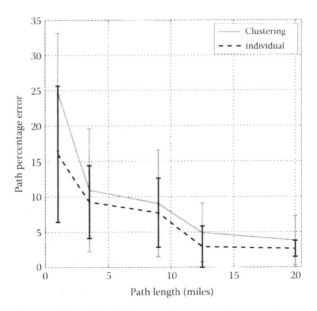

FIGURE 45.11 Mean path error when path length is varied for individual car models and cluster-based models.

45.6.2 Fuel Savings

In this section, we evaluate the fuel savings achieved when using the GreenGPS system. As we outlined in the implementation section, we are integrating the street parameters such as the stop signs, traffic lights, and average speed information into the OSM database. To evaluate fuel savings, we chose landmarks in the city of Urbana–Champaign that the authors visit in their daily commutes, such as work, gym, frequently visited restaurants, and shopping complexes. To eliminate subjective choice of routes between the selected landmarks, each of the authors selected a pair of landmarks then looked up both the shortest route and fastest route between these landmarks on MapQuest. The person then drove eight round trips (of approximately 20–40 min each) between their selected pair of landmarks; four on the shortest route and four on the fastest route, recording actual fuel consumption for each round trip. The landmarks

TABLE 45.4 Table Showing Fuel Consumptions for the Various Roundtrips between Different Landmarks

Car Details	Landmarks	Route Type	Fuel Consumption (gal)				GreenGPS Prediction
Honda Accord 2001	Home 1 to mall	Shortest	0.19	0.16	0.19	0.16	Shortest
		Fastest	0.22	0.23	0.25	0.22	
	Home 1 to gym	Shortest	0.19	0.20	0.19	0.18	Shortest
		Fastest	0.21	0.23	0.22	0.25	
Ford Taurus 2001	Home 2 to restaurant	Shortest	0.24	0.23	0.23	0.22	Shortest
		Fastest	0.3	0.28	0.29	0.29	
Toyota Celica 2001	Home 2 to work	Shortest	0.18	0.16	0.18	0.17	Fastest
		Fastest	0.17	0.14	0.16	0.15	
Nissan Sentra 2009	Home 3 to CUPHD[a]	Shortest	0.14	0.15	0.15	0.15	Fastest
		Fastest	0.13	0.13	0.14	0.14	
Honda Civic 2002	Grad housing to work	Shortest	0.33	0.32	0.33	0.3	Fastest
		Fastest	0.25	0.28	0.27	0.24	

[a] Champaign Public Health Department.

TABLE 45.5 Table Summarizing Savings of the Fuel-Efficient Route for All Cars

Car Details	Landmarks	Savings %
Honda Accord 2001	Home 1 to mall	31.4
	Home 1 to gym	19.7
Ford Taurus 2001	Home 2 to restaurant	26
Toyota Celica 2001	Home 2 to work	10.1
Nissan Sentra 2009	Home 3 to CUPHD	8.4
Honda Civic 2002	Grad housing to work	18.7

together with the shortest and fastest routes are shown in Figure 45.12. We then used the GreenGPS system to predict which of the two compared routes for each pair of landmarks is the better route, which it did correctly in every case.

The fuel consumption data for each roundtrip on the shortest and fastest routes for all the cars in this experiment are shown in Table 45.4.

We observe from Table 45.4 that the fuel-optimal route for destinations of the Honda Accord and Ford Taurus was the shortest route, whereas, for the other three destinations, it was the fastest route. Hence, picking the shortest or fastest routes consistently is not optimal. To confirm that the differences in fuel consumption between the compared routes are not due to measurement noise, we tested the statistical significance of the difference in means using the two paired *t-test*. The test yielded that the differences are statistically significant with a confidence level of at least 90%. The average savings (by choosing the correct route over the alternative) for each pair of landmarks and car are summarized in Table 45.5.

Comparing the total fuel consumed on the optimal route to the average of that consumed on the shortest route and fastest route (assuming the driver guesses at random in the absence of GreenGPS), the savings achieved are roughly 6% over the shortest path and 13% over the fastest, which is consistent with data we reported earlier in the feasibility study.* This is by no means statistically significant, since only a handful of routes were used in the experiments mentioned previously, but it nevertheless shows promise as a proof of concept.

---

* The feasibility study used different routes from those reported earlier.

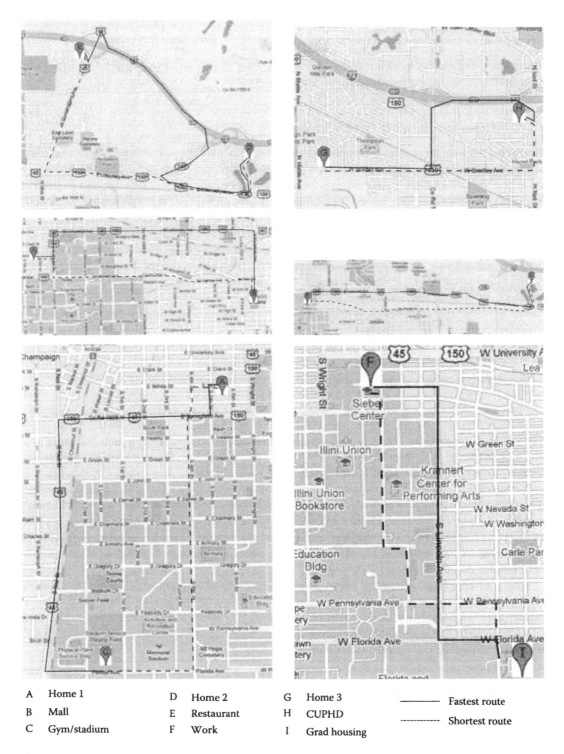

A	Home 1	D	Home 2	G	Home 3
B	Mall	E	Restaurant	H	CUPHD
C	Gym/stadium	F	Work	I	Grad housing

———— Fastest route

------------ Shortest route

FIGURE 45.12 Figure showing the landmarks and corresponding shortest and fastest routes.

45.7 Lessons Learned

This section presents, in its two respective sections, a brief discussion of our experiences with green navigation services and the limitations of the current study.

45.7.1 Experiences with Green Navigation

Several lessons were learned from GreenGPS, as an example of participatory sensing applications. First, we observed that data cleaning is an important problem and it is application dependent. We had several occasions when several fields were missing from the data. For example, the GPS sometimes failed to communicate with the DashDyno and the location fields were then empty. A simple scheme was used to filter complete data sets from those that were missing values. Another data-related issue was the presence of noise in the data. For example, in our setup, we observed that (in some car models) whenever the GPS communicated with the DashDyno, the *fuel rate* measurement had a large spike. This was likely due to improper use of sensor IDs, which led to data overwriting. Solutions have to be developed that filter the noise at the source. For example, we developed a simple filter (as a plugin to PoolView) that removes outliers from the data before storing it. An application-specific challenge was observed due to the slight variations in the OBD-II standards among different cars. For example, we observed that the Toyota Prius (by default) outputs the speed and fuel measurements in the metric system, rather than the Imperial system (which happens to be the default for the remaining cars in our data set). It is harder to propose generic solutions to such problems. They suggest, however, that unlike small embedded systems, participatory sensing applications involve a much larger number of heterogeneous components (e.g., different car types in GreenGPS). As such, components interact with each other or with aggregation services, subtle compatibility problems will play an increasing role. Troubleshooting techniques are needed that are good at identifying problems resulting from unexpected or bad interactions among different individually well-behaved components. This is to be contrasted, for example, with debugging tools that attempt to find bugs in individual components.

Next, privacy challenges come to the forefront in participatory sensing systems. A large class of participatory sensing systems monitor location information continuously, which poses significant privacy issues. Simple anonymization of data will not work in such situations, as the GPS traces can lead to privacy breaches (e.g., reveal the home location of the user and thus uncover their identity). Techniques such as the ones proposed in [17,31], which rely on data perturbation can be used to preserve privacy, while still allowing accurate modeling. In our current study, individual users simply switch off data collection devices when they feel the need for privacy.

Finally, another lesson learned relates to the initial experimental deployment of participatory sensing systems. A major hurdle in getting participatory sensing systems off the ground is to provide the right incentives to the individuals (who are part of the system) [32]. We believe that the initial deployment, which tends to be sparse, should be carefully designed in order to provide incentives for larger adoption. It should therefore be useful from the very early stages.

45.7.2 Limitations of Current Study

Apart from the limitations arising from the small size of the data set, discussed earlier, we also make the following observations. As expected, the main factors affecting fuel consumption of a vehicle on a path are the average speed, the speed variability (estimated by averaging the speed squared), and the engine idle time (estimated from the number of stop signs and stop lights on the path). A limitation of the study is that we did not explore the use of real-time traffic conditions for purposes of fuel estimation. Rather, we opted to use statistical averages of speed, speed variability, and idle time. It is easy to see how such statistical averages can be computed for different hours of the day and different days of the week given a sufficient amount of historical data, yielding expected fuel consumption (in the statistical sense of expectation). The outcome is that individual trips may differ significantly from the statistical expectation. However, by consistently following routes that have a lower expected fuel consumption, savings will accumulate in the long term. Drivers may think of GreenGPS as a long-term investment. Short-term results may vary, but long-term expectations should tend to come true.

In order to achieve the next level of optimization, a next generation of GreenGPS can take into account the real-time situation. Our experience reveals, not surprisingly, that the degree of congestion plays the largest role in accounting for fuel consumption variations among individual trips of the same vehicle. On lightly utilized streets, another main factor is the degree to which traffic lights are synchronized. Lack of synchronization accounted for up to a 50% increase in fuel consumption in our measurements.

Another limitation of the current service is that it does not properly account for turns. Turns on the path add fuel consumption, often because delays in the turn lane differ from those in the through lane. In particular, our measurements show that left turns may add a considerable amount of delay to a path. Hence, routing should account for the type of turn as well.

Finally, we expect that fuel savings in larger cities will be higher than those reported in this chapter, both due to the larger variability in traffic conditions that could be taken advantage of and because of the larger connectivity which offers more alternatives in the choice of route. With the aforementioned caveats, we believe that the study remains of interest in that it explores problems typical to many participatory sensing applications, such as overcoming conditions of sparse deployment, adjusting to heterogeneity, and living with large day-to-day errors toward estimating cumulative properties. The GreenGPS study could therefore serve as an example what to expect in building similar services, as well as a recipe for some of the solutions.

45.8 Related Work

We divide this section into two sections, Section 45.8.1 presents related work in participatory sensing and Section 45.8.2 examines fuel-efficiency-related literature.

45.8.1 Participatory Sensing

Our navigation service is an example of participatory sensing services that have recently become popular in networked sensing. The concept of participatory sensing was introduced in [8]. In participatory sensing, individuals are tasked with data collection which is then shared for a common purpose. A broad overview of such applications is provided in [1]. Several early applications have been published. Examples include CenWits [23], a participatory sensing network to search and rescue hikers; CarTel [24], a vehicular sensor network for traffic monitoring; BikeNet [11], a bikers sensor network for monitoring popular cyclist routes; and ImageScape [33], cellphone camera networks for sharing diet-related images. Our application, GreenGPS, introduces a novel example of this genre that enables individuals to compute fuel-efficient routes within a city.

45.8.2 Fuel Efficiency

A comprehensive study that provides optimal route choices for lowest fuel consumption is presented in [13]. In the paper, fuel consumption measurements are made through the extensive deployment of sensing devices (different from the OBD-II) in experimental cars. These fuel consumption measurements are then used to compute the lowest fuel consumption route. In contrast to the work in [13], our chapter uses a sparse deployment to build mathematical models for predicting fuel consumption for other streets and cars. In [7], the influence of driving patterns of a community on the exhaust emissions and fuel consumption were studied. Feedback was provided to the community regarding the driving patterns to cut back on the fuel consumption and exhaust. A driver support tool, FEST, was developed in [10]. FEST uses sensors installed in the car along with a software to determine the driving behavior of the driver and provide real-time feedback to the individual for the purpose of reduction in fuel consumption. An extension to FEST that includes more experiments and further evaluation can be found in [38]. A feedback control algorithm was developed in [34] that determines speed of automobiles on highways with varying terrain to achieve minimal fuel consumption. An extension to the work in [34] was developed in [22]. In [22], suggestions of driving style to minimize fuel consumption were made for varying road and trip types (e.g., constant grade road, hilly road). The problem was formulated using a control theoretic approach.

UbiGreen [15] is a mobile tool that tracks an individual's personal transportation and provides feedback regarding their CO_2 emissions.

In a separate study [25], it was shown that rising obesity has a significant impact on the total fuel consumption in the United States. Models were developed that studied the impact of obesity on the amount of fuel consumed in passenger vehicles.

Our work represents the first participatory sensing service that aims at improving fuel consumption. Using data collected from volunteer participants, models are built and continuously updated that enable navigation on the minimum-fuel route.

45.9 Conclusions and Future Work

In this chapter, we presented a proof of concept study that suggests the advantages of green vehicular navigation. A service was described that relies on OBD-II data collected and shared by a set of users via a participatory sensing framework to enable computing more energy-efficient routes. Lessons were described that extrapolate from experiences with this service to broad issues with green navigation design in general. This study shows that significant fuel savings can be achieved, which not only reduces the cost of fuel, but also has a positive impact on the environment by reducing CO_2 emissions. An important issue addressed was surviving conditions of sparse deployment. A hierarchy of models described to estimate the fuel consumption, shooting for models that are unbiased, if not accurate.

45.A Appendix: Deriving the Physical Model for Fuel Consumption

Assuming that the engine RPM is ωs^{-1}, the torque generated by the engine is $\Gamma(\omega)$, the final drive ratio is G, the kth gear ratio is g_k, and the radius of the tire is r, then F_{engine} is given by the following equation: $F_{engine} = \frac{\Gamma(\omega)Gg_k}{r}$.

The frictional force, $F_{friction}$, is characterized by the gravitational force acting on the car, given by $mgcos(\theta)$, where m is the mass of the vehicle and g is the gravitational acceleration and the coefficient of friction, c_{rr}. The equation for frictional force is: $F_{friction} = c_{rr}mgcos(\theta)$.

The gravitational force, F_g, due to the slope is given by the following equation: $F_g = mgsin(\theta)$.

Finally, the force due to air resistance, F_{air}, is given by the following equation: $F_{air} = \frac{1}{2}c_dA\rho v^2$.

In this equation, c_d is the coefficient of air resistance, A is the frontal area of the car, ρ is the air density, and v is the current speed of the car.

Assuming that the car is on an upslope, the final force acting on the car is given by the following equation: $F_{car} = F_{engine} - F_{friction} - F_{air} - F_g$.

In order to obtain a relation between the fuel consumed and the aforementioned forces, we note that the fuel consumed is related to the power generated by the engine at any instance of time t. If f_r is the fuel rate (fuel consumption at a given time instance) and P is the instantaneous power, then $f_r \propto P$. Power is related to the torque function, $\Gamma(\omega)$, and engine RPM, ω as follows: $P = 2\pi\Gamma(\omega)\omega$. Hence, we obtain $f_r = \beta\Gamma(\omega)\omega$.

In the aforementioned equation, β is a constant. Further, we also have the relationship $v = r\omega$ from rotational dynamics. From the aforementioned equations, we obtain the fuel consumption rate as a function of the forces acting on the car as follows:

$$F_{car} = ma$$

$$= \frac{f_r G g k}{r\omega\beta} - c_{rr}mg\cos(\theta) - \frac{1}{2}c_d A \rho v^2$$

$$- mg\sin(\theta)$$

$$mav = \beta'f_r - c_{rr}mg\cos(\theta)v$$

$$- \frac{1}{2}c_d A \rho v^3 - mgv\sin(\theta)$$

$$f_r = k_1 mav + k_2 mvc\cos(\theta)$$

$$+ k_3 A v^3 + k_4 mv\sin(\theta)$$

Finally, we can obtain the equation for the fuel consumed, f_c, by integrating the rate of fuel consumption with respect to time. We obtain the following equation:

$$f_c = \int_{t_1}^{t_2} f_r(t)\, dt$$

$$= \int_{t_1}^{t_2} (k_1 mav + k_2 mvc\cos(\theta) + k_3 A v^3$$

$$+ k_4 mv\sin(\theta))\, dt$$

In order to further derive a model that can be used for regression analysis, we will detail the various components that are part of the fuel consumption of a car. As shown in the aforementioned equation, a moving car at a constant speed on a straight road which does not encounter any stop lights or traffic will only need to overcome the frictional forces caused by the road, the air, and gravity. These are represented by $\int_{t_1}^{t_2} k_2 mvc\cos(\theta)$, $\int_{t_1}^{t_2} k_3 A v^3$, and $\int_{t_1}^{t_2} k_4 mv\sin(\theta)$, respectively. On the other hand, the first component $\int_{t_1}^{t_2} k_1 mav$ can be broken down further into two components, one is the extra fuel consumed due to encountering stop signs (ST) and traffic lights (TL) and the second one is the extra fuel consumed due to congestion. Hence, the previous equation now becomes the following:

$$f_c = \int_{t_1}^{t_2} (k_{11} mav(ST + vTL)$$

$$+ k_{12} mav) + k_2 mvc\cos(\theta)$$

$$+ k_3 A v^3 + k_4 mv\sin(\theta))\, dt$$

If we replace v with $\bar{v}$, the average speed, assume that θ remains constant, and we know $a = dv/dt$, we can further simplify the aforementioned integral to the following:

$$f_c = k_{11}m\bar{v}^2(ST + \nu TL) + \frac{k_{12}m\bar{v}^2}{2}$$

$$+ k_2 m\Delta d cos(\theta) + k_3 A\bar{v}^3 \Delta t$$

$$+ k_4 m\Delta d sin(\theta)$$

In the aforementioned equation, Δd is the distance traveled and Δt is the time traveled. Dividing the aforementioned equation by Δd gives us the metric *fuel consumed per mile (gpm)*, which is appropriate for our analysis purposes. Hence, our final model will now be

$$gpm = k_{11}m\bar{v}^2\frac{(ST + \nu TL)}{\Delta d} + k_{12}m\frac{\bar{v}^2}{2\Delta d}$$

$$+ k_2 m cos(\theta) + k_3 A\bar{v}^2 + k_4 m sin(\theta)$$

References

1. T. Abdelzaher et al. Mobiscopes for human spaces. *IEEE Pervasive Computing*, 6(2):20–29, 2007.
2. Actron. Elite autoscanner. http://www.actron.com/product_category.php?id=249
3. Auterra. Dashdyno. http://www.auterraweb.com/dashdynoseries.html
4. AutoTap. Autotap reader. http://www.autotap.com/products.asp
5. AutoXRay. Ez-scan. http://www.autoxray.com/product_category.php?id=338
6. D. M. Bevly, R. Sheridan, and J. C. Gerdes. Integrating INS sensors with GPS velocity measurements for continuous estimation of vehicle sideslip and tire cornering stiffness. In *Proceedings of American Control Conference*, Arlington, VA, pp. 25–30, 2001.
7. K. Brundell-Freij and E. Ericsson. Influence of street characteristics, driver category and car performance on urban driving patterns. *Transportation Research, Part D*, 10(3):213–229, 2005.
8. J. Burke et al. Participatory sensing. *Workshop on World-Sensor-Web, Co-located with ACM SenSys*, Boulder, CO, 2006.
9. Y. Chen et al. Regression cubes with lossless compression and aggregation. *IEEE Transactions on Knowledge and Data Engineering*, 18(12):1585–1599, 2006.
10. V. der Voort. Fest—A new driver support tool that reduces fuel consumption and emissions. *IEE Conference Publication*, 483:90–93, 2001.
11. S. B. Eisenman et al. The bikenet mobile sensing system for cyclist experience mapping. In *Proceedings of SenSys*, Sydney, New South Wales, Australia, November 2007.
12. EPA. Emission facts: Greenhouse gas emissions from a typical passenger vehicle. http://www.epa.gov/OMS/climate/420f05004.htm
13. E. Ericsson, H. Larsson, and K. Brundell-Freij. Optimizing route choice for lowest fuel consumption—Potential effects of a new driver support tool. *Transportation Research, Part C*, 14(6):369–383, 2006.
14. J. Farrelly and P. Wellstead. Estimation of vehicle lateral velocity. In *Proceedings of IEEE Conference on Control Applications*, Dearborn, MI, pp. 552–557, 1996.
15. J. E. Froehlich et al. Ubigreen: Investigating a mobile tool for tracking and supporting green transportation habits. In *Proceedings of Conference on Human Factors in Computing*, Boston, MA, pp. 1043–1052, 2009.

16. R. K. Ganti, N. Pham, H. Ahmadi, S. Nangia, and T. F. Abdelzaher. GreenGPS: A participatory sensing fuel-efficient maps application. In *Proceedings of the 8th International Conference on Mobile Systems, Applications, and Services*, MobiSys '10, San Francisco, CA, pp. 151–164, 2010.

17. R. K. Ganti, N. Pham, Y.-E. Tsai, and T. F. Abdelzaher. Poolview: Stream privacy for grassroots participatory sensing. In *Proceedings of SenSys '08*, Raleigh, NC, pp. 281–294, 2008.

18. Garmin eTrex Legend. www8.garmin.com/products/etrexlegend

19. Google. Google Maps. http://maps.google.com

20. GPS POI. Red light database. http://www.gps-poi-us.com/

21. J. Gray et al. Data cube: A relational aggregation operator generalizing group-by, cross-tab and sub-totals. *Data Mining and Knowledge Discovery*, 1(1):29–54, 1997.

22. J. N. Hooker. Optimal driving for single-vehicle fuel economy. *Transportation Research, Part A*, 22A(3):183–201, 1988.

23. J.-H. Huang, S. Amjad, and S. Mishra. Cenwits: A sensor-based loosely coupled search and rescue system using witnesses. In *Proceedings of SenSys*, San Diego, CA, pp. 180–191, 2005.

24. B. Hull et al. Cartel: A distributed mobile sensor computing system. In *Proceedings of SenSys*, Boulder, CO, pp. 125–138, 2006.

25. S. H. Jacobson and L. A. McLay. The economic impact of obesity on automobile fuel consumption. *Engineering Economist*, 51(4):307–323, 2006.

26. MapQuest. http://www.mapquest.com

27. National Aeronautics and Space Administration (NASA). Landsat data. http://landsat.gsfc.nasa.gov/data/

28. Nic Roets. Gosmore. http://wiki.openstreetmap.org/wiki/Gosmore

29. OpenStreetMap. Openstreet map. http://wiki.openstreetmap.org/

30. Owen Brotherwood. Symbtelm. http://sourceforge.net/apps/trac/symbtelm/

31. N. Pham, R. Ganti, Y. Sarwar, S. Nath, and T. Abdelzaher. Privacy-preserving reconstruction of multidimensional data maps in vehicular participatory sensing. In *LNCS Proceedings of EWSN*, Coimbra, Portugal, pp. 114–130, 2010.

32. S. Reddy, D. Estrin, and M. Srivastava. Recruitment framework for participatory sensing data collections. To appear in *Proceedings of International Conference on Pervasive Computing*, Helsinki, Finland, 2010.

33. S. Reddy et al. Image browsing, processing, and clustering for participatory sensing: Lessons from a Dietsense prototype. In *Proceedings of EmNets*, Cork, Ireland, pp. 13–17, 2007.

34. A. B. Schwarzkopf and R. B. Leipnik. Control of highway vehicles for minimum fuel consumption over varying terrain. *Transportation Research*, 11(4):279–286, 1977.

35. Traffic. Real-time traffic conditions. http://www.traffic.com/

36. H. E. Tseng. Dynamic estimation of road bank angle. *Vehicle System Dynamics*, 36(4–5):307–328, 2001.

37. U.S. Census Bureau. Tiger database. http://www.census.gov/geo/www/tiger/

38. M. van der Voort, M. S. Dougherty, and M. van Maarseveen. A prototype fuel-efficiency support tool. *Transportation Research, Part C*, 9(4):279–296, 2001.

46

Energy-Aware Mobile Multimedia Computing

Jianxin Sun
University of Nebraska–Lincoln

Dalei Wu
University of Nebraska–Lincoln

Jiucai Zhang
University of Nebraska–Lincoln

Xueyi Wang
University of Nebraska–Lincoln

Song Ci
University of Nebraska–Lincoln

46.1 Introduction

In recent years, mobile multimedia have become feasible and popular due to the rapid advances of semiconductor and portable diverse technology. Technology advances in video compression and transmission over wireless communication networks have enabled mobile multimedia on portable wireless devices, such as cellular phones, laptop computers connected to WLANs, and cameras in surveillance and environmental tracking systems. Video coding and streaming are also envisioned in an increasing number of applications in the areas of battlefield intelligence, reconnaissance, public security, and telemedicine. Present 3G and emerging 4G wireless systems and IEEE 802.11 WLAN/WMAN have dramatically increased transmission bandwidth and generated a great amount of users on video streaming applications. Although wireless video communications is highly desirable, a primary limitation in wireless systems is the basic design architecture that most mobile devices are typically powered by batteries with limited energy capacity. This limitation is of fundamental importance due to the high energy consumption rate in encoding and transmitting video bit streams during multimedia communications. Moreover, due to the relatively slow development of battery technologies, energy stored in a battery fitted in the limited size of mobile devices cannot catch up with the power consumption of the super multimedia processor continuously developed in the pattern of Moore's law. Thus, the gap between power consumption of the mobile multimedia application and the limited power source is becoming widened. Much work and research have been focused on energy-aware mobile multimedia communication and green computing of multimedia coding processes. Basically, there are three main directions of dealing with this problem. The first direction is to engage in hardware architecture improvement to optimize energy efficiency, which mainly depends on the technique and new design of microelectronics. Secondly, the nonlinear characteristics of different battery types, such as battery current effect, have been analyzed and adopted to achieve higher battery utilization. Piles of battery cells are operated to provide energy according to a dynamic sequence or pattern, which is called power management and scheduling. The third main direction looks into series of procedures applied to multimedia communication and wisely selects complexity control parameters in each procedure to secure the delivered quality and minimize energy consumption at the same time. It has been shown that achieving a satisfactory user experience needs a systematic consideration of both video source adaptation and network transmission adaptation, indicating that the core of mobile multimedia system design is how to achieve an ideal energy allocation balance between computation and communication. In other words, it depends on how to jointly select those computational complexity control parameters during video coding and transmission according to the real-time status and constraints, such as the video content characteristics, available network resources, underlying network conditions, battery capacity conditions, and distortion requirement.

46.2 Energy-Aware Wireless Video Communication

A conceptual power-aware mobile multimedia system is illustrated in Figure 46.1. At the transmitter, different video source adaptation methods, such as typical prediction and quantization, scalable video coding [24], transcoding [34], object-based video coding [32], and summarization [22], may provide a different video compression rate based on the video content to match the receiver capability. Then, the compressed data packets are transmitted over wireless links. To combat the lossy nature of a wireless channel, adaptive modulation and channel coding schemes as well as transmitter power at the physical layer can be adjusted based on the channel state information (CSI). Mobile multimedia receiver devices demodulate the received bit stream, perform error detection and correction, decode the received bit stream, and display reconstructed video clips. Mobile receiver devices may interact with the transmitter

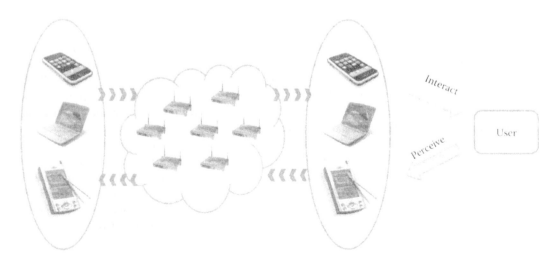

FIGURE 46.1   A conceptual power-aware mobile multimedia system.

devices to adaptively adjust the compression rate to provide differentiated services following the interactive activities of end users. Note that mobile multimedia devices are powered by battery, which is not an ideal energy source, since it tends to provide less energy at higher discharging currents. To minimize the total consumed battery energy for delivering a video clip with satisfied quality, joint rate–distortion–complexity optimization to prolong the battery operating time is necessary.

46.2.1 Power Management in Mobile Devices

Since battery technology cannot satisfy the growing power demand of mobile multimedia devices, power management technology is needed to increase system power efficiency. However, efficient use of the limited battery energy is one of the major challenges in designing mobile multimedia communication devices with limited battery energy supply. This is because (1) real-time multimedia is bandwidth-intense and delay-sensitive, and the battery may need to continuously discharge; (2) a wireless channel dynamically varies over time and space due to fast and large-scale channel variations; (3) different mobile devices have different limited processing power levels, limited memory and display capabilities, and limited battery energy supply due to the size and weight constraints; (4) video quality does not increase linearly as the complexity increases; and (5) battery discharge behaviors are nonlinear. Since the performance of each part in the mobile multimedia device is dynamic and heterogeneous, all these innate conflicts induce major research challenges in designing these mobile multimedia devices. Therefore, how and when to apply a particular power reduction technique is a challenging problem. However, there does not yet exist a systematic method for the power management of mobile devices in real-time multimedia applications.

46.2.1.1 Dynamic Voltage and Frequency Scaling

Multimedia content is time-varying as well as delay sensitive. As a result, to maintain a dynamic balance between the operating level of the processor and the QoS of multimedia application is challenging. In [5], an offline linear programming method has been proposed to determine the minimum energy consumption for processing multimedia tasks under stringent delay deadlines. In [2], an optimal frequency was assigned by a buffer-controlled DVS framework to optimally schedule active and passive states for a video decoding system. The work in [31] uses the workload of a video application to determine the frequency and voltage of a processor for playing streaming video with less power consumption while minimizing data losses. The proposed DVS algorithm has been implemented on a PXA270 processor with a Linux 2.6 kernel. In [1], both CPU and multimedia accelerator have been considered to reduce the power consumption of handheld systems. In [21], based on the statistical analysis of more than 600 processor load trace files, a novel interval-based DVS scheme has been proposed to handle the nonstationary behavior by using an efficient online change detector and important parameters, and thereby the penalty incurred by DVS can be efficiently controlled.

46.2.1.2 Maximizing Available Battery Capacity

Due to the nonlinear battery effects, the actual battery capacity of a fully charged battery is always less than its theoretical capacity. Battery capacity decreases as the discharging current increases and will recover the decreased capacity when the battery has been rested or discharged at a low current rate. Therefore, useable battery capacity is significantly affected by the discharge current shape. As a result, a minimum-power-consumption policy does not necessarily result in the longest battery operating time. Battery-aware scheduling schemes attempt to tailor the current of a device to match the optimal discharge rate of the battery. However, for the run-time and delay sensitive multimedia application, how to achieve the best trade-off between the discharge current shape and video quality is extremely challenging.

46.2.1.3 Power-Aware Transmission and Buffer Management

Wireless network interface cards (NICs) have multiple operation modes such as sleep, idle, transmit, and receive. Each mode has different power consumption. Thus, significant energy savings can be achieved by switching the operation mode from idle to sleep or even off during idle periods. However, an extra amount of power consumption is spent to activate or deactivate the electronic components for mode transition. For multimedia applications, how to optimally set the transition point of the NIC is a crucial problem. Bursty traffic could combine the short idle intervals into longer ones to reduce the number of mode transition [6,7]. Consequently, power consumption on mode transition is reduced. In [35], the minimum buffer size on the receiver side was determined to achieve the maximum energy saving under three cases: single-task, multiple subtasks, and multitask. In [23], a power-saving approach using a realistic network framework in the presence of noise has been analyzed. The transcoded video is buffered by proxy middleware buffers, and then the buffered video is transferred in bursts over a given time. Thus, the NIC modes are alternatively switched between active and idle to save power. In [4], a power-aware transmission scheme can switch off the card while frames are being played back until a low-threshold level is reached in the client buffer. In [10], the video data is queued in a buffer and sent by bursts at a longer interval. Consequently, much energy on transmission can be saved.

46.2.2 Power Consumption Trade-Off between Computation and Communication

The total power consumed by a mobile device is mainly composed of the power to code the source at the application layer and the power to transmit the coded bits at the physical layer. A high compression ratio will increase the encoding computational complexity and require more computational power. For desired compression introduced distortion, the computational power is a decreasing function of the coded bit rate. On the other hand, to maintain the desired bit error rate, adequate transmission power is needed. Therefore, the total power consumption on coding and transmitting video frame k is a convex function of the transmission rate, which can be denoted as

$$P_{total}^k = P_C^k(R) + P_T^k(R) \tag{46.1}$$

where
   P_C^k is the consumed power in coding the kth frame
   P_T^k is the consumed power in transmitting the coded bits at transmission rate R

Overall, all practical communication networks have limited bandwidth and are lossy by nature. Furthermore, wireless channel conditions and multimedia content characteristics may change continuously, requiring constant value updates of source and channel parameters. In addition, multimedia streaming applications typically have different quality of service (QoS) requirements with respect to packed loss probability and delay constraints. Therefore, the total power consumption of mobile multimedia devices can be minimized by taking advantage of the specific characteristics of the video source and jointly adapting video source coding, transmission power, and modulation and coding schemes.

Generally, to minimize the total power P_{tot}, the source coding parameters S, channel (transmission) parameters C, NIC setting N, and decoder parameters (e.g., error concealment strategy) Q have to be jointly considered to satisfy distortion and delay constraints. The goal of power consumption trade-off between computation and communication is to minimize the total consumed power. The problem can be stated as

$$minP(S, C, N, Q)$$

$$s.t. \begin{cases} D_{tot}(S, C, N, Q) \leq D_0 \\ T_{tot}(S, C, N, Q) \leq T_0 \\ C_{tot}(S, C, N, Q) \leq C_0 \end{cases} \tag{46.2}$$

where

D_0 is the maximum distortion to ensure the satisfied video quality

T_0 is the end-to-end delay constraint imposed by the given video application

C_0 is the maximum computational complexity that the mobile multimedia device can provide

The selection of S, C, N, and Q will affect the end-to-end distortion D_{tot}, delay T_{tot}, and computational complexity C_{tot}.

To solve the problem in (46.2), we need to understand (1) how source adaptation at a video codec affects the computational complexity and the achieved video quality; (2) how transmission adaptation affects the power consumption on transmission and the obtained video quality. We will discuss these two issues in the following sections in detail.

46.3 Power-Aware Video Coding

Video coding achieves high compression efficiency, and enables high resolution videos to be played by mobile multimedia devices. However, the high coding efficiency of video coding is achieved at the cost of high computational complexity. As a result, a significant burden is put on the processor, which is challenging for mobile multimedia devices with limited processing capabilities and battery energy.

46.3.1 Estimation of Codec Power Consumption by Its Predictable Computational Complexity

The video encoding and decoding flexibility provides a variety of multimedia implementation platforms, and enables significant tradeoff between video coding quality and computational complexity. In order to optimally select the optimal operating point of a multimedia application for a specific system, the rate distortion and the complexity characteristics of the operational video coders should be accurately modeled. For example, the computational complexity of an H.264/AVC baseline decoder is mainly determined by two major components: time complexity and space (or storage) complexity. The computational complexity of each module can be found in [13]. A tool for the complexity analysis of reference description has been proposed in [3]. In [27], a generic rate–distortion–complexity model has been proposed to generate digital item adaptation descriptions for image and video decoding algorithms running on various hardware architectures. The model can estimate average decoding complexities as well as the transmission bit-rate and content characteristics. As a result, the receiver can negotiate with the media server/proxy to have a desired complexity level based on their resource constraints. Based on operational source statistics and off-line or online training to estimate the algorithm and system parameters. An analytical rate–distortion–complexity modeling framework for wavelet-based video coders has been proposed in [12].

46.3.2 Computational Complexity Reduction by Approximation and the Corresponding Challenges

The computational complexity can be scalable in various aspects. Based on the observations (1) not every round of local refinement of fast motion search algorithms can achieve an equally good sum of absolute difference operations; (2) motion estimation of smaller block modes is often redundant. A joint rate–distortion–complexity optimization framework has been proposed to balance the coding efficiency and the complexity cost of the H.264 encoder in [14]. The method can cut off the complexity-inefficient motion search rounds, skip redundant motion search of small block modes, and terminate motion search at the optimal rate–distortion–complexity points. In [29], scalable memory complexity reduction has been considered via recompressing I- and P- frames prior to motion-compensated prediction. A simple rate–distortion–complexity adaptation mechanism for wavelet-based video decoding based on the number of decoded nonzero coefficients used prior to the inverse discrete wavelet transform has been proposed in [28].

In addition, choosing the right set of encoder parameters results in efficiently coded video. However, joint rate–distortion–complexity analysis of H.264 is complex due to the large number of possible combinations of encoding parameters. As a result, exhaustive search techniques are infeasible in encoder parameter selection. Several heuristic algorithms have proposed to reduce the computational complexity in video coding. In [16], a subset of coding parameter choices are selected and algorithmic simplifications are enforced, and then the effect of each parameter choice and simplification on both performance and complexity reduction is quantified. Rate–distortion–complexity optimization of integer motion estimation in H.264 has been discussed in [26]. In [30], the computational complexity and distortion are estimated based on the encoding time and mean squared error measurement. Furthermore, the generalized Breiman, Friedman, Olshen, and Stone (GBFOS) algorithm has been used to efficiently obtain parameter settings so that obtained distortion-complexity points are close to optimal. In [9], a nonheuristic nonprobabilistic approach based on nonadditive measure quantitatively captures the dynamic interdependency among system parameters under uncertainties, which is a possible method to effectively and efficiently optimize codec parameters.

Furthermore, the quality performance does not increase linearly as the complexity increases. There is a saturation point of quality improvement. Beyond that point, significant computational effort may get little performance improvement, which makes joint rate–distortion–complexity analysis significantly challenging. Moreover, the video content and its characteristics can be very different sequence by sequence, or even frame by frame. The video content can be slow motion such as head-and-shoulder video, fast motion such as sport videos, or global motion. A video frame may contain a simple scene with a few object motions or a complex scene with many object motions.

The goal of power-aware codec design is to optimally select codec modes to minimize the power consumption on computation with the desired visual quality and delay constraints. However, joint rate–distortion–complexity optimization makes our optimization framework even more challenging as the state space increases significantly if more options are considered in the optimization. In that sense, designing deterministic power-aware codec algorithms is extremely challenging. Therefore, various heuristic approaches have been proposed to design power-aware codes. Selection of different video compression algorithms will bring about different levels of video quality and power consumption. Content-aware algorithms can reduce the power consumption with the lossless user perception, while lossy fast algorithms can adaptively trade off the user perception with power consumption. Consequently, a power-aware codec can dynamically select video compression algorithms to reduce power consumption based on user satisfaction, video content characteristics, as well as battery states. In [17], a configurable video coding system is proposed, which uses an exhaustive search and the Lagrangian multiplier method to optimize the performance and computational complexity. Power-aware concepts and considerations of specific conditions such as different battery status, signal content, user preferences, and operating environments have been proposed. The proposed system can dynamically set the codec mode based on different battery situations to prolong the battery operating time. In [20], an embedded compression algorithm and VLSI architecture with multiple modes for a power-aware motion estimation has been presented, which reduces external access caused by video content and further reduces the power consumption of the codec. The architecture adaptively performs graceful tradeoffs between power consumption and compression quality. The methodology of power-aware motion estimation has also been addressed in literature. In [8], hardware-oriented algorithms and corresponding parallel architectures of integer ME and fractional ME have been proposed to achieve memory access power reduction and provide power scalability and hardware efficiency, respectively.

46.3.3 Scalable Video Coding

Scalable video coding (SVC) provides the capability to easily and rapidly fit a compressed bit stream with the bit rate of various transmission channels and with the display capabilities and computational resource constraints of various receivers. This is achieved by structuring the data of compressed video bit streams into layers. The base layer bit streams correspond to the minimum quality, frame rate, and resolution, which provide basic video quality and must be transferred, and determines the minimum power needed to drive the codec. The enhancement layer bit streams represent the same video at gradually increased quality and/or increased resolution, and/or increased frame rate, which provide a flexible coding structure for temporal, spatial, and quality scalability. Properly enabling the enhancement layer allows balancing of the video quantity and computational complexity so as to provide a power-aware feature for codec design.

For mobile devices, throughput variations and varying delay depend on the current reception conditions and need to be considered. Scalability of a video bit stream provides various media bit rates to match device capability without the need of transcoding or re-encoding. SVC can intelligently thin a scalable bit-stream. Bit rate scalable media may combine with unequal error protection, selective retransmission, or hierarchical modulation schemes to strongly protect the important part of the scalable media for overcoming worst-case error scenarios and give less protection to the enhancement layer in order to overcome the most typical error situations. Thus, video quality may gracefully degrade to adapt to channel conditions. In [59], a video bit rate adaptation method relying on a scalable representation drastically reduces computational requirement in the network element.

46.4 Power-Aware Video Delivery

Transmitting video over wireless channel faces a unique challenge. Due to the shadowing and multipath effects, wireless channel gain varies over time, and thus signal transmission is significantly unreliable. Therefore, constant power cannot lead to the best performance. Although the reliability of signal transmission can be increased by increasing the transmitter power, most mobile multimedia devices are powered by battery with a limited power source, making it an unpractical solution. How to achieve satisfied QoS over a fading channel with the minimum power consumption is critical for mobile multimedia device design. In this section, we examine and review the most popular techniques for power-aware video delivery in mobile multimedia applications, i.e., joint source-channel coding and power adaptation, and cross-layer design and optimization. We present a general framework that takes into account multiple factors, including source coding, channel resource allocation, and error concealment, for the design of power-aware wireless video delivery systems.

46.4.1 General Framework

Since video encoding and data transmission are the two dominant power-consuming operations in wireless video communication, we focus on how to jointly optimize source coding parameters S (e.g., prediction mode and quantization step size) and channel parameters C (e.g., channel codes, modulation modes, transmission power levels, or data rates) in a power-aware video communication system to achieve a targeted video quality or energy usage. Moreover, the delay performance is more crucial than the computational complexity in real-time video delivery. Therefore, from (46.2), the problem of power-aware wireless video delivery can be formally stated as

$$\min_{\{S,C\}} E_{tot}$$

$$s.t. \begin{cases} D_{tot}(S, C) \leq D_0 \\ T_{tot}(S, C) \leq T_0 \end{cases} \tag{46.3}$$

where

E_{tot} is the total energy consumption

T_0 is the end-to-end delay constraint imposed by the application

D_0 is the end-to-end distortion constraint

For video delivery over a lossy channel, the distortion at the receiver is a random variable from the sender's point of view. Thus, the expected end-to-end distortion (averaged over the probability of loss) is usually used to characterize the received video quality and guide the source coding and transmission strategies at the sender.

The end-to-end distortion $E[D_k]$, the end-to-end delay T_{tot}, and the total energy E_{tot} in are all affected by parameters S and C. We use $D_{tot}(S, C)$, $T_{tot}(S, C)$, and $E_{tot}(S, C)$ to explicitly indicate these dependencies. The expected distortion for the kth packet can be written as

$$E[D_k] = (1 - p_k)E\left[D_k^r\right] + p_k E\left[D_k^l\right], \tag{46.4}$$

where

p_k is the probability of loss for the kth packet

$E\left[D_k^r\right]$ is the expected distortion if the packet is received correctly, which accounts for the distortion due to source coding as well as error propagation caused by interframe coding

$E\left[D_k^l\right]$ is the expected distortion if the packet is lost, which accounts for the distortion due to concealment

The probability of packet loss p_k depends on the channel state information (CSI), transmitter power, modulation, and channel coding used. Given transmission rate R, the transmission delay needed to send a packet of L bits is $T = L/R$. The energy needed to transmit the packet with transmission power P is given by $E = PL/R$.

46.4.2 Joint Source-Channel Coding and Power Adaptation

In the literature, joint source-channel coding and power adaptation is a critical technique to achieve power-aware video delivery. In this section, we consider several examples to show how the source coding and channel parameters, including the transmission power, can be jointly selected to achieve energy efficient video coding and transmission.

A joint source-channel coding and power adaptation system is a scheme, in which source coding parameters at the encoder and channel parameters at the transmitter are jointly selected by the controller based on the source content, the error concealment strategy, and the available CSI. In power-aware wireless video delivery systems, transmitter power adaptation and channel coding are two powerful techniques to overcome bit errors caused by unreliable wireless network links. Taking advantage of the specific characteristics of video source and jointly adapting video source coding decisions with transmission power, modulation, and coding schemes can achieve substantial energy efficiency compared with nonadaptive transmission schemes. The authors in [36] proposed a framework where source coding, channel coding, and transmission power adaptation are jointly designed to optimize video quality given constraints on the total transmission energy and delay for each video frame. In addition to the used rate-compatible punctured convolutional (RCPC) codes, transmission power of each packet is also adapted to decrease the loss probabilities of packets. The work in [11] jointly considered optimal mode and quantizer selection with transmission power allocation.

To illustrate the performance gain of joint adaptation of the source coding and transmission parameters in power-aware mobile video systems, we present some experimental results, which are discussed in detail. In the experiment, a joint source coding and transmission power allocation (JSCPA) approach is compared with an independent source coding and power allocation (ISCPA) approach in which S and C are independently adapted. It is important to note that both approaches use the same transmission energy and delay/frame. In addition, the generalized skip option is used by the JSCPA approach to improve efficiency. The idea is that if the concealment of a certain packet results in sufficient quality, then the algorithm can intentionally not transmit this packet in order to allocate additional resources to packets that are more difficult to conceal. Due to the independent operation between the video encoder and the transmitter in the ISCPA approach, the relative importance of each packet, i.e., their contribution to the total distortion, is unknown to the transmitter. Therefore, the transmitter treats each packet equally and adapts the power in order to maintain a constant probability of packet loss. The JSCPA, on the other hand, is able to adapt the power/packet and, thus, the probability of loss, based on the relative importance of each packet. For example, more power can be allocated to packets that are difficult to conceal.

To sum up, power-aware joint source-channel coding usually should implement the following three tasks: (1) finding an optimal power adaptation scheme and bit allocation between source coding and channel coding for given channel loss characteristics; (2) optimizing the source coding to reduce the computational complexity and achieve the target source rate; and (3) optimizing the channel coding to achieve the required robustness. Although, these three tasks are separately mentioned, they are essentially correlated. Properly choosing the mode and coding rate of codec, channel coding schemes, and transmission power can reduce the total power of the system.

46.4.3 Power-Aware Cross-Layer Design and Optimization

Due to limited adaptation to dynamic wireless link conditions and interaction between layers, traditional layer-separated protocols and solutions fail to provide QoS for mobile multimedia applications. Therefore, more efficient adaptation requires cross-layer design, not only from the video application's side, but also from the network protocol's side. Cross-layer design for power-aware multimedia aims to improve the overall performance and energy efficiency of the system by jointly considering the video encoder and multiple protocol layers. A cross-layer controller is designed at the sender (the source node) to provide the following functionalities: (1) interact with each layer and obtain the corresponding managerial information, such as the expected video distortion from the encoder and the network conditions from lower layers; (2) perform optimization and determine the corresponding optimal values of control variables residing in various layers. The control variables may include, but not limited to, source coding parameters S at the application layer and channel parameters C at the lower layers that include the sending rate at the transport layer, transmission path at the network layer, retransmission limit and channel coding at the data link layer, and modulation and transmitter power at the physical layer. In this cross-layer framework, network conditions, such as CSI, packet loss rate, network throughput, network congestion status, etc., are all assumed to be available to the controller. How to timely acquire and deliver these network condition information to the controller still remains a challenging task.

Power-aware cross-layer design and optimization for mobile multimedia has received many research efforts. Various design techniques and optimization methods have been developed, including almost all the work of joint source-channel coding and power adaptation, where video source coding and communication decisions have been jointly considered. Besides source coding adaptation, other video source adaptation techniques can also be considered in power-aware cross-layer optimization, such as scalable video stream extraction [24], transcoding [34], object-based video coding [32], and summarization [15]. Cross-layer optimization for resource allocation and scheduling is another interesting research topic in power-aware mobile multimedia. Plenty of research has focused on multiuser wireless video streaming systems [25], where the assignments of the transmission power as well as other network resources among multiple users were discussed.

46.5 Challenges and Future Directions

Based on the earlier discussion, there has been a dramatic advance in the research and development of mobile multimedia systems. However, due to the limited energy supply in mobile device batteries, unfriendly wireless network conditions, and stringent QoS requirement, current research on mobile multimedia still faces several major challenges. In this section, we will list these challenges and point out the corresponding future research directions.

Power management in mobile devices: Efficient use of the limited battery energy is challenging due to (1) nonlinear discharge behavior of battery, (2) high QoS requirement of real-time multimedia applications, (3) dynamic wireless network conditions, and (4) interactive activities of mobile endusers. Therefore, developing efficient methods for scheduling battery discharge under different battery capacity status and different workload is imperative to prolong the battery operating time.

Rate–distortion–complexity analysis of video codecs: Joint rate–distortion–complexity analysis for advanced video codecs, such as H.264 codec, is complex due to the large number of possible combinations of encoding parameters. It becomes more challenging because, (1) the quality performance does not increase linearly as the complexity increases, and (2) different videos with different contents and characteristics have different rate–distortion–complexity results. Therefore, developing efficient, accurate, and content-aware rate–distortion–complexity analysis models for different video codecs is another challenging research task.

Computational complexity: Many source parameters (e.g., prediction mode, quantization step size, and summary choice) and channel parameters (e.g., transmitter power level, modulation, channel coding, and scheduling) could be considered as the control variables for the optimization of mobile multimedia systems. In order to achieve global optimality, we need to consider control variables and the interactions among them as much as possible. Moreover, the size of the state space of an optimization problem normally increases exponentially with the number increasing of the selected control variables and their operating points. Therefore, to make the best trade-off between the system performance and the computational complexity, how to reduce the computational complexity and how to determine the suitable control variables and their operating points still remain challenging.

Network information feedback and cross-layer signaling: To perform the best adaptation of control variables to the underlying network conditions, power-aware cross-layer optimization for mobile multimedia requires both accurate and timely feedback of network status information (e.g., CSI and availability information of network resources), as well as more effective communications between network layers. However, in the literature, perfect channel state information is usually assumed available at the controller, which is not real in practice. Therefore, how to manage the cost of acquiring and transmitting the necessary network conditions and how to design cost-effective and time-efficient cross-layer signaling architectures are still the challenging issues.

46.6 Mobile Multimedia Communication Energy Footprinting and Optimization

We develop a systematic optimization framework for wireless video delivery under the constraint of the predefined video distortion. We first discuss the experimental methods and models to footprint the energy consumption in video codec and transmission. Based on the derived model, the problem of battery-aware wireless video codec and delivery is formulated to jointly select the video codec and transmission parameters to minimize the battery consumption under the constraint of expected received video distortion. Our major contributions made in this paper are (1) footprinting and modeling of battery energy consumption based on real measurements; (2) developing a systematic optimization framework of video codec and wireless communications parameters from the perspective of the battery remaining capacity and the desired video quality.

46.6.1 Energy Consumption Profiling

46.6.1.1 Encoder Energy Consumption Profiling

Video coding or compression is a basic technology that enables the storage and transmission of a large amount of digital video data. Many standard video encoder systems, including MPEG-1/-2/-4, H.26x, and H.264/AVC employ a hybrid coding architecture based on DCT and the Motion Estimation Compensation (ME/MC) scheme. Existing processes during wireless video communication include modules of ME (motion estimation), PROCODING (including DCT, inverse DCT, quantization, inverse quantization, and reconstruction) and ENC (entropy encoding). During the video coding process, with the help of electric quantity measure equipment, the energy used in the battery-powered multimedia processor can be derived from the operating time of the CPU, which enables the hardware resources to accomplish a series of steps for video coding. On a specific video coding platform, the operating time of the CPU depends on the characteristics of the specific video clip and complexity parameters selected in the steps of video coding. In the process of PROCODING, the quantization parameter (QP), a parameter that controls the quality and bit rate of video compression, is a key factor that affects the number of nonzero MBs (NZMB) in one video frame that are needed to be coded. While the computation for quantization is independent of the bit rate, with a smaller quantization step size, more computation for variable length coding (VLC) is needed due to the increased number of nonzero coefficients. Many experiments have shown that all the steps together in PROCODING employ a large proportion of CPU occupancy and eat up more than 50% of the total energy consumption on the encoder. Compared with other modules in source coding, PROCODING energy consumption is almost twice the energy consumed in ME and six times the energy consumed in ENC. Quantization is the key complexity control parameter in the processes of PROCODING, therefore, it is reasonable for us to set quantization step q as the main optimal complexity control parameter in the video coding processes to calculate the energy used to encode a specific video clip on a certain hardware platform. The total time used in coding the whole video clip depends on the CPU running time spent on encoding every frame of this video. Denote the total number of frames in a video clip as n and the time used to encode kth frame as t_k. Then the total operating time of CPU T_e^{tot} to encode the a video clip can be expressed as

$$T_e^{tot} = \sum_{k=1}^{n} t_e^k, \tag{46.5}$$

On the other hand, total energy used in coding the whole video clip also depends on the energy E_e^k, which is used to encode the kth frame of this video. So the total operating energy of CPU E_e^{tot} used for coding can be written as

$$E_e^{tot} = \sum_{k=1}^{n} E_e^k, \tag{46.6}$$

It is possible for us to directly find the amount of electric quality used to encode every frame of a video by referring the corresponding experiment test result from equipment of energy measurement. In this way, once the hardware platform is decided, the energy in the encoder for one frame is a function of the CPU running time t_k spent on coding this frame and the key complexity control parameter q_k chosen to compress the same frame. The function of total energy used in encoder can be given by

$$E_e^{tot} = \sum_{k=1}^{n} f_e(t_k, q_k). \tag{46.7}$$

where $f_e(\cdot)$ is the uniform way to calculate the energy consumption of one frame during coding processes.

46.6.1.2 Transmission Energy Consumption Profiling

The total transmission energy can be calculated by adding together the energy used on transmitting every frame of a video. This is given by

$$E_t^{tot} = \sum_{k=1}^{n} E_t^k, \tag{46.8}$$

where E_t^k is the energy consumption used for transmitting the kth frame. The energy used to transmit a frame depends on the compressed bits of this frame and channel transmission rate. Compressed bits of one frame is the size of a stream which is generated after coding processes of a video frame and mostly decided by the characteristics of the input video and the quantization step q applied on this frame. Channel transmission rate depends on the transmission bandwidth and adaptive modulation and coding (AMC) scheme. Different choice of an AMC scheme applied in the transmitter will result in different transmission rates and spectral efficiency. Let W be the underlying channel bandwidth, and K_i be the transmission rate of AMC scheme i. Then, the resulting transmission rate when data is transmitted by using the ith AMC scheme is

$$R_i = K_i \cdot W, \tag{46.9}$$

If we use F_i to represent the compressed bits size of the ith frame after coding processes, then the energy used for kth frame can be denoted as

$$E_t^k = P \cdot \frac{F_i}{R_i} = P \cdot \frac{F_i}{K_i \cdot W}, \tag{46.10}$$

where P is the transmission power. F_i can also be determined by referring the corresponding experiment test result. In general, the energy in transmission for one frame is a function of time spent on coding processes t_k, AMC scheme i, and the complexity control parameter q_k chosen to compress the current frame. Therefore, the total energy consumption in transmitting a video clip can be written as

$$E_t^{tot} = \sum_{k=1}^{n} f_t(t_k, q_k, i), \tag{46.11}$$

where $f_t(\cdot)$ is the function to calculate the energy consumption used to transmit one frame.

46.6.2 Expected End-to-End Distortion

During the wireless video communication process, the total expected end-to-end distortion is caused during source coding and transmission. In order to acquire an accurate result of the distortion through the whole wireless video transmission system, in this work, we consider and calculate the overall end-to-end distortion instead of just simply adding the coding-introduced distortion and the transmission-introduced distortion together. Because a robust error concealment technique is necessary to avoid significant visible error in the reconstructed frames at the decoder, we consider a simple but efficient temporal concealment scheme used in our previous research [33]: a lost macroblock is concealed using the median motion vector candidate of its received neighboring macroblocks (the topleft, top, and top-right) in the preceding row of macroblacks. The candidate motion vector of a macroblock is defined as the median motion vector of all 4×4 blocks in the macroblock. If the preceding row of macroblocks is also lost, then the estimated motion vector is set to zero and the macroblock in the same spatial location is the previously reconstructed frame used to conceal the current loss. Although some straightforward error concealment strategies do not cause packet dependencies, as a generic framework, the more complicated scenario is considered here as a superset for the simpler cases. Due to the difficulty in computing the actual video quality perceived by the end users, in this work, the received video quality is evaluated as the expected end-to-end distortion by using the ROPE method. The expected distortion is accurately calculated in real-time at the source node by taking all related parameters into account, such as source codec parameters (e.g., quantization, packetization, and error concealment) and network parameters (e.g., packet loss rate and throughput). Therefore, given the dependencies introduced by the aforementioned error concealment scheme, the expected distortion of slice/packet π_i can be calculated at the encoder as

$$
\begin{aligned}
E[D_i] = (1 - p_i)E\left[D_i^R\right] + p_i(1 - p_{i-1})E\left[D_i^{LR}\right] \\
+ p_i P_{i-1} E\left[D_i^{LL}\right],
\end{aligned} \tag{46.12}
$$

where

p_i is the loss probability of packet π_i

$E\left[D_i^R\right]$ is the expected distortion of packet π_i if received

$E\left[D_i^{LR}\right]$ and $E\left[D_i^{LL}\right]$ are respectively the expected distortion of the lost packet π_i after concealment when packet π_{i-1} is received or lost

The expected distortion of the whole video frame which contents m packets, denoted by $E[D]$, can be written as

$$
E[D] = \sum_{i=1}^{m} E[D_i], \tag{46.13}
$$

Generally, multiple modulation and coding schemes are available to wireless stations in a wireless data network to achieve a good trade-off between the transmission rate and transmission reliability. Modulation schemes that allow a larger number of bits per symbol have symbols closer to each other in the constellation diagram, which may result in more error in decoding. Varying code rates can be employed with each modulation scheme to adapt to changing channel conditions by allowing more redundancy bits for channel coding (lower code rate k/n) as channel conditions deteriorate. As the code rate decreases, the effective data rate is reduced, and hence the achievable throughput decreases. We set the term scheme i to refer to a specific choice of AMC scheme. The probability of error in a packet of L bytes, for a given AMC scheme i, as a function of the bit error probability $p_{b,i}$, can be expressed as $p_{e,i}(L) = 1 - (1 - p_{b,i})^{8L}$. Moreover, $p_{e,i}$ can also be approximated with sigmoid functions [18,19] in the form of

$$
p_{e,i}(L) = \frac{1}{1 + e^{\lambda(x-\delta)}}, \tag{46.14}
$$

where x is the signal-to-interference-noise-ratio (SINR). Table 46.1 shows the sigmoid parameters (λ, δ) for the eight AMC schemes in modeling packet transmissions over an 802.11a WLAN network. From this table and (46.14), it is easy to see that $p_{e,i}$ depends on the specific AMC scheme i, and so the overall distortion since the end-to-end distortion is the function of $p_{e,i}$. Once the packet error probability is calculated, the expected end-to-end distortion can be derived based on (46.12) and (46.13).

We have noted that, except for the characteristics of the input video, the QPs applied in source coding procedure play another critical role in contributing the total distortion since the larger the quantization step size is, the more small DCT coefficients will be lost. Thus, from (46.13), different levels of distortion will be achieved under different levels of QP. In other words, the value of quantization step q needs to be considered as another parameter to control the total distortion. Therefore, for a specific platform, the total expected distortion associated with AMC scheme i and QP q can be denoted as

$$
E[D]_{tot} = D(q, i). \tag{46.15}
$$

46.6.3 Optimized Battery Capacity Framework under Distortion Constraint

In order to solve the formulated optimization problem, two profiles need to be established in advance. The first profile is about how the pattern of the expected received frame distortion changes according to different set of choices on QPs q and AMC scheme i. The second is how the battery capacity used for the delivery of one frame is decided by the same set of choices on the QP and AMC schemes. Therefore, in our framework, we choose QPs q and AMC scheme i to form a two-dimensional (2-D) independent vectorial variable. Let Q be the total options of QP, I the total optional AMC schemes. So every video frame has $Q \cdot I$ options of this vector. Because both parameters are the key control variables to determine the working pattern in both coding and transmission, we name this vector as the control vector and denote it as (q, i) according to the definitions of previous sections.

Video is fundamentally different from other multimedia resources, for it basically comprises a group of separated video frames. When we deal with the optimization of battery capacity under the constraint of video distortion, it is not reasonable to select only one set of optimal control vector to process all the video frames uniformly, because the dynamics in video content and channel conditions make it necessary to adjust the QP and AMC parameters frame by frame. Therefore, we apply the optimization to the best control vector for each frame to minimize the battery capacity used to deliver each frame under the constraint of expected received frame distortion. We can observe that the total expected distortion and total energy consumption are based on the choice of control vector used to code and transmit each video frame. For a specific video, after the execution of experiment and model applications, the expected received video distortion and battery capacity used for delivering the whole video clip can be calculated by referring to the received frame distortion and battery capacity used for delivering each frame.

46.6.4 Experimental Results

We conducted experiments to show the performance of the proposed framework. Four video sequences with varied contents (Carphone, Foreman, Coastguard, Mobile) in QCIF format were considered in our work. An Imote2 wireless sensor node which applies PXA271 XScale processor was used in the experiment. The Arbin measurement system was in charge of monitoring and recording all the desired electronic data. The Y-component of the first 50 frames of each video sequence was encoded with the H.264 codec (JVT reference software, JM 16.2). We chose the quantization step size (QP) and AMC schemes listed in Table 46.1 as the tunable source coding and transmission parameters. The permissible QP values are $[9, 12, 15, \ldots, 36]$. According to Table 46.1, the permissible AMC scheme i values are $[1, 2, 3, \ldots, 8]$ because different QP results in different bit rate. To maintain a smooth date rate and thereby a relatively constant power consumption on transmission to extend the battery lifetime, the difference of the selected QP for neighboring slices was limited within a threshold of 3. All frames, except the first one, were coded as inter frames. To reduce error propagation due to packet loss, 10 random I macroblocks

TABLE 46.1 Approximation of Packet Error Probability for Different AMC Schemes

Modulation Scheme	δ (dB)	λ (dB^{-1})	Code Rate (Bits/Symbol)	AMC Scheme (i)
BPSK	2.3	0.640	0.5	1
BPSK	6.1	0.417	0.75	2
QPSK	5.3	0.461	1	3
QPSK	9.3	0.444	1.5	4
16-QAM	10.9	0.375	2	5
16-QAM	15.1	0.352	3	6
64-QAM	18.2	0.625	4	7
64-QAM	21.2	0.419	4.5	8

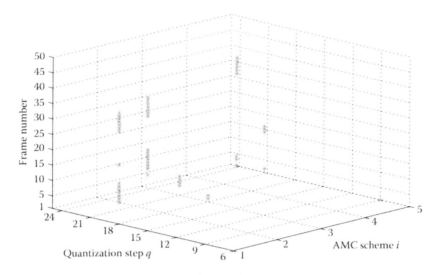

FIGURE 46.2 The optimal solution achieving the minimized distortion without the battery capacity constraint.

were inserted into each frame. The frames were packetized such that each packet/slice contains one row of MBs, which enables a good balance between error robustness and compression efficiency.

For the experiment using the video clip Foreman, Figure 46.2 shows the condition that no capacity constraint for video delivery was applied, and the optimization process chose the optimal solution which has the minimized distortion in each frame. In this case, the solution has an average PSNR of 44.9 *dB* for each frame, battery capacity consumption is 0.0119 *Ah*. From the figure, we can see that the control vector of each frame concentrates in the lower range of AMC scheme *i* and quantization step *q*. Figure 46.3 shows the scenario where no distortion constraint is applied for video delivery, and the optimization process chose the optimal solution which has the minimized battery capacity consumption in each frame.

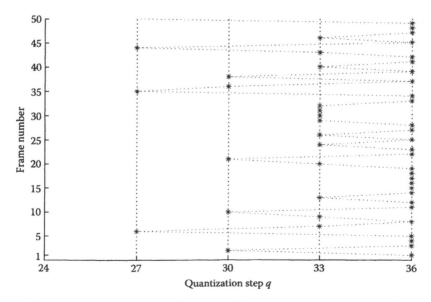

FIGURE 46.3 The optimal solution achieving the minimized battery capacity consumption without the distortion constraint (AMC scheme $i = 8$).

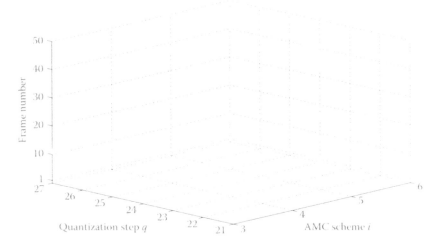

FIGURE 46.4   The optimal solution under the constraint of 36 *dB* average frame distortion.

In this case, the solution has an average PSNR of 29.2 *dB* for each frame, the battery capacity consumption is 0.0101 *Ah*. We can also see that the control vector of each frame concentrates in the higher range of the quantization step *q*, and all the frames are transmitted under the AMC scheme of number 8.

The optimization framework proposed in this paper was tested by three experiments under different values of distortion constraint. In the first experiment, we applied the optimization toward the first 50 frames of the Foreman video clip and set the average frame distortion constraint as 36 *dB*. After executing our framework, the control vector for each frame can be decided to minimize the battery capacity consumption resulting in a received frame which has a distortion under 36 *dB*. Figure 46.4 shows the 50 control vectors corresponding to the first 50 frames of the tested video clip. Every point in the space represents an optimized control vector of one frame to satisfy the constraint. All the optimized control vectors of these 50 frames have formed an optimal solution for this video clip. By applying our optimization, this solution can still be derived as the number of the video frames grows. Figures 46.5

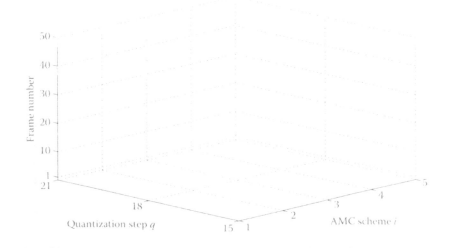

FIGURE 46.5   The optimal solution under the constraint of 42 *dB* average frame distortion.

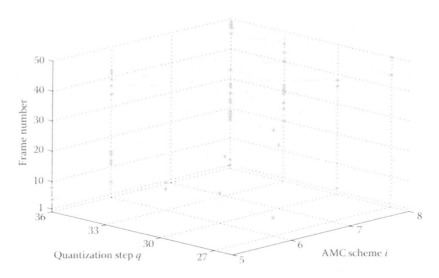

FIGURE 46.6   The optimal solution under the constraint of 29 *dB* average frame distortion.

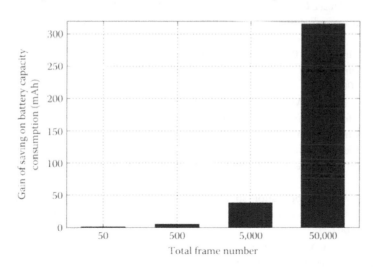

FIGURE 46.7   Comparison of battery capacity saving in delivering 50, 500, 5,000, and 50,000 frames.

and 46.6 show the other two testing results of establishing the optimized solutions under the average frame distortion constraints of 42 and 29 *dB*.

Figure 46.7 represents how much battery capacity can be saved according to the total number of video frames with the proposed framework. In the figure, the gain of saving on battery capacity increases in an exponential fashion when the total number of video frames needed to be delivered increases. As a result, the proposed optimization can save a considerable amount of battery capacity if it is applied to a relatively long video delivering case.

46.7 Conclusion

Because of the large number of video source and transmission adaptation parameters, time-varying video content characteristics, uncertain wireless channel conditions, and nonlinear battery effects, enabling energy-aware mobile multimedia is extremely challenging. In this chapter, we studied the energy-aware mobile multimedia computing in each component of power-aware mobile multimedia systems with focus on energy-aware video coding and energy-aware video delivery. Research challenges in the field of energy-aware mobile multimedia were introduced. In order to apply the idea of energy-aware mobile multimedia computing, we also proposed an analytical framework for mobile wireless video communication systems driven by battery. A method to optimize the battery capacity consumption under a constraint of expected received video distortion was proposed. Based on analytical results, the video coding and transmission were jointly considered to minimize the battery capacity used for one video frame under the constraint of expected received frame distortion. Experimental results verified the efficiency and effectiveness of the proposed optimization framework.

References

1. J. Ahn, J. hi Min, H. Cha, and R. Ha. A power management mechanism for handheld systems having a multimedia accelerator. In *Proceedings of the Sixth Annual IEEE International Conference on Pervasive Computing and Communications*, Hong Kong, China, pp. 663–668, 2008.
2. E. Akyol and M. Van der Schaar. Compression-aware energy optimization for video decoding systems with passive power. *Proceedings of the IEEE*, 18(9):1300–1306, 2008.
3. H. Ates and Y. Altunbasak. Rate-distortion and complexity optimized motion estimation for h.264 video coding. *IEEE Transactions on Circuits and Systems for Video Technology*, 18(2):159–171, 2008.
4. D. Bertozzi, L. Benini, and B. Ricco. Power aware network interface management for streaming multimedia. In *Proceedings of the IEEE Wireless Communications and Networking Conference*, Orlando, FL, pp. 926–930, 2002.
5. Z. Cao, B. Foo, L. He, and M. Van der Schaar. Optimality and improvement of dynamic voltage scaling algorithms for multimedia applications. In *Proceedings of the 45th Annual Design Automation Conference*, Anaheim, CA, pp. 179–184, 2008.
6. S. Chandra. Wireless network interface energy consumption implications of popular streaming formats. In *Proceedings of the Multimedia Computing and Networking*, San Jose, CA, pp. 185–201, 2002.
7. S. Chandra and A. Vahdat. Application-specific network management for energy-aware streaming of popular multimedia formats. In *Proceedings of the Usenix Annual Technical Conference*, Monterey, CA, pp. 329–342, 2002.
8. Y.-H. Chen, T.-C. Chen, Y.-H. Tsai, S.-Y. Chien, and L.-G. Chen. Fast algorithm and architecture design of low-power integer motion estimation for H.264/AVC. *IEEE Transactions on Circuits and Systems for Video Technology*, 17(5):568–577, 2007.
9. S. Ci and H.F. Ge. Significance measure with nonlinear and incommensurable observations. In *Proceedings of the IEEE Global Communications Conference*, Washington, DC, pp. 26–30, 2007.
10. R. Cornea, A. Nicolau, N. Dutt, and D. Bren. Annotation based multimedia streaming over wireless networks. In *Proceedings of the IEEE/ACM/IFIP Workshop on Embedded Systems for Real Time Multimedia*, Washington, DC, pp. 47–52, 2006.
11. Y. Eisenberg, C. Luna, T. Pappas, R. Berry, and A. Katsaggelos. Joint source coding and transmission power management for energy efficient wireless video communications. *IEEE Transactions on Circuits and System for Video Technology*, 12(6):411–424, 2005.
12. B. Foo, Y. Andreopoulos, and M. Van der Schaar. Analytical rate-distortion-complexity modeling of wavelet-based video coders. *IEEE Transactions on Signal Processing*, 56(2):797–815, 2008.
13. Z. He and S. K. Mitra. A unified rate distortion analysis framework for transform coding. *IEEE Transactions on Circuits and Systems for Video Technology*, 11(12):1221–1236, 2001.
14. Y. Hu, Q. Li, S. Ma, and C.C.J. Kuo. Joint rate-distortion-complexity optimization for H.264 motion search. In *Proceedings of the IEEE International Conference on Multimedia and Expo*, Toronto, Ontario, Canada, pp. 1949–1952, 2006.

15. J. Huang, Zi. Li, M. Chiang, and A.K. Katsaggelos. Joint source adaptation and resource allocation for multi-user wireless video streaming. *IEEE Transactions on Circuits and Systems for Video Technology*, 18(5):582–595, 2008.

16. I.R. Ismaeil, A. Docef, F. Kossentini, and R.K. Ward. A computation-distortion optimized framework for efficient DCT-based video coding. *IEEE Transactions on Multimedia*, 3:298–310, 2001.

17. P. Jain, A. Laffely, W. Burleson, R. Tessier, and D. Goeckel. Dynamically parameterized algorithms and architectures to exploit signal variations. *Journal of VLSI Signal Processing Systems*, 36(1):27–40, 2004.

18. D. Krishnaswamy. Game theoretic formulations for network-assisted resource management in wireless networks. In *IEEE Vehicular Technology Conference*, Vancouver, British Columbia, Canada, pp. 1312–1316, December 2002.

19. D. Krishnaswamy. Network-assisted link adaptation with power control and channel reassignment in wireless networks. In *3G Wireless Conference*, San Francisco, CA, pp. 165–170, May 2002.

20. D. Kwon, P. Driessen, A. Basso, and P. Agathoklis. Performance and computational complexity optimization in configurable hybrid video coding system. *IEEE Transactions on Circuits and Systems for Video Technology*, 16(1):31–42, January 2006.

21. M. Li, Z. Guo, R.Y. Yao, and W. Zhu. A novel penalty controllable dynamic voltage scaling scheme for mobile multimedia applications. *IEEE Transactions on Mobile Computing*, 5(20):1719–1733, 2006.

22. Z. Li, F. Zhai, and A. Katsaggelos. Joint video summarization and transmission adaptation for energy-efficient wireless video streaming. *EURASIP Journal on Advances in Signal Processing*, 2008:1–12, 2008.

23. S. Mohapatra, R. Cornea, N. Dutt, A. Nicolau, and N. Venkatasubramanian. Integrated power management for video streaming to mobile handheld devices. In *Proceedings of the 11th ACM International Conference on Multimedia*, Irvine, CA, pp. 582–591, 2003.

24. J. Ohm. Advances in scalable video coding. *Proceedings of the IEEE*, 93(1):42–56, 2005.

25. P. Pahalawatta, R. Berry, T. Pappas, and A. Katsaggelos. Content-aware resource allocation and packet scheduling for video transmission over wireless networks. *IEEE Journal on Selected Areas in Communications*, 25(4):749–759, 2007.

26. J. Stottrup-Andersen, S. Forchhammer, and S.M. Aghito. Rate-distortion-complexity optimization of fast motion estimation in H.264/MPEG-4 AVC. In *Proceedings of the IEEE International Conference Image Process*, Brondby, Denmark, pp. 111–114, 2004.

27. M. Van der Schaar and Y. Andreopoulos. Rate-distortion-complexity modeling for network and receiver aware adaptation. *IEEE Transactions on Multimedia*, 7(3):471–479, 2005.

28. M. Van der Schaar, D. Turaga, and V. Akella. Rate-distortion-complexity adaptive video compression and streaming. In *Proceedings of the IEEE International Conference Image Processing*, Davis, CA, pp. 2051–2054, 2004.

29. M. Van der Schaar and P.H.N.D. With. Near-lossless complexity scalable embedded compression algorithm for cost reduction in DTV receivers. *IEEE Transactions on Consumer Electronics*, 46(4):923–933, 2000.

30. R. Vanam, E.A. Riskin, S.S. Hemami, and R.E. Ladner. Distortion-complexity optimization of the H.264/MPEG-4 AVC encoder using the GBFOS algorithm. In *Proceeding of the IEEE Data Compression Conference*, Snowbird, UT, pp. 303–312, 2007.

31. H.M. Wang, H.S. Choi, and J.T. Kim. Workload-based dynamic voltage scaling with the QoS for streaming video. In *Proceedings of the Fourth IEEE International Symposium on Electronic Design, Test and Applications*, Hong Kong, People's Republic of China, pp. 236–239, 2008.

32. H. Wang, F. Zhai, Y. Eisenberg, and A. Katsaggelos. Cost-distortion optimized unequal error protection for object-based video communications. *IEEE Transactions on Circuits and Systems for Video Technology*, 15(12):1505–1516, 2005.

33. D. Wu, S. Ci, H. Luo, H. Wang, and A.K. Katsaggelos. Application-centric routing for video streaming over multi-hop wireless networks. In *IEEE Communications Society Conference Sensor, Mesh and Ad Hoc Communications and Networks*, Rome, Italy, pp. 1–9, June 2009.

34. J. Xin, C. Lin, and M. Sun. Digital video transcoding. *Proceedings of the IEEE*, 1(1):84–97, 2005.

35. W. Yuan and K. Nahrstedt. Buffering approach for energy saving in video sensors. *Proceedings of the IEEE International Conference on Multimedia and Expo*, Baltimore, MD, pp. 289–292, 2003.

36. F. Zhai, Y. Eisenberg, T.N. Pappas, R. Berry, and A.K. Katsaggelos. Joint source-channel coding and power adaptation for energy efficient wireless video communications. *Signal Processing: Image Communications*, 20(4):371–387, April 2005.

47

Ultralow-Power Implantable Electronics

Seetharam
Narasimhan
Case Western Reserve
University

Swarup Bhunia
Case Western Reserve
University

47.1 Introduction

Implantable microsystems have emerged as important and highly effective instruments for monitoring and manipulating internal activities within the body. With great advances in electronics and electrode technology over the past decades, it has become possible to implement complex implantable systems, which interface with the biological organisms to monitor internal body signals and manipulate the activity of body parts using electrical/chemical/optical stimulation. One of the success stories in the field of implantable biomedical devices is the cardiac pacemaker, which has been successfully implanted in a large population of patients. With emerging requirements of biomedical devices to interface with different body parts for therapeutic usage, pervasive implantable devices are rapidly becoming a reality. Figure 47.1a shows some example applications of bioimplantable devices. These devices are used to recognize and treat symptoms of various diseases like epilepsy, heart disease, Parkinson's disease, blindness, urinary incontinence, etc. Several applications of biomedical devices are mentioned in [1]. Examples of implantable medical devices produced commercially by Medtronic, a leading medical technology company, are presented in [2].

A typical implantable system contains sensing circuitry for recording biological signals, circuits for signal conditioning and analysis, and transceiver circuits for external communication, as shown in Figure 47.1b. The front-end electronics are usually analog/mixed-signal circuits for signal conditioning (preamplifier, filter, data converter, etc.) of the recorded biological signal. In order to be untethered to wires for power and data transfer, the implantable systems tend to have radio frequency (RF) transceivers and portable energy sources either within the implant unit or as wearable modules connected to the implanted sensing system. The back-end transceiver circuits [3] usually deal with wireless data communication, typically at low radio frequencies, which are more effective in penetrating skin and other tissue, without

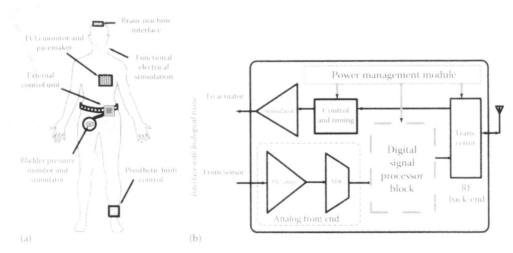

(a)   (b)

FIGURE 47.1 (a) Pervasive nature of bioimplantable systems, which are used for monitoring and/or controlling activities of different body parts. (b) Overall block diagram of a typical bioimplantable system, highlighting the DSP block.

causing damage. The energy source can be a rechargeable battery that is powered through inductive or RF links or using alternative sources of energy, commonly termed as *energy harvesting* or *energy scavenging*. The digital unit is used to provide control of different tasks and for digital signal processing (DSP) for on-board real-time data analysis. This block is often implemented using a commercial microcontroller or DSP unit, which is software-configurable during the initial testing phase when the algorithms are being developed.

Figure 47.2 shows the amplitude range and frequency range of various types of bioelectric signals that are sensed by typical biomedical interfaces [4]. According to the frequency and amplitude range of different signals, the design specifications for the corresponding signal conditioning electronics vary. As biological signals typically have very-low-frequency content (in the range of fraction of Hz to a few kHz), the computational speed requirement of the processing electronics is very low compared to state-of-the-art microprocessors, which have maximum operating frequencies in the order of few GHz. It has been shown in [5] that the system power is dominated by the wireless transmission and that can be reduced drastically by using complex signal processing on-chip to reduce the amount of data to be transmitted. However, the power budget for the electronics is also extremely low. Moreover, the transmission frequencies of the implantable transceivers are limited to a few MHz, according to the

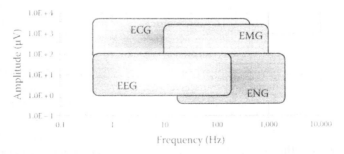

FIGURE 47.2 Frequency range and amplitude range of various biological signals of interest. ECG: electrocardiogram (extracellular action potentials from cardiac tissue), EMG: electromyogram (extracellular action potentials in muscle fibers), EEG: electroencephalogram (neural signals from the brain measured outside the skull), and ENG: electroneurogram (extracellular neural signal).

regulations by the Federal Communications Commission (FCC) [1]. As the number of recording sites increases and more complex data analysis needs to be performed online, the DSP unit in the implantable system needs to have increased computational power. The DSP block serves two main purposes: (1) it recognizes meaningful patterns to trigger appropriate measures, for example, neuromodulation or drug delivery; and (2) it performs data compression to reduce the amount of raw data to be transmitted wirelessly to the external control unit. For example, a neural recording from an array of 100 electrodes sampled at 25 kHz per channel with 10 bit precision yields an aggregate data rate of 25 Mbps, which is well beyond the reach of state-of-the-art wireless telemetry. Also, if the system keeps transmitting raw sampled data, it will drain the power source, which seriously hampers the freely operating lifetime of the device.

A computational task for real-time online multichannel neural signal analysis requires special purpose, low-power, robust, and area-efficient hardware, since conventional microprocessor or DSP chips would dissipate too much power and are too large in size for an implantable device. For hardware implementation of this signal processing block, nanoscale technologies offer great potential due to their terascale integration density, low switching power, and high performance. However, this also brings in number of design challenges, such as exponential increase in leakage power [6], reduced yield and lack of robustness due to reduced noise margin [7]. It is important to develop circuit/architecture level design solutions tailored to the computational algorithms for neural signal processing (NSP) that can leverage the benefits of nanoelectronics while addressing its limitations.

In this chapter, the requirement of DSP hardware design for implantable signal processing systems is analyzed and the emerging design challenges are identified. The conventional design requirements are in terms of power, area, and real-time performance. It should be noted that long-term reliable operation of such systems is becoming an important design challenge in the nanoscale technology regime due to the reduced robustness of the nanoscale transistors. Besides, programmability of the system is also important to adapt to individual patient conditions and/or to any temporal change. Various circuit/architecture design solutions for implementing ultralow power and robust on-chip signal processing in implantable systems are discussed. In summary, it can be seen that conventional design solutions for implementation of energy-efficient hardware can be combined with novel solutions that exploit the nature of the application for an implantable system and its signal processing algorithm in order to achieve efficient solutions in the target design parameter space. Finally, energy procurement for the implantable system is described, including harvesting from alternative energy sources and transcutaneous power telemetry for battery recharging. Section 47.2 gives a background about implantable systems that specifically deal with neural control.

47.2 Implantable Systems for Neural Control

Microsystems that interface with the central or peripheral nervous system are a growing class of implantable systems. A major limitation of our current understanding of the nervous system comes from the limitations of existing techniques for monitoring its electrochemical activity. Currently, it is possible to precisely measure the activity of small numbers of neurons (e.g., using intracellular electrodes), or to obtain an overall measure of the activity of large neural assemblies [using techniques such as electroencephalogram (EEG) or functional magnetic resonance imaging (fMRI)]. But precisely recording and analyzing the activities of a large number of neurons (on the scale of hundreds to thousands) in a normally behaving animal is not currently feasible. This is a significant limitation, because many important aspects of neural control occur due to the activities of populations of nerve cells at this scale.

Neural interface systems typically require monitoring the activities of the neural network at the cellular level using microelectrodes capable of sensing neural action potentials. Considering that a human nervous system typically consists of $\sim 10^{11}$ neurons each with diameters 4–100 µm, a realistic implantable microsystem should be capable of monitoring a large number of sites (in the range of hundreds or thousands) simultaneously with closely spaced electrodes and transmit the multichannel recorded data wirelessly for untethered measurements [8]. Similarly, microstimulation by insertion of electronic signals requires interfacing with the neural networks at the cellular level and controlling a large number of sites simultaneously. Researchers are increasingly using these implantable devices as interfaces to the central and peripheral nervous system to achieve better understanding of the mechanisms of neural communication and control. By studying simple organisms with tractable nervous systems, one can gain insight into the correlation between patterns of neural activity at the level of individual neurons and the resultant behavior of the organism [9]. Such behaviorally meaningful patterns can range from single spikes in a single neuron to timed bursts of neural spikes from a population of neurons, depending on the granularity of the behavior being studied.

Numerous efforts have been made to use arrays of electrodes and associated electronics for understanding the signals in a complex nervous system [8,10]. Implantable neural interfaces have been explored in diverse contexts including neural stimulation, as in cardiac pacing and functional electrical stimulation (FES). FES of nerves or muscles is used to assist patients in grasping, standing, or urination, while deep brain stimulation (DBS) has been shown to be an effective treatment for Parkinson's disease. Cochlear implants are commercially available for treating deafness, while visual prostheses have had preliminary success in creating sensations of vision in human beings. Extensive research has been done on developing brain–computer interfaces (BCI) [11,12] in which a tetraplegic person (whose limbs are paralyzed typically due to spinal cord injury) can control the movement of a computer cursor or a robotic arm. Current implementations of these systems, however, do not perform in situ signal processing, although some of them use simple control algorithms based on external sensor data.

In the algorithm-development stage, use of powerful external computers for real-time signal processing in conjunction with simple signal acquisition hardware is common. After the algorithm is stabilized, it needs to be tested for robustness under real operating conditions. Typical experiments start with in vitro preparations where the relevant neurons and associated muscles are extracted and used for interfacing with a prototype system. The next stage is deployment of a miniaturized prototype system in vivo (i.e., within the animal body). In order to record biological signals in the real setting, the animal needs to recuperate after surgical insertion of the implant and allowed to move untethered to any power or data wires. Recordings from freely behaving animals in their natural environment are crucial to final validation of the algorithm under real environmental noise conditions. In order to allow chronic implantation and operation, the system needs to be more independent in terms of energy resource, data telemetry, and control for decision making. As implantable systems enter the clinical realm where they are inserted in human beings for supervisory/assistive/preventive healthcare, factors like reliability, biocompatibility, energy efficiency, and self-sufficiency become more and more important.

The need for a closed-loop neural system, which records from multiple neurons, analyzes the neural activity and stimulates some neurons based on the analysis, has been emphasized before [13]. However, most of the current neural interface systems employ sophisticated data analysis performed on an external computer. Real-time closed-loop neural control can greatly benefit from in situ signal processing using low-power miniaturized hardware. Such in situ processing is more important for chronic implantations as well as to facilitate ambulatory movements of a patient. Although intense research has been carried out on designing the analog front-end circuitry [8,14,15] and algorithms for off-line signal analysis, the design of algorithms and digital circuits for online signal processing inside the implantable system is comparatively new.

The neurons are the communication pathways for the neural signals to travel to and from the central nervous system. However, these neurons are more than simple conducting wires for neural signals. The electrical signal within a neuron occurs as a spike (where a spike means a concentrated short-time voltage change with a high-frequency band) in the intracellular membrane potential due to precisely defined electrochemical processes. These spikes are known as action potentials, which travel along the axon and provide electrochemical stimuli to other neurons in their vicinity. The occurrence of a spike as well as its particular shape and frequency of occurrence, along with the considerations of simultaneous activity in related neurons, determine the propagation of a particular neuron's signal and its effect on a neuron's presynaptic terminal, at which signals are generated that propagate (either electrically or chemically) to other neurons. The electrochemical activity can be sensed using microelectrodes to pick up variations in electrical potential or chemical concentrations of neurotransmitters. Neuroelectric electrodes can be of two types—intracellular, which are invasive but can pick up the voltage profile inside individual neurons, and extracellular, which are noninvasive (causing minimal neural tissue damage) but pick up neural signals of different strengths from different neurons in the vicinity along with lots of background noise. A major component of this noise is the biological noise (e.g., action potentials from neural cells far from the measuring electrodes) and other electrical noise (e.g., body movements). Figure 47.3 shows the activity recorded from the same neuron using intracellular and extracellular electrodes. The latter recording has lower amplitude and is corrupted by background noise.

In order to use the recorded information containing neuronal action potentials mixed with background noise, one needs to perform spike detection and alignment followed by classification of the action potential signals. From the neuroscientist's point of view, it is important to retain the spike topology and shape in the transmitted data, since they convey important information regarding the activity inside a neuron, and can be critical for identifying neuronal subtypes. In the context of NSP, major tasks to be performed by the DSP unit are denoising of the recorded signal, detection of spikes, spike sorting, detection of bursts, efficient representation of spike data (for data compression) and recognition of meaningful patterns (that relate to the overall behavior of an animal) from the multichannel recorded signal. Previous efforts for spike detection have been aimed at using simple thresholding schemes, with or without an adaptive threshold for detecting spikes in the recorded neural signal. A simple thresholding scheme for on-chip spike detection using analog comparators was proposed in [15]. An on-chip data compression circuit that detects spikes using a simple adaptive thresholding scheme and transmits their amplitudes is described in [16]. More complicated spike detection (energy-filter-based) and sorting (maximum–minimum clustering-based) algorithms have been proposed and implemented in silicon [17]. Various spike sorting architectures were compared for energy efficiency in [18]. Wavelet-based spike detection [9] and custom hardware implementation of discrete wavelet transform (DWT) for multichannel data compression [19,20] have also been investigated before. The main design challenges for implantable systems are described in the following section.

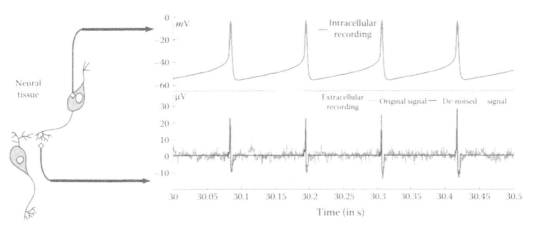

FIGURE 47.3 Extracellular and intracellular recordings from the same neuron in *Aplysia californica* (sea-slug). Denoising and spike detection in the extracellular signal are performed using a wavelet-based adaptive thresholding algorithm. (From Narasimhan, S. et al., Wavelet-based neural pattern analyzer for behaviorally significant burst pattern recognition, in *Proceedings of the 30th International Conference of the IEEE Engineering in Medicine and Biology Society*, 2008.)

47.3 Design Challenges for Implantable Systems

The major design challenges for implantable systems are highlighted in Figure 47.4. The first two parameters are the most significant and have been known to implantable system designers for ages. The implantable system needs to fit within a small area and hence, the chips inside the system should have a small form factor as well. The second design parameter is power. There are two reasons behind the quest for ultralow power. First, the implantable systems are equipped with a small battery with limited capacity for supplying energy to all the active circuits. If the circuits consume too much power, the lifetime of the system becomes reduced. Nowadays, rechargeable batteries with RF [4] or inductive powering [21] techniques, as well as methods for harvesting energy [1] from within the body are used to increase the availability of power to the implantable system without increasing the area too much. But, the second reason for using low-power circuits is to limit the power dissipation and avoid the associated problems like tissue damage due to overheating or temperature-induced circuit reliability degradation.

Typically, circuit designers tend to trade-off performance to achieve ultralow power and area. As shown in Figure 47.5, performance is typically not a constraint when considering biosignal processing, because

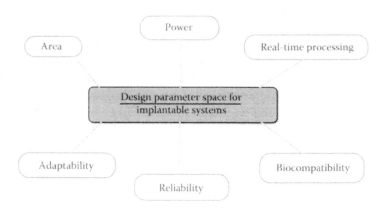

FIGURE 47.4   Design parameter space for designing DSP hardware in implantable systems.

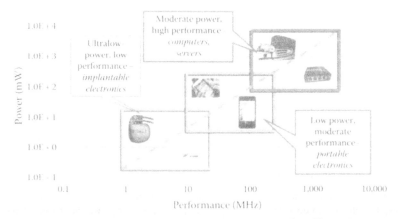

FIGURE 47.5   Power-performance spectrum for consumer electronics. Implantable systems occupy the ultralow power, low-performance part of the entire range; hence, their design requirements are different from high-performance microprocessors.

of the inherently low-frequency content and hence, low-sampling frequency of these signals. However, the processing circuits need to perform their computations in real-time, so that the feedback loop can be closed without too much latency. As the number of recording channels increases, the requirement for time multiplexing of the signal processing hardware increases their operating frequency requirements to tens of MHz. The other three issues in bioimplantable systems are adaptability, reliability, and biocompatibility. It should be noted that design solutions for addressing these issues should not come at the expense of increase in area or power consumption. Hence, overhead is an important design parameter when considering different solutions for these challenges.

47.3.1 Reliability

As the number of channels increases, the requirement for complex signal processing circuitry necessitates the use of nanoscale technologies that give better performance, high transistor density, and low area and power. However, to fully leverage the potential of nanoscale devices in implantable system design, some major design challenges need to be addressed, such as the exponential increase in standby leakage, lack of robustness due to reduced noise margin, and yield loss with process fluctuations. Further, temporal parameter variations (due to environmental variations and device aging effects) also affect robustness of operation. The various techniques [6] for reduction of dynamic and standby power tend to have adverse effects on the robustness of the design at nanoscale because of process-induced parameter variations. Variations in manufacturing process can lead to wide variations in circuit parameters, causing loss in yield. Conventionally, one needs to follow a worst-case design approach with a wide design margin in order to avoid failures caused by process and temporal variations, which are intrinsic to nanometer technologies. However, such a design approach considerably compromises power dissipation and die area.

Parameter variations can cause a loss in both yield and reliability. Yield is defined as the probability of failure of an as-processed device, while reliability is defined as functional failure of the device during its operation. A process with low yield (due to various extrinsic defects) is unacceptable to begin with, but even a process with high yield (low initial defects) but relatively large degradation rates (poor reliability) is unacceptably expensive in the long term. For bioimplantable systems, reliability of various components is an issue of major interest since the implanted chips are expected to function without failure for a long period of time (e.g., 10 years) under harsh operating conditions. There are two types of temporal variations—short term and long term, which can lead to degradation in robustness of operation. Short-term variations can be caused by environmental changes like temperature and voltage fluctuations. Long-term variations can be caused by aging-induced effects like negative bias temperature instability (NBTI), hot carrier injection (HCI), and time-dependent dielectric breakdown (DDB).

47.3.2 Adaptability

Another significant design issue for bioimplantable systems is the introduction of reconfigurability. The various parameters that are kept tunable in a software implementation need to be assigned fixed values in hardware. However, variations in the nature of biological signals from patient-to-patient and temporal variations in signal and noise characteristics even for the same patient necessitate regular calibration and tuning of the various parameters. It has been observed by various researchers that the shape and features of biological signals measured with the same sensors tend to change over time due to biomechanical factors like electrode drift, building-up of scar tissue over the sensors, etc. To achieve adaptability, one can use software reconfigurability, where the algorithm is coded in an embedded microprocessor [2,22–24]. However, compared to a custom implementation, this can be extremely poor in terms of area and power requirements. Also, certain signal processing algorithms require special instructions that might require the use of special DSP chips, which are also area and power-hungry. One might use special custom integrated circuits implementing special functions as coprocessors or accelerators for supporting

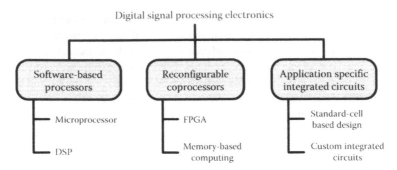

FIGURE 47.6 Different ways of implementing the DSP electronics for implantable systems. The design parameters to be traded-off are area, power, performance, reconfigurability, and design-time/effort.

the general-purpose computing electronics. However, the microprocessors still consume much area and standby power, which can drain the battery to a large extent, especially for implantable systems where energy is a scarce resource.

Another alternative is to consider hardware reconfigurable platforms. Existing field programmable gate array (FPGA) based reconfigurable systems are also beyond the area and power budget affordable by implantable systems. This necessitates the investigation of alternative reconfigurable architectures like memory-based computing (MBC) [25], which can be custom-designed to implement a particular algorithm within the area, power, and performance bound, but keeps the option of reconfigurability in order to suit patient-to-patient and temporal variability. Such a reconfigurable MBC framework that uses a dense memory array as the underlying computing element, leading to nearly $2\times$ power reduction at iso-performance along with significant improvement in resource usage compared to state-of-the-art FPGA has been presented in [25].

Figure 47.6 shows the different design options for implementing digital circuits for implantable systems. On one end of the spectrum, we have off-the-shelf software-programmable components like microprocessors and DSPs, whereas on the other end, we can design application-specific integrated circuits (ASICs), which are tailor-made for the application under consideration. These can be designed using a standard-cell based approach or a completely custom-design where each computing element and gate can be optimized for meeting the design specifications. In order to introduce hardware reconfigurability, one can use off-the-shelf FPGAs or incorporate MBC elements within the custom-designed ASIC. In order to satisfy the area and power constraints of implantable systems, one should use a judicious mix of reconfigurable MBC and custom logic, which can give the required adaptability. This is similar to the concept of hardware–software codesign used in embedded systems for achieving programmability as well as efficiency in terms of speed and energy. Again, one can use the properties of the biosignal processing algorithm [26] to perform this division between custom logic implementation and reconfigurable hardware.

47.3.3 Biocompatibility

Biocompatibility of the implantable system is typically achieved using proper packaging techniques that allow normal functioning of the electronics inside a harsh biological environment. The packaging techniques usually involve multiple coatings with biocompatible polymers or silicone glue, which allow hermetic sealing of the system [27] and at the same time prevent tissue damage. The tissue damage could be due to mechanical means caused by displacement of the implantable system internal to the body or due to thermal damage caused by overheating. The power dissipated by the implantable electronics as well as the energy involved in data and power telemetry through the tissue could cause damage or discomfort to the biological organism. Biocompatible packaging is not directly related to energy-efficient computing within the implantable system and hence, beyond the scope of this chapter.

Next, a circuit–architecture codesign approach for achieving ultralow-power and robust signal processing electronics for bioimplantable systems is described. The choice of an algorithm and its architectural implementation is motivated by the application domain and the system specifications. It is shown that contrary to conventional thought, the optimal operating voltage and frequency for each system is determined by the various trade-offs between area, power, and performance along with new design requirements of reliability and adaptability.

47.4 Circuit–Architecture Codesign Approach to Achieve Ultralow-Power Operation

Miniature size and low energy consumption are two primary design requirements for implantable systems. Though the size of an implantable system is still limited by the size of battery, antenna, and off-chip components, the area of the signal processor can be significant with increased requirement for on-chip processing of signals from multiple channels. While nanoscale (sub 90 nm) technologies provide high integration density, faster switching speed, and lower switching power per transition, they also bring new challenges. These include exponential increase in leakage current [6] and reliability issues due to process-induced parameter variations as well as temporal parameter variations due to temperature and voltage variations (collectively termed as P–V–T variations) [7]. Hence, in order to use nanoscale technologies for implantable signal processing electronics, the power and parameter variation–induced reliability issues need to be addressed using appropriate design methodologies. Typically, designers use worst-case design methods to achieve ultra-reliable operation in presence of variations, at the expense of area and power. However, for implantable signal processing electronics, it seems more appropriate to adopt a nominal design approach and use circuit–architecture-level techniques that exploit the properties of the algorithm in order to further trade-off power and reliability.

To decrease power consumption, one can use popular techniques such as clock gating and supply voltage gating. Clock gating saves dynamic power in the clock line that drives the sequential elements such as flip-flops and latches. Supply gating saves leakage power due to the stacking effect [6,28]. Both approaches require identification of idle cycles for a logic block during which it can be "gated." Voltage scaling [29] along with commensurate scaling of frequency [23] is an effective power reduction approach due to quadratic dependence of power on supply voltage. Static or dynamic change in transistor threshold voltage (Vth) is another method of achieving low-power operation [28]. However, the effectiveness of dual-Vth designs diminishes at nanoscale technologies due to increase in band-to-band tunneling leakage current and impact of process variations on parametric yield.

To achieve ultralow-power operation, the supply voltage can be scaled below the transistor threshold voltage, when the circuit starts operating in the subthreshold region. This leads to a huge reduction in both dynamic and leakage power with a corresponding increase in path delays, leading to ultralow-frequency operation. Subthreshold design [1] is a well-researched design technique, especially suited for low-sampling rate applications like biomedical signal processing. However, it comes at the cost of increased vulnerability of a design to variation-induced failures leading to reduced reliability and yield loss. At scaled voltage, digital circuits can suffer from functional failure due to variations in circuit delays. For example, in low-power DWT hardware, the time required for computation of the most significant wavelet coefficients can be affected due to voltage scaling, causing the clocked storage elements (flip-flops) to latch wrong logic values. With increasing die-to-die and within-die parameter variations in nanoscale technologies, maintaining high yield and reliability of operation in subthreshold design is becoming a major challenge. On the other hand, superthreshold design, although more effective in terms of yield and reliability, usually dissipates much higher power. Hence, a design technique that merges the advantages of both worlds is most desirable. In this context, a novel near-threshold computing approach has been proposed in [30], which evaluates both subthreshold and superthreshold approaches and investigates design solutions to enable low-power and robust operation, by exploiting the nature of the NSP algorithm.

47.4.1 Subthreshold versus Superthreshold Operation

Results obtained by HSPICE simulations at 70 nm predictive technology model (PTM) [31] for the main processing element (PE) of the DWT algorithm [32] are shown in Figure 47.7. Voltage scaling causes the total energy consumption to decrease almost exponentially for a "nongated" nominal design, whereas a "gated" design has comparatively large decrease in overall energy requirement per operation. On the other hand, the critical path delay increases exponentially with voltage scaling, especially in the subthreshold region (below 0.2 V) causing the energy-delay product (EDP) to have a steep increase, especially for the gated design, which fails to meet the operating frequency requirements below 0.15 V. It should be noted that the minimum energy point (see Figure 47.7a) has a much lower supply voltage than the point with the minimum EDP (see Figure 47.7c), which corresponds to 0.6 V. Taking this as the nominal superthreshold voltage, one can operate the circuit at a relatively higher frequency than required and then turn on the supply-gating circuitry for the remaining idle time where the circuit is waiting for the next set of operands. The opportunity to increase the supply-gating time window increases with the increasing frequency of operation. The simulation results for total energy (Figure 47.7d), dynamic power (Figure 47.7e), and leakage power (Figure 47.7f) are obtained for the nominal supply voltage (0.6 V) with and without supply gating. Naturally, the reduction in leakage power with supply gating causes the total energy per operation to decrease with increasing frequency, in spite of the corresponding increase in dynamic power. In fact, one can obtain comparable energy savings by performing subthreshold operation at ultralow-frequency and near-threshold operation at moderate frequencies, by application of extensive power gating.

Since both subthreshold design and near-threshold design (with power gating) have similar energy consumption, one can opt for either technique to implement the design for implantable signal processing circuit. However, at nanoscale technologies, process variations can cause wide variations in circuit parameters like critical path delay, causing delay failures under extreme conditions. The detrimental effects of process variations are exacerbated in designs with low-power techniques applied. The impact of process variations (modeled as Vth variations) on the critical path delay of a PE is different at different

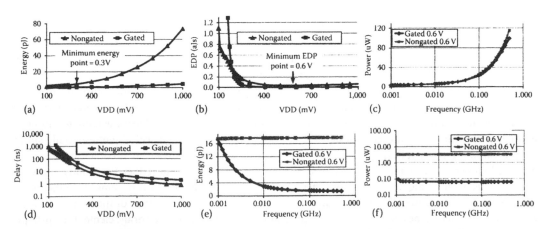

FIGURE 47.7 Finding the optimal operating voltage and frequency for the main PE of a DWT block. (a) Energy versus supply voltage (VDD). (b) EDP versus VDD. (c) Dynamic power versus frequency. (d) Delay versus VDD. (e) Energy versus operating frequency. (f) Leakage power versus frequency. The energy consumption for both gated and nongated designs are compared at different voltages and frequencies.

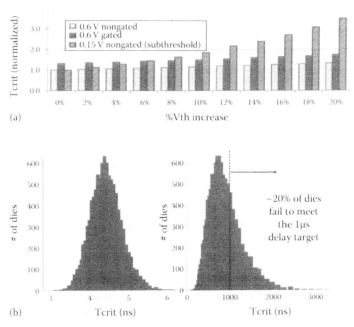

(a)

(b)

FIGURE 47.8 (a) Impact of process variations on critical path delay at subthreshold and superthreshold voltages with supply gating. (b) The impact on parametric yield.

supply voltages. As observed in Figure 47.8a, the delay increases exponentially with increasing process variations in the subthreshold regime, whereas the delay for the near-threshold design with and without supply-gating transistors (which are kept on during active mode, but still affect performance) show a linear increase. The sleep transistors are sized to have maximum power savings, allowing up to 30% performance degradation under nominal conditions. The increase in critical path delay, T_{crit} (normalized with respect to delay of nongated design at same supply voltage) is plotted for increasing Vth variations in Figure 47.8a. The values of T_{crit} for the three cases under nominal conditions are as follows:

1. Superthreshold nongated design at 0.6 V: 3.34 ns
2. Near-threshold gated design at 0.6 V: 4.39 ns
3. Subthreshold nongated design at 0.15 V: 732.40 ns

It is seen that under extreme process variations, the delay of a subthreshold design can increase by up to 3.5×, causing extreme yield loss, while a superthreshold design with supply gating, has a comparatively minimum impact under variation, even though the performance overhead under nominal conditions is relatively larger. The yield results, considering only inter-die variations, are obtained by assuming a Gaussian distribution with 30% standard deviation for the Vth. The resultant delay distributions for 10,000 dies (obtained using Monte Carlo simulations in HSPICE) for superthreshold (0.6 V) gated design and subthreshold (0.15 V) nongated design are shown in Figure 47.8b. Given a delay target of 1 μs, the yield in case of the subthreshold design is 79.94%, while all superthreshold designs pass the delay target of 30 ns, corresponding to an operating frequency of 33.33 MHz (which is the nominal frequency).

47.4.2 Preferential Design Approach to Achieve Further Power Reduction

One can achieve comparable energy reduction by subthreshold low-frequency operation and near-threshold operation with extensive power gating. However, the former can cause huge yield loss and decrease in robustness under large parameter variations. Moreover, operation in the superthreshold

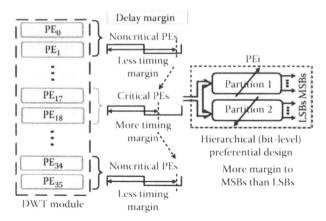

FIGURE 47.9   Preferential design methodology for a parallel DWT architecture.

domain opens avenues for architecture-level design techniques like preferential design [30] to further decrease energy consumption with graceful degradation in robustness. Low power and robustness of operation typically impose contradictory design requirements. In logic circuits, the principal failure mechanism under device parameter variations at nanoscale technologies is delay failure, which occurs when the maximum path delay of a circuit exceeds the clock period. Low-power design using voltage scaling accentuates delay failure probability under variations. To avoid these delay failures, conventionally, one needs to follow a worst-case design approach. However, such a design approach considerably compromises power dissipation and die area. On the other hand, if one opts for a nominal design, any variation-induced failure might cause drastic changes in the outputs, which are more critical in terms of signal quality, because of timing failure in the maximum delay paths.

In the preferential design approach [32], critical components in terms of performance (e.g., the computational blocks for most significant wavelet coefficients in DWT hardware) are designed with more relaxed timing margin than noncritical ones. In such an approach, possible variation-induced failures are confined to noncritical components of the system, thus allowing graceful degradation in performance under variations. Preferential design methodology for a parallel architecture of the DWT block is shown in Figure 47.9. Further, due to higher timing margin in critical components, the system becomes suitable for further voltage scaling to achieve ultralow-power operation. It is to be noted that the preferential design approach uses the nature of the signal processing algorithm to reduce the design overhead. This maintains relatively high reliability of operation without having to adopt a worst-case design approach by confining any variation-induced failures to insignificant components of the signal processing algorithm. The results of using preferential design in implementing a NSP algorithm are provided in Figure 47.10. The major design parameters are compared with other state-of-the-art implementations of NSP algorithms in Table 47.1 [30].

47.5 Energy Harvesting and Management

With much advancement in microelectronic circuits and micro-electro-mechanical systems (MEMS) sensors and actuators, the power source [1] is another great obstacle facing the development of implanted autonomous microsystems. As the computing requirement inside the implanted system increases, so does the power requirement. However, even the leakage power is substantial in nanoscale technologies causing rapid drainage of battery charge. On the other hand, small form factor and weight require the use of miniature batteries, which have reduced capacity. In the context of bioimplantable systems that can

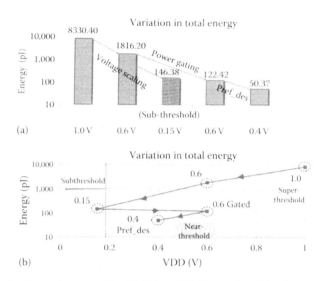

FIGURE 47.10 Variation in total energy of the DWT-based NSP algorithm following a circuit–architecture code-sign approach [30], which decides the optimal operating voltage and frequency for the circuit to achieve ultralow-power operation. The *preferential design* approach allows further reduction in power while limiting any variation-induced failures to least significant computations.

TABLE 47.1 Comparison of Design Parameters for NSP Implementations

References	Analog/Digital	Technology (μm)	Area (mm^2)	Power (μW)	Frequency (MHz)	Voltage (V)	No. of Channels
[15]	Analog	1.50	0.094	57	N/A	5.0	1
[16]	Digital	0.50	6.00	2600	2.5	3.0	32
[17]	Digital	0.35	63.36	100	5.12	1.65	128[a]
[19]	Digital	0.18	1.856	1530	16.0	1.8	16
[20]	Digital	0.18	0.221	76	6.4	1.3	32
[30]	Digital	0.07	0.214	1	33.33	0.6	20

[a] Results are presented for a single channel.

operate at reduced supply voltage, a new class of microsensors has been proposed. These sensors harvest energy from their environment to supply electrical power for operation of the implantable systems. Figure 47.11 shows several alternative sources of energy that can be exploited to procure power for a resource-constrained implantable system. Of particular interest are energy converters that scavenge energy from within the biological organism where the system is implanted. For example, vibration energy from heartbeats can be used to supply power to a heart-rate monitor. Similarly, chemical energy from excess glucose in the blood stream can be converted to electric energy in order to power a blood-sugar monitoring system for diabetic patients.

Magnetic induction and RF powering are often used for powering units implanted within the body. An alternative to conventional battery-operated systems is energy harvesting from natural sources like solar energy, thermal energy, kinetic energy, or vibrational energy. Muscle-actuated piezoelectric generation has been shown to produce 46 μW of power, which is certainly not adequate to constantly supply power to the electronics but can be used for powering rarely activated implantable systems. In general, the harvested energy will not be adequate for real-time operation of a wireless sensor, but a lithium-ion battery can be trickle charged during times of inactivity, then later used to operate the microsystem. Solar panels can be embedded on the clothing or headgear and used for scavenging energy from sunlight or other bright sources of light. Vibration-based energy harvesting systems, using a steel substrate with deposited lead–zirconate–titanate (PZT) piezoelectric thin films, can generate power densities ranging from 120 to 520 $\mu W/cm^2$ depending on different input vibration parameters. Novel techniques for recycling heat energy (dissipated in the implanted electronic chips using thermoelectric generators) or kinetic energy involved in motion are being developed and might be incorporated as new sources of energy or as shunting feedback paths to preserve energy in a usable form.

Each alternative source of energy has its own pros and cons and some are more suitable for a particular application than others. However, it is observed that a single "unconventional" source of energy may be unreliable and might not produce enough power to supply a multitude of circuits and sensors having varying energy requirements. Hence, there arises the need for an energy management system, which mixes the energy available from a variety of sources, stores the surplus energy, routes it to a variety of sinks, and efficiently controls the acquisition as well as the delivery to minimize power wastage. Such a system is illustrated in Figure 47.11, which acquires energy from various sources and stores it in a portable rechargeable battery or supercapacitor system. It is capable of supplying energy on-demand to the various energy sinks in the implantable system.

Currently, implantable systems have been designed to be powered by far-field electromagnetic radiation (or RF powering) or by near-field magnetic induction (Figure 47.12). Some of them are constantly powered by the external antenna and require the external system to be in close proximity (outside the skin). Often, these sources could be used for recharging a battery associated with the implant unit, which is capable of sustaining the implantable system for a whole day. The energy management unit in these systems typically includes power converters to convert the acquired AC power to a regulated DC supply voltage, which can be used by the electronics. These energy management units can also incorporate voltage converters to

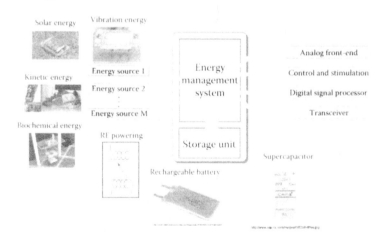

FIGURE 47.11 Energy management system for harvesting energy from alternative sources and storing it or supplying it to the implantable electronics.

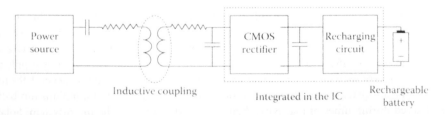

FIGURE 47.12 Inductive charging for transcutaneous powering of an implantable system.

produce different supply voltages for different parts of the implantable electronics. For instance, a high-frequency stimulation unit for pain block might need a high voltage for a short duration, whereas the neural recording and processing electronics might be operating at near-threshold conditions. The sleep control signals for powering down inactive circuits and reducing leakage power can also be generated by the energy management circuitry. The goal of the energy management unit is to get better efficiency, compared to usual energy delivering systems. There are four aspects that can be used to characterize the efficiency of such a system: (1) increasing the lifetime of energy sources and sinks, (2) minimizing the loss of energy in the routing path, (3) maximizing the availability (improving reliability), and (4) energy savings in the long run.

47.6 Summary

In this chapter, we have presented issues and solutions related to energy-efficient implantable system design. As these systems become pervasive, there is an increasing need to minimize their energy dissipation to enable chronic operation. We have described the design parameter space for an implantable system and presented low-overhead design techniques for the DSP component. With increasing amount of recorded data in these systems, there is a need for efficient DSP to analyze the data in order to take appropriate action and/or to compress it to transmit it wirelessly to an external receiver. As the data analysis algorithms get more sophisticated, the implantable systems can operate autonomously, without the need for external control. To address this increase in complexity, one can consider using nanoscale technologies, which provide the advantages of high-performance and high-integration density.

However, nanoscale technologies also represent new challenges in terms of exponential increase in leakage current and parameter variation-induced failures leading to degradation in yield and reliability. Instead of using conventional design principles to address these challenges, one can exploit the nature of the algorithm as well as signal acquisition characteristics to achieve ultralow power and robust operation. An efficient design solution that adequately addresses all the important design parameters can be achieved through an algorithm-circuit–architecture codesign approach. Moreover, the variability in biological environment and subject conditions necessitates circuit–architecture design styles, which allow for tunability of certain parameters in order to adapt to the specific needs of the patient. Adaptability can be in the form of software or hardware reconfigurability and can be built into the system as long as the power and area constraints are not violated.

Optimal operating voltage and frequency can be determined by extensive design-time simulations taking into account the effects of process variations. For some applications, optimal design point might be in the subthreshold regime, where the supply voltage is lower than the transistor threshold voltage. On the other hand, to address the robustness and performance issues in the subthreshold region, it may be necessary to adopt a near-threshold operating condition. In the later case, voltage scaling and power gating can be employed aggressively to achieve ultralow power and robust operation. Finally, appropriate choice of energy source, methods for energy transduction, and energy management need to be coupled with energy-efficient system design approaches to achieve implantable systems for chronic usage, which can operate reliably under extremely tight power budget.

References

1. Chandrakasan AP, Verma N, and Daly DC (2008) Ultralow-power electronics for biomedical applications. *Annual Review of Biomedical Engineering* 10:247–274.
2. Allan R (2003) Medtronic sets the pace with implantable electronics. *Electronic Design*, [Online] http://electronicdesign.com/article/components/medtronic-sets-the-pace-with-implantable-electronic.aspx (accessed 24 May, 2010).

3. Bradley PD (2006) An ultra low power, high performance medical implant communication system (MICS) transceiver for implantable devices. In *Proceedings of the IEEE Biomedical Circuits and Systems Conference*, IEEE, London, Nov 29–Dec 1, 2006.

4. Yazicioglu RF, Torfs T, Merken P, Penders J, Leonov V, Puers R, Gyselinckx B, and Hoof CV (2009) Ultra-low-power biopotential interfaces and their applications in wearable and implantable systems. *Microelectronics Journal* 40:1313–1321.

5. Chandler RJ, Gibson S, Karkare V, Farshchi S, Markovic D, and Judy JW (2009) A system-level view of optimizing high-channel-count wireless biosignal telemetry. In *Proceedings of the 31st International Conference of the IEEE Engineering in Medicine and Biology Society*, IEEE, Minneapolis, MN, Sep 3–6, 2009.

6. Roy K, Mukhopadhyay S, and Mahmoodi-Meimand H (2003) Leakage current mechanism and leakage reduction techniques in deep sub-micrometer CMOS circuits. *Proceedings of the IEEE* 91(2):305–327.

7. Borkar S, Karnik T, Narendra S, Tschanz T, Keshavarzi A, and De V (2003) Parameter variations and impact on circuits and microarchitecture. In *Proceedings of Design Automation Conference*, ACM/IEEE Anaheim, CA, June 2–6, 2003.

8. Wise KD, Anderson DJ, Hetke JF, Kipke DR, and Najafi K (2004) Wireless implantable microsystems: High-density electronic interfaces to the nervous system. *Proceedings of the IEEE* 92(1): 76–97.

9. Narasimhan S, Cullins M, Chiel HJ, and Bhunia S (2008) Wavelet-based neural pattern analyzer for behaviorally significant burst pattern recognition. In *Proceedings of the 30th International Conference of the IEEE Engineering in Medicine and Biology Society*, IEEE, Vancouver, British Columbia, Canada Aug 20–25, 2008.

10. Sawan M, Yamu H, and Coulombe J (2005) Wireless smart implants dedicated to multichannel monitoring and microstimulation. *IEEE Circuits and Systems Magazine* 5(1):21–39.

11. Black MJ, Bienenstocky E, Donoghue JP, Serruyay M, Wuz W, and Gao Y (2003) Connecting brains with machines: The neural control of 2D cursor movement. In *Proceedings of the First International IEEE EMBS Conference on Neural Engineering*, IEEE, pp. 580–583, Mar 20–22, 2003.

12. Nurmikko AV, Donoghue JP, Hochberg LR, Patterson WR, Song YK, Bull CW, Borton DA, Laiwalla F, Park S, Ming Y, and Aceros J (2010) Listening to brain microcircuits for interfacing with external world—Progress in wireless implantable microelectronic neuroengineering devices. *Proceedings of the IEEE* 98(3):375–388.

13. Nicolelis MAL (2001) Actions from thoughts. *Nature* 409:403–407.

14. Chi B, Yao J, Han S, Xie X, Li G, and Wang Z (2007) Low-power transceiver analog front-end circuits for bidirectional high data rate wireless telemetry in medical endoscopy applications. *IEEE Transactions on Biomedical Engineering* 54(7):1291–1299.

15. Harrison R (2008) The design of integrated circuits to observe brain activity. *Proceedings of the IEEE* 96(7):1203–1216.

16. Olsson III RH, and Wise KD (2005) A three-dimensional neural recording microsystem with implantable data compression circuitry. *IEEE Journal of Solid-State Circuits* 40(12):2796–2804.

17. Chae MS, Yang Z, Yuce MR, Hoang L, and Liu W (2009) A 128-channel 6 mW wireless neural recording IC with spike feature extraction and UWB transmitter. *IEEE Transactions on Neural Systems and Rehabilitation Engineering* 17(4):312–321.

18. Zumsteg ZS, Kemere C, O'Driscoll S, Santhanam G, Ahmed RE, Shenoy KV, and Meng TH (2005) Power feasibility of implantable digital spike sorting circuits for neural prosthetic systems. *IEEE Transactions on Neural Systems and Rehabilitation Engineering* 13(3):272–279.

19. Gosselin B, Ayoub AE, Roy JF, Sawan M, Lepore F, Chaudhuri A, and Guitton D (2009) A mixed-signal multichip neural recording interface with bandwidth reduction. *IEEE Transactions Biomedical Circuits and Systems* 3(3):129–141.

20. Kamboh AM, Raetz M, Oweiss KG, and Mason A (2007) Area-power efficient VLSI implementation of multichannel DWT for data compression in implantable neuroprosthetics. *IEEE Transactions Biomedical Circuits and Systems* 1(2):128–135.

21. Ghovanloo M, and Atluri S (2007) A wide-band power-efficient inductive wireless link for implantable microelectronic devices using multiple carriers. *IEEE Transactions on Circuits and Systems I* 54:2211–2221.

22. Chestek CA, Samsukha P, Tabib-Azar M, Harrison RR, Chiel HJ, and Garverick SL (2006) Microcontroller-based wireless recording unit for neurodynamic studies in saltwater. *IEEE Sensors Journal* 6:1105–1114.

23. Sun Y, Huang S, Oresko JJ, and Cheng AC (2010) Programmable neural processing on a smartdust for brain-computer interfaces. *IEEE Transactions on Biomedical Circuits and Systems* 4(5):265–273.

24. Yeager DJ, Holleman J, Prasad R, Smith JR, and Otis BP (2009) NeuralWISP: A wirelessly powered neural interface with 1-m range. *IEEE Transactions on Biomedical Circuits and Systems* 3(6):379–387.

25. Paul S, and Bhunia S (2009) Computing with nanoscale memory: Model and architecture. In *Proceedings of IEEE/ACM International Symposium on Nanoscale Architectures*, IEEE/ACM, San Francisco, CA, July 30–31, 2009.

26. Rapoport BI, Wattanapanitch W, Penagos HL, Musallam S, Andersen RA, and Sarpeshkar R (2009) A biomimetic adaptive algorithm and low-power architecture for implantable neural decoders. In *Proceedings of the 31st International Conference of the IEEE Engineering in Medicine and Biology Society*, IEEE, Minneapolis, MN, Sep 3–6, 2009.

27. Najafi K (2007) Packaging of implantable microsystems. In *Proceedings of IEEE Sensors Conference*, IEEE, Atlanta, GA, Oct 28–31, 2007.

28. Mutoh S, Douseki T, Matsuya Y, Aoki T, Shigematsu S, and Yamada J (1995) 1-V power supply high-speed digital circuit technology with multithreshold-voltage CMOS. *IEEE Journal of Solid-State Circuits* 30(8):847–854.

29. Gonzalez R, Gordon BM, and Horowitz MA (1997) Supply and threshold voltage scaling for low power CMOS. *IEEE Journal of Solid-State Circuits* 32(8):1210–1216.

30. Narasimhan S, Chiel HJ, and Bhunia S (2011) Ultra low-power and robust digital signal processing hardware for implantable neural interface microsystems. *IEEE Transactions on Biomedical Circuits and Systems* 5(2):169–178.

31. Predictive Technology Model [Online], http://www.eas.asu.edu/~ptm/

32. Narasimhan S, Chiel HJ, and Bhunia S (2009) A preferential design approach for energy-efficient and robust implantable neural signal processing hardware. In *Proceedings of the 31st International Conference of the IEEE Engineering in Medicine and Biology Society*, IEEE, Minneapolis, MN, Sep 3–6, 2009.

Energy-Adaptive Computing: A New Paradigm for Sustainable IT

Krishna Kant
George Mason University

48.1 Introduction

The rapid proliferation of information technology (IT) continues to increase the number of computing devices, their supporting infrastructure, and the overall energy consumption. At the chip level, a dramatic increase in wire resistance, increasing device density, and the emerging 3-D integration is conspiring to make power densities unsustainable and heat removal very difficult [1,2]. Because of increasing miniaturization and computing power, similar issues arise at higher levels as well. For example, the tight form factors of blade servers, notebooks, and PDAs make heat dissipation very challenging to manage. Cooling limitations essentially limit the amount of power that the device can consume. For mobile devices, the increasingly sophisticated processing stresses the battery life further. At the global level, it is expected that in spite of aggressive efforts at enhancing power efficiency of computing systems, total IT power consumption is likely to continue to go up [3] and so is the electricity cost associated with running large data centers.

Many of these issues have been well recognized and have resulted in substantial improvements in energy efficiency at a variety of levels—from low-power HW design to aggressive use of available power modes to intelligent load and activity management (e.g., see [4] and references therein). Coordinated power and thermal management have been examined at multiple levels [5–8]. These efforts are expected to continue in the foreseeable future. Yet, the sheer increase in the computing base and its environmental impact have raised a number of sustainability concerns. Although important, energy efficiency is only one of many dimensions of sustainable computing. Sustainability also relates to such diverse subjects as minimization of toxic chemicals used in chip fabrication, design for reuse, and easy end-of-life disposability, operation with renewable energy sources that are often variable, and overall minimization of materials and natural resources used. In this chapter, we address the last two aspects and show that they are related.

The fundamental paradigm that we consider is to replace the traditional overdesign at all levels with rightsizing coupled with smart control in order to address the inevitable lack of capacity that may arise occasionally. In general, such lack of capacity may apply to any resource; however, we only consider its manifestation in terms of energy/power constraints. Note that power constraints could relate to both real constraints in the power availability as well as the inability to consume full power due to cooling/thermal limitations. Power consumption limitation indirectly relates to capacity limitation of other resources as well, particularly the dominant ones such as CPU, memory, and secondary storage devices. We call this *energy adaptive computing* or EAC.* The main point of EAC is to consider energy related constraints at all levels and dynamically adapt the computation to them as far as possible.

It is important to note that the operational energy consumption does not matter much if the energy comes from renewable source. The main issue instead is life-cycle impact of the infrastructure required to generate and distribute the renewable energy [9,10]. A direct use of locally produced renewable energy could reduce the distribution infrastructure, but must cope with its often variable nature. This again comes back to the need for coping with occasional limitations in available power. Thus, better adaptation mechanisms allow for more direct use of renewable energy.

The outline of the rest of the chapter is as follows. Section 48.2 discusses how energy adaptive computing can help enhance sustainability of computing infrastructure. Section 48.3 discusses issues of end to end energy adaptation. Section 48.4 then discusses the challenges posed by EAC. Section 48.5 discusses some concrete results on energy and thermal adaptation in data centers. Finally, Section 48.6 concludes the chapter.

48.2 Sustainability and Energy Adaptive Computing

It is well recognized by now that much of the power consumed by a data center is either wasted or used for purposes other than computing. In particular, up to 50% of the data center power may be used for purposes such as chilling plant operation, compressors, air movement (fans), electrical conversion and distribution, and lighting. Furthermore, the operational energy is not the only energy involved here. Many of these functions are quite materials and infrastructure heavy, and a substantial amount of energy goes into the construction and maintenance of the cooling and power conversion/distribution infrastructures. In fact, even the "raw" ingredients such as water, industrial metals, and construction materials involve considerable hidden energy footprint in form of making those ingredients available in usable form.

It follows that from a sustainability perspective, it is not enough to simply minimize operational energy usage or wastage; we need to minimize the energy that goes into the infrastructure as well [10]. This principle applies not only to the supporting infrastructure but to the IT devices such as clients and servers themselves. Even for servers in data centers, the increased emphasis on reducing operating energy makes the nonoperational part of the energy a larger percentage of the life-cycle energy consumption and could almost account for 50% [11]. For the rapidly proliferating small mobile clients such as cell-phones and PDAs, the energy used in their manufacture, distribution, and recycling could be a dominant part of the lifetime energy consumption.

Toward this end, it is important to consider data centers that can be operated directly via locally produced renewable energy (wind, solar, geothermal, etc.) with minimal dependence on the power grid or large energy storage systems. Such an approach reduces carbon footprint not only via the use of renewable energy but also by reducing the size and capacity of power storage and power-grid related

---

* Here "energy adaptation" implicitly includes power and thermal adaptation as well.

infrastructure. For example, a lower power draw from the grid would require less heavy-duty power conversion infrastructure and reduce its cost and energy footprint. The downside of the approach is more variable energy supply and more frequent episodes of inadequate available energy to which the data center needs to adapt dynamically. This issue can be addressed via large energy storage capacity; however, energy storage is currently very expensive and would increase the energy footprint of the infrastructure.

In large data centers, the cooling system not only consumes a substantial percentage of total power (up to 25%) but also requires significant infrastructure in the form of chiller plants, compressors, fans, plumbing, etc. Furthermore, chiller plants use a lot of water, much of which simply evaporates. Much of this resource consumption and infrastructure can be done away with by using ambient (or "free") cooling, perhaps supplanted with undersized cooling plants that kick in only when ambient temperature becomes too high. Such an approach requires the energy consumption (and hence the computation) to adapt dynamically to the available cooling ability. The energy available from a renewable source (e.g., solar) may be correlated with the temperature (e.g., more solar energy on hotter days), and such interactions need to be considered in the adaptation mechanisms.

The power supply and distribution infrastructure can also be significant energy wasters at all levels. For example, a large data center fed by a 33 kV power supply requires a number of stages of voltage step-down, conditioning, storage, and distribution that together can consume up to 10% of the incoming power, which could be in multiple megawatts. Thus, reducing this infrastructure enhances energy efficiency. On the side, the power supplies that go into individual servers and other assets can also be big power wasters. For example, a server consuming 500 W and sporting a high efficiency power supply with 85% efficiency still wastes 75 W of power. (Since this waste is in form of heat, additional power is wasted in removing the resulting heat.) Often, servers run at rather low utilization levels (5%–15%), and the power supply efficiency is typically much poorer at lower utilizations. Even worse, for redundancy, the servers may employ two concurrent load sharing power supplies, which means that each of them will never exceed more than 50% load. Smart *phase shedding* power supplies address this problem by providing a number of "phases" [12]. As the server utilization dips, more and more phases can be turned off, thereby keeping the power supply utilization and efficiency high. For example, a power supply with 8 phases may have all phases active at 90% server utilization, but at server utilization of 45%, only 4 phases need to be active and will still provide the same power supply efficiency. Similar approaches apply to on-board voltage regulators (VRs). An intelligent mechanism for phase changes that properly accounts for other energy adaptation mechanisms, in effect, could significantly increase energy efficiency of the data center assets.

Yet another sustainability issue is the overdesign and overprovisioning that is commonly observed at all levels of computer systems. In particular, the power and cooling infrastructure in servers, chassis, racks, and the entire data center is designed for worst-case scenarios that are either rare or do not even occur in realistic environments. Realistic workloads rarely stress more than one resource at a time—for example, it is easy to saturate the CPU when it executes primarily out of the caches, but if a significant memory access activity is involved, the CPU will not be able to get the required data quickly enough and will stall. Therefore, taxing CPU and DRAM simultaneously may not even be feasible. It may be possible to saturate both CPU and NIC simultaneously via one or more steady streams of packets of a suitable size, but this may be possible only when it is possible to deposit incoming packets directly into the cache (without a DMA to the memory).

Although data centers are beginning to "derate" specified power and cooling requirements to address this lack of realism, derating alone is inadequate for two reasons: (a) it still must provide a significant safety margin, and (b) derating is used only for sizing up the data center power distribution and cooling capacities, not in server design itself. Instead, we argue for much leaner design of all components having to do with power/thermal issues: heat sinks, power supplies, fans, voltage regulators, power supply capacitors, power distribution network, uninterrupted power supply (UPS), air conditioning equipment, etc. This leanness of the infrastructure could be either static (e.g., lower capacity power supplies and heat sinks, smaller disks, DRAM, etc.), or dynamic (e.g., phase shedding power supplies, hardware resources

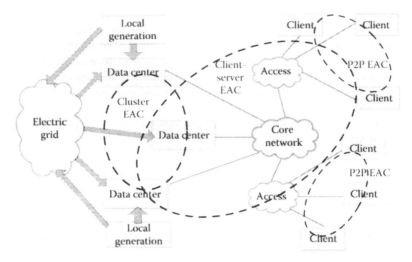

FIGURE 48.1   Illustration of energy adaptation loops.

dynamically shared via virtualization). In either case, it is necessary to adapt computations to the limits imposed by power and thermal considerations. We assume that in all cases the design is such that limits are exceeded only occasionally, not routinely.

48.3 Distributed Energy Adaptive Computing

It is clear from the previous discussion that many advanced techniques for improving energy efficiency of IT infrastructure and making it more sustainable involve the need to dynamically adapt computation to the suitable energy profile. In some cases, this energy profile may be dictated by energy (or power) availability, in other cases the limitation may be a result of thermal/cooling constraints. In many cases, the performance and/or QoS requirements are malleable and can be exploited for energy adaptation. For example, under energy challenged situations, a user may be willing to accept longer response times, lower audio/video quality, less up-to-date information, and even less accurate results. These aspects have been explored extensively in specific contexts, such as adaptation of mobile clients to intelligently manage battery lifetime [13]. However, complex distributed computing environments provide a variety of opportunities for coordinated adaptation among multiple nodes and at multiple levels. In general, there are three types of distributed energy adaptation scenarios: (a) cluster computing (or server to server), (b) client–server, and (c) peer to peer (P2P) (or client to client). These are shown pictorially in Figure 48.1 using dashed ovals for the included components and are discussed briefly in the following. Notice that in all cases, the network and the storage infrastructure (not shown) are also important components that we need to consider in the adaptation.

Although we discuss these three scenarios separately, they generally need to be addressed together because of multiple applications and interactions between them. For example, a data center would typically support both client–server and cluster applications simultaneously. Similarly, a client may be simultaneously involved in both P2P and client-server applications.

48.3.1 Cluster EAC

Cluster EAC refers to computational models where the request submitted by a client requires significant computation involving multiple servers before the response can be returned. That is, client involvement in the service is rather minimal, and the energy adaptation primarily concerns the data center infrastructure.

In particular, a significant portion of the power consumed may go into the storage and data center network and they must be considered in adaptation in addition to the servers themselves.

In cluster EAC, the energy adaptation must happen at multiple levels. For example, the power capping algorithms may allocate a certain power share to each server in a chassis or rack, and the computation must adapt to this limit. In addition, there may be a higher level limit as well—for example, the limit imposed by the power circuits coming into the rack. At the highest level, energy adaptation is required to conform to the power generation (or supply) profile of the energy infrastructure. The limits placed at the lower level generally need to be more dynamic than at higher levels. Translating higher level limits into lower level limits is a challenging problem and requires a dynamic multilevel coordination [8]. A related issue is that of energy limitation along the software hierarchy (e.g., service, application, software modules, etc.) and corresponding multilevel adaptation.

48.3.2 Client–Server EAC

The client–server EAC needs to deal with both client-end and server-end adaptation to energy constraints in such a way so that the client's QoS expectations are satisfied. A coordinated client–server energy adaptation could even deliver benefits beyond adaptation per se. As the clients become more mobile and demand richer capabilities, the limited battery capacity gets in the way. The client–server EAC can provide better user satisfaction and service by seamlessly compensating for lack of client resources such as remaining battery, remaining disk space, operating in a hot environment, etc. In fact, such techniques can even help slow down client obsolescence and thus enhance the goal of sustainability. Reference [14] considers middleware for reconfiguring software in pervasive computing scenarios in order to adapt to dynamically changing environment and user requirements. Similar adaptation can be done for energy as well.

Client–server EAC can be supported by defining client energy states and the QoS that the client is willing to tolerate in different states and during state switching. This information could be communicated to the server side in order to effect appropriate adaptation actions as the client energy state changes. For a more comprehensive adaptation, the intervening network should also be communicated the QoS requirements and should be capable of exploiting them. The situation here is similar to but more complex than the contract based adaptation considered in [15]. The major challenge is to decide an optimal strategy to ensure the desired end-to-end QoS without significant overhead or increase in complexity. In client–server computing, the client adaptation could also be occasionally forced by the server-side energy adaptation. Since server-side adaptation (such as putting the server in deep sleep state and migrating the application to another server) can affect many clients, the server-side adaptation decisions become quite complex when interacting with a large number of geographically distributed and heterogeneous clients. Involving the network also in the adaptation further complicates the problem and requires appropriate protocol support.

48.3.3 Peer-to-Peer EAC

In a P2P setting involving increasingly mobile clients, energy consumption is becoming an important topic. Several recent papers have attempted to characterize energy consumption of P2P content sharing and techniques to improve their energy efficiency [16–18]. Energy adaptation in a P2P environment is quite different from that in a client–server setting. For simple file-exchange between a pair of peers, it is easy to consider the energy state of the requesting and serving peers and that of their network connections; however, a collective adaptation of a large number of peers can be quite complex. Furthermore, it is important to consider the fundamental P2P issue of get–give in this adaptation. In particular, if a peer is in a power constrained mode, it may be allowed to be more selfish temporarily (i.e., allowed to receive the appropriate low-resolution content that it needs without necessarily supplying content to others). In a more general situation, such as BitTorrent where portions of file may come from different clients, deciding and coordinating content properties and assembling the file becomes more challenging. In particular, it might be desirable to offload some of these functions to another client (i.e., not in energy constrained mode). In general, addressing these issues requires defining appropriate energy related metrics relative to the content requester, all potential suppliers (or "servers"), transit nodes and the intervening network. A framework that allows minimization of global energy usage while satisfying other local performance and energy requirements can be quite challenging.

48.4 Challenges in Distributed EAC

In the preceding section, we made little distinction between "energy" and "power"; however, there are subtle differences with respect to both their minimization and adaptation to limits. Energy (or equivalently average power over long periods) can be minimized by deliberately increasing power consumption over short periods. For example, it may be possible to minimize energy for a given set of tasks to be executed by running these tasks at the highest speed and then putting the machine in a low-power mode. This "race-to-halt" policy has traditionally been suboptimal because of the possibility of significant voltage reductions at low speeds in the traditional DVFS (dynamic voltage frequency switching). However, as voltages approach minimum threshold for reliable operation, the voltage differences between various available speeds become rather small. At the same time, the idle power consumption continues to go up due to increased leakage current. In such a situation, race-to-halt increasingly becomes an increasingly attractive scheme. Adaptation to an energy constraint can use the race-to-halt strategy effectively by deliberately batching up work so that the utilization during the active period and the length of inactive period are both maximized. Notice that a power limit may still require using operation at lower speeds.

There are many situations where simultaneous energy and power constraints may be required. For example, in a data center, the capacity of power circuits (e.g., chassis or rack circuit capacity, or capacity of individual asset power supply) must be respected while at the same time abiding by energy constraints (or constraint in terms of average power over longer periods).

The real energy or power limitation usually applies only at a rather high level—at lower levels, this limitation must be progressively broken down and applied to subsystems in order to simplify the overall problem. For example, in case of a data center operating in an energy constrained environment, the real limitation may apply only at the level of the entire data center. However, this limitation must be broken down into allocations for the physical hierarchy (e.g., racks, servers, and server components) and also along logical hierarchy (e.g., service, application, and tasks). While such a recursive break-down allows independent management of energy consumption at a finer-grain level, an accurate and stable allocation is essential for proper operation. We address this issue in detail in Section 48.5.

Good energy allocation or partitioning requires an accurate estimation of energy requirements at various layers. Although a direct measurement of energy consumption is straightforward and adequate for making allocations, it only allows reactive (or after the fact) estimation. For example, if additional workload is to be placed on a server, it is necessary to know how much power it will consume *before* the placement decision is made. This is often quite difficult since the energy consumption not only depends on workload and hardware configuration but also on complex interactions between various hardware and software components and power management actions. For example, energy consumed by the CPU depends on the misses in the cache hierarchy, type of instructions executed, and many other micro-architectural details and how they relate to the workload being executed.

A fairly standard method for estimating power is to compute power based on a variety of low-level performance monitoring counters that are available on-chip and becoming increasingly sophisticated. The trick usually is to find a small set of counters that can provide a good estimate of power [19,20]. While quite accurate, such a scheme does not have much predictive power since the relationship between high-level workload characteristics and performance counters is nontrivial. If multiple tasks are run on the same machine, they can interact in complex ways (e.g., cache working set of one task affected by presence of another task). Consequently, neither the performance nor the power consumption adds up linearly, e.g., the active power for two VMs running together on a server does not equal the sum of active powers of individual VMs on the same server. It may be possible to come up with an estimation method that can account for such interference. The situation is similar to that for estimating total bandwidth consumption of multiple applications communicating over a single link. The notion used in this context is "effective bandwidth" [21,22]. A natural question is whether it is possible to develop a similar notion for power so that it is possible to use simple arithmetic in deciding migration of task and VMs.

As available energy (or power) dips significantly below that required for normal (unconstrained) operation, good energy allocations become progressively more difficult to achieve. These complications arise from the fact that the optimal operating point depends on a variety of factors including the hardware configurations, nature and importance of the workload, and how frequently the workload characteristics change and interactions between various hardware and software components. The interdependence between various hardware and software components may make their relative energy consumption to change quite substantially as the energy budgets shrink. For example, if CPUs and memory are allocated only 1/2 of their normal power for a workload, the changed workload behavior could make this proportion significantly suboptimal. Thus a continuous monitoring and adjustment to energy needs is required in order to keep energy allocation close to optimal.

When energy availability is restricted, certain applications—particularly those involved in background activities—do not even need to run. Others may run less frequently, with fewer resources, or even change their outputs, and still provide acceptable results. For applications that are driven by client requests and must run, the treatment depends on a variety of factors such as SLA requirements, level of variability in the workload characteristics, latency tolerance, etc. For example, if the workload can tolerate significant latencies and has rather stable characteristics, the optimal mechanism at the server level is to migrate the entire workload to a smaller set of servers so they can operate without power limitations and shut down the rest. In this case, a trade-off is necessary with respect to additional energy savings, SLA requirements, and migration overheads.

A comprehensive trade-off requires accounting for not just the servers but also for the storage and networking infrastructure. As within a single platform, the relative energy consumption behavior between servers, storage and network could change significantly under severe energy constraints and needs to be considered carefully. As the workload becomes more latency sensitive, the latency impact of reconfiguration and power management actions must be taken into account. In particular, if firing up a shutdown server would violate latency and response time related SLA, it is no longer possible to completely shut down the servers, and instead one of the lower latency sleep modes must be used. A less stable workload may also require use of less severe power management actions.

Power management techniques typically take advantage of the low utilization of resources so that idle energy consumption can be minimized. This is done either by putting the devices into inactive low-power mode when idle (including complete shutoff), or running them at lower frequencies and voltages so as to raise the device utilization (i.e., the traditional DVFS controls) [4,23]. Traffic batching can help reduce the overhead of entering and exiting sleep states [24] at the expense of adding additional latencies. In the past, much of the work has focused mostly on a rather narrow application of these techniques, such as DVFS control of CPUs or nap states for DRAM, but more complex scenarios involving coordinated control of multiple subsystems are beginning to be analyzed [25].

It is important to note that in the case of EAC, often the problem is not inadequate work, but rather inadequate energy to process the incoming work. Obviously, in order to reduce the average power consumption, we need to slow down processing, except that this slowdown is not triggered by idling. The basic techniques for slowing down the computation still remain the same and may involve either forcing the device into low-power sleep modes or lower DVFS states. Reference [26] compares the effectiveness of the two methods. However, unlike the situation where the goal is to minimize wasted energy, an energy constrained environment requires careful simultaneous management of multiple subsystems in order to make the best use of the available energy. For example, it is necessary to simultaneously power manage CPU, memory, and IO adapters of a server in order to ensure that the energy can be delivered where most required.

In addition to power management, the inability to process all of the incoming workload may require some additional load management actions to avoid build up of long queues. In a high-performance computing type of environment driven by long running jobs, delaying completion of running jobs or startup of new jobs usually has no further consequences. In a transactional system driven by user requests, further actions in form of dropping requests, redirecting them to another facility, migrating away entire applications, or reducing processing requirements at the cost of degraded output quality may be necessary. All of these actions require an accurate mechanism for evaluating "before" and "after" energy requirements for making intelligent decisions. The difficulty here is that because of interference between workloads, power consumptions do not necessarily add up, as already mentioned earlier.

The admission control, migration, and power management of a large number of resources at multiple levels raises a lot of interesting issues in terms of the stability and optimality of the control in addition to the issues of the overhead and lag associated with information exchange. A comprehensive control theoretic framework is required in order to address these issues [7,27]. When the control extends over multiple physical facilities, perhaps each with differing energy costs, the problem becomes even more complex.

Although much of the earlier discussion concerns servers, similar issues apply to clients and their subsystems. For example, the partitioning and control of power among CPU, memory, storage, and other portions of a client involves the same set of issues as servers. However, the P2P interaction between clients involves some unique issues as already mentioned earlier.

Although much of our discussion has focused on servers and clients, storage and network also need serious consideration in energy adaptive computing because of increasing data intensiveness of most applications. Energy management of rotating magnetic media often involves long latencies (in spinning down or spinning up the drives) and reliability issues resulting from RPM changes or repeated starts and stops. The emerging solid state storage (SSD) can be helpful in this regard. The energy management of network devices such as switches and routers is inherently difficult because of its nonlocal impact. For example, if a router/switch port is placed in a low power mode, every application and endpoint whose traffic goes through this port will be affected. Reference [28] discusses the notion of "shadow ports" to allow queuing of packets coming into ports that are in low power mode, but lower level hardware techniques can also ensure that packets arriving into an inactive port are not lost.

When the network power management is triggered by shortage of available power (as opposed to idle or low traffic conditions), the impact is much more severe, since the energy management will result in accumulation of packets and significantly increase flow latencies. The end-to-end admission control required to manage the traffic needs to carefully manage these latencies, performance impact on various applications with varying latency sensitivity, and application timeouts. A significant amount of work remains to be done to address these issues adequately.

While the topic of applications changing their behavior in the face of energy limitations has been explored in several specific contexts such as audio/video streaming, rendering a Web page, P2P content sharing [13,18], and mechanisms to specify and manage the adaptation have been proposed [15,29], there is scope for considerable further work on how and when to apply various kinds of adaptation mechanisms (e.g., lower resolution, higher latency, control over staleness and/or accuracy, etc.) under various kinds of power/thermal limitation scenarios.

The main theme in EAC has been to cut down "fat" at all levels and thereby lower not only the direct energy consumption but also the entire life-cycle energy costs that are essential to examine from a sustainability perspective. This leanness has a downside: the increased fragility in the system that can be exploited by attackers. In particular, just as current systems can be victimized by denial of service (DoS) attacks, the systems proposed here can be further victimized by denial of energy (DoE) attacks. For example, it is possible to craft "power viruses" whose aim is to consume as much power as possible. A carefully planned attack using such viruses can significantly disrupt a distributed EAC scheme and lead to instabilities and poor performance. Protection mechanisms against such energy attacks are essential to realize the EAC vision.

48.5 Energy Adaptation in Data Centers

In this section, we briefly describe results for client–server EAC considered earlier. The details of the scheme, called Willow, are discussed in [30] and covered here only briefly. Willow provides a hierarchical energy adaptation within data centers in response to both demand and supply side variations. Although a comprehensive energy adaptation scheme could include many facets, the current design of Willow is geared toward load consolidation and migration. These mechanisms may need to be augmented with workload alteration such as shutting down of low-priority tasks, change of codecs, or batching of event handling; however, since these aspects are very much workload dependent, they are not considered explicitly. Willow does consider adaptation to thermal constraints by considering them in energy-equivalent terms.

48.5.1 Multilevel Power and Thermal Control

As stated in Section 48.3.1, power/thermal management is required at multiple levels from individual devices within an asset up to the entire data center. For example, it is possible to "shift" power within a server among, say, CPU, memory, and NIC, depending on the low-level workload phase [31]. This can be accomplished by a coordinated control of available operational states of these devices so that the power efficiency can be maximized. The higher levels in this hierarchy include individual assets (servers, switches, routers, etc.), chassis and/or racks, clusters, and the entire data center. In a power limited situation, each level will need to operate under a suitably determined and dynamically varying "power budget." Figure 48.2 shows such a structure along with a power management unit (PMU) at each level.

In the hierarchical power control model that we have assumed, the power budget in every level gets distributed to its children nodes in proportion to their demands. The component in each level $l + 1$ has configuration information about the children nodes in level l. For example, the rack level power manager requires knowledge of the power and thermal characteristics of the individual assets in the rack. The components at level l continuously monitor the demands and utilization levels and report them to level $l + 1$. This helps level $l + 1$ to continuously adjust the power budgets. Level $l + 1$ then directs the components in level l as to what control action needs to be taken.

Because of the fluctuating demand and workload characteristics, the power consumption will also fluctuate and any demand estimation and corresponding adaptations need to be discretized over a suitable time granularity, henceforth denoted as Δ_{DI}. It is assumed that this time granularity is sufficiently coarse to accommodate accurate power measurement and its presentation, which can be quite slow.

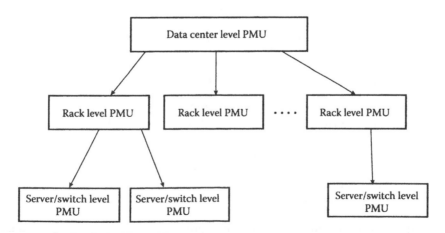

FIGURE 48.2 A simple example of multilevel power control in a data center.

Henceforth, we denote consumed power as CP_l at level l. Typically, appropriate time granularity at the level of individual servers are of order of 100s of ms. It may be desirable to use an even coarser granularity at higher levels such as rack or cluster. The consumed power estimate can be further smoothed by using a simple exponential smoothing over successive "slots" of duration Δ_{Dl}.

Since the supplied power could also fluctuate in general, a similar discretization applies to the supply side power estimation as well. The time-constant for supply changes, and hence the discretization period, Δ_{Sl} is typically much larger than Δ_{Dl}. This is typically a result of temporary energy storage in UPS batteries or capacitors in the supply path. For convenience, we assume $\Delta_{Sl} = \eta_1 \Delta_{Dl}$, where η_1 is suitably chosen integer >1.

Energy efficient operation is an essential aspect of coping with energy deficient scenarios, and they are also valuable during energy surplus periods. In view of this, our Willow system also performs workload consolidation when the demand in a server is low. This allows some servers to be put in a deep sleep state such as S3 (suspend to memory) or even S4 (suspend to disk). Since the activation/deactivation latency for these sleep modes can be quite high, the corresponding time constant, denoted Δ_{Al}, is even higher. In particular, we assume that $\Delta_{Al} = \eta_2 \Delta_{Dl}$, where $\eta_2 > \eta_1$ is an integer constant.

48.5.2 Supply and Demand Side Adaptations

Let BP_{l+1}^{old} be the overall budgeted power at level $l+1$ during the last period and BP_{l+1} the overall power budget at the end of current period. If the difference between the two is too small to have much of a performance impact, we can either leave the level l budgeted powers unchanged or simply alter the budget of the node with highest allocation. In other cases, we need to reallocate the power budgets of nodes in level l. In doing so, we consider both *hard* and *soft* constraints:

1. Hard constraints are imposed by the thermal and power circuit limitations of the individual components. In our control scheme the thermal limits play an important role in deciding the maximum power that can be supported in the component.
2. Soft constraints are imposed by the division of power budget among the other components in the same level.

The reallocation of budgets at level l is somewhat different depending upon whether we have a decrease or increase at level $l+1$. In case of a decrease, the following actions are initiated:

1. If there are any overprovisioned nodes their surplus power is taken away to the extent of satisfying the deficit. If the overall surplus exceeds the deficit, no further action is necessary.
2. In case of unsatisfied deficit, any potential workload modification (e.g., shutting down of low priority tasks) is attempted.
3. If the deficit still persists, a reallocation is attempted following the method described in the following.

In case of budget increase, the following actions are initiated:

1. If there are any underprovisioned nodes, they are allocated just enough power budget to satisfy their demand. (Note that if an admission control is currently in effect to reduce the demand, it needs to be removed or loosened.)
2. If the surplus power still exists, any actions such as freezing of background tasks or workload modifications in effect (e.g., lower resolution) are undone.
3. If surplus is still available at a node, then the surplus budget is allocated to its children nodes proportional to their demand.

The demand side adaptation to thermal and energy profiles is done systematically via migrations of the demands. We assume that the fine grained power control in individual nodes is already being done so that any available idle power savings can be harvested. Our focus in this paper is on workload migration strategies to correct any imbalances in energy supply and demand. For specificity, we consider only those types of applications in which the demand is driven by user queries and there is minimum or no interaction between servers, (e.g., transactional workloads). The applications are hosted by one or more virtual machines (VMs) and the demand is migrated by migrating these virtual machines. A finer grain control is possible by controlling the query fraction directed toward multiple virtual machines serving the same application; however, such an approach is not always feasible and is not considered here.

There are a few considerations in designing a control strategy for migration of demands:

1. *Error Accumulation*: Because of the top-down subdivision of power budgets, any errors and uncertainties get more pronounced at lower levels.
2. *Ping-Pong Control*: A migration scheme that migrates demand from server A to B and then immediately from B to A due to erroneous estimations leads to a ping-pong control. A ping-pong wastes energy and introduces latencies.
3. *Imbalance*: The inaccuracies in estimation could leave some servers power deficient while others have surplus power.

We avoid these pitfalls by a variety of means including *unidirectional control*, explained in the following, and allowing sufficient margins both at the source and the destination to accommodate fluctuations after the migrations are done.

Unidirectional control initiates migrations only when power constraints are tight ended. Furthermore, the migrations are initiated in a bottom up manner. If the power budget $BP_{l,i}$ of any component i is too small then some of the workload is migrated to one of its sibling nodes. We call this a local migration. If the sibling nodes of component i do not have sufficient surplus to accommodate its excess demand, then the workload is migrated to one of the children of another $l + 1$ component. We call this a nonlocal migration. Local migrations are always preferred to nonlocal migrations for two reasons:

1. The overheads involving networking resources are reduced when the migrations are local rather than nonlocal.
2. The VMs might have some local affinity with common resources like hard disks or SANs, and a nonlocal migration might affect this affinity.

The migration decisions are made in a distributed manner at each level in the hierarchy starting from the lowermost level. The local demands are first satisfied with the local surpluses and then those demands that are not satisfied locally are passed up the hierarchy to be satisfied nonlocally. The final rule in the unidirectional control scheme is that the migrations are destined to a node only if the power availability of the node is not reduced by the event that caused the migration. For instance, if the power availability of a rack has gone down, no migrations are allowed into that rack. Similarly if the power availability of the entire data center has reduced no migrations can happen immediately. (Migrations may still happen if as a result of budget reductions we decide to shut down certain tasks and consequently increase the power availability.)

In order to mitigate against random short-term fluctuations in supply and demand, a migration is done if and only if the source and target nodes can have a surplus of at least P_{min}. Also, a migration does not have the secondary effect of splitting the demand between multiple nodes. Finally Willow also does some consolidation. When the utilization in a node is really small, the demand from that node is migrated away from it and the node is deactivated.

The problem of migration coupled with load consolidation can be formulated as a bin packing problem with variable sized bins. Here the bin size represents the current. The variable sized bin packing problem is a NP-hard problem as such, and numerous approximation schemes are available in literature [32]. We choose one such simple scheme called FFDLR [32]. The FFDLR successively packs the bins in the order of largest bin first. The net result is packing in a minimum number of bins.

FFDLR solves a bin packing problem of size n in time $O(n \log n)$. The optimality bound guaranteed for the solution is $1.5^* \text{OPT} + 1$ where OPT is the solution given by an optimal bin packing strategy. We chose this algorithm for two reasons. First, it is simple to implement with guaranteed bounds. Second, the repacking into the fewest (and largest) bins means that the workload is migrated to the fewest servers so that the others can be shut down.

48.5.3 Experimental Results

In this section, we report some experimental results for Willow. The results are obtained primarily via simulation but with parameters determined via actual experiments. Figure 48.4 shows the measured server power consumption as a function of CPU utilization for a CPU bound workload. Not surprisingly, the idle power consumption is quite substantial and increases monotonically with utilization.

To illustrate the impact of the proposed scheme, a tiny data center consisting of 18 servers was simulated in MATLAB®. The servers are arranged in four levels as shown in Figure 48.3. Each server runs a random mix of four different application types with relative processing requirements of 1, 2, 5 and 9, respectively. We assume that all the applications are transactional, driven by user queries and the query arrival process is Poisson. We assume the multiplicative constants $\eta_1 = 4$ and $\eta_2 = 7$. We assume an average server power consumption of 450 W, typical ambient temperature to be 25 °C and thermal limit of 70 °C. We assume, for simplicity, that the slot size Δ_{DI} is large enough (typically 100s of ms) to accommodate CPU temperature changes associated with demand changes. Figure 48.5 shows adaptation to situation where servers 1–14 are in a "cool" zone of 25 °C ambient temperature, whereas, the rest are in a "hot" zone of 40 °C ambient temperature. The figure shows the average power consumption under various utilization settings as a function of server number. As expected, the control effected to keep the

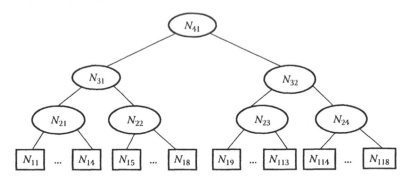

FIGURE 48.3 Configuration of the simulated data center. (From Kant, K., Murugan, M., and Du, D., Willow: A control system for energy and thermal adaptive completing, *Proceedings of IPDPS*, Anchorage, AK, 36, 2011. ©2011 IEEE.)

Utilization (%)	Average Power (W)
0	160
20	175
40	190
60	215
80	230
100	245

FIGURE 48.4 Measured server power consumption vs. utilization.

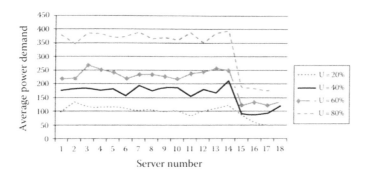

FIGURE 48.5 Power consumption at various utilization levels. (From Kant, K., Murugan, M., and Du, D., Willow: A control system for energy and thermal adaptive completing, *Proceedings of IPDPS*, Anchorage, AK, 36, 2011. ©2011 IEEE.)

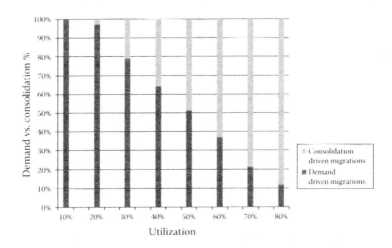

FIGURE 48.6 Relative fractions of demand vs. consolidation migrations. (From Kant, K., Murugan, M., and Du, D., Willow: A control system for energy and thermal adaptive completing, *Proceedings of IPDPS*, Anchorage, AK, 36, 2011. ©2011 IEEE.)

server temperatures below the acceptable limit results in much lower allowable power consumption for servers 15–18.

Since Willow attempts to both consolidate the workload as well as abide by power constraints, the migrations are driven by both factors. Figure 48.6 shows that the migrations are driven by consolidation at lower utilizations and by power constraints at higher utilizations. Figure 48.7 shows the proportion of migration traffic as a fraction of normal job arrival rate. This can be interpreted as the probability that a job would need to be migrated. It is seen that the migration probability increases with utilization at first due to more activity in the system and reaches a maximum around 50% utilization level. Beyond this, the opportunities for migration begin to dry up due to lack of servers with adequate surplus to accept the migrated jobs. At very high utilization levels, it is no longer possible to cope with low power availability periods via migration, and load shedding (not shown) becomes essential.

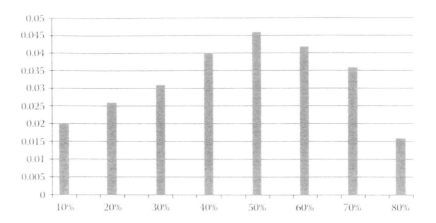

FIGURE 48.7 Migration probability vs. server utilization. (From Kant, K., Murugan, M., and Du, D., Willow: A control system for energy and thermal adaptive completing, *Proceedings of IPDPS*, Anchorage, AK, 36, 2011. ©2011 IEEE.)

48.6 Conclusions

In this chapter, we discussed the notion of energy adaptive computing that attempts to go beyond achieving energy savings based on the minimization of device idling and energy wastage. One of the goals of energy adaptive computing is to make IT more sustainable by minimizing overdesign and waste in the way IT equipment is designed, built, and operated. As pointed out in this chapter, such an approach brings multiple new challenges in the energy management of IT systems that need to be explored more fully.

As an illustration, we considered energy adaptation in a data center and devised a simple algorithm, i.e., stable, low-overhead and able to handle significant variations in energy availability. The stability in the decision making directly translates into reduced networking impact since there are no flip-flops in migrations.

The area of energy adaptive computing can be explored in many different directions, including the issues pointed out among the challenges. The data center adaptation reported here can be refined further to consider applications with different QoS requirements running together. Since migrations are often necessary because of normal resource availability issues such as computing power or memory, it is necessary to coordinate such migrations along with those that are done as a result of power/energy limitations so that the overall QoS objectives can be satisfied.

Acknowledgment

The author wishes to acknowledge the contributions of Muthukumar Murugan in the data center energy adaptation work reported in Section 48.5.

References

1. P. Emma and E. Kursun, Opportunities and challenges for 3D systems and their design, *IEEE Design & Test of Computers*, 26(5):6–14, 2009.
2. B.P. Wong, A. Mittal, Y. Cao, and G. Starr, *Nano-CMOS Circuit and Physical Design*, Chapter 1, John Wiley, Hoboken, NJ, 2005.
3. J.G. Koomey, Worldwide electricity used in data centers, *Environmental Research Letters*, 3:1–8, 2008.

4. K. Kant, Data center evolution: A tutorial on state of the art, issues, and challenges, *Elsevier Computer Networks Journal*, 53(17):2939–2965, December 2009.

5. J. Moore, J. Chase, P. Ranganathan, and R. Sharma, Making scheduling cool: Temperature-aware workload placement in data centers, *Proceedings of USENIX Annual Technical Conference*, Berkeley, CA, pp. 61–74, 2005.

6. R. Nathuji and K. Schwan, VirtualPower: Coordinated power management in virtualized enterprise systems, *Proceedings of SOSP'07*, Stevenson, WA, pp. 265–278, 2007.

7. X. Wang and Y. Wang, Coordinating power control and performance management for virtualized server clusters, *IEEE Transactions on Parallel and Distributed Systems*, 22(2):245–259, 2010.

8. R. Raghavendra, P. Ranganathan, V. Talwar, et al., No power struggles: Coordinated multi-level power management for the data center, *Proceedings of 13th ASPLOS*, Seattle, WA, pp. 48–59, March 2008.

9. S. Boyd, A. Horvath, and D. Dornfeld, Life-cycle energy demand and global warming potential of computational logic, *Environmental Science & Technology*, 43(19):7303–7309, 2009.

10. A.J. Shah and M. Meckler, An energy-based framework for assessing sustainability of IT systems, *Proc. of ASME Conf.* 2009, pp. 823–832.

11. J. Chang, J. Meza, P. Ranganathan et al., Green server design: Beyond operational energy to sustainability, *Proceedings of HotPower*, Vancouver, British Columbia, Canada, 2010.

12. D. Freeman, Digital power control improves multiphase performance, *Power Electronics Technology*, December 2007 (www.powerelectronics.com).

13. J. Flinn and M. Satyanarayanan, Managing battery lifetime with energy-aware adaptation, *ACM Transactions on Computer Systems*, 22(2):137–179, May 2004.

14. A. Corradi, E. Lodolo, S. Monti, and S. Pasini, Dynamic reconfiguration of middleware for ubiquitous computing, *Proceedings of 3rd International Workshop on Adaptive and Dependable Mobile Ubiquitous Systems*, London, U.K., pp. 7–12, 2009.

15. V. Petrucci, O. Loques, and D. Moss, A framework for dynamic adaptation of power-aware server clusters, *Proceedings of 2009 ACM Symposium on Applied Computing (SAC'09)*, Honolulu, HI, pp. 1034–1039, 2009.

16. S. Gurun, P. Nagpurkar, and B.Y. Zhao, Energy consumption and conservation in mobile peer-to-peer systems, *Proceedings of International Conference on Mobile Computing and Networking*, Los Angeles, CA, pp. 18–23, September 2006.

17. I. Kelenyi and J.K. Nurminen, Energy aspects of peer cooperation measurements with a mobile DHT system, *Proceedings of ICC*, Beijing, China, pp. 164–168, 2008.

18. I. Kelenyi and J.K. Nurminen, Bursty content sharing mechanism for energy-limited mobile devices, *Proceedings of 4th ACM Workshop on Performance Monitoring and Measurement of Heterogeneous Wireless and Wired Networks*, Tenerife, Canary Islands, Spain, pp. 216–223, 2009.

19. Y. Cho, Y. Kim, S. Park, and N. Chang, System-level power estimation using an on-chip bus performance monitoring unit, *Proceedings of ICCAD*, San Jose, CA, pp. 149–154, 2008.

20. G. Contreras and M. Martonosi, Power prediction for intel XScale processors using performance monitoring unit events, *Proceedings of ISLPED*, San Diego, CA, pp. 221–226, 2005.

21. C.-S. Chang and J.A. Thomas, Effective bandwidth in high-speed digital networks, *IEEE Journal on Selected Areas in Communications*, 12(6):1091–1100, August 2002.

22. A.I. Elwalid and D. Mitra, Effective bandwidth of general Markovian traffic sources and admission control of high speed networks, *IEEE/ACM Transactions on Networking*, 1(3):329–343, August 2002.

23. V. Venkatachalam and M. Franz, Power reduction techniques for microprocessors, *ACM Computing Surveys*, 37(3):195–237, September 2005 (http://www.ics.uci.edu/vvenkata/finalpaper.pdf).

24. A. Papathanasiou and M. Scott, Energy efficiency through burstiness, *Proceedings of the 5th IEEE Workshop on Mobile Computing Systems and Applications (WMCSA'03)*, Monterey, CA, pp. 44–53, October 2003.

25. S. Mohapatra and N. Venkatasubramanian, A game theoretic approach for power aware middleware, *Proceedings of 5th ACM/IFIP/USENIX International Conference on Middleware*, Toronto, Canada, pp. 417–438, October 2004.

26. K. Kant, Distributed energy adaptive computing, *Proceedings of ACM Hotmetrics, 2009, ACM Performance Evaluation Review*, Vol. 37, No. 3, Seattle, WA, pp. 3–7, December 2009.

27. Y. Wang, K. Ma, and X. Wang, Temperature-constrained power control for chip multiprocessors with online model estimation, *SIGARCH Computer Architecture News*, 37(3):314–324, 2009.

28. G. Ananthanarayanan and R.H. Katz, Greening the switch, *Proceedings of HotPower*, San Diego, CA, pp. 7–7, 2008.

29. J.P. Sousa, V. Poladian, D. Garlan et al., Task based adaptation for ubiquitous computing, *IEEE Transactions on Systems, Man and Cybernetics*, 36(3):328–340, 2006.

30. K. Kant, M. Murugan, and D. Du, Willow: A control system for energy and thermal adaptive computing, *Proceedings of IPDPS*, Anchorage, AK, pp. 36–47 2011.

31. P. Ranganathan, P. Leech, D. Irwin, and J. Chase, Ensemble-level power management for dense blade servers, *Proceedings of ISCA*, Boston, MA, pp. 66–77, 2006.

32. D.K. Friesen, M.A. Langston, Variable sized bin packing, *SIAM Journal of Computing*, 15(1):222–230, 1986.

Social and Environmental Issues

49

Evolution of Energy Awareness Using an Open Carbon Footprint Calculation Platform

Farzana Rahman
Marquette University

Sheikh Iqbal
Ahamed
Marquette University

Casey O'Brien
Marquette University

He Zhang
Tsinghua University

Lin Liu
Tsinghua University

49.1 Introduction

The term "GHG footprint" refers to the total greenhouse gas (GHG) emissions that an individual or company is responsible for emitting into the atmosphere. Defining GHG footprint is the first step in reducing GHG footprint or going carbon neutral. Reducing GHG emissions requires measuring the emissions in specific areas and identifying the major GHG emitting activities. Based on the knowledge of the major emitting activities and their impact on the environment, domain experts could build quantitative models and propose solid mitigation strategies and plans.

In the computing community, personal carbon footprint calculators have emerged to assist individuals in determining their carbon footprint—the "total greenhouse gas emissions caused directly and indirectly by a person, organization, event or product" [Liu09]. On the Internet, there are a variety of products that claim to calculate an individual's carbon footprint. Produced by government agencies, nongovernmental organizations, and private enterprises, these calculators draw on household activities and transportation to arrive at an estimated amount of carbon dioxide per year. For example, Google PowerMeter [GooglePM] receives information from utility smart meters and energy management devices and provides customers with access to their home electricity consumption right on their personal iGoogle homepage. The Carbon Diem [CarbonD] project utilizes global positioning system (GPS) and mobile phones to give accurate usage of CO_2 for fixed travel pattern. However, the recent rise in carbon calculators has been accompanied by inconsistencies in output values given similar inputs for individual behavior. A comparative study is analyzed in [Padgett08].

There are two issues with these calculators. First, while the intent of the calculators brings awareness to the issue of carbon emissions, they produce vastly inconsistent results given similar inputs. A comparative study of representative calculators was performed by Padgett et al. [Padgett08]. Second, these calculators are static and fail to take into account the dynamic behavior of human beings. In other words, these calculators need to respond to the current activities of an individual.

Despite all these facts, people are interested in making personal changes to reduce their contribution to climate change. We focus our efforts on people who are actively seeking to reduce their carbon footprint. These users have questions about how best to direct their efforts, such as how much additional electricity does increasing the thermostat on the air conditioner by one degree consume? For such users, tracking energy usage for each dynamic activity (electricity usage, driving a long route, etc.) is as important as getting suggestions to reduce carbon footprint by rational decision making. This approach allows users to prioritize among the many possible ways they can reduce their environmental impact. These two features are not addressed by any of the calculators developed so far. However, the advances in mobile technology provide a unique opportunity to produce dynamic carbon footprint calculations that allow individuals to make informed, environmentally responsible decisions. Here, we present the design of an open carbon footprint platform (OCFP) that includes the aforementioned two important features. OCFP is a platform-agnostic standard open framework that will provide the necessary interfaces for software developers to incorporate the latest scientific knowledge regarding climate change into their applications.

49.2 Significance of Ubiquitous Carbon Footprint Calculation

Recently, a growing number of companies have started to collect the emission data that results from their production and service provision. GHG or carbon dioxide accounting methods may not only be applied to countries or organizations but also to individual persons. But, there are two issues with these calculators. First, while the intent of the calculators brings awareness to the issue of carbon emissions, they produce vastly inconsistent results given similar inputs. Second, these calculators are static and fail to take into account the dynamic behavior of human beings. In other words, these calculators need to respond to the current activities of an individual. Pervasive computing and sensor technologies offer great potential to estimate instant and ubiquitous CO_2 consumption of each individual or an organization as a whole. The use of carbon footprint calculator applications will increase the awareness of individuals to reduce their own CO_2 contribution in the environment. The use of ubiquitous technology to reduce CO_2 consumption can be better understood by the following examples.

Example 49.1

Bob is a salesperson who needs to go to different places everyday because of the nature of his job. He uses his GPS to determine the route to his destination place. However, by using a simple smartphone application Bob can reduce his carbon footprint immensely. Using the GPS on his smartphone, the application can track his travel and can build a velocity profile figuring out how efficiently he is driving. The application allows him to find the most fuel efficient routes for his vehicle between arbitrary end points. As a result, it helps Bob to reduce his carbon footprint while he is driving. The application lets Bob to review his trips with details that allow him to better understand his practice to reduce his carbon footprint. The application allows Bob to join an online carpool community. It can also show the results of people in his local area going the same way. This helps Bob to participate in the carpool and reduce his own carbon footprint even further.

49.3 Ubiquitous Data Collection Techniques and Challenges

The usability of an automated carbon footprint calculator application is dependent on the data collection methods as well as on the collected data itself. The advances in mobile technology provide a unique opportunity to perform dynamic data collection that allows individuals to make environmentally responsible decisions.

There are different techniques for obtaining relevant condition monitoring information that may be used in the estimation of carbon footprint. Sensors can either be simply associated directly with the items of interest or they can simply report back any environmental situation. There is also a wide range of sensing options available from simple disposable time–temperature indicators (TTI) that require a visual inspection to data loggers having either a fixed wired connection or using a wireless link, such as IEEE 802.15.1 (Bluetooth), IEEE 802.15.4 (used by ZigBee), or IEEE 802.11 (Wi-Fi) to transfer data to a datastore or real-time monitoring system. There are also some newer technologies such as wireless identification and sensing platform (WISP) [Sample07,Buettner08] that can be used to collect environmental or human activity data. GPS on a mobile device can also be used to track velocity of car or human location. Moreover, sensor-enabled RFID technology is also a good technique to track/trace through the monitoring of conditions such as temperature, contamination, shock, humidity, etc.

With current technologies, the commercial justification for using sensor-enabled RFID tags is no longer limited by traditional constraints such as size and price, and they are now commercially available as active or semipassive tags. Passive versions can be anticipated in the future if the power consumption of the sensing circuitry becomes sufficiently low or nonelectrical sensors are used. Moreover, some important sensing capabilities of recent smart mobile devices can be used for different monitoring purposes. In the context of sensing and environmental or behavioral data collection, currently available technologies can be classified into two broad categories, which are briefly discussed next.

49.3.1 Continuous Monitoring

A continuous monitoring sensor provides time-related data samples that can be recorded at known intervals. Since the data can be collected at either regular or irregular intervals, any exceptional instances that occur during the course of the monitoring process can thus be stored. These sensors are ideally suited to goods that can degrade over time (e.g., as a result of temperature, humidity, etc.). Continuous monitoring has its pros and cons. On the positive side, the rich information is useful for evaluating the impact of condition changes in key processes for more detailed analysis. However, on the negative side, continuous data generally requires a power source, extra memory, and additional components, which add to the cost, reliability, life, and size of the tag.

49.3.2 Discrete Monitoring

The discrete monitoring sensor provides a single binary state regarding the condition characteristic of an item or an event. For example, it detects whether the item has been subjected to any discrete event such as an occurrence of a damaging shock that exceeds acceptable limits. Discrete monitoring sensors have their advantages and disadvantages. On one hand, the sensor does not require additional memory space due to the "true/false" binary characteristic, thus making it a cheap and simple solution. On the other hand, since there might not be any timing information as typically associated with continuous data collection, tracking down the origin of the event might be limited to corresponding object read points, provided that the state of the sensor was also interrogated at each read point.

49.4 Definition and Properties of Ubiquitous Carbon Footprint

A carbon footprint is the total amount of GHG emissions caused by an organization, event, or product. However, in this chapter, we introduce the notion of ubiquitous carbon footprint *(UCF)* that can be addressed through several quantifiable *context contributors*. By context contributors, we refer to those dynamic human activity or environmental changes that are responsible for significant CO_2 emission. A UCF consists of various context contributors and most of them evolve to address the unique characteristics found in ubiquitous computing environments. Thus, we define the properties of UCF, in this chapter, from their scope in ubiquitous scenarios. In our model, total UCF, UCF_t, depends on the number of context contributors $(cc_1, cc_2, \ldots, cc_k)$ at time t. We consider the following notations:

$$UCF_t(cc) = \sum_{i=1}^{k} UCF_t(cc_i) \tag{49.1}$$

where k = number of context contributors.

Context contributors can be categorized into two types: *explicit context contributor (ecc)* and *implicit context contributor (icc)*. Explicit context contributors are those contexts that directly contribute toward CO_2 emission (e.g., burning fuel, electricity consumption, etc.). Whereas implicit context contributors depend on the *ecc* to contribute in CO_2 emission. For example, using air conditioner or heater, increasing or decreasing room temperature are examples of *icc*. Therefore, we can redefine the Equation 49.1 as follows:

$$UCF_t(cc) = UCF_t(ecc) + UCF_t(icc) \tag{49.2}$$

where

$$UCF_t(ecc) = \sum_{i=1}^{m} UCF_t(ecc_i)$$

$$UCF_t(icc) = \sum_{j=1}^{n} UCF_t(icc_j)$$

where
  m is the number of explicit context contributors
  n is the number of implicit context contributors

Details of the equation for the overall UCF calculation have been described in later sections of the chapter. Throughout this chapter, we use the term "mobile device/device" to refer to a "person" as we assume that each individual wishing to determine personal carbon footprint will have a mobile device (e.g., cell phone, personal digital assistant (PDA), smart phone, etc.). A UCF has following properties and relationships:

1. **Context dependent:** A mobile device will have different carbon footprint values for different contexts.

$$UCF_t(cc_i) \neq UCF_t(cc_j) \quad \text{where, } i \neq j$$

However, the carbon footprint value of some aggregated contexts may be similar to that of a single context or some other combination of aggregated contexts. For example,

$$UCF_t(cc_m) + UCF_t(cc_n) = UCF_t(cc_i) = UCF_t(cc_x) + UCF_t(cc_y) + UCF_t(cc_z)$$

2. **Dynamic:** A mobile device may have different carbon footprint values at different times provided that the context contributor has been changed.

$$UCF_{t+1}(cc_i) \neq UCF_{t+2}(cc_j) \quad \text{where, } i \neq j$$

However, if the context contributor is same over a period of time, the carbon footprint can be same:

$$UCF_{t+1}(cc_i) = UCF_{t+2}(cc_i)$$

3. **Partially transitive:** If device A's carbon footprint depends on a particular *icc*, which is dependent on another *ecc*, then this chain relation between *icc* and *ecc* implies that a carbon footprint value can be inferred for A on *icc*.
4. **Nonsymmetric:** In general, two devices A and B having the same carbon footprint value do not necessarily indicate that both devices depend on the same context contributor. Let, $UCF_t(cc_i)$ is the carbon footprint of device A and $UCF_t(cc_j)$ is the carbon footprint of device B. Therefore,

$$UCF_t(cc_i) \neq UCF_t(cc_j) \Rightarrow cc_i \neq cc_j$$

5. **Context relation dependent:** UCF is dependent on relations among different context contributors.

49.5 Unique Characteristics of Open Carbon Footprint Platform

The OCFP has the following characteristics: (1) users' capability to develop various carbon footprint monitoring applications based on a standard framework, (2) a privilege for the developers to customize and adapt the interfaces of the framework based on their requirements, (3) a component-based computational model that can be adapted from application to application, (4) release and publication of some cloud-based Web services that can perform some core calculation for various applications, and (5) allowing the end user to benefit from the use of various carbon footprint monitoring applications that will increase energy awareness. The research will benefit the future green computing areas and applications, having a standard and open framework for application development, requiring the developers only to customize the interfaces and allowing the end users to benefit from such applications by contributing less CO_2.

49.6 Impact of Using Open Carbon Footprint Platform

The use of OCFP may have a wide range of impacts in the fields of green computing, eco-friendly software development, green applications, environment friendly computer technologies and systems, energy efficient design of algorithms and systems, green supply chain process, various industries involving monitoring process, and many more applications that involve environmental and behavioral monitoring.

One of the major contributions of the use of OCFP is that it will provide the necessary interfaces for software developers to incorporate the latest scientific knowledge regarding climate change into their applications. This is possible since OCFP is developed by consulting the latest climate science and computational models. The use of OCFP for the development of various personal carbon footprint applications can allow users to increase awareness regarding their own contribution of GHG gasses that will eventually lead to less energy consumption.

49.7 Using Open Carbon Footprint Platform for Carbon Footprint Application Development

By consulting the latest climate science and computational models, we propose a platform-agnostic open carbon footprint framework that will provide the necessary interfaces for software developers to incorporate the latest scientific knowledge regarding climate change into their applications. OCFP will maintain a clouded knowledge base that will give developers access to a dynamic source of computational information that can be brought to bear on real-time sensor data.

OCFP aims at providing a UCF calculation and carbon footprint abatement consulting service for individual users. With the sensors installed inside the portable devices, the system can collect the information needed for the pervasive CO_2 emission. OCFP is supported by a calculation platform, which contains quantitative computation models of human–environment interaction. Experts can use collective intelligence to update the models and offset knowledge on this platform. In order to develop a flexible carbon calculation platform, we need to establish and manage a set of flexible human–environment interaction models and to find a solution for model integration. The result obtained from an individual user's impact on the environment could then be used to match the relevant knowledge and suggestions in the knowledge base and, therefore, promote awareness of reducing carbon emission and facilitate GHG offset knowledge sharing among individuals. Our system has two major parts:

- *Portable devices:* It finds major parameters in the human behavior model. It finds data collection methods from sensors and portable devices.
- *Model management and calculation:* It finds a way of managing a set of quantitative models for calculating personal carbon footprint. It can build up a customized carbon footprint calculator using the applicable quantitative models.

49.8 Architecture of Open Carbon Footprint Platform

One major property of OCFP is that it can manage a set of simple footprint calculation models and can integrate them to calculate the overall carbon footprint of an individual user. Figure 49.1 shows the system architecture of the personal carbon footprint calculation platform. The system includes three layers: the *model layer*, the *middleware layer*, and the *sensor layer*. The sensor layer collects raw data from the registered sensors and smart meters and sends the data to the middleware layer. Raw data are then analyzed by the middleware to remove the noise and recognize some current conditions of the user (for instance, whether he/she is currently traveling by car or bike). When the middleware layer finishes analyzing the data, the processed data along with the deducted current conditions of the user is served as the model input for the model layer for footprint calculation.

The basic architecture of OCFP framework is shown in Figure 49.2. OCFP is designed in such a way so that it can be modified or customized by the application developers. We designed the architecture of OCFP to facilitate the application developer to get all kinds of core services to develop their application in semitransparent manner.

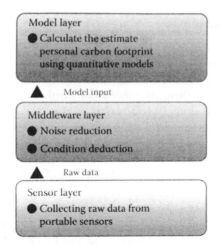

FIGURE 49.1 Data flow in different layers of OCFP.

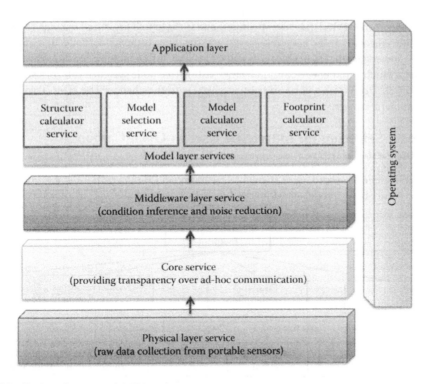

FIGURE 49.2 Basic architecture of OCFP.

49.9 Calculation Models in Open Carbon Footprint Platform

The OCFP framework uses a set of quantitative models to calculate the user's personal carbon footprint. All of these quantitative models could get their input data from the sensors and their respective middleware, and having GHG emission estimation as their calculation output. Each of the models calculates a certain category of the user's carbon footprint under some restraints.

49.9.1 Model Storage Structure

After learning about the representation of a single model, the following questions still lie ahead: in what kind of data structure can we store all these models? And how can these models work together to calculate the overall carbon footprint for personal users?

For the model storage structure, there are concerns of whether it is easy to update and search and whether it is effective for model selection and calculation. We use a decomposition tree structure to store the models. Figure 49.3 shows an example of the hierarchical model collection tree (HMCT) in which we store the quantitative models used for calculation. The HMCT is a decomposition structure of the personal user's carbon footprint with the root node representing the user's overall carbon emission. The root node and its descendent nodes can be divided into several smaller subcategories, each forming a subnode of the current node.

In the example shown in Figure 49.4, the user's personal carbon footprint is divided into direct carbon emission from electricity consumption, fossil fuel consumption, natural resources consumption, and other indirect emissions. Each of the tree nodes could also be divided into even smaller categories. For example, the electricity consumption node could be divided into two subcategories: direct carbon emission from electricity consumption caused by lighting and direct carbon emission from electricity consumption caused by using office equipment.

Each tree node of the HMCT has two attributes: one is the actual measured average carbon emission value of that subcategory that the node represents; and the other is the model collection of all the models that fall to the subcategory that tree node represents. The HMCT does not need to be exhaustive at first. As the new models, which could not be covered by the existing tree node specified categories, are introduced to the system, new tree nodes will be added to comply with the change. The HMCT will be managed manually by domain experts.

49.9.2 Model Selection for Individual User in Open Carbon Footprint Platform

When the user is defined according to the data availability and model restraints, a set of models along with the substructure of the model collection tree could be selected to form a customized model tree. The

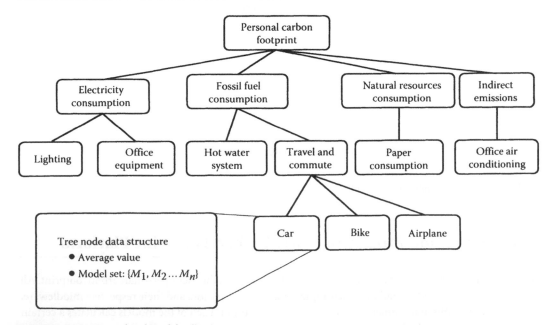

FIGURE 49.3 Hierarchical model collection tree.

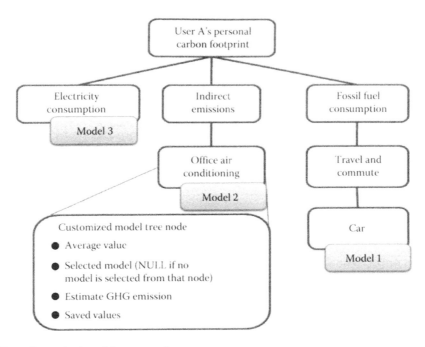

FIGURE 49.4 Customized model tree example.

customized model tree is used not only for storing the calculation models for specific user, but also for the calculation of the overall carbon emission for the user. To avoid duplicate calculation, we select the model according to the following rules:

- At most, one model can be selected from a certain HMCT tree node's model set.
- If one model is selected from a certain HMCT tree node, no model could be selected from its descendent nodes.

Under such rules, there is usually more than one set of models that could be selected from the HMCT. So there are several strategies that could be applied during selection according to the user's preference, for example, the user could select the model set with the minimum calculation error or the maximum model coverage. We use the minimum error strategy, which uses a deep first search to select a set of models with the largest model calculation error minimum. After the model set being selected from the HMCT, the customized model tree also keeps the substructure of the HMCT from the tree root to the nodes with selected models.

Although the customized model tree is much like an HMCT after branch cut, it has a slightly different data structure for tree nodes. The tree node in the customized model tree has four attributes: Average value is the same as the one in the HMCT tree nodes; selected model is the model being selected from its corresponding HMCT tree node (null if there is no model selected); estimate GHG emission, which represents the accumulative total carbon emission calculation result from the branch formed by the node and its descendent nodes; and saved values is a set of constant values collected from the user, which will be used as part of the calculation input for selected model in the node.

Figure 49.5 shows an example of a typical customized model tree. According to user A's given data and sensors that user A could access, three models, out of all the models, which restrains could be satisfied, could be selected from the HMCT according to the aforementioned method. The customized model tree could then be built on basis of the selected models and the substructure of the HMCT. For the node of the customized model tree, node electricity consumption, office air conditioning, and car have the selected model saved in their selected model attribute, respectively. From the user input data, the car model and

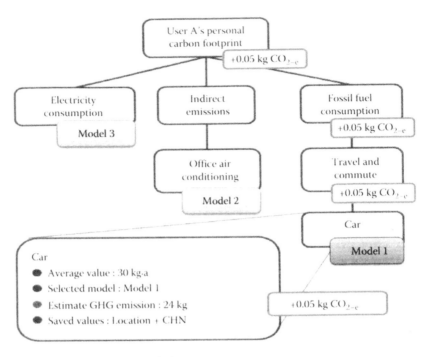

FIGURE 49.5   Personal carbon footprint calculation.

the location of user A's house and office is then being saved in the saved values attribute of the three tree nodes. When the node first appeared in the user's customized model tree, the estimate GHG emission will be set to $0\,kg\,CO_{2-e}$.

49.9.3 Carbon Footprint Calculation in Open Carbon Footprint Platform

After the customized model tree is built, the user's personal carbon footprint could be calculated using the tree. The system will calculate using the following five steps:

1. Building a multidimensional index tree using the trigger restraints of the selected models.
2. Collecting data automatically from the data sources according to the preset time step.
3. When the data are collected, the system will check the trigger restraints index tree and find the active models in the customized model tree.
4. For each active model, it will calculate GHG emission increase according to the input.
5. The increase of the carbon emission will then be added to the estimate GHG emission attribute of the active model's node and also to the node's ancestor nodes in the customized model tree.

In Figure 49.6, an example of calculation is given to illustrate this process. After the trigger restraints index tree is built and a set of input data is collected, the system checks the index tree and finds the trigger restraints for Model 1 are satisfied. After Model 1's general restraints are checked, Model 1 is activated, and an increase of carbon emission could be calculated using the input data and the saved values of the node car. The calculation results in an increase of $0.05\,kg\,CO_{2-e}$. This increase is added to the estimate GHG emission attribute of car and its ancestors.

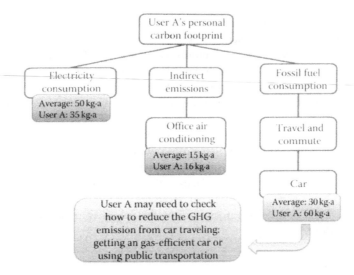

FIGURE 49.6 Calculation result analysis.

49.10 UCFC: A UCF Calculator Application Developed Using OCFP

In order to demonstrate the use of OCFP, we developed a ubiquitous carbon footprint calculator application (UCFC) that determines UCF based on a users' dynamic activity. UCFC is different from the already existing calculators, as it uses the built-in sensors of the portable devices to determine user activity. UCFC also allows the users of a lab/room to be aware of their carbon footprint based on their ubiquitous activity, heating, and cooling system of the room. A computational model is used in a UCFC application that uses various sensor data estimating CO_2 consumption due to a person's activity within the lab/room or room's environment. Using the collected sensor data, UCFC calculates an approximate carbon footprint by summing up the results of the applicable models. An extensive survey has been conducted before the design of UCFC to perform requirement analysis. The survey was conducted on 30 people to understand their concern regarding increased carbon footprint, their requirements, and finally their practices and desire to reduce personal carbon footprint. The motivation behind this survey was to incorporate the inference and findings of the survey results in the design and implementation of UCFC prototype application. According to the model management and calculation methods, a management tool has also been developed for the domain experts to manage the stored quantitative models. Finally, to evaluate the feasibility of UCFC application, a usability study has been performed. Some of the important features of UCFC application are described next.

49.11 Features of UCFC Application

OCFP provides an important tool for individuals to assess their CO_2 emissions. It also allows application developers to develop such types of applications on top of the UCFC platform. As this field continues to expand, accurate and transparent values will be needed to educate users and motivate effective responses on individual and policy levels. However, UCFC provides some unique features as follows:

- *Ubiquitous:* It is a mobile and ubiquitous application, that is, it can be used at any time and any place if the user has a cell/ mobile/smart phone or PDA.

- *Adaptive:* It focuses to address a generalized solution to CO_2 measurement problem. However, it can be adapted to be used with other sensors too.
- *Sensor based:* Our proposed carbon footprint calculator is different from the already existing calculators. Since this system could use the built-in sensors in the portable devices as a basic source of input, this sensor data will help the user keep track of carbon footprint with minimum effort.
- *User friendly:* It requires minimal end-user interaction. Users of UCFC do not have to be technology experts.
- *Low cost:* It uses a low cost and reusable sensor, so that it can be cost effective. It has a simple Web interface that can be used by the user to monitor values of different sensors.
- *Ease of mobility:* It does not need Internet connectivity all the time. It needs Wi-Fi/Internet connectivity only when it wants to upload data to the server. This allows it to be used anytime and anywhere.
- *Customizable:* UCFC is designed in such a way that it can be modified or customized by the application developers.

49.12 Conclusion

Carbon footprint calculators are important tools for estimating CO_2 emissions and for providing information that can lead to behavioral and policy changes. However, these calculators produce estimates of carbon footprints that can vary by as much as several metric tons per annum per individual activity. Although most calculators are relatively new, their numbers are growing, and their methods are proliferating. Calculators can increase public awareness about CO_2 emissions and ways to reduce them. They can also affect the type and magnitude of emissions reduction efforts and offset purchases. However, a lack of a dynamic and UCF calculator has been observed lately. Therefore, our proposed OCFP can determine a user's carbon footprint based on their dynamic activity. Given the prevalence and potential influence, OCFP can provide greater public benefit by providing greater consistency and ubiquity.

References

Buettner08 M. Buettner, B. Greenstein, A. Sample, J. R. Smith, and D. Wetherall. (2008). Revisiting smart dust with RFID sensor networks. In: *Proc. ACM Workshop on Hot Topics in Networks*, Alberta, Canada.

CarbonD Carbon Diem, http://www.carbondiem.com/ (accessed 8th Aug 2011).

GooglePM Google PowerMeter, www.google.org/powermeter/howitworks.html (accessed 8th Aug 2011).

Liu09 L. Liu, H. Zhang, and S. I. Ahamed. (2009). Some thoughts on climate change and software engineering research. In: *First International Workshop on Software Research and Climate Change*, Orlando, FL, USA.

Padgett08 J. Padgett, S. Paul, C. Anne, J. H. Clarke, and M. P. Vandenbergh. (2008). A comparison of carbon calculators. *Environmental Impact Assessment*. 28: 106–115.

Sample07 A. P. Sample, D. J. Yeager, P. S. Powledge, and J. R. Smith. (2007). Design of a passively-powered, programmable platform for UHF RFID systems. In: *IEEE International Conference on RFID March 26–28, 2007*. Texas, USA, pp. 149–156.

50

Understanding and Exploiting User Behavior for Energy Saving

Vasily G.
Moshnyaga
Fukuoka University

50.1 Introduction

Energy consumption is one of the most critical aspects of today's society. According to the report published recently by the International Energy Agency (IEA), the world marketed energy consumption will grow by 49% in the next 25 years rising from 495 quadrillion British thermal units (Btu) in 2007 to 590 quadrillion Btu in 2020 and 739 quadrillion Btu in 2035 [1]. In spite of growing popularity of green and renewable energy sources, such as wind, solar, bio, hydro, nuclear, etc., burning of fossil fuels (petroleum, coal, and gas) will remain the world's biggest source of energy production, contributing to greenhouse and climate warming. Worldwide efforts to reduce carbon dioxide emission and increase energy security are in trouble if nothing is done to reduce the energy gobbled by both information and communication technologies and consumer electronics. According to the IEA, the energy used by computers and consumer electronics will not only double by 2022, but increase threefold by 2030. Even with improvements foreseen in energy efficiency, consumption by electronics in the residential sector is set to increase by 250% by 2030 and most likely will become the largest end-use category in many countries before 2020, unless effective steps are taken [1].

Reducing energy consumption of electronic systems, however, represents the biggest challenge that designers will have to face in the next decade. Energy-efficient design requires systematic optimization at all levels of design abstraction: from process technology and logic design to architectures and algorithms. Existing techniques include those at the algorithmic level (such as, strength reduction [2], algorithmic [3] and algebraic transformations [4], and retiming [5]), architectural level (such as pipelining [6] and parallel processing [7]), logic (logic minimization [8,9], precomputation [10]), circuit [11], and technological level [12]. We refer to these optimizations as static because they are applied during the design phase, assuming a worst-case scenario, and their implementation is time invariant.

An alternative approach is to adjust the system operation and energy consumption to the application workload dynamically, that is, during system operation. Methods include the following:

1145

1. *Application-driven system reconfiguration* [13]. The idea is to map a wide class of signal-processing algorithms to an appropriate architectural template and then reconfigure the system to deliver just right amount of computational resources for an application.
2. *Dynamic voltage-frequency scaling* [14]. The trade-off is that at lower voltages, circuits become slower, and the maximum operating frequency is reduced. Significant energy savings can be achieved by recognizing that peak performance is not always required, and therefore, the operating voltage and frequency of the processor can be scaled dynamically. The frequency can be lowered by performing operations in parallel to maintain the same overall performance.
3. *Energy-quality scaling* [15,16]. The idea is to obtain the minimum energy consumption at required computational quality. If the computational order of a system is varying based on signal statistics (known a priori), processing the most significant (from the point of precision) computations first achieves required computational quality at less energy. The implementation requires dynamically reconfigurable architectures that allow energy consumption per input sample to be varied with respect to quality. Applications include digital signal processing (noise reduction, filtering), image processing (discrete cosine transform [DCT], inverse DCT [IDCT] Fourier transform), video processing (vector quantization, video decoding), etc.
4. *Network-driven optimizations* [17]. The distributed network resources are exploited to reduce energy consumption in portable (wireless) devices by transferring large amounts of computation typically performed at the portable devices to the high-powered servers of the network.

Despite differences, these dynamic methods exploit the same idea; namely, to keep the system in the lowest power mode whenever there is no activity on inputs, and activate the system whenever the input signals change. To implement the idea, the system incorporates an extra unit that constantly monitors the input activity or workload (X) and based on it determines the new operational mode (S) for the system, as shown in Figure 50.1a. Depending on the application, the workload (X) can be measured by different metrics (e.g., the average rate, at which events arrive at the processor [16], and the idling time per sample interval [14–18]). If the workload overcomes a given threshold, the controller activates/deactivates the system modules, reconfigures the system, or changes frequency and voltage to deliver the required performance at the minimum energy consumption per sample.

Unfortunately, putting new functionality into a system to monitor the workload does not come for free; it drains energy and consumes extra area and cost. Also, switching the system between the modes introduces delays that affect performance. Therefore, solutions are usually linked to predicting the workload in advance based on various heuristics [15]. However, it is not always possible to make correct predictions due to peculiarities of application, operational environment, and/or the user demands, which are varying in time.

In general, there are two sources of energy losses in a device: *intrinsic* losses and *user-related* losses. The intrinsic energy losses are caused by the engineering design, technology, and materials used in construction of the device. For example, a plasma TV intrinsically dissipates more energy than a LCD TV; poor insulation of doors and walls results in intrinsic energy losses in fridges, etc. The user-related losses are associated with varying and inefficient device usage. Keeping a TV on when nobody watches

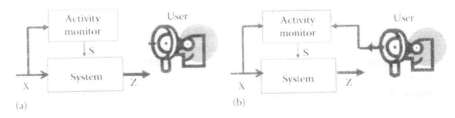

FIGURE 50.1 Energy management schemes: (a) existing and (b) proposed.

it or leaving a refrigerator door unnecessarily open, for example, cause energy losses associated with bad device usage.

Existing energy management policies are device centric; that is, they either ignore the user, assuming unchangeable operational environment for the device, or rely on very simplified policies, which eventually lead to large energy losses. Take a TV for example. A variety of methods has been proposed to reduce energy consumption of TVs. The majority of them, however, target the intrinsic energy losses without considering the viewer. As result, the television sets produce bright and high-quality pictures independently whether there is a viewer or not. According to [19], the energy consumption of consumer electronic devices can differ by a factor of two due to usage. Some experts estimate that 26%–36% of the total domestic energy losses are due to unreasonable usage of appliances [20]. Clearly, to reduce the losses, we must make the device energy management user centric, that is, adaptable to varying user behavior. No energy should ever be spent uselessly.

In this chapter, we present a novel user-centric approach to energy management, which monitors not only the system workload but also the user behavior and the environment.

50.2 User-Centric Energy Management

The main idea of our approach is to extend controller functionality to monitor the demands on system operation imposed by the user and adjust the system performance to the variation in these demands. Figure 50.1b illustrates the idea. The user monitoring is done by nonintrusive sensors (e.g., temperature sensors, motion sensors, video camera or image sensors produced by complementary metal-oxide semiconductor [CMOS] technology and known as CMOS image sensors, acoustic sensors, and radio-frequency identification [RFID] tag readers). Having the sensor readings, the controller evaluates the system operation (power consumption mode, output quality, active resources, etc.) and the user status (location, position, movement, eye gaze, etc.), estimates the user demands for the system at that time, and based on them generates signals that deliver the required functionality without losing energy. For example, when a user watches TV, the TV is kept bright. If there are several people in the room but nobody looks at the TV screen, the screen is dimmed to save energy. When nobody is present in the room, the TV is powered down or put into a sleep mode that keeps only the audio on.

In our research, we have investigated several applications of this novel energy management approach. Examples include personal computers (PCs), television sets, personal media gadgets, and home automation system. All these systems are user centric; they receive inputs from the user and deliver services to the user. Obviously, their energy management must be user centric. In the following sections, we discuss some applications of the approach in detail.

50.2.1 Personal Computer

The PC is a classic user-centric system that interactively receives commands from its user (through keyboard, mouse, microphone, etc.) and delivers him or her information through a display, printer, speaker, or other output device. According to the SMART2020 report [21], the number of PCs (desktops and laptops) globally is expected to increase from 592 million in 2002 to more than 4 billion in 2020. Due to this large growth, their share in the total Information and Communications Technology (ICT) CO_2 emission footprint will rise by 2.7 times in comparison to 2002 and reach 45% in 2020. Therefore, new technologies that would transform how PCs use energy are very important.

In a typical PC, the display accounts for 1/3 of the total PC energy consumption [22]. To reduce the energy, the OS-based advanced configuration and power interface (ACPI) [23] sets the display to low-power modes after specified periods of inactivity on the mouse and/or keyboard. The ACPI efficiency strongly depends on inactivity intervals set by the user. On one hand, if the inactivity intervals are improperly short, for example, 1 or 2 min, the ACPI can be quite troublesome by shutting the display off when it must be on. On the other hand, if the inactivity intervals are set to be long, the ACPI efficiency decreases. Because modifying the intervals requires system setting, a half of the world's PC users never adjust the power management of their PCs for fear that it will impede performance [24]. Those who do the adjustment, usually assign long intervals. HP inspected 183,000 monitors worldwide and found that almost a third were not set to take advantage of the energy-saving features. Just enabling these features after 20 min of inactivity can save up to 381 kWh for a monitor per year [25]. Evidently, to prevent such a problem, the PC energy management must employ more efficient user monitoring.

In contrast to the ACPI, which "senses" the user through keyboard and/or mouse, we "watch" the user through camera or CMOS image sensor [26]. Our user-centric energy management is based on the following assumptions:

1. PC is equipped with a color video camera (CMOS image sensor). The camera is located at the top of display. When the user looks at display, it faces the camera frontally.
2. The target object is a single-PC user. The user's motion is slow relatively to the frame rate. The background is stable and constant.
3. The display has a number of backlight intensity levels with the highest level corresponding to the largest power consumption and the lowest level to the smallest power, respectively. The highest level of backlight intensity is enabled either initially or whenever the user looks at the screen.

Figure 50.2 shows the proposed display energy-management scheme. The user's eye-gaze detector receives an RGB color image, analyzes its content, and outputs two logic signals, u_1 and u_0. If a face pattern is detected in the image, the detector sets u_0 to 1; otherwise, $u_0 = 0$. The zero value of u_0 enforces the voltage converter to shrink the backlight supply voltage to 0 V; dimming the display off. If the gaze detector determines that the user looks at the screen, it sets $u_1 = 1$. When both u_0 and u_1 are 1, the display operates as usual in the high-power mode. If the user's gaze has been off the screen for more than N consecutive frames, u_1 becomes 0. When $u_0 = 1$ and $u_1 = 0$, the input voltage (V_b) of the high-voltage inverter is decreased by ΔV. This voltage drop lowers backlight luminance and so shrinks the power consumption of the display. Any on-screen gaze in this low-power mode reactivates the initial backlight luminance and moves the display onto the high-power mode. However, if no on-screen gaze has been detected for N consecutive frames and the backlight luminance has already reached the lowest level, the display is put to sleep. Returning back from the sleep mode requires pushing the on button.

To detect eye gaze, we use a low-complexity eye-tracking algorithm [27] that scans the six-segment rectangular filter (SSR) [28] over the integral representation of the input image to locate the between-the-eyes (BTE) pattern of human face (see Figure 50.3). At each location, the SSR filter compares the integral sums of the segments as follows:

$$\text{Sum}(1) < \text{Sum}(2) \text{ and } \text{Sum}(1) < \text{Sum}(4), \tag{50.1}$$

$$\text{Sum}(3) < \text{Sum}(2) \text{ and } \text{Sum}(3) < \text{Sum}(6). \tag{50.2}$$

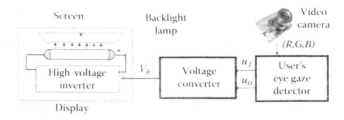

FIGURE 50.2   The display monitoring scheme.

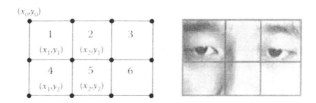

FIGURE 50.3 Illustration of the six-segment rectangular filter.

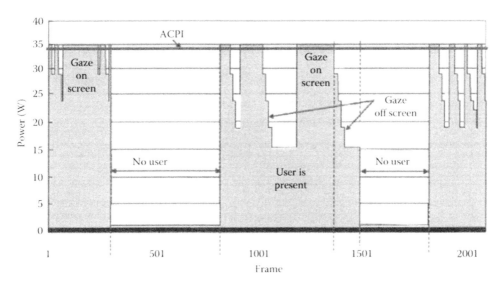

FIGURE 50.4 Power consumption profile of the tested display.

If the aforementioned criteria are not satisfied, the user is assumed to be absent in front of the camera. Otherwise, the SSR is considered to be a candidate for the BTE pattern (i.e., face candidate) and two local minimum (i.e., dark) points each are extracted from the regions 1 and 3 of the SSR for left and right eyes, respectively. If both eyes are located, the user is assumed to be looking at the screen. The details can be found in [29].

The detector was implemented in hardware [30] based on a single XILINX FPGA board and Video Graphics Array (VGA) CMOS camera connected to the board through parallel I/O interface. It operates at 48 MHz frequency and 3.3 V voltage and provides user-gaze detection at 20 fps rate, 84% detection accuracy while consuming in only 150 mW of power.

Figures 50.4 and 50.5 illustrate the power savings achieved by the proposed technology for 17″ thin film transistor liquid crystal display (TFT LCD) (35 W on peak, 0.63 W in sleep mode, four different levels of screen brightness) in comparison to the existing ACPI power management (20 min inactivity setting) [29]. Even though our technology takes a little more power than the ACPI for the user monitoring, it saves 36% of the total energy consumed by display on this short (100 s) test. For a longer test, the energy savings provided by our approach are much larger.

50.2.2 Television Set

Despite the wide acceptance of "green energy" regulations, TVs have become more energy consuming. Recently emerged plasma televisions, which are 50% bigger than their cathode-ray tube equivalents, consume about four times more energy. A 50 in. flat-screen plasma HD TV now burns over 500 W

FIGURE 50.5   Screenshots of PC display and the corresponding power consumption: when the user looks at screen, the screen is bright (power: 35 W); else the screen is dimmed (power: 15.6 W).

of power, thus consuming the same energy as a dishwasher or in-room air conditioner. Dimming the screen brightness is one of the most effective energy-saving techniques proposed for TV. Sensing light is already a feature of many TVs that enables dimming based on ambient light level. Also, users can set the brightness by selecting one of three operation modes: the "standard mode" delivers the highest level of brightness, the "saving mode" refers to the dimmed screen, and "no brightness mode" reflects the dark screen. The brightness level in the saving mode can also be changed. Unless the user changes the mode, the TV maintains the same brightness.

Like PC users, the majority of TV viewers do not change brightness for energy savings, fearing that it affects picture quality. Besides, users usually watch TV while doing other activities: reading books, working on the PC, preparing food, chatting with friends, etc. As a result, the TV wastes much energy to produce high-quality pictures when nobody is watching them.

To save energy, leading TV producers have embedded "user sensors" into TVs to adjust performance according to the user behavior. For example, the Panasonic VIERA® plasma TV senses the user through the remote controller. If the time from the last use of the remote controller exceeds a predefined duration (e.g., 1, 2, or 3 h), the TV automatically powers off. Additionally to this "remote button sensor," the latest Sony Bravia® HD TV incorporates an infrared motion sensor that switches the TV off when no motion has been detected in front of it over a period of time (e.g., 5 min, 30 min, or 1 h) preset by the user. Also Hitachi and Toshiba use hand gesture sensors to control TVs.

As we mentioned in Section 50.2.1, "sensing" the viewer explicitly (through fingers, hands, or body motion) has several problems: it is either incorrect (a moving dog or a tree in the window can keep the TV on) or troublesome, that is, requires the user either to push the remote control frequently to prevent

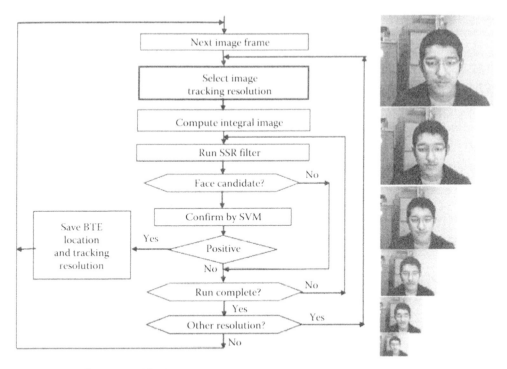

FIGURE 50.6 An illustration of the face tracking algorithm and image resolution used in the search.

shutting the TV down when someone is watching it or to enlarge the allowed duration of inactivity interval. Our approach is free of such problems because it "watches" the viewer through a camera and detects precisely whether he or she looks at the screen or not. In contrast to a PC, the TV viewer monitoring unit searches for any face in the image that looks at the camera frontally. The search is done by scanning the SSR filter (Figure 50.3) over the integral representation of green image plane produced by the camera. Because the size of a face decreases with the distance from camera, and large faces require large SSR filters, we repeat the scan for a scaled-down image (at most six times) ending the search whenever a viewer is detected. Figure 50.6 illustrates the flowchart of the searching process. The right column shows the scale-down images used in the search (the scales are 2/3, 1/2, 1/3, 1/4, and 1/6 from top to bottom).

The user-centric TV energy management works as following. Turning the TV on automatically sets the standard (brightest) mode and starts the camera-based viewer monitoring. If the monitoring cannot find faces in the current image or finds that none of the faces look at the screen, the TV is switched to the "energy-saving mode," by dimming the screen. If it has already been in the "energy-saving mode" longer than a predefined time, T_1, and still no viewers found, the TV is turned to the "no-screen" mode, that is, a dark screen (see Figure 50.7). Any face gazing at the screen reactivates the power-up mode and the process repeats. However, if the no-screen mode lasts longer than a preset time (T_2), the TV, camera, and monitoring unit are powered down.

Figure 50.8 shows the relative energy-saving figures achieved by the proposed "user-centric" energy management in comparison to the standard mode and the motion-based screen-off mode (set to 5 and 30 min of inactivity) estimated on 40″ KDL-40V5 Sony Bravia TV. The test lasted 40 min and covered three different parts. In the first part (10 min), two viewers were moving sporadically in front of the camera (TV) but not looking at the TV screen. In the second part (20 min), one of the viewers left the room and the second watched the TV while working with a PC. In the last part (10 min), the user read a book while seated motionless in front of the TV. In this test, the existing motion-based energy management activated the "no-screen" mode only in the last 5 min. In contrast, our "user-centric" management lowers the brightness

FIGURE 50.7   TV screenshots when the viewer (a) looks at screen and (b) when the viewer does not.

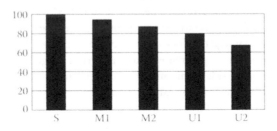

FIGURE 50.8   Results of TV energy saving.

each time the TV is not watched. The bar S in Figure 50.8 shows the relative energy dissipated in the standard mode; the bars M1 and M2 depict the relative energy consumed by the motion-based energy management in standard mode and the energy-saving mode, respectively; the bars U1 and U2 reflect the relative energy consumed by the proposed user-centric energy management in the standard mode and the saving mode, respectively. As one can see, our approach reduces the TV energy consumption by 21% and 32% respectively, while the motion-based scheme reduced only by 6% and 12%, respectively. Obviously, for other scenarios the results will be different.

50.2.3 Home Energy Management

Home is another application for which the user-centric energy management is profitable. In modern homes, electric and electronic devices account for 1/3 of the total household energy consumption [31]. Separate studies in U.S., Dutch, and British homes have reported that 26%–36% of domestic energy is "behavioral"—determined by the way we use machines, not the efficiency of the hardware itself [32]. Recently developed home systems [33–36] allow users to manage home appliances remotely and switch them on or off from an in-wall control panel or a PC. However, we humans have two amazing features that undermine such control. First, we are quite lazy to routinely perform manual control of dozens of appliances every day. Second, we frequently forget about the controls, leaving appliances on at night of even after going away from home. What we need is a system that not only monitors home appliances and reports their status to the user but automatically adjusts the work of the appliances to the user's needs. Obviously, such a system must be also user centric, that is, capable of monitoring not only electric appliances but also their users.

Unlike existing systems, our user-centric home energy management is a network of sensors, which monitors all users in a home, determines their location, behavior, and control instructions if any, and based on them automatically sets all home appliances to an appropriate mode that fully satisfies the current user's needs while maximally reducing the energy consumption. The heart of the system is the home control center (HCC), which is built on top of a standard gateway architecture, shown in Figure 50.9. The HCC is based on an open Linux platform enabling the home owner to build a technology-neutral smart home that can be controlled with a mobile phone, using a unified user interface. The HCC receives information from multiple sensors that monitor users and appliances and, once processed, generates the control signals for the actuators that control device operations. The user sensors include motion sensors, CMOS cameras or vision sensors, RFID, etc. The device sensors combine functions of conventional sensors and actuators with transmitters and receivers [37]. Each device sensor is attached to a corresponding appliance and the power line (Figure 50.10). When activated, it measures the value of electric current in the corresponding device, transmits the result to the base station, receives the control commands from the HCC and based on them connects or disconnects the device to/from the power line. Based on the data received, the location of users, their behavior, and instructions, the system can switch off those appliances that are active but not in use or activate those he or she needs. Unlike existing home

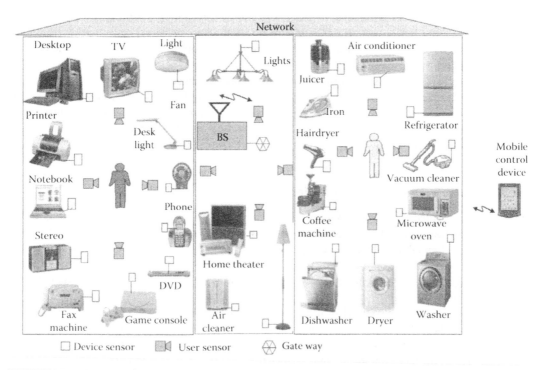

FIGURE 50.9 System architecture.

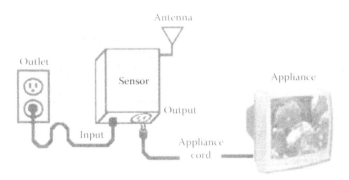

FIGURE 50.10   The device sensor.

automation approaches, we teach our system to understand the contextual cues of home occupants. The concept is similar to the aware home [38] with the difference that it aims at home energy saving. Our goal is to develop a user-centric pervasive system that coordinates operational modes of home appliances with the user's location, status, and performance to form a comfortable and pleasant home environment with low-energy consumption.

Based on sensor reading and the energy usage patterns of home occupants, the system sets the usage of all room devices, lighting, heating, ventilation and air conditioning (HVAC), TV, music, and also motorized shades to control sunlight entering room areas. It turns off lights in unoccupied areas, dims lights in corridors and hallways, and separates circuits, creating energy-saving solutions. Only when someone walks in the room, the system turns on lights or air conditioner. During mornings and afternoons, it automatically opens shutters and drapes, turning lights off when not needed, but at noon, it closes the drapes to decrease solar effects and save energy at the same time. The system turns off air conditioning in vacant areas, monitors doors or room windows to avoid air leaks, monitors filter cleanliness status, and other factors that help decrease energy consumption. It prevents overrunning of devices, pumps, water heaters, and HVAC and directly alerts when high-level consumption thresholds are reached. Obviously any user instruction (say switching the room lights on) interrupts the HCC to adjust its operation.

In our architecture, the user sensors are wired, while device sensors and HCC control are wireless. The user interface with the HCC is envisioned through touch screens and wall-mounted control panels. The panels can have from 2 to 16 buttons and are provided for each room. The buttons on each control panel are programmed for specific commands in each room creating moods and activating preprogrammed operation or device control. You can change the mood at the touch of a button.

50.2.4 Challenges and Open Problems

The research presented here is a work in progress and the list of things to improve it is long. In the current work on user-centric PC energy management, we restricted ourselves to a simple case of a single user. However, when talking about the user-gaze monitoring in general, some critical issues arise. For instance, how to handle several users? The main PC user might not look at the screen, while the others do. Concerning this point, we believe that a feasible solution is to keep the display active while there is someone looking at the screen.

User-centric TV management also has several challenging issues. First, the viewers can be positioned quite far from the TV set. Monitoring the viewers located far away from the camera is far from trivial. Second, the viewers can watch TV lying on a bed, so the viewer's face can rotate on a large angle. Third, the face illumination may change from a very bright to a complete darkness. In these conditions, the correct real-time face monitoring with low-energy overhead becomes really difficult.

User-centric home energy management is a typical task of context-aware computing, which requires solutions related to situation recognition, pervasive computing, advanced human–computer interface (HCI), ambient control, etc. At this time, we assume touch panels and keyboards as HCI input devices. In the future, there will be many possible types of input that a system might be capable of. In this case, it will be helpful to indicate to the user what is available for the selection of energy-saving options, particularly for those such as touch, voice, or gesture that may not be readily visible in the way that a keyboard is. Some possible input devices are virtual touchscreen, presence, voice, gestures, virtual displays, etc. Providing the user with "virtual" devices that project the selection menu or an image of a keyboard onto any flat surface nearby the user is another challenging problem. Odd as it sounds, using one of these soon becomes quite natural. Another possible alternative input for a home system is voice recognition. High-quality voice recognition requires substantial computing power, so handheld devices are still a few years away from being able to take dictation well enough to substitute for a keyboard, but sooner or later they will.

Finally, due to the shortage of energy resources, the cost of energy will eventually increase in the near future. We hope that making our environment more intelligent and user centric will help us consume less energy; make our lives less expensive and more enjoyable.

Acknowledgment

The work was sponsored by The Ministry of Education, Culture, Sports, Science and Technology of Japan under Regional Innovation Cluster Program (Global Type, 2nd Stage) and Grant-in-Aid for Scientific Research (C) No.21500063.

References

1. International Energy Agency, *Gadgets and Gigawatts—Policies for Energy Efficient Electronics*, IEA Press, Paris, France, p. 424, 2009.
2. A.P. Chandrakasan, M. Potkonjak, R. Mehra, J. Rabaey, and R.W. Brodersen, Minimizing power user transformations, *IEEE Transactions on Computer-Aided Design of Integrated Circuits and Systems*, 14, 12–31, January 1995.
3. K.K. Parhi, Algorithm transformation techniques for concurrent processors, *Proceedings of the IEEE*, 77, 1879–1895, December 1989.
4. M. Potkonjak and J. Rabaey, Fast implementation of recursive programs using transformations, *Proceedings of the 1992 IEEE International Conference on Acoustics, Speech and Signal Processing (ICASSP)*, San Francisco, CA, pp. 5629–5672, 1992.
5. C. Leiserson and J. Saxe, Optimizing synchronous systems, *Journal of VLSI Computing Systems*, 1, 41–67, 1983.
6. H. Loomis and B. Sinha, High speed recursive digital filter realization, *Circuits, Systems, and Signal Processing*, 3(3), 267–294, 1984.
7. K. Parhi and D.G. Messershmitt, Pipeline interleaving and parallelism in recursive digital filters, Part I & II, *IEEE Transactions on ASSP*, 37, 1099–1134, July 1989.
8. S. Malik and S. Devadas, A survey of optimization techniques targeting low power VLSI circuits, *Proceedings of the 32nd ACM/IEEE Design Automation Conference (DAC'95)*, San Francisco, CA, 1995.
9. S. Iman and M. Pedram, *Logic Synthesis for Low Power VLSI Designs*, Kluwer, Boston, MA, 1997.
10. M. Alidina, J. Monteirio, S. Devadas et al., Precomputation-based sequence logic optimization for low-power, *IEEE Transactions on Very Large Scale Integration (VLSI) Systems*, 2, 398–407, December 1994.

11. M.-C. Chang, C.-S. Chang, C.-P. Chao et al., Transistor- and circuit-design optimization for low-power CMOS, *IEEE Transactions on Electron Devices*, 55(1), 84–95, January 2008.

12. K. Chen and C. Hu, Device and technology optimizations for low power design in deep sub-micron regime, *Proceedings ACM/IEEE 1997 International Symposium on Low-Power Electronic Design (ISLPED)*, Monterey, CA, pp. 312–316, 1997.

13. J. Rabaey, Reconfigurable processing: The solution to low-power programmable DSP, *Proceedings of the 1997 IEEE International Conference on Acoustics, Speech and Signal Processing (ICASSP)*, Vol. 1, Munich, Germany, pp. 275–278, 1997.

14. T.D. Burd, T.A. Pering, A.J. Stratakos, and R.W. Brodersen, Dynamic voltage scaled microprocessor system, *IEEE Journal of Solid-State Circuits*, 35(11), 1571–1580, 2000.

15. M. Goel and N.R. Shanbhag, Dynamic algorithm transformations (DAT)—A systematic approach to low-power reconfigurable signal processing, *IEEE Transactions on Very Large Scale Integration (VLSI) Systems*, 7(4), 463–476, December 1999.

16. A. Sinha, A. Wang, and A. Chandrakasan, Energy scalable system design, *IEEE Transactions on Very Large Scale Integration (VLSI) Systems*, 10(2), 135–145, April 2002.

17. A. Chandrakasan, R. Amirtharajah, J. Goodman, and W. Rabiner, Trends in low power digital signal processing, *1998 IEEE International Symposium on Circuits and Systems (ISCAS)*, Vol. 4, Monterey, CA, pp. 604–607, 1998.

18. S. Lee and T. Sakurai, Run-power control scheme using software feedback loop for low-power real-time applications, *Proceedings of the Asia-South Pacific Design Automation Conference (ASPDAC)*, Yokohama, Japan, pp. 381–386, 2000.

19. K. Gram-Hansen, Domestic electricity consumption—Consumers and appliances, *Nordic Conference on Environmental Social Sciences (NESS)*, Turku/Abo, Finland, 2003.

20. E.W.A. Elias, E.A. Dekoninck, and S.J. Culley, The potential for domestic energy savings through assessing user behaviour and changes in design, *Fifth International Symposium on Environmentally Conscious Design and Inverse Manufacturing*, Tokyo, Japan, 2007.

21. SMART2020: Enabling the low carbon economy in the information age, The Climate Group, 2008, Available: http://www.smart2020.org/publications/ (accessed August 10, 2011).

22. A. Mahesri and V. Vardhan, Power consumption breakdown on a modern laptop, *Proceedings of the Power Aware Computer Systems, LNCS*, 3471, 165–180, 2005.

23. Advanced configuration and power interface specification, Rev. 3.0, September 2004, Available: http://www.acpi.info/spec.htm

24. Energy savings with personal computers, Fujitsu-Siemens, Available: http://www.fujitsu-siemens.nl/aboutus/sor/energy_saving/prof_desk_prod.html (accessed November, 2007).

25. 2006 Global citizenship report, Hewlett-Packard Co., Available: www.hp.com/hpinfo/globalcitizenship/gcreport/pdf/hp2006gcreport.pdf (accessed August 10, 2011).

26. V.G. Moshnyaga and E. Morikawa, LCD display energy reduction by user monitoring, *Proceeding of 2005 IEEE International Conference on Computer Design (ICCD)*, San Jose, CA, pp. 94–97, 2005.

27. S. Yamamoto and V.G. Moshnyaga, Algorithm optimizations for low-complexity eye-tracking, *Proceeding of the IEEE Systems, Man and Cybernetics*, San Antonio, TX, pp. 18–22, 2009.

28. S. Kawato, N. Tetsutani, and K. Osaka, Scale-adaptive face detection and tracking in real time with SSR filters and support vector machine, *IEICE Transactions Information & Systems*, E88-D(12), 2857–2863, December 2005.

29. V.G. Moshnyaga, The use of eye tracking for PC energy management, *Proceedings of the ACM Symposium on Eye-Tracking Research and Applications (ETRA 2010)*, Austin, TX, pp. 113–116, March 22–24, 2010.

30. V.G. Moshnyaga, K. Hashimoto, T. Suetsugu, and S. Higashi, A hardware system for tracking eyes of computer user, *Proceedings of the 2009 International Conference on Computer Design (CDES 2009)*, Las Vegas, NV, pp. 125–130, July 13–16, 2009.

31. V.G. Moshnyaga, K. Hashimoto, T. Suetsugu, and S. Higashi, A hardware implementation of the user-centric display energy management. In *Integrated Circuit and System Design: Power, Timing Modeling and Simulation*, J. Monteiro and R.V. Leuken (Eds.), Selected Papers, 19th International Workshop, PATMOS 2009, Delft, the Netherlands, September 2009, Springer LNCS 5953, pp. 56–65, 2010.

32. P. Owen, The rise of the machines: A review of energy using products in the home from 1970 to today, Energy Saving Trust, U.K., Available: http://www.energysavingtrust.org.uk/uploads/documents/aboutest/Riseofthemachines.pdf (accessed August 10, 2011)

33. T. Simonite, Innovation: Can technology persuade us to save energy? *New Scientist*, Available: www.newscientist.com, November 2009.

34. Energy Conservation and Home NETwork, ECHONET Consortium, Available: http://www.echonet.gr.jp/english/1_echo/index.htm (accessed August 10, 2011)

35. Insteon. The details, Available: http://www.insteon.net/pdf/insteondetails.pdf (accessed August 10, 2011)

36. C.D. Kidd, R. Orr, G.D. Abowd, C.G. Atkeson, I. Essa, B. MacIntyre, E. Mynatt, T.E. Starner, and W. Newsletter, The aware home: A living laboratory for ubiquitous computing research, in *Cooperative Buildings. Integrating Information, Organizations and Architecture*, N.A. Streitz et al. (Eds), LNCS 1670, Springer, pp. 191–198, 1999.

37. V.G. Moshnyaga, A wireless system for energy management of household electric appliances, *Proceedings of the Fourth International Conference on Computing, Communications and Control Technologies*, Orlando, FL, July 20–23, 2006.

38. S.S. Intille, Designing a home of the future, *IEEE Pervasive Computing*, April–June, 80–86, 2002.

51

Predicting User Behavior for Energy Saving

Mingsong Bi
The University of Arizona

Igor Crk
*IBM Systems and
Technology Group*

Chris Gniady
The University of Arizona

51.1 Introduction

Computer system designers face two challenges: energy and performance. The spectrum of system design considerations ranges from ultrahigh performance on one side, in which energy consumption is not a limiting factor, all the way to ultralow power design, where performance expectations are very low. The middle of the range is occupied by systems where both power and performance play a role in system design. Portable system designers are faced with a user demand for performance, functionality, and better user interfaces along with a longer battery life. Designers of servers and desktop systems focus almost entirely on performance, since energy usually is not a constrained resource. However, this trend has started to change since researchers have realized the positive financial and environmental implications of energy conservation for stand-alone servers and server clusters [1–4]. The challenge of designing energy-efficient systems lies in understanding the role of user interaction in energy consumption and in providing an energy-performance schedule that adequately accommodates user demand. Furthermore, system performance can be tailored to a user's pattern of interaction and its energy-performance schedule optimized.

User interaction can be as simple as launching a batch job or much more sophisticated, as in the case of interactive applications, where a user may continuously interact with an application. The goal is the same in both cases: meeting user demand. Users usually demand high performance; however, that does not preclude energy optimizations. Performance and energy consumption are tightly coupled, where higher performance is usually achieved at the cost of an increased power demand. However, the key observation is that higher power demand does not necessarily translate into an increase in energy consumption. For instance, hardware in a higher performance state may complete a particular task faster than the same hardware operating at a lower performance state. This reduces the time during which the entire system has to be on. On the other hand, a particular device may not be required by all tasks and so may be operated in a low performance state without a significant impact on the performance of the executing application. Similarly, the low performance state of a particular device does not impact performance if the demand placed on the device is sufficiently low.

Existing work that addresses performance and energy optimizations ranges from hardware optimizations to application transformations. Figure 51.1 shows a typical organization of computer systems and the potential for optimization at each level. Optimizations at lower levels are usually complementary to higher-level optimizations. The lower levels also provide the interfaces for managing both energy and performance at a higher level. The key observation is that the resulting behavior of all system layers is a response to the combination of user actions and application behavior in interactive environments.

Understanding user interaction can provide valuable information about which resources will be needed ahead of time. This leads to performance optimizations such as better resource allocations for applications that can utilize a given resource more productively, and transitioning devices to the more appropriate energy-performance state before the demand arrives. The last optimization refers to the situation where the device is turned off to reduce energy consumption. The challenge is to provide a performance/energy schedule that best matches the task at hand, since keeping the device in one performance level is not energy efficient as the demand placed on the device varies.

Figure 51.2 shows an example of utilization of user interactions in predicting device power states. In response to a user's interaction, the application initiates some I/O activity. Figure 51.2a shows a behavior of a reactive mechanism that evaluates initial I/O requests to verify the need for providing higher device performance, and as a result higher bandwidth. In other words, the I/O device is switched to the high-performance mode when the need for high I/O bandwidth is detected. We observe that the

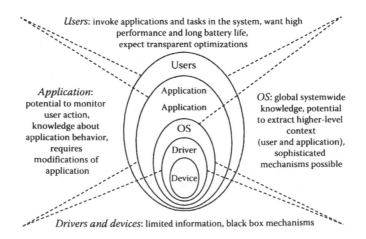

FIGURE 51.1 Computer system organization and optimization levels.

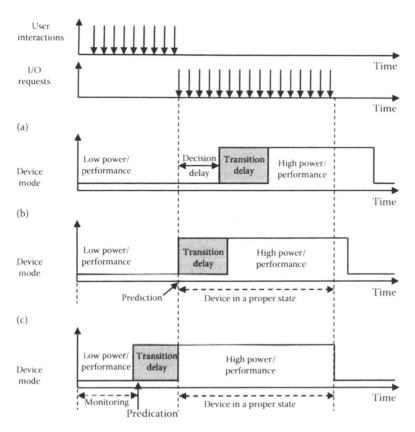

FIGURE 51.2 Prediction example. (a) Reactive approach. (b) I/O call–based prediction. (c) Interaction-based prediction.

decision evaluation period and the following mode transition period both impact the performance of high-bandwidth transfers and may lengthen the time spent serving the I/O requests [5].

Waiting for the request to arrive before generating a prediction and powering up the device degrades performance, resulting in longer time to service I/O requests and potentially increasing the system's overall energy consumption. Alternatively, we can utilize prediction mechanisms that predict I/O bandwidth demand based on the current I/O system call being invoked as shown in Figure 51.2b. We observe that decision time is eliminated, since the prediction mechanism uses the knowledge of past events to immediately generate a prediction based upon a given I/O system call. In this case, we can get an accurate prediction and eliminate the decision time, but we are still left with the transition delay. Predicting a transition to the high-power state to hide the transition delays proved to be challenging since we needed hints ahead of time to transition a device to a higher energy state before I/O requests arrived. The key to solving this problem is observing user interactions and inferring from user behavior when the increased performance demand will arrive. Therefore, interaction-aware prediction mechanisms that monitor user activity are able to predict the transition to a higher power level in time to meet the increased performance demand, as shown in Figure 51.2c.

51.2 Capturing User Context

Since their origin in the early 1980s, graphical user interfaces (GUIs) have become the de facto standard for interfacing with applications. Direct manipulation of graphical elements provides an intuitive means for a user to interact with the computing environment. The most common systems that provide a GUI environment are Microsoft Windows, Mac OS X, and the X Window System. Each windowing system provides programmers with common interface components that are the functional entities associated with specific features of an application, such as a button or menu option to save a file. As such, most application functionality can be accessed through mouse-driven interaction with the displayed visual elements, and as pointed out by Dix et al., virtually all actions take only a single click of the mouse button [6].

GUI organization emphasizes simplicity, where each interface component exposes a limited set of functions to the user. By interacting with the various interface components with the mouse or keyboard, the user triggers a series of actions. However, low-level application monitoring alone is insufficient to discern the exact intent of the user's interaction. For example, the monitoring of a system call occurring when a file is read, lacks the context necessary to distinguish user-driven application activity from automated application behavior. Similarly, global coordinates and type of mouse event, button up/down, is easily captured, but alone gives no indication about the interactive elements that are causing application or system activity. So, the additional context provided by simple monitoring of mouse events such as a button press or movement can provide the additional information necessary to make the distinction sought by the first example, but this is the limit of what this additional context provides.

In order to make the context provided by mouse input useful, the collected information must consist of more than simple mouse button events or movements. System-level activity that is caused by interaction with a particular element can be more accurately described and even predicted by the unique correlation of mouse activity with specific GUI elements. The transparent user context monitoring (TUCM) mechanism described in this chapter allows precise extraction of interactions context using existing technologies without the need for application modifications to monitor execution context.

51.2.1 User Activity Monitoring Systems

Accurate and detailed monitoring of user activity is the basis for continuing to improve performance and energy efficiency of computing systems. Most interactive applications are driven by simple point-and-click interactions. All operating systems targeted for consumers offer a GUI to facilitate uniform interfaces for interactions between users and application. As a result, virtually all interactions can be accomplished through mouse clicks [6]. Users interact with an application to accomplish specific tasks, like opening or saving a file. Many tasks can be accomplished by a single point and click, but other tasks require a sequence of interactions. For example, to save a new file, the user commonly clicks on the File menu then selects the Save option and is presented by a directory selection menu, once the filename is entered and the user clicks OK, the file is saved and disk I/O may be requested. It is argued that all GUI interactions resulting in disk I/O activity can be accurately captured and correlated to the activity they initiate; however, it should be noted that this does not always account for all occurring I/O activity.

One key challenge in inferring high-level notions of context is moving from low-level signals (e.g., mouse clicks or keyboard events) to meaningful high-level contextual information (e.g., chatting on the Internet or streaming video via Webcam). One of the earliest efforts in this area was the Lumiere Project [7]. Forming a foundation for Microsoft's Office Assistant, the Lumiere project used Bayesian models to reason about the time-varying goals of computer users from observed interactions and queries. More specifically, Lumiere intercepts events and maps them into observation variables in a Bayesian network. This network is then used to provide assistance to end users. Other promising methods for mapping low-level events to high-level contexts include hidden Markov models (HMMs) [8], or Bayesian classifiers [9]. Although, such methods often require specifying the type of activities to be inferred before training.

Another application of higher-level context derived from low-level events is often seen in resource and energy management, where user interactions and application behavior can provide contextual hints necessary for inferring the desirable system state. These hints can be derived from the application's user interface events [10] or from the applications themselves [11]. However, all of these existing mechanisms do not provide detailed and accurate context. As a result, the focus of this chapter is on transparent, nonintrusive approaches to gathering the necessary information that can be applied to both resource management and increasing the understanding of user interaction with GUI-based applications.

51.2.2 Need for Automatic User Interaction Capture

User activity can be captured at various levels of detail. The simplest method is to capture the coordinates of mouse clicks relative to the entire display or application window. Alone, the coordinates provide the lowest level of detail, since it is unclear what visual elements the user is interacting with. With this approach, several clusters of activity correspond to the application buttons or windows. As an example, consider a typical instant messaging application, where users interact by requesting audio/video streaming, typing and sending messages, or transferring files. One cluster may correspond to file transfer and Webcam interactions, while another may correspond to sending messages through clicking in a text box. The key challenge of inferring these hints is moving from low-level signals (e.g., mouse clicks or keyboard events) to meaningful high level contexts (e.g., viewing a streaming video). Differentiating functional clusters of mouse clicks using a K-means clustering technique is a possible approach, and has the desirable effect of identifying classes of mouse clicks that correspond to mouse-driven functions of the software. Once clustering converges, new mouse clicks can be classified with respect to the identified clusters.

Clustering is necessary when the only available information about user interactions are the relative coordinates of mouse events. K-means clustering is adequate to show the potential of dynamically capturing user behavior. However, it suffers from several shortcomings that are addressed in this dissertation:

- K-means require significant offline processing to compute the clustering. The number of clusters varies from one application to the next, requiring additional processing before mouse clicks can be correlated with network I/O patterns.
- The clustering technique used to detect the layout of the user interface is an approximation. Interactions with nearby interface buttons may introduce unrelated data points to existing clusters resulting in misclassifications. The technique is also sensitive to changing window locations and sizes and, as such, may introduce inaccuracies as the clustering has to be recomputed.
- The implementation considered a single application with a relatively simple interface and a single interactive window. Other applications may have a more complicated interface that not only requires multiple submenu selection but also supports interaction between multiple windows.

51.3 Automatic Interactive Context Capturing

To address the shortcomings of the coarse-grained interaction capture described earlier, we explore a complete set of low overhead mechanisms that transparently integrate with GUI and capture user behavior exactly without any uncertainty associated with clustering. As a result, the proposed system achieves a more accurate correlation and prediction. Improved capture accuracy translates to better prediction accuracies for interactive applications.

The key to the design of the TUCM system is that the graphics management subsystem in modern operating systems can be leveraged to transparently monitor, record, and utilize user interactions to improve system performance or energy efficiency. Subsequently, TUCM design has several requirements:

- User interactions have to be captured transparently without any modification to the application.
- Monitoring and capture should be efficient to prevent excessive energy consumed by the CPU, which would reduce system energy efficiency.

- The system should handle multiple applications in a graphically rich environment.
- User behavior correlation and classifications should be performed online and without direct user involvement.

All of those items have to be successfully addressed to provide efficient user context monitoring and management that can be applied for energy or performance management in the emerging computer systems.

51.3.1 Context Capture in Linux

Recent interface design trends emphasize simplicity, where designers break complex tasks into simpler ones that are represented by autonomous interface components (e.g., icons, menus, and buttons). Each interface component exposes a limited set of functions to the end user. The user invokes these functions by interacting with different interface objects using input devices (e.g., mouse, keyboard) which in turn trigger a series of actions. GUI toolkits and widget libraries provide programmers with abstracted building blocks that hide the specifics of physical interaction devices, manage the details of interaction for the application by abstracting it to callbacks, and provide encapsulation for the GUI appearance and behavior [12].

On Unix-like systems, the X Window System provides a common display protocol built on the client-server model [13]. The X Server is responsible for accepting graphical output requests from clients and reporting user input to clients. The stream of data from the client to the server contains low-level information regarding the window layout, such as window size and parent and children windows, while the data sent from the server to the client applications contains information about user interactions, such as mouse button events, the windows they occurred in, and relative coordinates. Section 51.3.1 addresses the implementation of context capture within the X Window System in a Linux environment. Similar context capture mechanisms can also be implemented for other operating systems, such as Microsoft Windows.

The basic architecture of an X application is shown in Figure 51.3. As shown, the application defines the widgets that are to be used to make up the user interface. A widget is an interactive interface element, such as a window, text box, button, etc. The application is also responsible for receiving and processing user input via callbacks. The user can only interact with the application by use of the graphical output drawn by the widgets and input device events, such as mouse button presses processed by the widgets.

By adding an intermediary layer, as shown in Figure 51.4, between the server and attached clients, the exact sequence of requests and events is observed. This layer provides a means of transparently monitoring user behavior. No modification of applications is necessary. Furthermore, user interactions are captured exactly, eliminating both the excessive computational overhead of computing a clustering and the inaccuracies associated with the clustering present in the previously described solution that relied

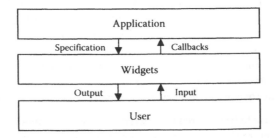

FIGURE 51.3 Environment overview in the X Window System.

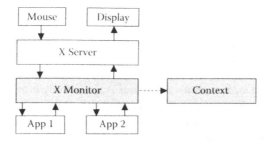

FIGURE 51.4 Augmenting the X Window system hierarchy with a shim layer enables context capture.

on K-means clustering. Since the need for cluster formation and behavior detection is eliminated, the offline processing needed in the clustering approach is eliminated, fully allowing for detection, correlation, and prediction to be performed online.

51.4 Applying Interaction-Based Prediction to Managing Delays in Disks

Decreasing energy consumption by decreasing the performance level of a component can significantly increase interactive delays. This is particularly apparent in the case of hard disk drives, where the retrieval of data from a spun-down disk results in a significant delay when platters are spun up to operational speed and during which the system may become unresponsive. Keeping the disk spinning and ready to serve requests eliminates interactive delays, but wastes energy. Stopping or slowing the rotation of disk platters during periods of idleness, i.e., periods during which I/O requests are absent, is the most effective means of reducing the energy consumed by a hard drive. While prior research has focused on predicting the upcoming idle periods in order to place the disk in a lower power mode, little has been done in predicting the arrival of I/O activity, especially in the arena of interactive user applications, where user-generated I/O requests alone do not generally exhibit discernible patterns.

Timeliness of power mode transitions affects not only the system's performance and the overall system's energy consumption but also user perception of the system's responsiveness. Significant delays are associated with the transition to a higher-performance state. For example, waiting for I/O requests to arrive before switching to a higher power level may degrade system performance, keep the system processing the task longer, and, as a result, increase overall energy consumption. Switching too early wastes energy, since the demand for high performance is not present. Therefore, timely transitions to the appropriate performance level are critical for achieving both best performance and energy efficiency. Monitoring user behavior provides not only the necessary context of execution that was previously unavailable to the predictors, but enables timely predictions before the need for high performance arrives.

So far we have seen that user interactions can be monitored efficiently. Next, we will consider the interaction-aware spin-up predictor (IASP) that correlates captured user interactions to reducing the interactive delays of hard disk power management. The mechanisms gather contextual information from the user's mouse interactions within a GUI and use them in predicting an upcoming I/O request. The idea is motivated by the observation that with a majority of common interactive applications, the user fully interacts with the application through its GUI. In this context, a simple action such as opening a file requires a sequence of mouse events. By correlating the sequence of steps to the resulting I/O, future I/O occurrences can be predicted when the user initiates the same set of operations again.

51.4.1 Aggressive Delay Reduction

The observation that user interactions are responsible for the majority of disk I/O in the interactive applications leads us to consider a simple mechanism that spins up the disk upon mouse clicks. The intuition dictates that if the user actively interacts with an application, which may require disk I/O, the disk should stay on to satisfy user requests. If the user is not actively interacting with the application, the likelihood that the disk will be needed drops and the disk can be shut down. Therefore, the naïve all-click spin-up mechanism (ACSU) spins the disk up upon each mouse click and keeps it spinning as long as the user is interacting with the application. Once the user stops interacting, the disk shuts down after a timeout period.

ACSU mechanisms act on all mouse clicks and spin up the disk as soon as possible, with the downside of unnecessary spin-ups for clicks that are not followed by any disk I/O. It is important to note that user interactions that require disk I/O are a small subset of all user interactions with the application. Therefore, ACSU mechanisms have the greatest potential for reducing spin-up delays at the expense of energy consumption caused by unnecessarily spinning the disk up and keeping the disk spun up without serving any disk I/Os. ACSU mechanisms set a lower bound on the spin-up delays for the proposed IASP and also illustrate the need for more intelligent prediction schemes that decide when the disk should be spun up to improve energy efficiency.

51.4.2 Monitoring and Correlating I/O Activity

Each application is monitored individually for mouse clicks and file I/O. This allows a more accurate correlation of file I/O activity to user interactions with an application. Two levels of correlation allow file I/O and disk I/O to be distinguished. First, the application's file I/O activity is captured by the kernel in the modified I/O system call functions that check for file I/Os. For example, a modified *sys_read* checks if the I/O call that entered *sys_read* is indeed file I/O since *sys_read* can be used for many types of I/O. This stage does not consider buffer cache effects since file I/O activity is captured before the buffer cache. As a result, a more accurate correlation between file I/O and mouse interactions is obtained. Second, once potential file I/O activity is detected, the call is followed to determine if it resulted in an actual disk I/O or it was satisfied by the buffer cache. This information is used to correlate the user interactions that invoke file I/O to the actual disk I/O. The access patterns in the buffer cache will also correlate to the user interactions, since user behavior is repetitive. Hence, it will be shown that IASP is able to predict actual disk I/Os with a high degree of accuracy.

51.4.3 Correlating File I/O Activity

Correlation statistics are recorded in the prediction table, which is organized as a hash table indexed by the hash calculated using the mouse event IDs. The click IDs are unique to the window organization and therefore do not result in aliasing between different applications and windows as explained earlier. The data stored by the prediction table contains only the unique event ID, the number of times the event was observed, and the number of times I/O activity followed. The counts are a simple, but efficient means of computing an empirical probability for future predictions. The table resides globally in a daemon and is shared among processes to allow table reuse across multiple or concurrent executions of the application. Furthermore, the table can be easily retained in the kernel across multiple executions of the application due to its small size.

In addition to the global prediction table, IASP records the history of recent click activity for each process in the system. Consider a typical-usage scenario shown in Figure 51.5 where a user is editing a file in a word processor. After a while, the user clicks through a file menu to change properties of the edited file. The recorded history of clicks is C1, C2, C3, C4. At this point, the user decides to work on the file again. If the time is long enough the clicks are considered to be uncorrelated and the history of clicks is

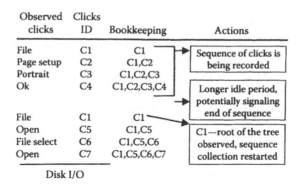

Observed clicks	Clicks ID	Bookkeeping	Actions
File	C1	C1	Sequence of clicks is being recorded
Page setup	C2	C1,C2	
Portrait	C3	C1,C2,C3	
Ok	C4	C1,C2,C3,C4	Longer idle period, potentially signaling end of sequence
File	C1	C1	
Open	C5	C1,C5	C1—root of the tree observed, sequence collection restarted
File select	C6	C1,C5,C6	
Open	C7	C1,C5,C6,C7	

Disk I/O

FIGURE 51.5 Example of possible interaction sequences.

cleared. Alternatively, the user may immediately proceed to open a new file with click sequence of C1, C5, C6, C7. In this case, the history is also reset when the user clicks on C1. Since all menus are organized as trees in the application, clicking on C1 signifies return to the root of the File menu tree. Therefore, when IASP detects a repeated click ID in the history, the history is restarted with the current click. It is still possible to record uncorrelated clicks in the history. For example, the user interacts with the Edit menu and subsequently opens a file. In this case, the history will contain clicks for edit menu interactions and the file open interactions. However, the uncorrelated clicks will have low probability and will eventually be made insignificant to the predictor with further training.

IASP uses very simple training where the observed count is updated every time a particular click is detected. In order to correlate file I/O activity, the history of clicks that lead to file I/O is traversed and the I/O count for every click present in the history is incremented. The ratio of both counts gives us the probability of file I/O following the particular click.

51.4.4 Correlating Disk I/O Activity

File I/Os issued by the application can be satisfied by the buffer cache and as a result may not require any disk I/O and the disk can remain in a power-saving state. Since not all file I/Os result in disk I/O, an additional correlation step is necessary to relate the mouse click to the disk I/O. A history of file I/Os generated by the particular click IDs is used to predict the future disk I/O generated by the particular click. An additional 2-bit history table with a 2-bit saturating counter records the history of file I/Os that resulted in disk I/O after a given mouse click was observed. The prediction table is updated using the history of file I/Os and the resulting disk I/O and the current outcome of the file I/O. Combination of the file I/O probability and the resulting I/O prediction results in a final decision about disk spin-up.

The proposed mechanisms rely on file I/O prediction and disk I/O prediction to separate application behavior from the file cache behavior. By considering both the probability of a particular click being followed by file I/O and the behavior of the resulting I/O in the buffer cache, IASP can accurately discern whether that click will result in actual disk I/O. Separating the predictor's training into file I/O and buffer cache behavior allows accurate correlation of clicks to the application's file I/O, which is the fundamental goal of this chapter. Mouse interactions with the application's GUI are strongly correlated to file I/O, so, intuitively, the goal of the described implementation is to filter all uncorrelated clicks first before the buffer cache impact on disk I/O is considered. Finally, a simple 2-bit history is used to predict buffer cache behavior, which provides sufficient accuracy. However, more sophisticated buffer-cache behavior prediction can be potentially employed to further improve IASP accuracies.

51.4.5 To Predict or Not to Predict

The critical issues that are addressed in this design are timeliness and accuracy, which turn out to be competing optimizations. Many application functions can be invoked with just a single click; however, certain operations may require several steps. In case of multiple clicks, the last click initiates a system action that is a response to the user's interaction. More specifically, if only the last click just before disk I/O occurred is used and correlated to the particular disk behavior, it results in high accuracy. Note that while this approach is very accurate, it is not very timely.

Correlating disk I/O to the last click occurring before the I/O request was observed does not provide adequate time before the I/O arrives to offset a significant portion of the spin-up latency, and so has a negligible impact on reducing the associated interactive delays. This scenario is illustrated in Figure 51.5. Clicking on C7, which is the final Open button in file open sequence, will result in an I/O system call. However, the click is immediately followed by I/O, and waiting for prediction until last click will provide little benefit in reducing delays exposed to users. Spinning up the disk upon a C1 click provides sufficient time to reduce delays; however, it may result in erroneous spin-ups, since the user may perform other operations that do not lead to file I/O.

51.4.6 Evaluation Methodology

A trace-based simulator is used to fully evaluate the effectiveness of ACSU and IASP mechanisms. An implementation on an actual system, replaying the traces in real time on a disk was used to validate the simulation results. Since the focus is on predicting spin-ups and, as a result, the shutdown mechanism is a simple timeout set to 20 s, it is comparable to the breakeven time of both disks. This means that the disk is shut down after 20 s of idleness. The specifications of the simulated disks belong to Western Digital Caviar WD2500JD and Hitachi Travelstar 40GNX hard drives and are shown in Table 51.1. The WD2500JD has a spin-up time of about 9 s from the sleep state, the surprising duration of which appears to be remarkably common for high-speed commodity drives. The 40GNX is designed for portable systems and as such has much lower energy consumption and spin-up time than the WD2500JD.

We use traces of six popular desktop applications for evaluation: Firefox, Writer, Impress, Calc, GIMP, and Dia. Firefox is a Web browser with which a user spends time reading page content and following links. In this case, I/O behavior depends on the content of the page and user behavior. Impress (presentation editor), Writer (word processor), and Calc (spreadsheet editor), are part of the Open Office suite of applications. All three are interactive applications with both user-driven I/O and periodic automated I/O, i.e., autosaves. GIMP is an image manipulation program used to prepare and edit figures, graphs, and photos. Finally, Dia is an application used for drawing diagrams for papers and presentations.

TABLE 51.1 Disk Energy Consumption Specifications

State	WD2500JD	40GNX
Read/write power	10.6 W	2.5 W
Seek power	13.25 W	2.6 W
Idle power	10 W	1.3 W
Standby power	1.8 W	0.25 W
Spin-up energy	148.5 J	17.1 J
Shutdown energy	6.4 J	1.08 J
State transition		
Spin-up time	9 s	4.5 s
Shutdown time	4 s	0.35 s

51.4.7 File I/O Correlation Accuracy

Accurate prediction ensures that the disk is not spun up needlessly, when no activity is forthcoming. First considered is the accuracy of correlating mouse clicks to file I/O at the application level, before it is filtered by the buffer cache. Figure 51.6 shows the breakdown of correct and incorrect spin-ups, i.e., hits and misses, for ACSU and IASP that result from predicting file I/O both with and without buffer caching. Hits are counted when the prediction to spin up the disk is made and it is followed by file I/O. Misses are those spin-ups which were not followed by any I/O, and missed opportunities are periods for which the mechanism failed to provide a prediction, but a spin-up was needed. Each missed opportunity results in the disk being spun up on demand, essentially spinning up when an I/O request arrives. ACSU mechanisms keep the disk powered up while a user is interacting with the application, minimizing the interactive delays. While it provides an upper bound for the number of I/O periods that may be predicted by clicks (i.e., the number of periods covered by IASP can equal but not exceed the number of periods covered by ACSU), it naïvely spins up the disk for all clicks, resulting in excess misses.

ACSU on average covers 81% of all file I/O periods, while IASP correctly covers an average of 79% of periods. ACSU shows the greatest number of misses for all applications, 52% of spin-ups are misses. This miss rate reflects the number of existing mouse clicks that do not correlate to any I/O. When the disk is spun up in ACSU, it will remain spinning as long as new clicks are observed and the idle threshold is not reached between any two clicks. IASP consistently results in the fewest misses, averaging 2%, which mostly occur while the predictor is warming up.

Coverage by the ACSU and IASP mechanisms is contingent on the availability of mouse events preceding I/O activity. Firefox, Writer, and Calc show the greatest number of misses and missed opportunities for both ACSU and IASP, meaning that there is a good deal more ambiguity in the mouse events available for prediction generated by these applications than the others. In the case of Firefox, most mouse activity occurs within the window displaying the visited Web pages. As such, the constantly changing structure of the window increases the number of mouse IDs that are encountered resulting in a high misprediction rate for ACSU. IASP, on the other hand, does not spin up the disk for these clicks, since their IDs are not observed as often as those that belong to the static part of the GUI. In the case of Writer and Calc, the relatively low coverage by both ACSU and IASP mechanisms is caused by lower availability of clicks preceding I/O. While most or all functionality of these applications is accessible through the GUI, the interaction is made simpler through the use of keyboard shortcuts. As only mouse events are considered, any I/O that occurs in response to a keyboard shortcut is not predicted by the proposed mechanisms. The applicability of keyboard events to I/O prediction will be explored in future research.

The applications for which both ACSU and IASP perform best are Impress, Dia, and GIMP. These applications have more complex GUIs for which extensive keyboard shortcuts are not intuitive to an

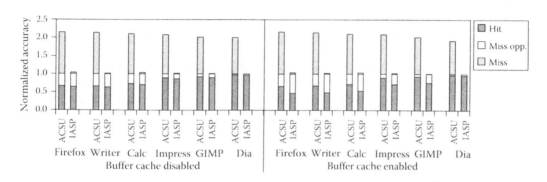

FIGURE 51.6 Prediction accuracy normalized to the total number of disk spin-ups with and without the buffer cache.

average user. This is the case with Dia and especially GIMP. All three of these applications also depend heavily upon the mouse, due to the graphical nature of their content and usage. Manipulating images, graphs, and figures is done most easily with the mouse and in the traces the user depended more heavily on the mouse for all interactions with these applications.

51.4.8 Impact of the Buffer Cache

The high file I/O prediction accuracy shown in Figure 51.6 represents the strong correlation between mouse clicks and file I/O. However, file I/O may also be satisfied by a buffer cache access, making a disk spin-up unnecessary. Hence, the impact of the buffer cache on prediction accuracy must be considered. From this point on, all figures show the mechanism with the buffer cache enabled. The buffer cache size is set to 512 MB, which is representative of current systems' capabilities. The buffer cache can satisfy many file I/Os resulting in fewer required disk accesses. In addition to introducing additional randomness into file I/O patterns, the buffer cache also increases the training time of prediction mechanisms due to both inclusion of the access history and fewer spin-ups encountered in the system. Figure 51.6 also shows hits, misses, and missed opportunities, but those metrics now realistically reflect the actual required disk I/O.

The aggressiveness of the ACSU mechanism again makes it a top performer when the amount of periods covered with correct spin-ups is considered. This behavior can be expected since the mechanism just keeps the disk on no matter if the buffer cache satisfies the request or not. Therefore, this mechanism's behavior is not impacted by the presence of the buffer cache. The different fraction of period misses, and hits, as compared to Figure 51.6, are due to a change in the periods' composition since there are fewer and longer periods due to filtering of I/O by the buffer cache. In this case, ACSU is able to spin up the disk ahead of time for 66% of required periods, while incurring misprediction rates as high as 54% with the average of 52%.

The impact is more pronounced in the case of IASP, since IASP uses contextual prediction selectively to predict what user activity will result in disk I/O. Therefore, the introduction of any randomness by the buffer cache affects the accuracy of the history-based IASP disk I/O prediction. IASP remains the most accurate mechanism, resulting in only 2% mispredictions while achieving 65% of correct spin-ups, on average. Low misprediction rate indicates that the randomness introduced by the buffer cache is insignificant and the history-based prediction is able to capture correctly the behavior of the disk I/O. Lower coverage indicates that the fewer I/O periods increase the fraction of learning time. It is worth noting that the significance of learning time decreases the longer the system stays on.

In the case of all applications except for Dia, the lower coverage of the IASP mechanism as compared to the coverage of the uncached I/O is due to learning, since predictions followed by an absence of disk I/O due to caching result in fewer learning opportunities. Interaction with GUI elements results in the requisite file data being stored in the cache. In the absence of a cache, even the infrequently used elements would generate disk I/O, but not so with the cache. In general, IASP greatly reduces the number of unnecessary spin-ups that are present in ACSU, at the cost of lower coverage, due to more energy-efficient spin-up policies.

In the case of Dia, the type of interactions encountered during tracing were limited to very simple actions, such as opening, creating, and saving a number of files containing various simple figures, meaning that the availability or absence of I/O was quickly learned by IASP. Creating even the simplest diagrams may require a large number of clicks. ACSU therefore exhibits a large number of mispredictions in this case, while IASP easily filters out the events that cause the program to, e.g., draw a triangle rather than open a file.

51.4.9 Delay Reduction

Figure 51.7 shows the average spin-up delays that are exposed to the user in the case of each of the two disks. Each missed opportunity seen in Figure 51.7 results in delay equal to the average spin-up time, 9 s

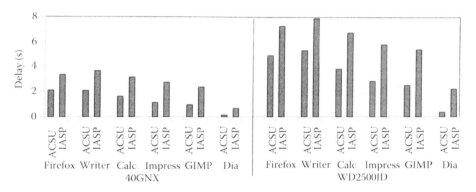

FIGURE 51.7 Average delay in seconds for 40GNX and WD2500JD disks.

in the case of WD2500JD and 4.5 s in the case of 40GNX. On the other hand, hits described in Figure 51.6 are predictions that result in the disk spinning up correctly and may arrive either early enough to allow the disk to spin up before the I/O arrives, resulting in no delay, or late, when the disk is in the process of spinning up when I/O arrives. The demand-based spin-up exposes full spin-up delay to the application during every spin-up, and therefore the average delay in the demand-based system is a full 9 s in the case of WD2500JD and 4.5 s in the case of 40GNX.

ACSU is very aggressive in reducing spin-up delays at the expense of increased energy consumption. High coverage of file I/O periods in Figure 51.6 results in an average spin-up delay reduction from 9 to 3.3 s, which is only 37% of the spin-up delay exposed by the demand-based spin-up for the WD2500JD. In the case of the 40GNX disk, the average spin-up delay is reduced to 1.36 s, which is 30% of demand-based spin-up delay.

IASP is able to shorten interactive delays exposed to the users down to 5.89 and 2.67 s for WD2500JD and 40GNX, respectively, while maintaining high accuracy and low energy consumption. IASP exposes 2.6 and 1.3 s more delay than ACSU for WD2500JD and 40GNX, respectively. ACSU sets the lower bound on the spin-up delay for mechanisms that utilize mouse interaction since it spins up or keeps the disk on for all mouse interactions as shown by the higher coverage in Figure 51.6. ACSU not only captures more I/O periods, but also does so earlier than IASP, since it is not governed by the confidence requirement set in IASP to prevent erroneous spin-ups. ACSU is therefore most effective in situations where low delay is desired, assuming that, of course, energy efficiency is also a desired attribute, but to a lesser extent. The higher accuracy of IASP makes it the most desirable choice when energy efficiency is important and users are willing to tolerate slightly higher delays than ACSU provides, which are still much lower than delays exposed by the demand-based spin-ups.

Highest delay reduction is present in Dia, where the delay is reduced by 93% and 85% by ACSU and IASP for 40GNX, and 96% and 85% for WD2500JD, indicating that there is plenty of user think time to overlap spin-up delays. On the other hand, Writer shows the lowest reduction in spin-up delays. The most significant factor contributing to the low reduction in spin-up delays in case of Writer is single-button interaction with toolbars, which results in I/O activity. For example, if the user clicks on the spell-check button in the toolbar rather than finding spell-check in the Tools menu, the resulting activity arrives quickly following the single mouse event that predicted it.

Reduction in delay is generally accompanied by increase in energy consumption, since the disk must remain on in order to minimize the delay. For example, if the disk is allowed to remain spinning for the entirety of an application's run, the interactive delays are eliminated, but at the cost of vastly increased disk energy consumption. On the other hand, simple demand-based mechanisms are often the lowest energy solution, due to the fact that they do not have extraneous spin-ups, but they incur delays each time the disk is spun up. What has been shown is that delay can be significantly improved using the proposed ACSU and IASP mechanisms.

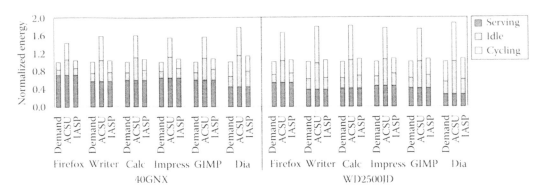

FIGURE 51.8 Energy consumption for 40GNX and WD2500JD disks.

51.4.10 Energy Savings

Figure 51.8 shows the details of energy consumption of the two disks. The energy consumption is divided into I/O serving energy, power-cycle energy, and idle energy. I/O serving energy is consumed by the disk while reading, writing, and seeking data. I/O serving energy is the same for all mechanisms, since the amount of I/O served is the same. Power-cycle energy is consumed by the disk during spin-up and shutdown and is directly related to the number of spin-ups, which also include erroneous spin-ups. Finally, idle energy is the energy consumed by the disk while it is spinning but not serving any I/Os. Idle energy is dependent on the number of I/O periods and the timeout before the disk is shutdown after an I/O period, additional idle energy consumption occurs in ACSU and IASP due to mispredictions, during which the disk idles before shutting down when I/O does not arrive. In addition, early spin-ups result in additional energy being consumed by the disk between the time when the disk is ready to serve data and the arrival of the first I/O, which is most prevalent in ACSU.

Due to a large number of mispredictions, ACSU consumes significantly more idle and power-cycle energy than IASP. On average, IASP consumes 30% less energy idling than ACSU and 40% less energy cycling power modes when using WD2500JD. In 40GNX's case, IASP consumes 27% less idle energy than ACSU and 25% less cycling energy. On average, with the WD2500JD, IASP consumes 6% more energy than the on-demand mechanism due to waiting after early spin-ups and the few mispredictions that result in the consumption of energy not present in the on-demand mechanism. Similarly, in the 40GNX case, IASP consumes 7% more energy than the on-demand mechanism. Keeping the disk always on has the effect of increasing idle energy consumption to levels that are prohibitively large for energy constrained systems. Overall, the energy consumed by WD2500JD using ACSU is 49% lower than keeping the disk always on and 70% lower in case of IASP. The energy consumed by 40GNX when using IASP is 64% lower. Differences in relative energy consumption result from the different power profiles of the two disks in question.

51.5 Applying Interaction-Based Prediction to Improve Energy Efficiency in CPUs

Today's CPUs are manufactured with support for energy management through dynamic voltage scaling (DVS). DVS allows software to dynamically reduce CPU voltage, which in turn reduces the CPU's energy consumption, since it is proportional to the square of the voltage, i.e., $E \propto V^2$. However, as the voltage is decreased, the maximum operating frequency is also reduced, resulting in a reduction in performance. Fortunately, maximal CPU performance is usually unnecessary for meeting performance expectations. For example, the performance of memory- or I/O-bound tasks may not be noticeably degraded when the

CPU's operating frequency is reduced. Furthermore, the perceived performance of real-time applications such as video players, games, or teleconferencing applications is not affected by varying CPU performance as long as the CPU provides the minimum performance required to maintain perceptual continuity for the user. Therefore, the key to transparently providing energy-efficient CPU operation is in accurate prediction of upcoming CPU demand and in providing the minimum required performance level that will meet the upcoming demand without introducing observable delays.

The challenge of efficient energy management lies in increasing the energy efficiency of the entire system rather than just an individual component. Introducing any execution delays through component-wise energy management is usually detrimental to the entire system. If the CPU runs slowly, the entire system may stay on longer, nullifying the energy saved by the CPU, or even increasing the energy consumption of the entire system. However, executing all tasks at the highest performance level is not necessary and may not reduce system delays, as is the case with interactive and real-time tasks. Therefore, it is critical to distinguish between the tasks that do not prolong program execution and those tasks that impose delays on the whole system when executed at lower performance levels. The former tasks can be executed at lower CPU performance settings to save energy, while the latter must be executed at the highest performance setting to minimize system-wide performance degradation and energy consumption.

To accurately correlate user interactions with performance demand, it is necessary to obtain a fine-grained interaction and execution context in complex GUI environments. In this chapter, we consider interaction-aware dynamic voltage scaling (IADVS), a highly accurate mechanism for matching CPU frequency to task demands and users' performance expectations. Compared to the existing coarse-grained approaches to interaction capture [10], IADVS's fine-grained interaction capture yields highly accurate predictions of upcoming performance levels demanded by tasks invoked by interactions with specific elements of the GUI. Ultimately, the high prediction accuracy results in energy-efficient executions of the upcoming tasks.

51.5.1 Interactive Task Classes

The majority of system actions in mobile and desktop systems are direct responses to user interactions. At the highest level, tasks can be classified into (1) those that require the highest performance level and (2) those that can be executed at lower performance levels to improve energy efficiency without impacting the system performance. Slowing down performance-oriented tasks prolongs execution, exposes delays to the user, and potentially increases energy consumption since the entire system spends additional time processing a given task [14]. However, users spend a majority of time performing low-level tasks, such as typing or reading, that do not require high CPU performance.

The user's response time while interacting with an application is dictated by the perception threshold. It has been shown that the average perception threshold ranges between 50 and 100 ms [15], a significant length of time for modern CPUs. Interactive task execution times shorter than the perception threshold are likely to be imperceptible to the user. Further, when a task completes execution early, the system must idly wait for the user to initiate the next task. Tasks that complete before the perception threshold is reached do not impact the speed with which the user interacts, and therefore do not prolong the application's execution time. Subsequently, we treat the perception threshold as the deadline for processing interactive tasks. Tasks with the execution time below the perception threshold can be executed at a lower CPU performance level, extending their execution time up to, but not beyond, the perception threshold. The lower CPU performance level improves energy efficiency while meeting the user's expectations of interactive performance. By preserving the perception deadline, we prevent the users from noticing any impact on the system from energy management mechanisms. Subsequently, user behavior is unaltered and we can focus on performance and energy metrics without the need for controlled user studies.

51.5.2 Detecting and Predicting UI-Triggered Tasks

The fundamental issue in energy management for CPUs is the frequency of power state switching. Frequent switching can often carry significant transition overheads, while infrequent switches may result in disparate tasks being executed at a CPU frequency that either does not meet performance demand or does not provide the desired energy savings. Subsequently, we select tasks triggered by user interactions as the main switching granularity. We define a task to be the sequence of operations by one or more threads that accomplish the same goal [10,11]. Specifically, when a thread processing an event causes a new event to occur, the new event is said to be dependent on the first event and is counted as a part of the current task. A trigger event is the first event in a series of events that is not dependent on any other events. Finally, a task is defined as the collection of all CPU processing, including its trigger event and all of the dependent events. The UI monitoring system identifies the application from user interactions, isolating the tasks initiated by the running application shortly following an interactive event.

We define UI-triggered tasks as those tasks that are preceded by UI events that occur when the UI controller (the X Window server in Linux) receives a mouse click or a keystroke. UI-triggered tasks include not only the handling of the device interrupt, but also the processing required to respond to the UI event. For example, when a user clicks the "blur" button in GIMP, the triggered task includes both the processing of the associated application functionality as well as the processing of the mouse interrupt event. The task ends when the operating system's idle process (swapper process in Linux) begins running, but the UI-triggered task is not blocked by I/O, or when the application receives a new user interface event [10]. It should be noted that the time a task spends waiting for I/O is not included in the total task duration in the analyses that follow. Likewise, task duration excludes preemption periods due to background daemons, such as the Name Server or NFS, since these processes are not dependent on the task that is being monitored.

When a UI event initiates a task, IADVS begins monitoring the system activities, recording the duration of CPU time for each relevant process. Once task completion is detected, the task's CPU demand is represented by the task length at the CPU's maximum frequency, which is computed as the sum of the processing time of each event during the task execution. To accurately measure the length of a task, we utilize the high-resolution Time Stamp Counter CPU register. At task completion, the corresponding entry in the prediction table is updated.

Each time a UI event occurs, IADVS performs a prediction table lookup using the captured interaction ID and predicts upcoming task demand. The CPU's frequency setting that can best meet the demand is set along with the corresponding CPU voltage. To minimize performance degradation due to mis-predictions, where a task continues past the perception threshold at a frequency lower than maximum, IADVS sets the CPU to the maximum frequency immediately upon reaching the perception threshold deadline.

51.5.3 Evaluation Methodology

To evaluate the IADVS mechanism, we developed a trace-driven simulation of IADVS and the following DVS mechanisms:

- **STEP**. An interval-based mechanism that runs each task by starting at the minimum frequency, stepping up the next available frequency following a fixed time interval. A 10 ms interval length is chosen to meet the average task demand.
- **ORACLE**. A task-based mechanism that utilizes future knowledge to select the optimal operating frequency for the task, fitting it to the deadline or running at maximum frequency if the task is longer than the deadline.

We assume that each DVS mechanism transitions the CPU to the minimum frequency immediately following a task, when the CPU begins idling.

TABLE 51.2 Phenom II X4 940 p-States with the Associated CPU Voltages (VDD), Current Operating Frequencies (COF) and the Power Consumptions (W)

	VDD (V)	COF (MHz)	CPU (W)		System (W)	
			Idle	Busy	Idle	Busy
P0	1.425	3000	29.2	40.8	110.7	124.2
P1	1.325	2300	20.8	28.0	99.8	109.1
P2	1.225	1800	15.3	20.3	92.6	99.2
P3	1.150	800	10.8	12.8	86.6	89.3

The traces used in the simulation were collected using a modified Linux kernel (2.6.27.5-117) running on an AMD Phenom II X4 940 with 4 GB RAM. All traces contain data from a large number of typical-usage sessions of a single user in the GNOME 2.20.1 environment. We use six applications commonly executed on desktop or mobile systems: AbiWord—a word processing application, Gnumeric—a spreadsheet application, Scigraph—scientific data plotting software, Eclipse—an integrated development environment for C/C++ and Java development, GIMP—image processing software, and LiVES—an integrated video editing and playback toolkit.

We used the AMD Phenom II X4 CPU [16] to study the implementation of the proposed mechanisms (the CPU specifications are summarized in Table 51.2). Similar to the previous findings [10,11,17], experiments are constrained by the small number of available CPU operating states. The four available power states are insufficient to efficiently match the CPU performance to task demand. Therefore, we simulated a custom CPU model based on the AMD Phenom II with the frequency range from 800 MHz to 3.0 GHz, supporting the intermediate frequencies in 250 MHz increments. We will refer to this model as the full frequency range CPU (FFRCPU).

51.5.4 Delay Reduction

Performance is an important factor for evaluating the effects of DVS mechanisms on interactive applications, since users do not tolerate excessive delays for the UI-triggered tasks. Figure 51.9a shows the total runtime for each application and for each mechanism normalized to ORACLE's. Any task that runs longer than its runtime in ORACLE exposes additional delays to the users.

Frequency stepping from the minimum in STEP results in long delays due to the time taken to ramp up to the maximum frequency. On average, STEP incurs 4% performance degradation, with the largest

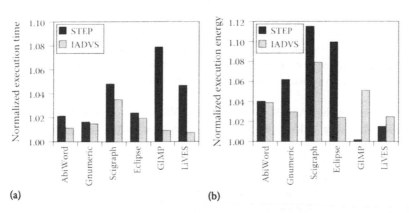

(a) (b)

FIGURE 51.9 (a) Runtime and (b) energy consumption normalized to ORACLE.

degradation being 8% in GIMP. By accurate prediction of the performance levels demanded by upcoming tasks, IADVS improves STEP's performance significantly. The most improvement occurs in GIMP and LiVES, where the majority of tasks are longer than the perception threshold. IADVS reduces STEP's delays in those two applications by 6% and 4%, respectively.

51.5.5 Energy Savings

Figure 51.9b compares the energy consumption of the studied mechanisms, normalized to ORACLE's energy consumption. Each DVS mechanism compared here is made to transition the CPU to the minimum frequency immediately when the CPU begins idling. Our focus is on energy consumed by the CPU when it is engaged in performing some task related to the application being interacted with. Subsequently, the energy consumed during CPU idling is the same across all mechanisms, and is excluded from the total energy consumption in Figure 51.9b.

Due to starting from minimum frequency and the gradual stepping up, STEP has to use higher than required frequency to catch up the processing deadline. In AbiWord, Gnumeric, Scigraph, and Eclipse, this behavior renders STEP to consume 8% more energy than ORACLE. IADVS reduces the energy consumption of STEP by 4% in these applications by accurate prediction mechanism. In GIMP and LiVES, STEP consumes comparable energy as IADVS. However, STEP trades energy efficiency for a significantly degraded execution performance, as shown in Figure 51.9a, which may result in increased energy consumption by other components since the entire system has to stay on longer.

51.6 Conclusions

The key observation in this chapter is that users are responsible for the demand placed on an interactive system, and they signal their intentions through interactions with the applications. Therefore, monitoring a user's interaction patterns with applications provides the opportunities for predicting I/O requests that follow the interactions and use this to spin-up the disk ahead of time, reducing delays, and predicting the upcoming performance levels on the CPU, closely matching expected demand to actual performance.

Section 51.4 proposes two disk spin-up mechanisms: ACSU that simply keeps the disk powered when users are interacting with the application and IASP that accurately and efficiently monitors user behavior. Both mechanisms reduce interactive delays exposed to the users due to energy management in hard disk drives. Hard disks contribute significantly to the overall energy consumption in computer systems. Therefore, aggressive energy management techniques attempt to maintain the hard drive in a low power state as much as possible, exposing long latency spin-up delays to the users. Reducing the spin-up delays provides a twofold benefit. First, the users are less irritated by constant lags in the responsiveness of the system due to disk spin-ups. Second, shorter delays allow the system to accomplish tasks quicker, resulting in less energy being consumed by other components that have to wait for the disk spin-up. Addressing the problem of the lengthy delays opens the possibility of more aggressive energy management mechanisms to be applied to disks.

Section 51.5 proposes the IADVS mechanism. IADVS accurately matches the CPU's operating frequency to the upcoming task's demand and the user's performance expectation. Slowing down performance-oriented tasks lengthens their execution time, exposes delays to the user, and may increase the amount of energy consumed due to the lengthened duration of a task. However, lengthening a task by less than the human perception threshold time introduces no perceptible performance degradation, exposing an opportunity for optimization that has not been previously explored.

Sections 51.4 and 51.5 describe specific applications of user monitoring for energy-efficient design of interactive systems. The described mechanisms were designed to be generic solutions, applicable to any application with a graphical interface, and requiring no modification of the applications themselves. Other resource management solutions could potentially perform better with hints from modified applications, but for unmodified applications, they must ultimately rely on monitoring some low-level activity. Application modifications are impractical due to the additional burden that is placed on programmers; therefore, the proposed mechanisms were designed to provide a transparent solution that provides high energy efficiency without the need for application modifications. More importantly, the proposed mechanisms are readily implementable in existing systems due to (1) the simplicity of the proposed mechanism, which monitors and correlates user behavior with system activity; (2) quantifiable, low computational, and storage overheads; and (3) online monitoring and prediction that does not require application modification or offline processing for the analysis of user interactions.

Acknowledgment

This chapter is based upon work supported by the National Science Foundation under grant no. 0844569.

References

1. R. Bianchini and R. Rajamony. Power and energy management for server systems. Technical Report DCS-TR-528, Department of Computer Science, Rutgers University, Green Ridge, NJ, June 2003.
2. J. Chase, D. Anderson, P. Thackar, A. Vahdat, and R. Boyle. Managing energy and server resources in hosting centers. In *Proceedings of the 18th Symposium on Operating Systems Principles*, Banff, Alberta, Canada, October 2001.
3. M. Elnozahy, M. Kistler, and R. Rajamony. Energy conservation policies for web servers. In *Proceedings of the Fourth USENIX Symposium on Internet Technologies and Systems*, Seattle, WA, March 2003.
4. E. Pinheiro, R. Bianchini, E. V. Carrera, and T. Heath. Load balancing and unbalancing for power and performance in cluster-based systems. In *Proceedings of the Workshop on Compilers and Operating Systems for Low Power*, New Orleans, LA, September 2001.
5. M. Anand, E. B. Nightingale, and J. Flinn. Self-tuning wireless network power management. *Wireless Networks*, 11(4):451–469, July 2005.
6. A. Dix, J. Finley, G. Abowd, and R. Beale. *Human–Computer Interaction* (3rd edn.). Prentice-Hall, Inc., Upper Saddle River, NJ, 2003.
7. E. Horvitz, J. Breese, D. Heckerman, D. Hovel, and K. Rommelse. The Lumiere project: Bayesian user modeling for inferring the goals and needs of software users. In *The 14th Conference on Uncertainty in Artificial Intelligence*, Madison, WI, 1998.
8. N. Oliver, A. Garg, and E. Horvitz. Layered representations for learning and inferring office activity from multiple sensory channels. *Computer Vision and Image Understanding*, 96(2):163–180, 2004.
9. E. M. Tapia, S. S. Intille, and K. Larson. Activity recognition in the home using simple and ubiquitous sensors. In *Second International Conference on Pervasive Computing*, Vol. 3001, pp. 158–175, Vienna, Austria, 2004.
10. J. R. Lorch and A. Jay Smith. Using user interface event information in dynamic voltage scaling algorithms. In *MASCOTS*, Orlando, FL, 00:46, 2003.
11. K. Flautner, S. Reinhardt, and T. Mudge. Automatic performance setting for dynamic voltage scaling. In *Proceedings of the Seventh ACM International Conference on Mobile Computing and Networking (MOBICOM 2001)*, Rome, Italy, 2001.
12. A. K. Dey. Providing architectural support for building context-aware applications. PhD thesis, 2000. Director-Gregory D. Abowd.
13. Inc. X Consortium. X Window system protocol. Online Technical Standard Specification, 2006. ftp://ftp.x.org/pub/X11R7.0/doc/PDF/proto.pdf (accessed on January 7, 2008)

14. D. Snowdon, S. Ruocco, and G. Heiser. Power management and dynamic voltage scaling: Myths and facts, *Proceedings of the 2005 Workshop on Power Aware Real-Time Computing*, New Jersey, USA, September 2005.

15. B. Schneiderman. *Designing the User Interface: Strategies for Effective Human-Computer Interaction*. Addison-Wesley, Menlo Park, CA, 1998.

16. AMD. AMD family 10h desktop processor power and thermal data sheet, February 2009. http://www.amd.com/us-en/assets/content_type/white_papers_and_tech_docs/GH_43375_10h_DT_PTDS_PUB_3.18.pdf

17. J. R. Lorch and A. J. Smith. Improving dynamic voltage scaling algorithms with pace. In *SIGMETRICS '01: Proceedings of the 2001 ACM SIGMETRICS International Conference on Measurement and Modeling of Computer Systems*, pp. 50–61, ACM Press, New York, 2001.

52

Toward Sustainable Portable Computing

Vinod Namboodiri
Wichita State University

Siny Joseph
Newman University

52.1 Motivation for Sustainable Portable Computing

Computing devices are playing different roles in server farms, data centers, office equipment, among others. With the increased awareness in how the world consumes energy, and its impact on the planet, it is natural to thus think about the impact of computing on global energy consumption. There have been many studies that document this impact looking specifically at information and computing technology [15,19,43].

The world is, however, changing the way it accesses the Internet, and computing in general. There is increasing relevance of portable, battery-operated devices in how we handle computing and communication tasks. The first phone call over a GSM cellular phone occurred in 1991. By the end of 2007, half the world population possessed such phones. This phenomenon is similar to the growth of computing devices in general where CPU processing power and capacity of mass storage devices doubles every 18 months. Such growth in both processing and storage capabilities fuels the production of ever more powerful portable devices. Devices with greater capabilities work with more data, and thus subsequently require greater capabilities to communicate data. This has resulted in a similar rapid growth of wireless communication data rates as well to provide adequate quality of service as shown in a study done by Novarum [33] and reproduced in Figure 52.1.

The increased role of portable devices has thus resulted in calls to look at sustainability in mobile computing [35,39]. The work in [39] finds that computing devices including data centers, server farms, desktops, and portable devices (laptops and mobile phones) account for about 3% of the global electricity usage. Surprisingly, portable devices were responsible for close to 20% of this share, which is expected to grow as power-hungry smart phones proliferate the market. The study also found that

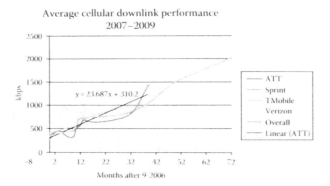

FIGURE 52.1 Cellular downlink performance through various U.S. carriers (From Novarum, Guidelines for successful large scale outdoor wi-fi networks, 2010.)

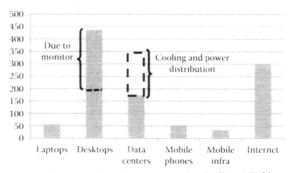

Sectorwise electricity consumption (million MWh)

FIGURE 52.2 Energy consumed by personal computing segment including front-end devices for users and back-end devices that provide the infrastructure.

computing devices (including portable devices) were responsible for 270 million tons of CO_2 emission annually—this number is equivalent to the emissions of 4.5 million vehicles on the road when driving 10,000 miles per year considering an average car giving 21 miles per gallon. The share of CO_2 emissions by the computing sector was found to be much larger than what was expected based on its share of global energy consumption. A big reason for this is the fact that most of the computing sector runs on electricity, most of which is produced from coal.

When it comes to looking at energy efficiency and the concept of sustainability in computing, the focus has invariably been data centers and mobile infrastructure like cell towers, as they were considered the power hogs within the computing sector. However, the study in [39] dispels this notion and points out that portable devices consume comparable, if not greater, energy than the devices/infrastructure believed to be power-hogs (reproduced in Figure 52.2). Thus, over the next few years, it is opportune to research ways to move toward sustainable portable computing.

52.2 Elements of Sustainable Portable Computing

Before we can move toward sustainable portable computing, it is important to define what it entails in terms of portable devices.

- *Reduction in electronic waste*: A recent study by Ofcom (a UK-based broadcast and telecom regulator) has concluded that a fundamental balance must be struck between the increasing environmental impacts of systems and services, and the social, economic, and commercial benefits delivered by such systems and services. Achieving this balance is considered to be a key challenge for the communications industry going forward. This study also highlights that consumer equipment, where devices have a small individual impact, often have a very substantial impact overall due to the large volumes involved and shorter product life compared to infrastructure systems. Thus, significant efforts are being put into the manufacturing and recycling phases of mobile computing devices. An increasing number of portable devices touted as green are emerging made from recycled material [5]. A recent European law known as the WEEE Regulations (Waste Electrical and Electronic Equipment) assigns financial responsibility to producers of electronic and electrical devices for the collection, treatment, and recycling of their WEEE. A global recycling survey highlighted that only 3% of handsets are recycled [32]. In response to legislation and lack of recycling awareness by mobile phone users, handset manufacturers have started promoting free-of-charge take-back programs. In spite of all these efforts, a major form of reducing electronic waste has gone largely unnoticed—reducing the rate at which these devices are replaced by increasing usage lifespan.
- *Energy-efficient operation*: As mentioned earlier in this section, portable devices consume energy that is not insignificant anymore due to the large number of such devices in use globally. Thus, an effort for sustainability in this area does require more attention from manufacturers in providing better information and "green" options to users and better usage behavior by individual users. It is also the responsibility of researchers to improve existing devices through improvements in hardware and software that allow portable devices to do more with less energy, and reduce energy wastage. Energy-efficient operation of these typically battery-life-constrained devices would further improve user satisfaction by providing greater operating lifetimes per battery charge, increasing the utility and thus the number of applications of such devices.

52.2.1 Importance of Software-Based Approaches

Typically, computer equipment manufacturers rely on periodic hardware updates to achieve more energy-efficient operation. Unfortunately, achieving this one facet of sustainable portable computing comes at the cost of the other facet: reduced electronic waste. Periodic hardware updates result in consumers dumping their existing hardware and using new ones. Take, for example, the rate at which cellular phones are upgraded every 2–3 years. Each upgrade results in possibly improved performance for the consumer, but results in the old hardware becoming a waste. As mentioned earlier, very little hardware is actually recycled—a study by the U.S. Environmental Protectional Agency (EPA) found that only about 10% of such waste is recycled in the United States, while a recent Nokia study showed that *only 3% of cell phones were recycled globally* [32]. Thus, it is imperative to focus more on nonhardware, software approaches to achieve energy-efficient operation as they reduce the dumping of existing hardware. Note that we do not imply that energy inefficiency is the cause of e-waste. Rather, we are stressing on how energy efficiency should be preferably achieved to get reduced e-waste as a by-product.

52.2.2 Organization of the Rest of the Chapter

Based on the discussion earlier, we believe that for portable computing to be energy-efficient and sustainable, we must look at the dual objectives of *energy-efficient operation* to reduce energy consumed and *increase in lifespan* to reduce electronic waste. The first category deals with how to design and/or operate a device more efficiently taking as little energy from the power grid as possible.* The second category

---

* This results in reducing carbon emissions as well, the figures for which were quantified in [39].

deals with how to decrease electronic waste by approaches that could help increase the period of time a consumer uses a device while maintaining or improving performance. Software approaches are preferred for this goal based on the discussion earlier in this section.

We begin by looking at an approach to increase device lifespan by relying on cloud computing and cognitive radios (CRs) in Section 52.3, an approach first put forward in [30]. This approach attempts to rely on more capable remote servers to provide improved performance and reducing the need for individual devices to be upgraded often. Next, in Section 52.4 we look at a novel, emerging approach to reduce energy consumed by portable devices by using small form-factor solar panels. We will answer interesting questions such as "How fast can a solar panel charge a portable device," and "How much energy benefit can be realized from such approaches"? Finally, in Section 52.5 we describe how firms and regulators could make an impact by providing economic incentives to promote sustainable portable computing.

52.3 Cloud Computing and Cognitive Radio Approach

52.3.1 Motivating Example

Consider a smart phone being used to play the game of chess by a person who is traveling. The game could be played locally on the device itself, or it could be played online. In the former version, all the computing required to make a move by the computer (the game opponent in a two-player format) is done using the device's resources. In the latter, online version, all the computation is done through a powerful remote server and conveyed through communication to the device. It is easy to see that more communication is required in the latter scenario by the portable device, but possibly significantly less computation.* The latter scenario allows the user to play higher-level games without having to upgrade his/her device in the future. All the resources needed for improved performance is already at the server, or could be by an upgrade at only this one location. Thus, by just focusing on the energy efficiency of the communication done, we can significantly improve the sustainability of smart phones for this application scenario.

52.3.2 Cloud Computing

The term, *cloud*, is used as a metaphor for the Internet, where resources, information, software, and hardware are transparently shared. The main advantages for using cloud computing are scalability, ubiquitous availability, and maintenance costs. The cloud computing concept is being furthered by a lot of major companies today like Google, IBM, Microsoft, etc. Many of us are using cloud computing in our daily lives without actually noticing it. For example, you go on a trip and have an album of digital pictures to share on your local computer. Photo sharing Web sites (e.g., Flickr or Picasa [2,4]) make it is easy to upload the pictures online and just send the link of the location as a URL for others to look at. This is cloud computing, where the album is stored in a data center online in a cloud and easily accessible from any location connected to the Internet. Another common cloud-based application that is emerging is office productivity tools. Google's Docs, and Microsoft's Office Live are two examples. Such tools, apart from increased availability, also make collaboration easier.

Cloud computing is typically a client–server architecture, where the client can be any portable device like a laptop, phone, browser, or any other operating system–enabled device. Users of portable devices these days want to share documents, check e-mail, and surf the Internet on the fly, and they represent an increasing segment of the population. A main issue with these portable devices is the constraints they present in terms of storage, memory, processing, and battery lifetime. By storing information on

---

* The amount of computation required could vary, for example, based on the size of the protocol stack employed and degree of encryption used.

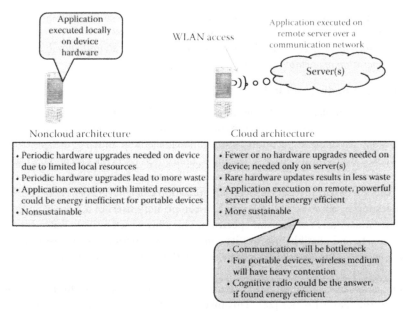

FIGURE 52.3 Motivation for pursuing energy-efficient usage of CRs in cloud computing scenarios.

the cloud, and interacting with the cloud through communication, all these constraints can be easily met. An interesting point to note is that all these constraints of portable devices are based on hardware, and when using the cloud paradigm, hardware upgrades do not always present obvious advantages. For example, a higher-powered processor and faster memory on the portable device do not seem as attractive anymore.* Frequent hardware updates are now done at the cloud's servers, which are much fewer in number. Thus, many portable devices can be possibly utilized for many more years than what is typical now, and hence reduce the manufacturing and recycling costs associated with this large-scale segment of mobile computing. This scenario is summarized in Figure 52.3.

52.3.2.1 Communication Challenge

The only possible downside is that all the burden is shifted onto the communication interface and the techniques it uses. With most of the portable devices currently using the wireless medium for communication, it is expected that the wireless spectrum will be highly congested, especially if cloud-based applications become more popular [8].

52.3.2.2 Why Cognitive Radios?

The state-of-the-art solution to wireless spectrum congestion is the CR paradigm as discussed next [27]. This paradigm can be implemented mostly with software techniques, maintaining our vision of sustainable mobile computing. The CR paradigm has its challenges as well, and one of those challenges (energy consumption) will be the focus of this chapter so as to move toward sustainable mobile computing.

52.3.3 Energy-Efficient Operation of Cognitive Radios

With an increased communication burden likely when portable devices rely on a cloud computing paradigm, next-generation wireless communication radios will likely work on the software-defined, CR

---

* On the other hand, improving graphics processors and more demanding playback of media could still necessitate hardware upgrades. However, as a recent article explains, running graphics based on capabilities on the cloud is an increasing trend [37].

paradigm where the communication medium and parameters are continuously optimized to maximize performance. The central question that needs answering is as follows:

Are CRs energy efficient in a cloud computing scenario, and how should they be configured to allow energy-efficient operation of portable devices?

52.3.3.1 Cognitive Radios and Related Work

Ever-increasing spectrum demands of emerging wireless applications and the need to better utilize the existing spectrum have led the Federal Communications Commission (FCC) to consider the problem of spectrum management. Under conventional spectrum management, much of the spectrum is exclusively allocated to licensed users. In the new CR paradigm, unlicensed users (aka secondary users) opportunistically operate in unutilized licensed spectrum bands without interfering with licensed users (aka primary or incumbent users), thereby increasing spectrum utilization efficiency. CRs have been seen as the way to minimize congestion by allowing multiplexing between primary users of a piece of spectrum with other opportunistic secondary users of the same spectrum. This allows each radio to watch for a less-congested spectrum to move to and possibly improve its communication performance. Research on the CR technique has mainly dealt with how spectra can be sensed, the coexistence of primary and secondary users, and the channel access aspect.

The sensing aspect of CR mainly deals with finding the right spectrum to use for communication, as introduced in the seminal paper [28]. This involves finding a spectrum that provides the best communication possibilities for the node in terms of metrics such as throughput, fairness, interference, and utilization. The channel assignment or allocation problem in CRs has been studied through different optimization formulations in [11,12,18,25,34,38,41,44]. Further, the detection and avoidance of primary users (PU) of the spectrum is of utmost importance as they are the incumbent users who have priority on the spectrum. Secondary users must ensure that they detect PU signals at very low signal-to-noise ratio (SNR) and move to another channel or stop communication to avoid interference to PUs. This detection process of a PU receiver and/or transmitters on the spectrum has been of considerable interest to researchers [9,20–22]. Some important considerations include the determination of the duration to sense the channel [16,42] and the duration to communicate packets on each spectrum [24].

The channel access aspect of CR can be classified based on the type of network architecture: infrastructure or ad hoc. MAC protocols for CR in *infrastructure networks* make use of the centralized base station to synchronize and conduct node access operations. The carrier sense multiple access (CSMA) MAC protocol proposed in [23] for infrastructure CR networks is a random-access protocol, which relies on differentiated access to the medium for packets from or to PUs, with other CR nodes having a lower priority. IEEE 802.22 is a standard for wireless regional area network (WRAN) using white spaces in the TV frequency spectrum. The development of the IEEE 802.22 WRAN standard is aimed at using CR techniques to allow sharing of geographically unused spectrum allocated to the Television Broadcast Service, on a noninterfering basis [3]. The IEEE 802.22 standard for CR uses the notion of superframes and slots at the base station to control access to the medium. In the downstream direction, 802.22 MAC uses time-division multiplexing, while in the upstream direction, it uses demand-assigned time division multiple access (TDMA). In general, in an infrastructure network, the base station is in control of the network and dictates what frequency all nodes in its network should use. Nodes are, however, free to search for and associate with other base stations to satisfy communication requirements.

In ad hoc CR networks, spectrum sensing and medium sharing are distributed in nature, along with responsibilities of forming packet forwarding routes and time synchronization, if required. Proposed protocols in literature can be classified further based on whether nodes have one or multiple radios. The dynamic open spectrum sharing (DOSS) MAC [25] proposes the use of three radios per node; one for data communication, one for control packets, and one for sending a busy tone. By sensing the busy tone, nodes know about an active PU because each data channel is mapped to the busy tone channel as well. Even if a node does not recognize the signals on the data channel of the PU, the busy tone conveys its presence. Further reading on MAC protocols for CR can be found in the recent survey in [13].

The work in [45] focuses specifically on using CR techniques for WLANs to solve the performance degradation due to congestion. Like other work, energy consumption with regard to CR techniques is not considered. The work by [27] specifically points out that one of the biggest motivations for CR techniques is WLAN spectrum congestion and the continuing density increase of wireless devices. This is also our motivation to explore the issue of energy consumption of a CR beginning with the WLAN scenario as summarized in Figure 52.4.

52.3.3.2 Negatives of Obtaining Cognition

The increased attention to develop CR techniques to find and use a wireless spectrum has, however, resulted in researchers overlooking the importance of energy consumption in the devices that employ such techniques. Scanning for a wireless spectrum, and possibly switching between frequency channels, is power intensive and could result in a rapid depletion of the lifetime of energy-constrained devices like personal digital assistants (PDAs), laptops, smart phones, wireless sensors, among others. The fact that the success of the CR technique depends on such a power-intensive operation can undercut the very paradigm in such portable devices. Thus, research needs to be done to study the extent of energy consumption and its impact on device lifetimes.

52.3.3.3 Positives of Having Cognition

On the positive side, the CR technique could also reduce the energy consumed for communication by finding a spectrum that is less congested. This would enable communication with less contention for the medium, another major factor of energy consumption in wireless devices. Higher contention for the medium typically results in more packet collisions, more time spent backing off when using CSMA protocols, and more overheard packets from other nodes. Thus, the CR technique's positive impact on energy consumption needs to be studied and quantified as well to understand how energy-constrained devices would fare in terms of operating lifetime.

The work in [17] presents techniques for reducing energy consumption of a CR. Their work targets *only* the physical-layer adaptations involving the power amplifier, modulation, coding, and radiated power. We, on the other hand, look at the problem in a top-down fashion. We study the impact of higher layer parameters such as scanning time per channel, number of contending nodes on the medium, and node distribution across channels in conjunction with physical-layer constructs like channel conditions.

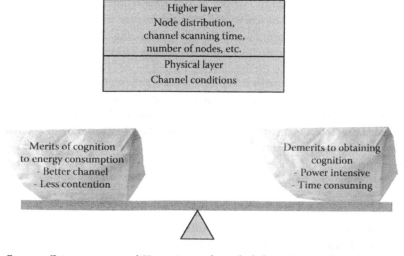

FIGURE 52.4 Energy-efficient operation of CR requires studying the balance between the positives and negatives of cognition. In this section we look at this balance by considering a subset of such, mainly upper network layer, factors.

52.3.4 Results

Here we look at some results on what can be expected when using CRs in cloud computing scenarios. The first set of results is of a comparison between a CR and a conventional radio in terms of energy consumption. This would indicate whether CRs are a good idea to start with when energy is a metric. The second set of results is for a comparative study of energy consumed by a commonly used application executed both locally as well as on a cloud. This would indicate whether cloud computing is more energy efficient than traditional, local computing. Based on these two preliminary, fundamental results, we point out the next steps to be taken toward sustainable mobile computing.

52.3.4.1 Cognitive Radio versus Conventional Radio

A CR differs from a conventional radio in that it can scan a predefined range of spectra and pick a channel for subsequent communication. The reason for moving from one channel to another could be due to several reasons: the arrival of a primary user, unfavorable channel conditions, or an increase in load because of growth of number of users on the former channel. All these factors would have led to increased energy consumption by the radio if it had stayed on the former channel. Thus, a CR can "hunt" for better channels that can reduce energy consumption. There are various possible scanning strategies that can be evaluated and compared. We base our results here on a simple strategy termed *optimal scan*. Doing a thorough, extensive scan of all available frequency spectra can guarantee a channel that is the best for communication in terms of minimizing energy consumption. Such a channel could minimize energy consumption for the following reasons: there are very few contending nodes on the chosen channel, or there exists a low-interference communication environment, or a combination of both that results in minimal energy wasted due to packet collisions, backoff time, or packet losses. Such an *optimal* strategy, however, can be very energy intensive in scanning all possible frequency spectra since the radio has to listen on each channel for some time, switch to another channel, and so on. Listening on a channel consumes almost as much power as communicating on a channel [14]. Interested readers can refer [30] for additional scanning schemes.

Preliminary work done with the optimal spectrum scanning techniques showed that results were highly dependent on the application scenario and configuration parameters [29]. For example, Figure 52.5 shows the role various parameters like number of channels, node distribution across channels, number of nodes on each channel, and time spent scanning each channel play in whether a CR can save energy over a conventional, nonscanning radio.

For the experiment results shown in Figure 52.5, node contention was chosen as the sole reason a CR changes channels. The number of nodes on a "current" channel was n_c and spectrum scanning was done to find a channel that had a fewer number of nodes contending. The conventional radio, of course, does not scan and stays put on the current channel. The energy savings shown thus compare the energy consumed by the conventional radio on the current channel versus the energy consumed by the CR on a newly found channel (including energy consumed to find this new channel). The parameter *node ratio* is the ratio of number of nodes active on the eventual chosen channel through the scanning process to that of the current channel. Thus, if there is a 10-fold reduction in the number of nodes on the chosen channel, the node ratio would be 0.1. The smaller the node ratio, the greater the reduction in contention by switching to this channel.

Further work needs to be done to examine the impact of application scenarios on various spectrum scanning schemes and configurations to better understand fundamental limits.

52.3.4.2 Cloud versus Noncloud

The goal of our preliminary work with cloud computing was to study the energy consumption of portable devices over a cloud-based application. For a representative device, we picked a laptop that we had experience doing energy measurements with previously. For cloud-based applications, we picked office productivity tools to begin with and later played video over remote servers using Web browsers.

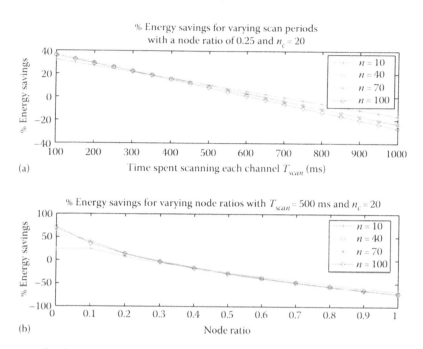

FIGURE 52.5 Results showing percentage energy savings over a conventional radio for varying parameters and $n_c = 20$ (a) The node ratio is kept at 0.25 and T_{scan} is varied from 100 to 1000 ms (b) The node ratio is varied from 0.01 to 1 while keeping T_{scan} fixed at 500 ms. (P. Somavat, S. Jadhav, and V. Namboodiri. Accounting for the energy consumption of personal computing including portable devices. In *First International Conference on Energy-Efficient Computing and Networking (e-Energy)*, Passau, Germany, April 2010.)

52.3.4.3 Methodology and Results

Our laptop used in the study was an IBM Lenovo SL400 with a 14 in. display. The study was done on the Windows 7 operating system. We initially studied the case of word processing applications and then the case of multimedia applications.

We used Microsoft Office 2007 as the desktop application, while Microsoft's Office Live and Google Documents were used as the cloud-based applications. Text was input into the word processing editor using an automated program at the same rate over a period of time on both versions. This was done multiple times to ensure reproducibility of results. The results obtained are shown in Figure 52.6. Six lines were plotted denoting how the battery lifetime decreases over time. The *Idle* line denotes how the laptop battery would have degraded over time if it was kept on and idle, with no applications running, and the WLAN interface switched off. The lines *C1* and *C2* denote the energy consumed when the cloud-based word processing application was executed with the WLAN interface on. The line *NC* denotes the energy consumed when the word processing application was executed locally. In this case, the Wi-Fi interface could be kept on or off as communication was not required. What is interesting is that cloud-based applications consume less energy over time, even compared to the noncloud version with Wi-Fi interface off, which is encouraging as a typical WLAN card, even in idle can consume about 10% of a laptop's power [26], and much more for a mobile phone or PDA [36]. This result confirms our intuition that cloud-based computing for portable devices can be more energy efficient, apart from reducing the need for hardware upgrades.

We also tested a multimedia application scenario by comparing energy consumed in playing video in both Flash video (FLV) and MPEG-II formats, locally and through a Web browser from remote servers. The results were computed locally and also over the cloud on both Windows 7 and Ubuntu 9.04. The FLV and MPEG-II video files had a 16:9 display ratio with 30 frames per second (fps) encoding.

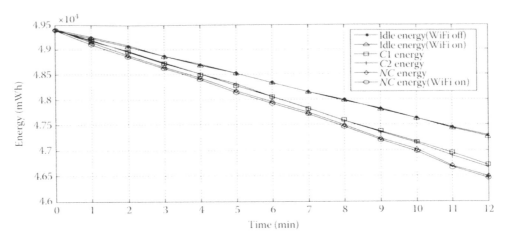

FIGURE 52.6 Comparison of battery lifetime reduction with a cloud-based office productivity application versus a similar noncloud application. Two different cloud applications were tested labeled C1 and C2. The wireless interface was kept switched on or off for the noncloud application and in idle mode.

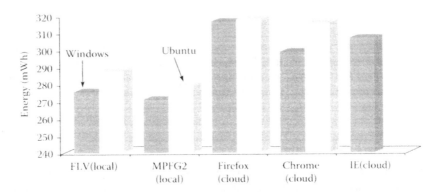

FIGURE 52.7 Energy consumption (MWh) of playing video on Windows and Ubuntu platforms, locally and remotely. The Wi-Fi interface was kept off for local video playback. Note that there is no version of Internet Explorer for Ubuntu.

From Figure 52.7, it can be seen that in the case of video, the cloud consumes more energy than local computation, but energy consumption over the cloud using Windows is better than when using Ubuntu. Playing an MPEG-II file over Windows saves 2.8% more energy than playing it over Ubuntu. Similarly, FLV files save 4.4% of energy on Windows than on Ubuntu. Among the browsers, Google Chrome has energy savings of 6% per minute on Windows platform. It can be observed that Windows provides energy-efficient cloud access over the Google Chrome browser when compared to Ubuntu.

52.3.5 Lessons

From the results, it is clear that greater reliance on communication by running applications in the cloud and using CRs can have benefits in terms of reduced energy consumption of portable devices depending on the type of application, along with the potential to reduce frequent hardware upgrades. For CRs, they can save energy over conventional radios under a specific set of configurations and application scenarios. When comparing cloud-based applications to those that are not, it is seen that energy could be saved at least for applications like word processing where very little local resources are used. For other applications like multimedia, local resources like graphics processors and audio hardware are used extensively requiring significant local energy consumption on top of the energy for communication.

52.4 Energy Harvesting Approach

The issue of energy consumption of computing devices can best be alleviated by all members of the computing community that includes researchers working on different aspects like displays, CPUs, network interfaces, memory, to name a few. Individual optimizations to each hardware component would help and could be improved by joint hardware optimizations. Similarly, software optimizations focusing on operating systems, data caching, data compression, to name a few, would be helpful as studied, and even put into practice over the last decade or so. Our task here is to pay particular attention on how to reduce energy consumed by portable devices. We believe that computer networking researchers can contribute the most in reducing portable device energy consumption due to the fact that an active network interface is likely to be the task that consumes the most power among all components [26]. For handheld devices like phones and PDAs, the share of power consumption due to the network interface(s) is likely to be greater than 50% [36].

52.4.1 Prior Approaches and Need for Awareness

a) Power management
Power management has been an active area of research over the years for computing devices as well as appliances [10]. Device components have been optimized to lower power consumption. An informal survey of 50 students at Wichita State University showed that most were unaware of power management modes of the network card, and used the default settings. A simple check of the default settings of 10 different laptops showed that power management was disabled in all of them. While controlling power management options might be becoming easier these days, user awareness is the single biggest factor to take advantage of such options.

b) Battery management
When batteries are necessary, as is the case when a user is mobile, better usage of batteries can get better efficiency. Better utilization of battery capacity can reduce power consumed of the electricity grid. In a similar vein, users must be made aware that leaving battery chargers plugged into power outlets can consume as much as 20%–30% of the operating power of these devices.

c) Adjust optimization metrics
Researchers in the area must be made aware that protocols and algorithms should be designed keeping total energy consumption as a metric, as opposed to individual node lifetimes when considering a collaborative network in specific scenarios. For example, consider the case of a static wireless mesh network. It may be possible to replace the batteries in such nodes periodically. In such an instance, protocols must be designed to minimize total energy consumption of the network as opposed to each node trying to minimize its energy consumption [31]. Another example is that of an ad hoc network in a conference room where a power outlet for each node is easily available and all nodes are plugged in. Again, in this scenario, the total power consumption by all nodes should be minimized as opposed to the common metric of individual node consumption.

52.4.2 Utilization of Energy Harvesting

From a networking researcher's perspective, there is a need to understand how to design protocols to utilize sources of energy harvesting effectively. For example, during periods when plenty of energy is available, portable devices could do more delay-tolerant tasks. Once the energy harvest is unavailable, the device can ramp down to minimal operating modes similar to the power management paradigm. It is more important to determine the feasibility of utilizing such harvesting for portable computing devices and, thus, reducing energy consumed from the grid.

Recently, there are many products on the market that allow the use of energy-harvesting techniques to power portable devices. For example, the SunlinQ product (cost about $320) [7] can charge laptops, while the Solar Gorilla (cost $220) can charge smaller devices like phones [6]. These devices can become more mainstream with time as prices drop due to scale, or through government subsidies to encourage more renewable energy use.

52.4.2.1 Experimental Results

For our experiments we used both a laptop device (a Lenovo SL400 running Ubuntu Linux) and a mobile phone (Apple iPhone 3GS), with the SunlinQ foldable solar panel for laptops and Solar Gorilla solar panel for mobile phones, [6,7]. A digital lux meter for accurate light level measurements was used for each of our experiments [1]. We show only the results for the phone as it is representative of those for laptops as well.

Figure 52.8 shows the comparison of charging time for an iPhone 3GS through a wall socket, and the two solar panels. The phone was in idle state throughout the charging process, which results in the transition to a low-power standby mode. The bigger, SunlinQ panel was found to have a similar rate of charging as a wall socket, a very promising result in terms of delay to charge. As these panels are foldable, they are easy to carry, but space restrictions may make it infeasible to use them in some situations. The small form-factor SunlinQ panel takes about 30% more time to charge the phone, but its small form-factor makes it easy to use in most situations. All experiments were done during the daytime between 10 a.m. and 4 p.m. where light intensity varied between 800 and 1200 lux. As laptops typically have a battery capacity (MWh) two to three times that of a smart phone, the time to charge them took about the same multiple of the results in Figure 52.8.

Our results open up possibilities such as charging portable devices during an outdoor lunch or while working in an office or coffee shop near a window, or while driving in a car during daytime. Experiments done with indoor fluorescent lighting, however, were not positive. With only about 100 lux of light intensity available, none of the solar panels could charge a phone.

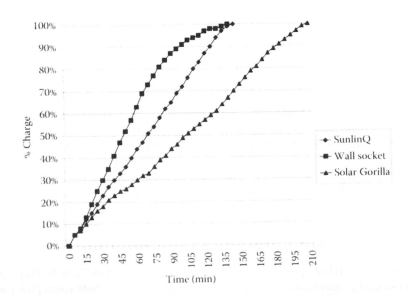

FIGURE 52.8   Comparison of charge time of an Apple iPhone 3GS between two solar panels and the wall socket.

52.5 Economic Incentive Approach

This approach is aimed at providing economic incentives to consumers, or at least modify the current incentives such that usage behavior is changed. For example, typical cellular phone carrier contracts in the United States last for 2 years after which customers typically upgrade to newer devices. These new devices offered are heavily subsidized conditional on the signing of a new 2 year contract. If customers would instead buy hardware without contracts and subsidies, they might be inclined to keep their devices longer. As reducing energy consumed in manufacturing devices and decreasing environmental waste provides environmental benefits, governments or regulators could provide incentives for customers to keep their devices. For example, a regulator could pay part of the carrier costs after a device has been used over a period of time. The incentive for the subsidy provider is that increasing lifespan has environmental benefits through reduced electronic waste and associated reduction in manufacturing and recycling energy costs, and carbon emissions as well. There has been some movement of the market in this direction by some "green" cellular carriers who offer customers the option to buy refurbished phones without contracts. A recent article in the *Wall Street Journal* discussed such an option by Sprint [40].

In the following, we develop a simple model for pricing that a firm can use to extend the length of a contract. The goals of an appropriate price are the following: (1) the firm is able to maintain its operating profit and (2) customers have an incentive to choose the longer plan. This exercise would be used to justify how there can be incentives for all parties involved to provide such options.

Let P_1 be the annual price charged to a customer for a mobile phone plan (say plan 1) of duration t_1 by a firm. Let P_2 be the annual price charged to the consumer for a similar plan (say plan 2) of duration t_2 by the firm. Let C be the annual per unit, average cost to the firm to provide service and the device to the customer. We assume this cost to be the same across both plans to keep the model simple.* We will assume a monopolistic model with a single firm and that the device is eventually recycled at the end of its term.

The profit for the firm from the customer for plan 1 is $(P_1 - C)t_1$. For plan 2, its profit would be $(P_2 - C)t_2$. Enforcing the requirement that the firm's profit remains the same under both plans, we can equate the two profit equations to get the price the firm should set for the new plan as

$$P_2 = \frac{P_1 t_1 + C(t_2 - t_1)}{t_2} \tag{52.1}$$

So, would the consumer adopt such a plan? From Equation 52.1 we can write an expression for percentage savings to a consumer for choosing a longer plan as

$$CostSavings = \frac{P_1 - P_2}{P_1} = \left(1 + \frac{C}{P_1}\right)\frac{t_2 - t_1}{t_2} \tag{52.2}$$

Figure 52.9 shows percentage cost savings to a consumer for durations of plan 2 varying from 1 to 7 years, with plan 1 duration, t_1 fixed at 1 year. For each value of t_2 we looked at four different ratios of $\frac{C}{P_1}$, which is the ratio of cost to revenue per year under the original plan 1. Higher values for the ratio imply smaller profit margins for the firm. The results indicate that even for an extension of 2 years, over the initial 1 year plan, the customer can save in the range of 100% in costs. Further, most of the cost savings are gained in the first few years of extension, implying little additional savings after the first 2–3 year extension. If the consumer feels that the reduction in utility per year as the phone ages can be offset by the increased savings, they will adopt these phones. Our thin-client or cloud computing approach

---

* It is easy to see that the service costs per year across both plans per unit should be similar for the carrier. The device costs per year for the longer plan should be smaller as it is amortized over a longer duration. However, if firms try to recuperate most of the device costs in the initial few years (duration of plan 1), the assumption of similar device costs across both plans can also be justified.

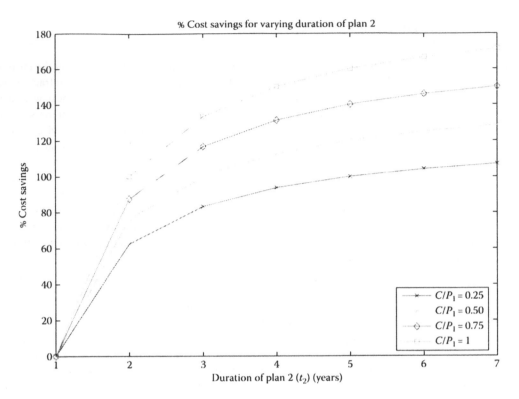

FIGURE 52.9 Percentage cost savings on varying the length of plan 2, keeping duration of original plan, plan 1, fixed at 1 year. Four different values of the cost to revenue ratio per year for the firm are considered.

described in Section 52.3 should further help maintain performance levels to some extent. The impact of any subsidies regulators can offer can be factored in to this model by defining a new term C_E for the environmental cost* through a combination of manufacturing and recycling of a device and any associated carbon emissions. By increasing lifespan, we are amortizing this cost over a longer duration that reduces the annual environmental costs. This reduction in annual cost can be passed along as a subsidy to the consumer to help improve the chances of adoption through additional cost savings. This subsidy can be calculated as follows:

For plan 1, the annual environmental cost per year is $\frac{C_E}{t_1}$, while that for plan 2 is $\frac{C_E}{t_2}$. Thus, annual environmental cost savings in moving to plan 2, which can be given as an annual subsidy S, is

$$S = C_E \frac{t_2 - t_1}{t_1 t_2}. \tag{52.3}$$

We can modify Equation 52.1 as well to incorporate the subsidy as

$$CostSavings = \frac{P_1 - P_2}{P_1} = \left(1 + \frac{C}{P_1}\right)\frac{t_2 - t_1}{t_2} + \frac{C_E}{P_1}\frac{t_2 - t_1}{t_1 t_2} \tag{52.4}$$

where P_2 was modified from Equation 52.1 as

$$P_2 = \frac{P_1 t_1 + C(t_2 - t_1)}{t_2} - S. \tag{52.5}$$

---

* This could be the monetary value attached to the emissions and fossil fuels used up. Carbon offset costs in terms of planting new trees could be one way to perceive this term.

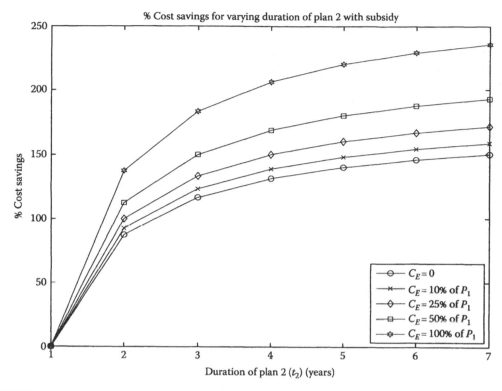

FIGURE 52.10 Percentage cost savings on varying the length of plan 2, keeping duration of original plan, plan 1, fixed at 1 year. The cost to revenue ratio per year for the firm is kept fixed. Five different values of the overall environmental cost to revenue for the firm for the first under plan 1 are considered.

Figure 52.10 shows percentage cost savings with the incorporation of a subsidy. We kept the ratio $\frac{C}{P_1}$ fixed at 0.75. For each value of t_2 we looked at five different ratios of $\frac{C_E}{P_1}$ (including the case with no subsidy), which is the ratio of overall environmental cost for a device to the revenue obtained from the customer in the first year under the original plan 1. Higher values for the ratio imply a higher value of subsidy provided. The results indicate how much subsidy would be required to achieve a desired level of savings. If the amount of cost savings required to convince a customer to adopt plan 2 is known, the subsidy could be adjusted accordingly to meet this goal.

52.6 Discussion

In this chapter, we defined the goals of reducing electronic waste and energy-efficient operation as the two tenets of sustainable portable computing. Three approaches toward achieving these tenets were then proposed aimed at increasing lifespan of portable devices and/or a reduction in energy consumed through the power grid.

The first approach evaluated the energy efficiency of using cloud computing and CRs as a way to improve device lifespan by relying on remote servers. Many cloud computing applications are increasing in popularity, signaling that they meet many performance expectations of users. Our evaluations showed that apart from benefits in reducing device replacement rate, this scenario could also reduce energy usage by devices for certain applications that use fewer local resources when relying on remote servers.

The second approach studied the feasibility of using solar power to charge device batteries, thus reducing energy consumed off the grid. It was shown that many solar panels currently available can deliver a similar charging performance to current wall outlet chargers under favorable light intensities.

Finally, the third approach studied what steps firms and regulators can take to provide economic incentives to customers to keep using their mobile phones longer. An approach for firms to set prices for longer duration contracts with customers was proposed, along with how subsidies from regulators could further incentivize the adoption of such plans.

This chapter thus proposed technical as well as economic solutions toward sustainable portable computing. Additional efforts will be needed in other areas like spreading greater awareness among users, looking at other forms of energy harvesting to charge batteries, and considering market competition in setting economic incentives.

References

1. Digital Lux Meter Product SKU : 001480-006. http://www.virtualvillage.com/digital-lux-meter-for-accurate-light-level-measurement-001480-006.html
2. Flickr. http://www.flickr.com
3. IEEE 802.22 Working group on wireless regional area networks. http://www.ieee802.org/22/
4. Picasa. http://picasa.google.com
5. Samsung Blue Earth mobile phone. http://www.samsung.com
6. Solar Gorilla solar charger. http://www.earthtechproducts.com/solargorilla-solar-charger.html
7. SunlinQ solar charger. http://www.earthtechproducts.com/p2008.html
8. A. Akella, G. Judd, S. Seshan, and P. Steenkiste. Self-management in chaotic wireless deployments. In *MOBICOM*, Cologne, Germany, pp. 185–199, 2005.
9. I.F. Akyildiz, W.-Y. Lee, M.C. Vuran, and S. Mohanty. Next generation dynamic spectrum access cognitive radio wireless networks: A survey. *Elsevier Computer Networks Journal*, 13(50):2127–2159, September 2006.
10. R. Brown, C. Webber, and J.G. Koomey. Status and future directions of the Energy Star program. *Energy*, 27:505–520, 2002.
11. L. Cao and H. Zheng. Stable and efficient spectrum access in next generation dynamic spectrum networks. In *IEEE INFOCOM*, Phoenix, AZ, pp. 870–878, April 2008.
12. K.R. Chowdhury and I.F. Akyildiz. Cognitive wireless mesh networks with dynamic spectrum access. *IEEE Journal on Selected Areas in Communications*, 26(1):168–181, January 2008.
13. C. Cormio and K.R. Chowdhury. A survey on MAC protocols for cognitive radio networks. *Elsevier Ad Hoc Networks*, 7(7):1315–1329, September 2009.
14. L.M. Feeney and M. Nilsson. Investigating the energy consumption of a wireless network interface in an ad-hoc networking environment. In *IEEE INFOCOM*, Anchorage, AK, 2001.
15. G. Fettweis and E. Zimmermann. ICT energy consumption—Trends and challenges. In *11th International Symposium on Wireless Personal Multimedia Communications (WPMC'08)*, Lapland, Finland, 2008.
16. A. Ghasemi and E.S. Sousa. Optimization of spectrum sensing for opportunistic spectrum access in cognitive radio networks. In *IEEE Consumer Communications and Networking Conference*, Las Vegas, NV, pp. 1022–1026, January 2007.
17. A. He, S. Srikanteswara, J.H. Reed, X. Chen, W.H. Tranter, K.K. Bae, and M. Sajadieh. Minimizing energy consumption using cognitive radio. In *IEEE IPCCC*, Austin, TX, pp. 373–377, December 2008.
18. Y.T. Hou, Y. Shi, and H.D. Sherali. Spectrum sharing for multi-hop networking with cognitive radios. *IEEE Journal on Selected Areas in Communications*, 26(1):146–155, January 2008.

19. K. Kawamoto, Y. Shimoda, and M. Mizuno. Energy saving potential of office equipment power management. *Energy and Buildings*, 36(9):915–923, September 2004.
20. H. Kim and K.G. Shin. Adaptive MAC-layer sensing of spectrum availability in cognitive radio networks. Technical Report CSE-TR-518-06, University of Michigan, Ann Arbor, MI, 2006.
21. H. Kim and K.G. Shin. Efficient discovery of spectrum opportunities with MAC-layer sensing in cognitive radio networks. *IEEE Transactions on Mobile Computing*, 7:533–545, May 2008.
22. W.Y. Lee and I.F. Akyildiz. Optimal spectrum sensing framework for cognitive radio networks. *IEEE Transactions of Wireless Communications*, 7(10):845–857, 2008.
23. S.-Y. Lien, C.-C. Tseng, and K.-C. Chen. Carrier sensing based multiple access protocols for cognitive radio networks. In *IEEE ICC*, Beijing, China, pp. 3208–3214, May 2008.
24. L. Luo and S. Roy. Analysis of search schemes in cognitive radio. In *IEEE Workshop on Networking Technologies for Software Defined Radio Networks*, San Diego, CA, pp. 647–654, June 2007.
25. M. Ma and D.H.K. Tsang. Joint spectrum sharing and fair routing in cognitive radio networks. In *IEEE CCNC*, Las Vegas, NV, pp. 978–982, January 2008.
26. A. Mahesri and V. Vardhan. Power consumption breakdown on a modern laptop. In *PACS*, Portland, OR, 2004.
27. N.B. Mandayam. Talk of cognitive radio networks & the future Internet. In *DIMACS Workshop on Next Generation Networks*, New Brunswick, NJ, August 2007.
28. J. Mitola and G.Q. Maguire. Cognitive radio: Making software radios more personal. *IEEE Personal Communication*, 6(4):13–18, 1999.
29. V. Namboodiri. Are cognitive radios energy-efficient?—A study of the wireless LAN scenario. In *IEEE IPCCC Workshop on Dynamic Spectrum Access and Cognition Radios*, Phoenix, AZ, December 2009.
30. V. Namboodiri. Towards sustainability in portable computing through cloud computing and cognitive radios. *Parallel Processing Workshops, International Conference on*, San Diego, CA, pp. 468–475, 2010.
31. V. Namboodiri, L. Gao, and R. Janaswamy. Power-efficient topology control for wireless networks with switched beam directional antennas. *Ad Hoc Networks*, 6(2):287–306, 2008.
32. Nokia. Global consumer survey reveals that majority of old mobile phones are lying in drawers at home and not being recycled, July 2009.
33. Novarum. Guidelines for successful large scale outdoor wi-fi networks. 2010. http://www.novarum.com/publications.php
34. C. Peng, H. Zheng, and B.Y. Zhao. Utilization and fairness in spectrum assignment for opportunistic spectrum access. *Mobile Networks and Applications*, 11(4):555–576, 2006.
35. R. Puustinen and G. Zadok. The green switch: Designing for sustainability in mobile computing. In *First USENIX International Workshop on Sustainable Information Technology*, San Jose, CA, February 2010.
36. V. Raghunathan, T. Pering, R. Want, A. Nguyen, and P. Jensen. Experience with a low power wireless mobile computing platform. In *ISLPED'04: Proceedings of the 2004 International Symposium on Low Power Electronics and Design*, Newport Beach, CA, pp. 363–368, 2004.
37. P.E. Ross. Cloud computing's killer app: Gaming. *IEEE Spectrum*, 46(3):14, March 2009.
38. Y. Shi and Y.T. Hou. A distributed optimization algorithm for multi-hop cognitive radio networks. In *IEEE INFOCOM*, Phoenix, AZ, pp. 1292–1300, April 2008.
39. P. Somavat, S. Jadhav, and V. Namboodiri. Accounting for the energy consumption of personal computing including portable devices. In *First International Conference on Energy-Efficient Computing and Networking (e-Energy)*, Passau, Germany, April 2010.
40. The Wall Street Journal. Sprint celebrates America recycles day, reminds consumers to recycle old unused wireless devices. http://www.thestreet.com/story/10922028/1/sprint-celebrates-america-recycles-day-reminds-consumers-to-recycle-old-unused-wireless-devices.html

41. M. Thoppian, S. Venkatesan, R. Prakash, and R. Chandrasekaran. MAC-layer scheduling in cognitive radio based multi-hop wireless networks. In *WoWMoM*, Buffalo, New York, pp. 191–202, April 2006.
42. P. Wang, L. Xiao, S. Zhou, and J. Wang. Optimization of detection time for channel efficiency in cognitive radio systems. In *Wireless Communications and Networking Conference*, Las Vegas, NV, pp. 111–115, March 2008.
43. C.A. Webber, J.A. Roberson, M.C. McWhinney, R.E. Brown, M.J. Pinckard, and J.F. Busch. After-hours power status of office equipment in the USA. *Energy*, 31(14):2823–2838, 2006.
44. Y. Yuan, P. Bahl, R. Chandra, T. Moscibroda, and Y. Wu. Allocating dynamic time-spectrum blocks in cognitive radio networks. In *ACM MobiHoc*, Montreal, Quebec, Canada, pp. 130–139, 2007.
45. Q. Zhang, F.H.P. Fitzek, and V.B. Iversen. Cognitive radio MAC protocol for WLAN. In *IEEE PIMRC*, Cannes, France, pp. 1–6, September 2008.

Index

Printed and bound by CPI Group (UK) Ltd, Croydon, CR0 4YY

21/10/2024

01777040-0016